MOLECULAR ASPECTS OF TRANSPORT PROTEINS

New Comprehensive Biochemistry

Volume 21

General Editors

A. NEUBERGER
London

L.L.M. van DEENEN
Utrecht

ELSEVIER
Amsterdam · London · New York · Tokyo

Molecular Aspects of Transport Proteins

Editor

J.J.H.H.M. DE PONT

*Department of Biochemistry, University of Nijmegen,
6500 HB Nijmegen, The Netherlands*

1992
ELSEVIER
Amsterdam · London · New York · Tokyo

Elsevier Science Publishers B.V.
P.O. Box 211
1000 AE Amsterdam
The Netherlands

Library of Congress Cataloging-in-Publication Data

Molecular aspects of transport proteins / editor, J.J.H.H.M. De Pont.
    p.   cm. -- (New comprehensive biochemistry ; v. 21)
  Includes bibliographical references and index.
  ISBN 0-444-89562-0 (alk. paper)
  1. Carrier proteins--Molecular aspects. 2. Ion pumps--Molecular
aspects. 3. Ion channels--Molecular aspects. 4. Sodium/potassium
ATPase--Molecular aspects.   I. Pont, J. J. H. H. M. de.
II. Series.
  [DNLM: 1. Biological Transport--physiology. 2. Carrier Proteins-
-metabolism. 3. Membrane Proteins--metabolism. 4. Protein Binding-
-physiology.   W1 NE372F v.21 / QU 55 M7145]
QD415.N48 vol. 21
[QP552.C34]
574.19'2 s--dc20
[574.87'5]
DNLM/DLC
for Library of Congress   92-18220
                                                                  CIP

ISBN 0 444 89562 0
ISBN 0 444 80303 3 (series)

© 1992 Elsevier Science Publishers B.V. All rights reserved.

No part of this publication may be reproduced, stored in a retrieval system or transmitted in any form or by any means, electronic, mechanical, photocopying, recording or otherwise, without the prior permission of the Publisher, Elsevier Science Publishers B.V., Copyright & Permissions Department, P.O. Box 521, 1000 AM Amsterdam, The Netherlands.

No responsibility is assumed by the publisher for any injury and/or damage to persons or property as a matter of products liability, negligence or otherwise, or from any use or operation of any methods, products, instructions or ideas contained in the material herein. Because of the rapid advances in the medical sciences, the publisher recommends that independent verification of diagnoses and drug dosages should be made.

Special regulations for readers in the USA – This publication has been registered with the Copyright Clearance Center Inc. (CCC) Salem, Massachusetts. Information can be obtained from the CCC about conditions under which photocopies of parts of this publication may be made in the USA. All other copyright questions, including photocopying outside of the USA, should be referred to the Publisher.

Printed on acid-free paper

Printed in The Netherlands

Preface

The first two volumes in the series New Comprehensive Biochemistry appeared in 1981. Volume 1 dealt with membrane structure and Volume 2 with membrane transport. The editors of the last volume (the present editor being one of them) tried to provide an overview of the state of the art of the research in that field. Most of the chapters dealt with kinetic approaches aiming to understand the mechanism of the various types of transport of ions and metabolites across biological membranes. Although these methods have not lost their significance, the development of molecular biological techniques and their application in this field has given to the area of membrane transport such a new dimension that the appearance of a volume in the series New Comprehensive Biochemistry devoted to molecular aspects of membrane proteins is warranted.

During the last decade hundreds of primary structures of membrane proteins have been published and each month several new sequences of transport proteins appear in the data banks. From these sequences global models for the structure of membrane proteins can be made using several type of algorithms. These models are very useful for a partial understanding of the structure of these proteins and may help us with understanding part of the mechanism of action. They do not, however, provide us with complete answers of how these pumps, carriers and channels actually function. The combination of biochemical (site-specific reagents), molecular biological (site-directed mutagenesis) and genetic approaches of which this volume gives numerous examples in combination with such biophysical techniques as X-ray analysis and NMR will eventually lead to a complete elucidation of the mechanism of action of these transport proteins.

It is clearly impossible to give a comprehensive overview of this rapidly expanding field. I have chosen a few experts in their field to discuss one (class of) transport protein(s) in detail. In the first five chapters pumps involved in primary active transport are discussed. These proteins use direct chemical energy, mostly ATP, to drive transport. The next three chapters describe carriers which either transport metabolites passively or by secondary active transport. In the last three chapters channels are described which allow selective passive transport of particular ions. The progress in the latter field would be unthinkable without the development of the patch clamp technique. The combination of this technique with molecular biological approaches has yielded very detailed information of the structure–function relationship of these channels.

Despite the limitation in the choice of membrane proteins, I hope that this volume will be useful for teachers, students and investigators in this field. Although only a limited number of transport proteins is discussed in this volume in detail, the approaches described here can be applied to other membrane proteins too and may lead to further progress in our understanding of this fascinating field.

Jan Joep H.H.M. De Pont
Nijmegen, The Netherlands,
January, 1992

List of contributors

Stephen A. Baldwin,
Departments of Biochemistry and Chemistry, and Protein and Molecular Biology, Royal Free Hospital School of Medicine (University of London), London NW3 2PF, U.K.

Rebecca M. Brawley,
Department of Pharmacology, Northwestern University Medical School, Chicago, IL 60611, U.S.A.

Chan Fong Chang,
Department of Pharmacology, Northwestern University Medical School, Chicago, IL 60611, U.S.A.

Jan Joep H.H.M. De Pont,
Department of Biochemistry, University of Nijmegen, 6500 HB Nijmegen, The Netherlands.

Rainer Greger,
Physiologisches Institut der Albert-Ludwigs-Universität, 7800 Freiburg i. Brsg., Germany.

M. Grenson,
Université Libre de Bruxelles, Faculté des Sciences, Département de Biologie Moléculaire, Laboratoire de Physiologie Cellulaire et de Génétique des Levures, B-1050 Bruxelles, Belgium.

Luis M. Gutierrez,
Department of Pharmacology, Northwestern University Medical School, Chicago, IL 60611, U.S.A.

M. Marlene Hosey,
Department of Pharmacology, Northwestern University Medical School, Chicago, IL 60611, U.S.A.

Peter Igarashi,
Department of Medicine, Yale University School of Medicine, New Haven, CT 06510, U.S.A.

Peter Leth Jørgensen,
Biomembrane Research Centre, August Krogh Institute, University of Copenhagen, 2100 Copenhagen OE, Denmark.

J.S. Lolkema,
The BIOSON Research Institute, University of Groningen, 9747 AG Groningen, The Netherlands.

Anthony Martonosi,
Department of Biochemistry and Molecular Biology, State University of New York

Health Science Center, Syracuse, NY 13210, U.S.A.
Cecilia Mundina-Weilenmann,
Department of Pharmacology, Northwestern University Medical School, Chicago, IL 60611, U.S.A.
O. Pongs,
Zentrum für Molekulare Neurobiologie, 2000 Hamburg 20, Germany.
G.T. Robillard,
The BIOSON Research Institute, University of Groningen, 9747 AG Groningen, The Netherlands.
Gene A. Scarborough,
Department of Pharmacology, University of North Carolina at Chapel Hill, Chapel Hill, NC 27599, U.S.A.
Tom J.F. Van Uem,
Department of Biochemistry, University of Nijmegen, 6500 HB Nijmegen, The Netherlands.

Contents

Preface v

List of contributors vii

Chapter 1. Na,K-ATPase, structure and transport mechanism
Peter Leth Jørgensen 1

1. Introduction 1
   1.1. The Na,K-pump 1
   1.2. Recent review articles on Na,K-ATPase structure and function 2
2. Structure of Na,K-ATPase 2
   2.1. Purified membrane-bound and soluble Na,K-ATPase 2
      2.1.1. Enzymatic properties 3
      2.1.2. Electron microscopy and crystal analysis 3
      2.1.3. Three-dimensional models 5
   2.2. Cytoskeletal associations 6
   2.3. Proteolytic dissection of membrane-bound Na,K-ATPase 7
   2.4. Membrane organization of the α subunit 7
   2.5. Structure of the β subunit of Na,K-ATPase 10
3. Nucleotide binding and phosphorylation 11
   3.1. The nucleotide binding domain in the α1β1 unit 12
      3.1.1. Comparison with the nucleotide binding sites in adenylate kinase 12
      3.1.2. Selective chemical labelling with ATP analogues 12
   3.2. Conformations of the nucleotide binding area 13
   3.3. The phosphorylation site, high- and low-energy phosphoforms, E_1P–E_2P ... 13
4. Cation binding and occlusion 15
   4.1. Capacity for binding and occlusion of Na^+ or $K^+(Rb^+)$ 16
   4.2. Isolation of the cation occlusion and transport path after tryptic digestion 17
   4.3. Transport stoichiometry and net charge of Na^+ and K^+ complexes with Na,K-ATPase 17
5. Structural transitions in the protein related to energy transformation and Na,K-transport 18
   5.1. Conformation dependent proteolytic cleavage of Na,K-ATPase 18
   5.2. Tryptophan fluorescence and secondary structure changes 19
   5.3. Cleaved derivatives; cleavage of bond 2 and the regulatory function of the N-terminus 20
   5.4. Effect of C_3 cleavage on E_1P–E_2P transition and cation exchange ... 20
   5.5. Mutagenesis in yeast H-ATPase and Ca-ATPase from sarcoplasmic reticulum 21
   5.6. Coupling to ion translocation 22
References 23

Chapter 2. Structure and function of gastric H,K-ATPase
Tom J.F. Van Uem and Jan Joep H.H.M. De Pont 27

1. Introduction 27
2. Tissue and cell distribution 28
3. Structure 28
   3.1. The catalytic α subunit 28
   3.2. The β subunit 31
   3.3. Molecular organization 34
   3.4. Conformations of H,K-ATPase 34
4. Kinetics of H,K-ATPase 36
   4.1. Overall reaction 36
   4.2. Phosphorylation from ATP 38
   4.3. Characteristics of ATP hydrolysis 39
   4.4. Hydrolysis of p-nitrophenylphosphate 40
   4.5. Phosphorylation from inorganic phosphate 41
5. Transport by H,K-ATPase 42
   5.1. The H^+–ATP stoichiometry 42
   5.2. Ion selectivity 42
   5.3. Electrogenicity of ion transport 43
6. Lipid dependency of H,K-ATPase 44
7. Solubilization and reconstitution 45
8. Inhibitors of H,K-ATPase 46
9. Conclusions 49
References 49

Chapter 3. The Ca^{2+} transport ATPases of sarco(endo)plasmic reticulum and plasma membranes
Anthony Martonosi 57

1. Introduction 57
2. The classification of Ca^{2+}-ATPase isoenzymes 58
   2.1. The Ca^{2+} transport ATPases of sarco(endo)plasmic reticulum (SERCA) ... 58
      2.1.1. SERCA1 58
      2.1.2. SERCA2 58
      2.1.3. SERCA3 59
      2.1.4. SERCA-type Ca^{2+}-ATPases from non-mammalian cells (SERCAMED) 59
   2.2. The plasma membrane Ca^{2+} transport ATPases (PMCA) .. 59
      2.2.1. rPMCA1 and rPMCA2 62
      2.2.2. rPMCA3 62
      2.2.3. rPMCA4 62
      2.2.4. hPMCA 62
3. The deduced amino acid sequences of the fast-twitch and slow-twitch isoforms of the sarcoplasmic reticulum Ca^{2+}-ATPase 62
4. The predicted topology of the Ca^{2+}-ATPases 64
   4.1. The Ca^{2+}-ATPase of the sarcoplasmic reticulum 64
      4.1.1. The cytoplasmic headpiece 65

		4.1.1.1.	The phosphorylation and nucleotide binding domains	65
		4.1.1.2.	The transduction or B domain.	67
		4.1.1.3.	The hinge domain.	67
		4.1.1.4.	The stalk region.	67
	4.1.2.		The transmembrane domain	68
	4.2.		The predicted domains of the plasma membrane Ca^{2+}-ATPase	68
5.	Reconstruction of Ca^{2+}-ATPase structure by electron microscopy			70
	5.1.		The vanadate-induced E_2-type crystals	70
		5.1.1.	Image reconstruction in three dimensions from negatively stained and frozen hydrated crystals	71
	5.2.		Crystallization of Ca^{2+}-ATPase by Ca^{2+} and lanthanides in the E_1 state	73
	5.3.		Crystallization of Ca^{2+}-ATPase in detergent-solubilized sarcoplasmic reticulum	73
6.	X-ray and neutron diffraction analysis of the Ca^{2+}-ATPase of sarcoplasmic reticulum			77
7.	Site specific mutagenesis of sarcoplasmic reticulum Ca^{2+}-ATPase.			78
	7.1.		The search for the Ca^{2+} binding site	78
		7.1.1.	Mutation of amino acids in the stalk sector	78
		7.1.2.	The probable location of Ca^{2+} binding sites in the transmembrane domain.	79
	7.2.		Mutations in the putative catalytic site	79
		7.2.1.	Mutations around Asp351	79
		7.2.2.	The mutations around Lys515.	80
		7.2.3.	The role of sequences 601–604 in ATP binding and Ca^{2+} transport	80
		7.2.4.	Mutations in the R616–K629 region of the Ca^{2+}-ATPase (Thr625, Gly626, Asp627)	81
		7.2.5.	Mutations in the 701–707 region	81
		7.2.6.	Mutations of Lys712.	81
		7.2.7.	The structure of the ATP binding site	81
	7.3.		The β strand sector. Conformational change mutants	82
	7.4.		The transmembrane segments of the Ca^{2+}-ATPase	83
8.	*In situ* proteolysis of Ca^{2+}-ATPase			84
	8.1.		Hydrolysis of Ca^{2+}-ATPase by trypsin	84
		8.1.1.	The T_1 cleavage	84
		8.1.2.	The T_2 cleavage	85
		8.1.3.	The cleavage of Ca^{2+}-ATPase by trypsin at the T_3 and T_4 sites	86
	8.2.		The effect of other proteolytic enzymes on the Ca^{2+}-ATPase.	87
		8.2.1.	Chymotrypsin	87
		8.2.2.	Thermolysin	87
		8.2.3.	Staphylococcal V8 protease.	87
	8.3.		Vanadate-catalyzed photocleavage of the Ca^{2+}-ATPase	87
9.	Monoclonal and polyclonal anti-ATPase antibodies.			88
	9.1.		Antibodies reacting with the N- and C-terminal regions of the Ca^{2+}-ATPase	89
	9.2.		Distribution of epitopes in the cytoplasmic domain of Ca^{2+}-ATPase.	89
	9.3.		Antibodies reacting with the putative luminal domain of the Ca^{2+}-ATPase	90
10.	Covalent modification of side-chain groups in the Ca^{2+}-ATPase			91
	10.1.		Sulfhydryl groups.	91

10.1.1. Identification of cysteine residues that react with *N*-ethylmaleimide (MalNEt)... 91
10.1.2. The reaction of iodoacetamide and its N-substituted derivatives with the Ca^{2+}-ATPase.. 92
10.1.2.1. Iodoacetamide (IAA) and 5-(2-acetamidoethyl)aminonaphthalene-1-sulfonate (IAEDANS)....................................... 92
10.1.2.2. The reaction of 6-(iodoacetamido)fluorescein (IAF) with the Ca^{2+}-ATPase... 92
10.1.3. Modification of Ca^{2+}-ATPase with 7-chloro-4-nitrobenzo-2-oxa-1,3-diazole (NBD-Cl).. 92
10.1.4. The disulfide of 3'(2')-*O*-biotinyl-thioinosine triphosphate (biotinyl-S^6-ITP_2)....................................... 93
10.2. Modification of lysine residues 93
10.3. Modification of arginine residues... 94
10.4. Modification of histidine 95
10.5. Modification of carboxyl groups 96
    10.5.1. The reaction of Ca^{2+}-ATPase with dicyclohexylcarbodiimide... 96
    10.5.2. The reaction of Ca^{2+}-ATPase with *N*-cyclohexyl-*N'*-(4-dimethyl-amino-α-naphthyl) carbodiimide (NCD-4) 97
    10.5.3. Reaction of Ca^{2+}-ATPase with the carbodiimide derivative of ATP 97
11. Spatial relationships between functional sites in the sarcoplasmic reticulum Ca^{2+}-ATPase 98
    11.1. Intramolecular distances determined by fluorescence energy transfer ... 98
        11.1.1. The location of the high-affinity Ca^{2+} binding site 100
        11.1.2. The ATP binding site 101
        11.1.3. The use of IAEDANS as reference point for distance measurements... 102
    11.2. Thermal fluctuations in the structure of the Ca^{2+}-ATPase........... 103
References 105

Chapter 4. The Neurospora crassa plasma membrane H^+-ATPase
Gene A. Scarborough 117

1. Introduction 117
2. Structural features of the H^+-ATPase molecule 118
    2.1. H^+-ATPase conformational changes....................................... 118
    2.2. The purified ATPase preparation 119
    2.3. Subunit composition of the H^+-ATPase 119
    2.4. The minimum functional unit 120
    2.5. Primary structure of the H^+-ATPase 121
    2.6. Secondary structure of the H^+-ATPase................................... 121
    2.7. Protein chemistry of the H^+-ATPase 122
    2.8. Chemical state of the H^+-ATPase cysteines 122
    2.9. Transmembrane topography of the H^+-ATPase 123
    2.10. A first-generation model for the tertiary structure of the H^+-ATPase .. 124
3. The molecular mechanism of transport 129
References 131

Chapter 5. The Enzymes II of the phosphoenolpyruvate-dependent carbohydrate transport systems
J.S. Lolkema and G.T. Robillard 135

1. Introduction 135
   1.1. PTS carbohydrate specificity 135
   1.2. PTS components 135
   1.3. PTS nomenclature 136
2. Enzyme II structure 138
   2.1. Sequence homology 138
   2.2. Domain structure 138
   2.3. Domain function 140
        2.3.1. The A domain 140
        2.3.2. The B domain 142
        2.3.3. The A and B domains of *E. coli* III^{Man} 143
        2.3.4. The C domain 143
   2.4. Domain interactions 143
        2.4.1. Association state of E-II 144
        2.4.1.1. E-II^{Mtl} 144
        2.4.1.2. E-II^{Glc} 145
        2.4.2. Kinetics of domain interaction 146
3. Binding studies 147
   3.1. General considerations 147
   3.2. Equilibrium binding to E-II 149
   3.3. Orientation of the binding site 149
   3.4. Kinetics of binding 151
4. The coupling between transport and phosphorylation 153
   4.1. General considerations 153
   4.2. Phosphorylation of free cytoplasmic carbohydrates 154
   4.3. Facilitated diffusion catalyzed by E-II 155
        4.3.1. Diffusion in uptake studies 155
        4.3.2. Diffusion in efflux studies 156
        4.3.3. Regulation of efflux 157
   4.4. Coupling in vectorial phosphorylation 158
5. Steady-state kinetics of carbohydrate phosphorylation 160
   5.1. General considerations 160
   5.2. The *R. sphaeroides* II^{Fru} model 161
   5.3. The *E. coli* II^{Mtl} model 163
References 164

Chapter 6. Mechanisms of active and passive transport in a family of homologous sugar transporters found in both prokaryotes and eukaryotes
Stephen A. Baldwin 169

1. Introduction 169
2. The kinetics of sugar transport in mammalian cells 170
   2.1. Substrate specificity 170
   2.2. Specific inhibitors of transport 172
   2.3. Kinetics of transport in the erythrocyte 174

		2.3.1.	General properties and methods of investigation	174

    2.3.1. General properties and methods of investigation 174
    2.3.2. Transport asymmetry and the effect of cytoplasmic ATP 176
  2.4. Kinetic models for the transport process 177
  2.5. Measurements of individual rate constants for steps in the transport cycle 179
3. Characterization of the isolated human erythrocyte transporter 182
  3.1. Purification and kinetic properties of the transporter protein 182
  3.2. Molecular properties of the isolated protein 184
    3.2.1. Polypeptide composition and glycosylation state 184
    3.2.2. Secondary structure 184
    3.2.3. Oligomeric state 185
4. The structure of the human erythrocyte glucose transport protein 185
  4.1. Amino acid sequence 185
  4.2. Arrangement in the membrane 186
    4.2.1. Topology 186
    4.2.2. Three-dimensional arrangement 189
  4.3. Location of the substrate-binding site(s) 189
    4.3.1. Insights from proteolytic digestion 189
    4.3.2. Photoaffinity labelling with cytochalasin B 189
    4.3.3. Photoaffinity labelling with bis-mannose derivatives 190
    4.3.4. Photoaffinity labelling with forskolin and its derivatives 191
    4.3.5. Photoaffinity labelling with miscellaneous inhibitors 191
5. Conformational changes and the mechanism of transport 192
  5.1. Influence of substrates and inhibitors on reactivity towards group-specific reagents 192
  5.2. Biophysical studies 194
  5.3. Differential susceptibility of conformers to proteolysis 195
6. Homologous transporters and their distribution in mammalian tissues 196
  6.1. GLUT-1 197
  6.2. GLUT-2 198
  6.3. GLUT-3 199
  6.4. GLUT-4 199
  6.5. GLUT-5 200
7. Homologous transporters in other organisms 200
  7.1. Fungal transporters 200
  7.2. Protozoan transporters 201
  7.3. The transporters of photosynthetic organisms 201
  7.4. Bacterial transporters 202
    7.4.1. The galactose, arabinose and xylose transporters of *E. coli* ... 202
    7.4.1.1. The D-xylose/H^+ transporter 202
    7.4.1.2. The L-arabinose/H^+ transporter 202
    7.4.1.3. The D-galactose/H^+ transporter 202
    7.4.2. The citrate and tetracycline transporters of *E. coli* 203
    7.4.3. The lactose transporter of *E. coli* 207
8. Clues to the mechanism of transport from comparison of the homologous transporters 208
9. Summary 210
References 211

Chapter 7. Amino acid transporters in yeast: structure, function and regulation
M. Grenson   219

1. Introduction   219
2. Physiological background: assimilation of exogenous nitrogen compounds used as a source of nitrogen or as building blocks   220
3. General characteristics of amino acid transporters in *Saccharomyces cerevisiae*.   222
   3.1. Accumulation of amino acids   222
   3.2. Multiplicity and specificity of amino acid transporters in *Saccharomyces cerevisiae*   222
   3.3. Functional specialization of amino acid transporters   223
   3.4. Irreversibility of amino acid accumulation   223
   3.5. Role of the vacuole in amino acid retention   224
   3.6. Efflux of amino acids   225
4. Identifying transport systems.   225
   4.1. Isolating mutants affected in uptake systems   226
   4.2. An example of transporter identification in a complex case: the three GABA transport systems of *Saccharomyces cerevisiae*.   226
5. Structure and evolution of amino acid transporters   227
   5.1. Molecular cloning and nucleotide sequencing of amino acid permease genes.   227
   5.2. A family of amino acid transporters with amino acid sequence homologies   231
6. Regulation of amino acid transport   232
   6.1. Regulation of permease activity.   232
   6.2. Regulation of permease synthesis.   234
      6.2.1. Case of the NCR-insensitive amino acid permeases   234
      6.2.2. Case of the NCR-sensitive amino acid permeases   234
      6.2.2.1. Constitutive expression of permease genes   234
      6.2.2.2. Inducible permeases   235
   6.3. Nitrogen-catabolite repression (NCR) and nitrogen-catabolite inactivation (NCI): two superimposed regulatory mechanisms affecting uptake systems for nitrogenous compounds   237
      6.3.1. Regulation of amino acid permease activity as a function of nitrogen availability   238
      6.3.1.1. Nitrogen-catabolite inactivation (NCI): negative control of GAP1 activity.   238
      6.3.1.2. Positive control of GAP1 activity.   239
      6.3.1.3. How is GAP1 activity regulated?   239
      6.3.2. Nitrogen-catabolite repression (NCR)   240
      6.3.2.1. NCR affects permease gene transcription or transcript accumulation.   240
      6.3.2.2. Glutamine as an effector of NCR   240
      6.3.2.3. The *URE2/GDHCR* gene product as a negative regulatory protein which participates in the repression of permease synthesis   240
      6.3.2.4. The *GLN3* gene product as a possible target for the *URE2/GDHCR* gene product.   240
      6.3.2.5. Double regulation of the ammonia-sensitive permeases   241
7. The *APF1* gene product, a common factor of unknown function which increases the activity of amino acid permeases.   241
8. Summary and prospects.   241

References 243

Chapter 8. Structure and function of plasma membrane Na^+/H^+ exchangers
Peter Igarashi 247

1. Introduction 247
   1.1. The Na^+/H^+ exchanger 247
   1.2. Functional heterogeneity 248
2. Biochemical properties of Na^+/H^+ exchangers 249
   2.1. 'Group-specific' modification 249
      2.1.1. Imidazolium 250
      2.1.2. Carboxyl 251
      2.1.3. Sulfhydryl 252
      2.1.4. Amino 253
      2.1.5. Carbohydrate 254
   2.2. Identification and characterization of candidate transport protein(s) . . . 255
      2.2.1. Covalent labeling 255
      2.2.2. Affinity chromatography 257
      2.2.3. Other 258
3. Molecular cloning of Na^+/H^+ exchangers 260
   3.1. cDNA cloning and primary structure 260
      3.1.1. Human Na^+/H^+ exchanger cDNA 261
      3.1.2. Other species 263
   3.2. Tissue and membrane localization 265
   3.3. Isoforms 267
   3.4. Genomic cloning 268
4. Summary and future directions 269
References 270

Chapter 9. Cl^--channels
Rainer Greger 273

1. Introduction 273
2. Different types of Cl^--channels 274
   2.1. Cl^--channels in the nervous system 275
   2.2. Cl^--channels of muscle and electric organ 276
   2.3. Cl^--channels in apolar non-excitable cells 276
   2.4. The problem of detecting small Cl^--channels 277
   2.5. Epithelial Cl^--channels 278
3. The structure and molecular basis of Cl^--channels 280
   3.1. The $GABA_A$-receptor and glycine-receptor channels 281
   3.2. The *Torpedo marmorata* Cl^--channel 281
   3.3. Muscle Cl^--channels 282
   3.4. Epithelial Cl^--channels 282
4. Pharmacological modulation of Cl^--channels 283
   4.1. Pharmacological modulation of $GABA_A$-receptor and glycine-receptor channels 283
   4.2. Inhibition of epithelial Cl^--channels 284

5.	Regulation of epithelial Cl^--channels		287
	5.1.	The Cl^--channel defect in cystic fibrosis	288
	5.2.	Mechanisms of Cl^--channel activation in epithelia	290
References			291

Chapter 10. Voltage-gated K^+ channels
O. Pongs 297

1.	Introduction		297
2.	Structure and biophysical properties of cloned voltage-gated K^+ channels		298
	2.1.	K^+ channels of the *Shaker* family	298
	2.2.	K^+ channels of the MBK/RCK/HBK family	301
	2.3.	K^+ channels of *Shaker* relatives	302
	2.4.	Pharmacology of K^+ channels	302
3.	Structure of K^+ channel genes		306
	3.1.	Genes in *Drosophila*	306
	3.2.	Vertebrate genes	307
4.	The basis of K^+ channel diversity		307
	4.1.	Properties of homo- and heteromultimers	307
	4.2.	Functional domains in K^+ channels	308
5.	General structural implications		309
References			311

Chapter 11. Structure and regulation of voltage-dependent L-type calcium channels
M. Marlene Hosey, Rebecca M. Brawley, Chan Fong Chang, Luis M. Gutierrez and Cecilia Mundina-Weilenmann 315

1.	Introduction			315
	1.1.	Subtypes of Ca^{2+} channels		315
	1.2.	Functions of L-type Ca^{2+} channels		317
2.	L-type Ca^{2+} channels			318
	2.1.	Pharmacology		318
	2.2.	Biochemical and molecular characterization		319
		2.2.1.	Isolation and purification of the multisubunit dihydropyridine-sensitive Ca^{2+} channels from skeletal muscle	320
		2.2.2.	Identification and purification of L-type channel proteins from other cells	321
		2.2.3.	DNA cloning and expression of channel proteins	322
		2.2.3.1.	Isoforms of the α_1 subunit	322
		2.2.3.2.	Cloning of DNAs for other putative channel subunits	323
		2.2.4.	Roles of subunits of L-type Ca^{2+} channels	324
		2.2.5.	Reconstitution of Ca^{2+} channels	325
	2.3.	Regulation of Ca^{2+} channels by protein phosphorylation and G-proteins		326
		2.3.1.	Phosphorylation by cAMP-dependent protein kinase	327
		2.3.2.	Regulation of L-type channels by PKC and other protein kinases	329
		2.3.3.	Regulation of L-type channels by phosphoprotein phosphatases	330

2.3.4. Regulation of Ca^{2+} channels by G-proteins 331
References 332

Index 337

CHAPTER 1

Na,K-ATPase, structure and transport mechanism

PETER LETH JØRGENSEN

Biomembrane Research Centre, August Krogh Institute, University of Copenhagen, 2100 Copenhagen OE, Denmark

1. Introduction

1.1. The Na,K-pump

The Na,K-pump is ubiquitous and located at the surface membrane of most animal cells. Primary active Na,K-pumping is a key process for the active uptake of nutrients, salts and water and for the regulation of fluid and electrolyte homeostasis in mammals. The pump maintains electrochemical gradients for Na^+ ($\Delta\mu Na$) for utilization in carrier mediated secondary active transport processes in kidney, intestine, lung and other epithelia. In coupling the hydrolysis of ATP to active transport of $3Na^+$ out and $2K^+$ into the cell, the pump is electrogenic and maintains ion gradients required for regulation of cell volume and the pump works as a battery for the electrical activity of excitable cells.

The Na,K-pump was first purified from the outer medulla of mammalian kidney [1,2] and the protein was characterized in membrane-bound form and in detergent solution [3,4]. This is a preparation of choice in studies of protein structure and membrane organization [5], conformational transitions coupled to ion translocation [6] and identification of sites for binding nucleotides and cations [7]. By incubation in vanadate medium, the proteins of the purified membrane-bound pump protein can be organized in crystalline arrays with the $\alpha\beta$ unit as the minimum asymmetric unit [5,8]. Low resolution models of the overall structure of the Na,K-pump molecule can be constructed on the basis of diffraction analysis of p1 and p12 crystals [9–11]. This knowledge of pump structure and function has been important for understanding the physiological function and regulation of the Na,K-pump in kidney [12,13].

Application of recombinant DNA techniques led to the primary structure of the α subunit [14,15] and β subunit [15,16] of the Na,K-pump in mammalian kidney and a number of tissues and species (for [17] and [18]). The unitary concept that the Na,K-pump was essentially the same protein in all tissues and cells has been abandoned as three structurally distinct α subunit isoforms were identified in several species [17,19]. Three β subunit isoforms have now also been disclosed and extensive studies of tissue specific and developmental expression of the genes encoding the isoforms and the hormonal regulation of their expression are now being reported. In functional and regulatory terms, the significance of the expression of the different combinations of the α1, α2, and α3 or β1, β2, and β3 subunits in brain, skeletal muscle, heart and other tissues remains obscure, mainly because so little is known about the Na,K-transport and enzymatic properties of isozymes other than the renal (α1β1) Na,K-pump.

The α1β1 isozyme of Na,K-ATPase remains the 'household' pump that is expressed in kidney, other epithelia and most other cells. This chapter is focused on the structural organization of this renal Na,K-pump and the molecular mechanisms behind the transformation of chemical energy to movement of Na^+ and K^+ across the membrane. Particular emphasis is placed on the organization of the proteins in the membrane, their interaction with cytoskeletal components, and identification of protein segments that are engaged in binding of nucleotides or cations and in conformational changes in the protein that bring about a reorientation of the cation binding sites.

1.2. Recent review articles on Na,K-ATPase structure and function

The expansion of our knowledge of the structure and function of Na,K-ATPase is reflected in a rapid succession of reviews on Na,K-ATPase genes and regulation of expression [17], subunit assembly and functional maturation [20], the isozymes of Na,K-ATPase [18], and the stability of α subunit isoforms during evolution [21], physiological aspects and regulation of Na,K-ATPase [22], reconstitution and cation exchange [23], chemical modification [24], and occlusion of cations [25]. Other valuable sources are the review articles [26] and recent developments [27] reported at the International Na,K-pump Conference in September 1990.

2. Structure of Na,K-ATPase

2.1. Purified membrane-bound and soluble Na,K-ATPase

The procedure for purification of Na,K-ATPase in membrane-bound form from the outer renal medulla of mammalian kidney offers the opportunity of exploring the structure of the Na,K-pump proteins in their native membrane environment. The protein remains embedded in the membrane bilayer throughout the purification procedure thus maintaining the asymmetric orientation of the protein in the basolateral membrane of the kidney cell in the purified preparation. This preparation has been particularly useful in studies of ultrastructure, protein conformation and for identification of sites for binding of ATP and cations [5–7]. A further advantage is that the preparation from outer medulla contains only the $\alpha 1\beta 1$ isozyme of Na,K-ATPase, while most other preparations consist of two or three α subunit and β-subunit isoforms.

The membrane-bound preparation from kidney is easily solubilized in non-ionic detergent and analytical ultracentrifugation shows that the preparation consists predominantly (80–85%) of soluble $\alpha\beta$ units with M_r 143 000 [28]. The soluble $\alpha\beta$ unit maintains full Na,K-ATPase activity, and can undergo the cation or nucleotide induced conformational transitions that are observed in the membrane-bound preparation. A cavity for occlusion of $2K^+$ or $3Na^+$ ions can be demonstrated within the structure of the soluble $\alpha\beta$ unit [29], as an indication that the cation pathway is organized in a pore through the $\alpha\beta$ unit rather than in the interphase between subunits in an oligomer.

2.1.1. Enzymatic properties

In an ideal pure preparation of Na,K-ATPase from outer renal medulla, the $\alpha 1$ subunit forms 65–70% of the total protein and the molar ratio of α to β is 1:1, corresponding to a mass ratio of about 3:1 [1,5]. Functionally the preparation should be fully active in the sense that each $\alpha\beta$ unit binds ATP, P_i, cations and the inhibitors vanadate and ouabain. The molecular activity should be close to a maximum value of 7 000–8 000 P_i/min. The highest reported binding capacities for ATP and phosphate are in the range 5–6 nmol/mg protein and close to one ligand per $\alpha\beta$ unit [29], when fractions with maximum specific activities of Na,K-ATPase [40–50 μmol P_i/min mg protein) are selected for assay.

2.1.2. Electron microscopy and crystal analysis

The purified membrane bound Na,K-ATPase consists of disc-shaped membrane fragments, 1 000–3 000 Å in diameter, with no tendency for vesicle formation. The densely packed protein particles with diameters of 30–50 Å represent $\alpha\beta$ units that can be visualized by negative staining with phosphotungstic acid or uranyl acetate. They are arranged in irregular clusters or strands and appear to be free to move in the plane of the membrane without formation of well defined oligomeric structures [30]. From negatively stained images similar to those shown in Fig. 1, the average density of protein in the membrane is estimated to be 12 000 $\alpha\beta$ units/μm^2. This corresponds to a concentration of α subunit in the lipid bilayer of about 7 mM or 0.5–1 g protein/ml of lipid phase. These are conditions for supersaturation and formation of crystalline arrays of the protein units in the membrane fragments is rapidly induced in the presence of vanadate that stabilize the protein in a state similar to the E_2P conformation [31]. In Na,K-ATPase, the predominant crystal form, shown in Fig. 2a, has the two-sided plane group symmetry, p1, and contains one protomeric $\alpha\beta$ unit per unit cell. Crystals with two-sided plane group symmetry, p21, with two $\alpha\beta$ units occupying one unit cell, are transient and less frequent

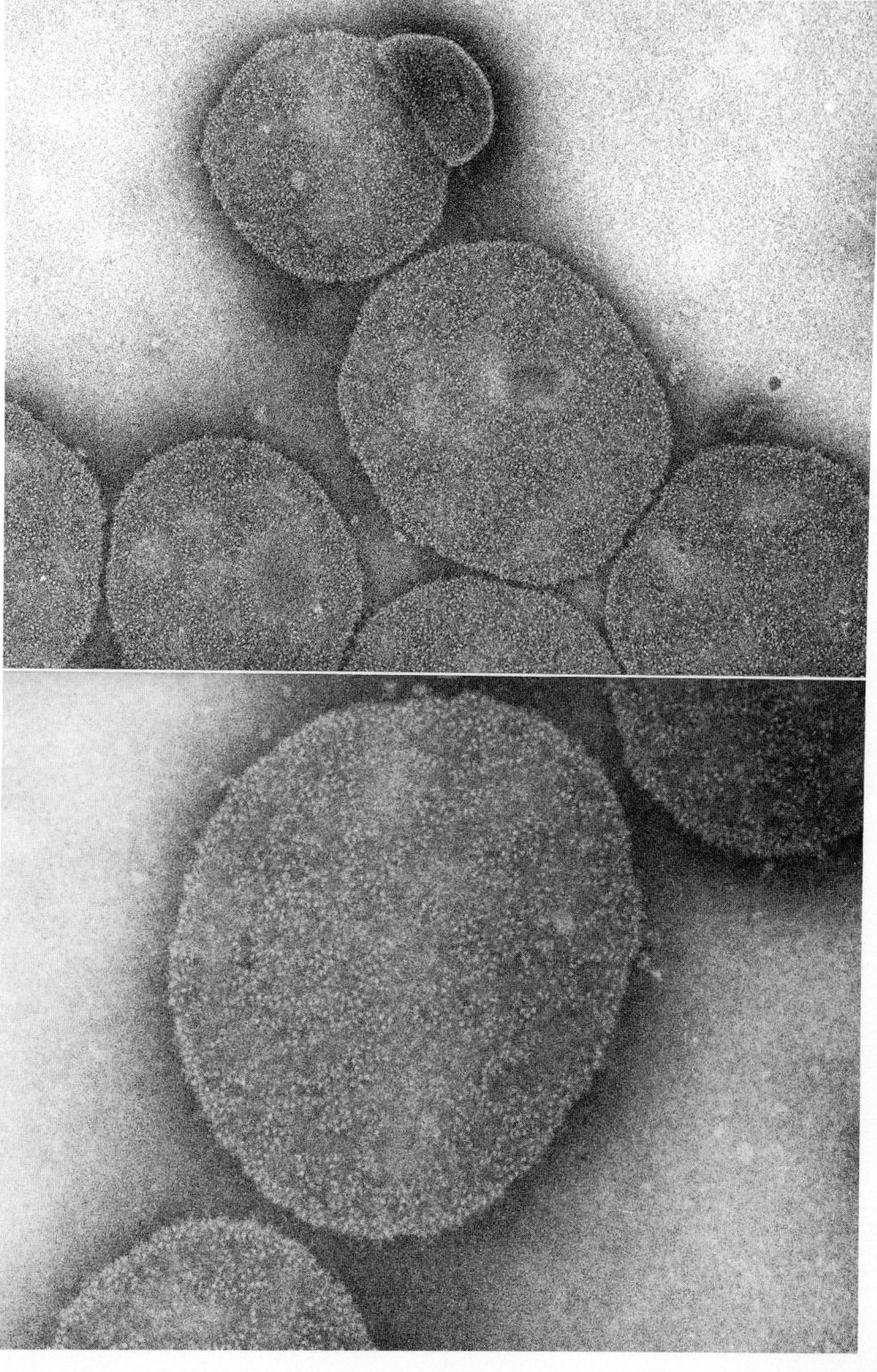

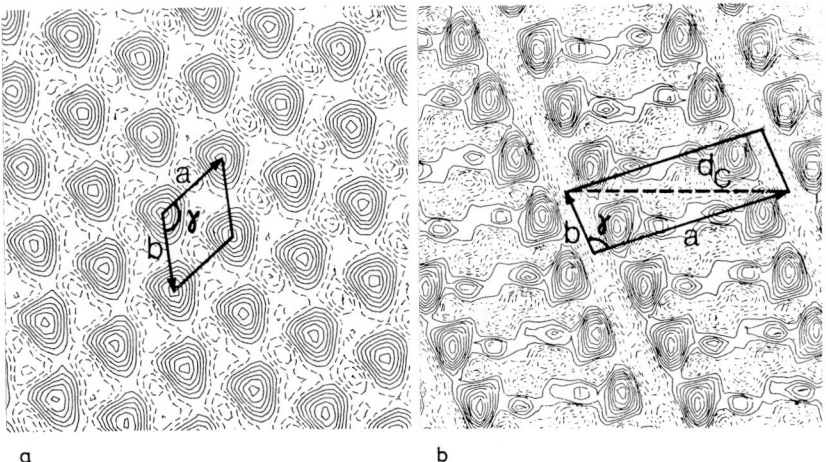

Fig. 2. Crystalline arrays of Na,K-ATPase in the membrane with (a) a protomeric $\alpha\beta$ unit as minimum asymmetric unit in a p1 crystal or (b) with an oligomeric $(\alpha\beta)_2$ unit in the unit cell of a p21 crystal. The p1 crystal was formed after incubation of purified membrane-bound Na,K-ATPase in 0.25 mM sodium monovanadate, 1 mM $MgCl_2$ at 4°C. For formation of the p21 crystal the purified membrane-bound Na,K-ATPase was incubated in 12.5 mM phosphate, 1 mM $MgCl_2$ and 10 mM Tris-HCl, pH 7.5 at 4°C. The membranes were negatively stained with uranyl acetate and micrographs were obtained at 235 000 × magnification. Images suitable for further analysis were densitometered at 20-μm intervals. Projection maps were calculated using the Fourier transform amplitudes and phases collected at the reciprocal lattice points. The protein-rich regions are drawn with unbroken contour lines, while negative stain regions have dashed lines. In the reconstructed images 1 mm corresponds to 2.8 Å. The unit cell dimensions are in a: $a = 53$ Å, $b = 51$ Å, $\gamma = 120°$; and in b: $a = 135$ Å, $b = 44$ Å, $\gamma = 101°$. From [33].

[32,33], Fig. 2b. The appearance of two crystal forms shows that the protein in the membrane exists in equilibrium between the protomeric $\alpha\beta$ unit and oligomeric $(\alpha\beta)_2$ forms. The high rate of crystal formation of the protein in vanadate solution shows that transition to the E_2 form reduces the difference in free energy required for self association of the protein. This vanadate-method for crystallization has been very reproducible [34–36] and it also leads to crystalline arrays of Ca-ATPase in sarcoplasmic reticulum [37] and H,K-ATPase from stomach mucosa [38].

2.1.3. Three-dimensional models
Low resolution models (20–30 Å) based on diffraction analysis of membrane crystals of Na,K-ATPase [34,35,39] and Ca-ATPase [40,41] show that the cytoplasmic protrusions of the proteins are remarkably similar. A notable difference is a 10–20 Å

Fig. 1. Negative staining by phosphotungstic acid of Na,K-ATPase purified in membrane-bound form. The membrane surfaces are covered by particles arranged in clusters between smooth areas. From [2] procedure as described by Deguchi et al. [30].

protrusion on the extracellular surface of the model for Na,K-ATPase while the Ca-ATPase model has a smooth extracytoplasmic surface.

The limitation of the resolution of these reconstructions is the internal order of the Na,K-ATPase crystals. At the given resolution of 20–25 Å some interactions and basic structure characteristics can be resolved, but there is no assignment of structural detail. Higher resolution has been obtained in studies of Ca-ATPase from sarcoplasmic reticulum in lamellar arrays consisting of sheets of protein arrays separated by lipid layers that are prepared from soluble Ca-ATPase in non-ionic detergent [42,43]. Thin three-dimensional crystals were grown by adding purified Ca-ATPase to appropriate mixtures of detergent, lipid and calcium [44]. They are rapidly frozen and maintained in frozen-hydrated state during electron microscopy. Electron diffraction extends to 4 Å and images contain phase data to 6 Å resolution. Based on these projections and the previously determined low-resolution structure of Ca-ATPase a packing diagram for the three-dimensional crystals is presented. A model with a specific arrangement for ten transmembrane α-helices is proposed [45].

2.2. Cytoskeletal associations

In the polarized tubule cells of mammalian kidney, the specific associations of Na,K-ATPase with cytoskeletal components and the cell–cell contacts appear to be important for the induction of polarity of the αβ unit between luminal and basolateral membranes. The epithelial cell adhesion molecule (CAM) uvomorulin (cadherin) functions as an inducer of cell polarity for the constitutively expressed α1β1 units through cytoplasmic linkage to the membrane cytoskeleton. Loss of polarity with incorrect localization of Na,K-ATPase to apical membranes has been associated with a number of diseases including polycystic kidney disease [46,47]. One link to the cytoskeleton is a high affinity binding site for ankyrin ($K_d = 10^{-8}$) that has been demonstrated in the purified renal Na,K-ATPase [48]. A fraction of the α1β1 units seems to associated in Na,K-ATPase–ankyrin–fodrin complexes with similarity to the capnophorin–ankyrin–spectrin complexes in the cytoskeleton of the human erythrocyte. Induction of cell–cell contact alters the properties and distribution of these proteins. Before contact between the epithelial cells, the ankyrin–fodrin tetramers form complexes with the membrane proteins, either Na,K-ATPase or uvomorulin. On cell–cell contact, uvomorulin seems to mediate redistribution so that the Na,K-ATPase–ankyrin–fodrin complexes accumulate at the sites of cell–cell contact [49,50]. It is proposed that the cell–cell contacts via uvomorulin induce the specific distribution at the cell surface of Na,K-ATPase during development of the polarized epithelial cells. Once polarity has been established, the proteins are replaced by targeting to the appropriate membrane from the Golgi complex.

Neuron–glial adhesion in nerve cell cultures is mediated by the β2 subunit AMOG (adhesion molecule on glia) in the α2β2 isozyme of Na,K-ATPase [51]. Antibodies to the β2 subunit dissociate cell–cell associations and also increase the rate of active transport suggesting that the β2 subunit may be part of a system for regulation of pump activity in the brain.

2.3. Proteolytic dissection of membrane-bound Na,K-ATPase

Proteolytic cleavage has proven to be an efficient tool for exploring the structure and function of the Na,K-ATPase. Exposure and protection of bonds on the surface of the cytoplasmic protrusion provides unequivocal evidence for structural changes in the α subunit accompanying E_1–E_2 transition in Na,K-ATPase [52]. Localization of the proteolytic splits provided a shortcut to identification of residues involved in E_1–E_2 transition [33,53,54] and to detection of structure–function correlations [33]. Further proteolysis identifies segments at the surface of the protein and as the cytoplasmic protrusion is shaved off all ATP-dependent reactions are abolished.

Proteases seem to be unable to shave away protein closer than 10–20 residues from the membrane embedded segments. Extensive proteolysis with trypsin or thermolysin in KCl medium removes 50–60% of the total αβ unit protein without reducing the content in the membrane of peptides that have been labelled with [^{125}I]-iodonaphthylazide (INA) [55] or [^{125}I]-trifluoromethyl-iodo-phenyldiazirine (TID) prior to proteolysis [3,56]. Table I gives a list of peptide fragments of the α subunit remaining in the membrane after extensive trypsinolysis [7] for comparison with the limit peptides after digestion of the denatured protein [57]. The β subunit is remarkably resistant to proteolysis probably because it is protected by the carbohydrate moieties. It is not clear to what extent the β subunit protects parts of the α subunit from digestion. Remarkably, the remnants of the αβ unit in the heavily digested membranes retain their ability for binding and occluding Na^+ or Rb^+ ions which indicates that the coordinating groups of the cation sites are contributed by residues in the intramembrane segments of the αβ unit [7].

2.4. Membrane organization of the α subunit

The presence of four transmembrane segments in the N-terminal half of the α subunit of Na,K-ATPase was predicted from the results of controlled proteolysis of three cytoplasmic sites (Fig. 3A) combined with selective chemical labelling with photosensitive ouabain, phosphorylation [5,59] and insertion of small hydrophobic probes, INA [55] or TID [56,60]. In contrast, neither chemical labelling experiments nor hydroplot analysis lead to a decision as to whether two, four, or six transmembrane segments are formed by the C-terminal part of the α subunit (residues 779–1016). The model in Fig. 3 has the maximum number of transmembrane segments and their localization in the sequence is given in Table I. Comparison of the hydrophobic labels [59] shows that TID provides a reliable random labelling of amino acid side chains that are in contact with the lipid bilayer [56] while INA or adamantanyl diazirine are prone to nucleophilic attachment. AD labels exclusively the C-

TABLE I

Transmembrane segments in α subunit and β subunit of pig kidney Na,K-ATPase in the model of Fig. 3.

Segment in model		Hydropathic index	Negatively charged residues	Hydrophobic labelling of peptide			Tryptic limit peptide in membrane
No.	Residues in sequence			TID	Adamantanyl diazirine	INA	N-terminal sequence and mass (kDa) [58]
		($w=19$)		[57]	[57]	[57]	

α subunit							
M1	89–109	2.1		Asp^{68}–Lys^{146}		Asp^{68}–Lys^{146}	68 DGPNAL–11 kDa
M2	123–143	2.0		Asp^{68}–Lys^{146}		Asp^{68}–Lys^{146}	68 DGPNAL–11 kDa
M3	287–306	2.5		Ile^{263}–Lys^{342}			263 IATLAS–9 kDa
M4	314–334	2.0	Glu^{329}	Ile^{263}–Lys^{342}			263 IATLAS–9 kDa
M5	771–791	2.1	Glu^{779}	Ser^{768}–Lys^{827}	Ser^{768}–Lys^{827}		737 QAADMI–8 kDa
M6	797–817	1.6	$Asp^{804,808}$	Ser^{768}–Lys^{827}	Ser^{768}–Lys^{827}		737 QAADMI–8 kDa
M7	845–865	1.5		Leu^{842}–Arg^{880}	Leu^{842}–Arg^{880}		830 NPKTDK–19 kDa
M8	909–930	1.6	Asp^{926}	Not labelled	Not labelled		830 NPKTDK–19 kDa
M9	946–966	1.6	$Glu^{953,954}$	Asn^{935}–Arg^{972}	Asn^{935}–Arg^{972}		830 NPKTDK–19 kDa
M10	974–995	1.2		Met^{973}–Arg^{998}	Met^{973}–Arg^{998}		830 NPKTDK–19 kDa
β subunit							
M1	34–58	2.2		Thr^{27}–Arg^{70}			14 FIWNSE–14 kDa

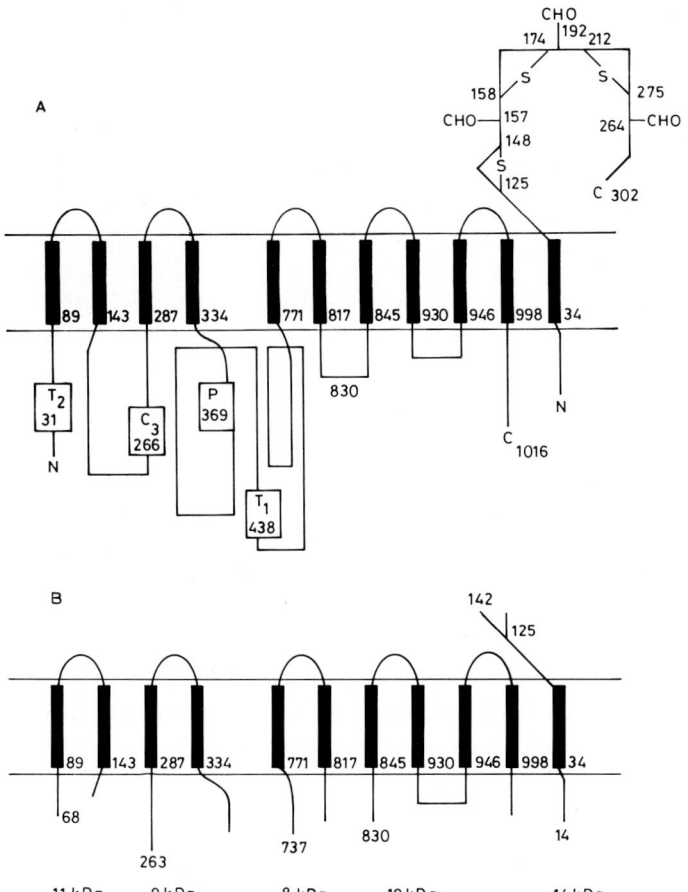

Fig. 3. (A) Disposition of αβ unit in the membrane, based on sequence information [14,15], selective proteolytic digestion of the α subunit [5,6] and hydrophobic labelling (Table I). The model for the β subunit is based on sequencing of surface peptides and identification of S–S bridges [64,65]. T_1, T_2 and C_3 show location of proteolytic splits. CHO are glycosylated asparagines in the β subunit. (B) Peptide fragments remaining in the membrane after extensive tryptic digestion of membrane-bound Na,K-ATPase from outer medulla of pig kidney as described by Karlish et al. [7,58].

terminal half of the α subunit and does not label the β subunit (see [59]). INA in low concentration labels only the N-terminal half of the α subunit probably in a reaction with Cys^{102} and Cys^{138} in M1 and M2, respectively. The INA-labelled intramembrane segment appeared as a 12-kDa fragment in SDS gels [59], and has now been identified in the Asp^{68}–Lys^{146} peptide which can also be labelled with TID [57].

The intramembrane segments consist of 21–25 amino acid residues with overrepresentation of the hydrophobic residues Phe, Ile, Leu, Val, Trp, Tyr, but also of Pro and Cys. This may suggest that S–S bridge formation is part of stabilizing intramembrane structures. Prolines or glycines break the continuity of membrane helices and the excess of proline is interesting in view of the demonstration that membrane-buried proline residues are found in transport proteins, while they are excluded from the membranous domains in non-transport proteins [61]. A few basic or acidic side chains are found in the transmembrane segments and the segments carry charged residues close to their cytoplasmic ends. These charges may react with headgroups of lipids to stabilize the structure in the membrane [62]. During biosynthesis the charged residues may have served as stop signals preventing transfer across the membrane.

One argument for a model with an equal number of eight or ten transmembrane segments in the α subunit is that the C-terminal seems to be exposed to the cytoplasm as it is in the Ca-ATPase of the sarcoplasmic reticulum [63]. Recent immunological studies suggest the presence of a cytoplasmic epitope near the C-terminus of the α subunit. In the model of the α1 subunit of Na,K-ATPase in Fig. 3, the transmembrane segments M1, M2, M3, M4, M5 are predicted by a hydropathic index of >2.0 and three segments M6, M8, and M9 have peak indices of >1.5 using a window of 19 residues or the hydroplot, Table I. Transmembrane segment M8 is included because it has a peak index of 1.6 in the hydroplot, but this segment does not seem to be labelled by the hydrophobic labels. In contrast the peptide Val^{545}–Arg^{589} is consistently labelled by hydrophobic labels [57]. This peptide has a maximum hydropathic index of 0.9 and possesses only short hydrophobic stretches. It may form a hydrophobic pocket in a β-sheet structure of the ATP binding area as proposed on the basis of homologies with adenylate kinase and other ATP binding proteins [6,78].

2.5. Structure of the β subunit of Na,K-ATPase

In the family of cation pumps, only the Na,K-ATPase and H,K-ATPase possess a β subunit glycoprotein (Table II), while the Ca-ATPase and H-ATPase only consist of an α subunit with close to 1 000 amino acid residues. It is tempting to propose that the β subunit should be involved in binding and transport of potassium, but the functional domains related to catalysis in Na,K-ATPase seem to be contributed exclusively by the α subunit. The functional role of the β subunit is related to biosynthesis, intracellular transport and cell–cell contacts. The β subunit is required for assembly of the αβ unit in the endoplasmic reticulum [20]. Association with a β subunit is required for maturation of the α subunit and for intracellular transport of the αβ unit to the plasma membrane. In the β1-subunit isoform, three disulphide bonds are formed by residues Cys^{125}–Cys^{148} Cys^{158}–Cys^{174}, and Cys^{212}–Cys^{275} [64,65] and their reduction is accompanied by loss of Na,K-ATPase activity [64,66]. The β2-subunit isoform seems to have a function as a recognition element for cell adhesion in the brain. This association may also mediate fine regulation of transport because antibodies to the β2 subunit (AMOG = association molecule of glia) cause dissociation of cell–cell adhesion and stimulate Na,K-transport [51].

The bulk of the hydrophilic residues of the β subunit are exposed on the extra-

TABLE II
Homology of β-subunit isoforms of Na,K-ATPase and H,K-ATPase corresponding to beta signatures no. 1 and no. 2 of the Prosite database [70]. Reference to sequences: β1 rat [71], β1 mouse [72], β2 rat [73], β2 mouse [51], β3 *Xenopus* [74], β subunit of H,K-ATPase [75]

Beta signature no. 1																			
Na,K-ATPase																			
β1 rat	11	W	K	K	F	I	W	N	S	E	K	K	E	F	L	G	R	T	G	G	S	W
β1 mouse	11	W	K	K	F	I	W	N	S	E	K	K	E	F	L	G	R	T	G	G	S	W
β2 rat	16	W	K	E	F	V	W	N	P	R	T	H	Q	F	M	G	R	T	G	T	S	W
β2 mouse AMOG	16	W	K	E	F	V	W	N	P	R	T	H	Q	F	M	G	R	T	G	T	S	W
β3 *Xenopus*	15	W	K	Q	F	I	Y	N	P	Q	K	G	E	F	M	G	R	T	A	S	S	W
H,K-ATPase beta rat	16	F	R	Q	Y	C	W	N	P	D	T	G	Q	M	L	G	R	T	P	A	R	W
Beta signature no. 2																						
Na,K-ATPase																						
β1 rat	145	R	K	V	C	R	F	K	L	D	W	L	G	N	C	S	G	L	N	D		
β1 mouse	145	R	K	V	C	R	F	K	L	D	W	L	G	N	C	S	G	L	N	D		
β2 rat	146	K	R	A	C	Q	F	N	R	T	Q	L	G	N	C	S	G	I	G	D		
β2 mouse AMOG	140	K	R	A	C	Q	F	N	R	T	Q	L	G	D	C	S	G	I	G	D		
β3 *Xenopus*	140	K	K	T	C	Q	F	N	R	T	S	L	G	I	C	S	G	I	E	D		
H,K-ATPase beta rat	148	K	F	S	C	K	F	T	A	D	M	L	Q	N	C	S	G	L	V	D		

cellular surface [67]. The β-subunit sequence has one hydrophobic segment (Ile^{34}–Leu^{58}) which is labelled from the bilayer by [^{125}I]-iodonaphthylazide [59] and [^{125}I]-trifluoromethyl-iodo-phenyldiazirine [60]. As an alternative to this model three transmembrane segments are proposed on the basis of papain digestion [68] and immunological studies [69]. It is remarkable that both the Phe^{14}–Arg^{142} and Gly^{143}–Ser^{302} segments remain associated with the membrane even after trypsinolysis and reduction of the Cys^{125}–Cys^{148} disulphide bridge [7], but there is no evidence for additional hydrophobic segments in the Gly^{143}–Ser^{302} segment.

The β subunit is not as well conserved as the α subunit, with 91% overall homology between the β subunit of sheep, pig, and human and 61% between the β subunit of human and *Torpedo*. Homologies between β-subunit isoforms and the β subunit of H,K-ATPase are moderate, 25–30%, but some segments are well conserved. As shown in Table II, beta signatures 1 and 2 are preserved in the β-subunit isoforms and β subunit of H,K-ATPase [70] as an indication that the positions of tryptophans and cysteines are well conserved elements.

3. Nucleotide binding and phosphorylation

A characteristic structural feature of the renal Na,K-pump protein is a cytoplasmic protrusion with approximate dimensions 45 × 65 Å in the plane of the membrane and a length of 50–60 Å in the plane perpendicular to the membrane. The bulk of the protrusion is formed by the large central domain (residues 340–780 in α subunit) that forms sites for ATP binding and phosphorylation. The second cytoplasmic domain (142–284 in α subunit) between M2 and M3 contains peptide bonds susceptible to proteolytic cleavage in E_1 forms and this domain is proposed to be involved in energy transduction [6,33]. The N-terminus is attached to M1 and involved in control of E_1–E_2 transition in Na,K-ATPase and of the rate of Na,K-pumping [6].

3.1. The nucleotide binding domain in the α1β1 unit

3.1.1. Comparison with the nucleotide binding sites in adenylate kinase

An indication of the functional groups required to form a nucleotide site may be obtained from examining sites in dehydrogenases, phosphofructo-kinase, and adenylate kinase [76–78]. The comparison with adenylate kinase is particularly interesting in view of the information available on the structure of the nucleotide binding area in that protein [78] and the homologies demonstrated earlier [6]. An important feature in the model of the ATP site in adenylate kinase is a hydrophobic pocket for accommodation of the adenine and ribose moieties which is formed by Ile, Val, His and Leu residues. The triphosphate moiety is flanked by a hydrophobic strand of parallel β-pleated sheet terminated by Asp [78]. This segment in adenylate kinase and in the β subunit of F1-ATPase shows some homology with respect to charges and hydrophobic residues to segments in α subunit of Na,K-ATPase (543–561). The observation that this segment is labelled by TID or AD [57] shows that it may also form a hydrophobic pocket in the nucleotide binding domain of Na,K-ATPase. Such a pocket may explain the high affinity of the reaction of the nucleotide site of Na,K-ATPase for reaction with compounds like trinitrophenyl-ATP [79] and tetrabromo-fluorescein (eosin) [80].

Transferred nuclear Overhauser effect (TRNOE) measurements complement the sequence data in the sense that part of the ATP moiety is organized in a manner similar to that in adenylate kinase [78], while there is a significant difference in the torsion angle of the bond between adenine and ribose. In Na,K-ATPase [81] the bound ATP adopts an anti conformation for the adenine ring with respect to the ribose ($\delta = 0 \pm 90°$) with a glycosidic torsion angle (δ) of 35°. The conformation of the ribose ring is N-type (C_2,-exo, E_3,-endo) with a torsion angle $\delta = 100°$. The orientation of O_5 relative to the ribose is determined by the torsion angle, $\delta = 178°$, a typical value for protein bound Mg-ATP [78].

3.1.2. Selective chemical labelling with ATP analogues

Labelling Na,K-ATPase with ATP analogues provides evidence for contribution from charged residues that are widely separated in the sequence of α subunit of Na,K-ATPase. The first indication came from ATP sensitive covalent insertion of fluorescein-isothiocyanate (FITC) into Lys^{501} in the α subunit [90]. The strong fluorescence signal provides a convenient probe for monitoring conformational transitions in the proteins. Site-directed mutagenesis of Lys^{501} reduces the activity of

Na,K-ATPase [82]. Asp^{710} in the highly conserved 704–722 segment is covalently labelled by ClR-ATP (α-[4-(N-2-chloroethyl-N-methyl-amino)]benzoyl-amide-ATP) [83]. Another conserved sequence, 470–487, is labelled by 8-azido-ATP and Lys^{480} in this segment is targeted by pyridoxal-ATP [84]. The lysine selective reagent (N-isothiocyano-phenyl-imidazole) also reacts in a conformation specific manner to block ATP binding, but the target residue has not been identified [85]. In these studies it has not been excluded that the chemical modification alters the conformational adaptability of the protein rather than blocking side chains that are contributing coordinating groups for the nucleotides.

3.2. Conformations of the nucleotide binding area

The segments contributing to nucleotide binding and phosphorylation domains undergo structural changes accompanying E_1–E_2 transition as the nucleotide binding region adapts for tight binding in the E_1 form with K_D 0.1 μM for ATP, while binding to the E_2 form is weak requiring millimolar concentrations of ATP for saturation.

The α subunit provides the necessary segments for formation of a nucleotide binding area, but the possibility of more than one nucleotide site per αβ unit is often raised. Two separate ATP sites with high and low affinity have been proposed on the basis of kinetic studies or inactivation experiments [86]. The different affinities for ATP can be explained [79] by the alternating E_1 and E_2 conformations with high and low ATP affinities, respectively. Since the maximum capacity of ATP binding in equilibrium experiments never exceeds one ATP bound per αβ unit, one may assume that a presumptive additional site has a low affinity and fewer coordinating groups than a site for high affinity binding.

3.3. The phosphorylation site, high- and low-energy phosphoforms, E_1P–E_2P

A sequence of ten amino acids (ICS–D–KTGTLT) around the phosphorylation site of Na,K-ATPase (Asp^{369}) is highly conserved among the Na,K-, H,K-, Ca-, and H-pumps [6]. There is also homology with the β subunit of F_1-ATP synthetase of mitochondria and chloroplasts (see [6]) except that Asp is replaced by Thr. Accordingly a covalent phosphorylated intermediate is not formed in F_1-ATPase. Mutagenesis of the phosphorylated aspartate residue in Na,K-ATPase [82], Ca-ATPase [87], or H-ATPase [88] completely blocks activity.

Transition from the 'high-energy' phosphoform $E_1P[3Na]$ to the K-sensitive $E_2P[2Na]$ of Na,K-ATPase are accompanied by conformational transitions in protein structure and changes of the capacity and orientation of cation sites. In the E_1 form of Na,K-ATPase, the exposure of Chy$_3$ (Leu^{266}) and Try$_3$ (Arg^{262}) to cleavage reflects that the cation sites of the phosphoprotein are in a conformation oriented towards the cytoplasm with a capacity for occlusion of three Na$^+$ ions. The E_2 form with exposed Try_1 (Arg^{438}) and protected Chy_3 and Try_3 occludes either $2Na^+$ or $2Rb^+(K^+)$ in the phosphoform or $2Rb^+(K^+)$ in the unliganded enzyme and it exchanges cations at the extracellular surface [6,33,89].

In the scheme in Fig. 4 [6,89], the E_1P–E_2P transition releases a single Na^+ ion at the extracellular surface and $E_2P[2Na]$ represents an occluded state in transition to E_2P–$2Na$ with Na^+ leaving the sites making them accessible for binding of K^+ from the extracellular phase. In a scheme involving two cycles the term E^*P was used for this intermediate [90] and it was observed that ouabain reacts with E^*P [91]. Studies of protein conformation and cation occlusion show that the correct notation for E^*P is $E_2P[2Na]$, since E_2P can occlude either $2Na^+$ or $2K^+$, without altering protein conformation as detected by proteolysis. In Na^+ medium, the $E_2P[2Na]$ intermediate is sensitive to ADP, because binding of one Na^+ allows it to return to the $E_1P[3Na]$ form for reaction with ADP and formation of ATP. After addition of K^+, exchange of Na^+ for K^+ at the extracellular surface would lead to dephosphorylation. The apparent ambiguity of the $E_2P[2Na]$ form with respect to reactivity to ADP and K^+ is therefore explained by the cation site occupancy, while the protein conformation of the $E_2P[2Na]$ intermediate is the same as that of other E_2 forms. With these properties of the $E_2P[2Na]$ complex, the ADP sensitive fraction of the phosphoenzyme comprises both E_1, and E_2 forms, namely the Na-occluded, $E_1P[3Na]$ and $E_2P[2Na]$. The conventional definition that the ADP-sensitive fraction of the phosphoenzyme corresponds to the amount of the E_1P form is therefore no longer valid. The redefinition of the ADP sensitive phosphoenzyme solves an objection that has repeatedly been raised towards the role of the phosphoenzyme in the Na,K-pump reaction, that the sum of the ADP- and K-sensitive phosphoenzymes exceed the total amount of phosphoenzyme (see [92]). Using the definitions above and

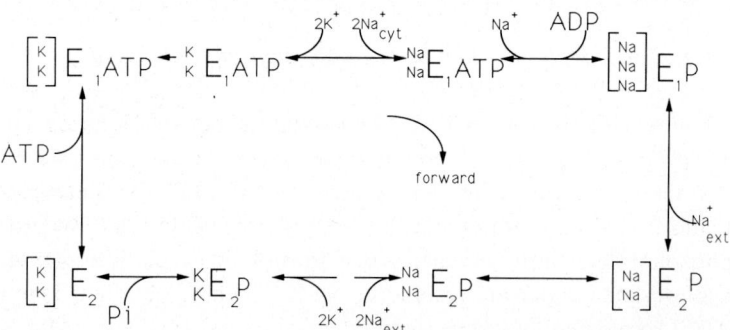

Fig. 4. E_1–E_2 reaction cycle of the Na,K-pump with four major occluded conformations and ping-pong sequential cation translocation. The phosphoforms can occlude Na^+ and dephosphoforms can occlude K^+ or Rb^+. Na^+ and K^+ without brackets are cations bound to an open form such that they can exchange with medium cations [Na^+] or [K^+] within brackets are occluded and prevented from exchanging with medium cations. It is proposed that release of Na^+_{ext} accompanies transition from $E_1P[3Na]$ to $E_2P[2Na]$, since the capacity for occlusion of Na^+ in the ouabain-stabilized E_2P form is lower than in the E_1P form prepared by incubation with CrATP [29] or oligomycin [89].

in Fig. 4, the sum of the amounts of ADP-sensitive and K-sensitive is equal to $E_1P[2Na]$ plus $2 \times E_2P[2Na]$ plus $E_2P[O]$ and thus exceeds the total phosphoenzyme by the amount of the $E_2P[2Na]$ intermediate.

Based on a series of studies of the effect of organic solvent on the reaction of Ca-ATPase with P_i and ATP synthesis, De Meis et al. proposed that a different solvent structure in the phosphate microenvironment in E_1 and E_2 forms the basis for existence of high- and low-energy forms of the aspartyl phosphate [93]. Acyl phosphates have relatively low free energy of hydrolysis when the activity of water is reduced, due to the change of solvation energy. The covalently bound phosphate may also reside in a hydrophobic environment in E_2P of Na,K-ATPase since increased partition of P_i into the site is observed in presence of organic solvent [6] in the same manner as in Ca-ATPase.

4. Cation binding and occlusion

To understand the Na,K-pump mechanism it is obviously important to identify the cation pathway and the sites for binding and occlusion of Na^+ and K^+ relative to the intramembrane portion of the protein. The groups coordinating the cations should be identified and it should be known if the pump has independent sites for Na^+ and K^+ or if the cations bind alternately to the same set of sites. With a stoichiometry of $3Na^+/2K^+$ per ATP split this would mean that two sites bind Na^+ and K^+ alternately, while one site only binds Na^+.

Known structures of other proteins or ionophores can be useful models for the cation sites in the Na,K-pump. In models for ionophores, cation binding sites consist of electrophilic carbonyl groups that are located in a fixed cavity to accommodate the size of the selected cations inside the ionophore. Transition from an open to a closed or occluded configuration of the cation binding sites in the ionophore involves only limited changes in conformation of carbonyl residues and a substitution of solvent molecules from the inner coordination sphere of the cation [94]. This arrangement allows for excellent cation specificities, e.g., the affinity of valinomycin for K^+ is about six orders of magnitude higher than for Na^+. The complex of a crown ether (cyclohexyl-18-crown-6) with Na^+ has five nearly coplanar oxygens surrounding the Na^+ with the ion coordination completed by the remaining oxygen of the hexaether and a water molecule, and with Na–O distances of 2.5–2.6 Å [95]. The corresponding complex with K^+ can be formed with minimal distortion of the cyclic ether and with bond lengths of 2.8 Å.

In the Ca-ATPase from sarcoplasmic reticulum, oligonucleotide-directed, site-specific mutagenesis has been applied to identify amino acids involved in Ca^{2+} binding. Mutation of 30 glutamate and aspartate residues, singly or in groups, in a stalk sector near the transmembrane domain has little effect on Ca^{2+}-transport. In contrast mutations to Glu^{309}, Glu^{771}, Asn^{796}, Thr^{799}, Asp^{800} or Glu^{908} resulted in loss of Ca^{2+}-transport [96]. As an indication that these mutations abolished Ca^{2+}-binding, phosphorylation from inorganic phosphate was observed even in the presence of Ca^{2+}. These data suggest that the carboxylate or carboxamide side chains or hydroxyl groups form coordinating groups in high affinity binding sites for Ca^{2+} near the centre of the transmembrane domain. The homology around these residues between Ca-ATPase and α subunit of Na,K-ATPase and other pump proteins suggest that their cation sites may have a similar location and overall structure.

4.1. Capacity for binding and occlusion of Na^+ or K^+ (Rb^+)

It is generally accepted that Na^+ ions can be occluded in E_1P forms. Occlusion of $3Na^+$ ions per EP has been demonstrated in chymotrypsin cleaved enzyme and in the Cr-ADP-E_1P[3Na] complex [29]. Three Na^+ ions can also be occluded per EP in a complex stabilized by oligomycin in the absence of Mg^{2+} or phosphate [97] while a maximum of two Na^+ ions are occluded per α subunit in the ouabain complex.

In preparations of high purity, the cation binding data above therefore correspond to $2Rb^+$ and $3Na^+$ occluded per αβ unit both in the soluble and in the membrane bound state. Fig. 5 shows that the capacity for occlusion is either $2Rb^+$ or $2Na^+$ per αβ unit. The apparent affinity of these E_2 forms for the cations varies over a wide range from $K_{1/2}(Rb) \sim 9\,\mu M$ to $K_{1/2}(Na) \sim 1.7\,mM$, but without changes in proteolytic cleavage patterns. The apparent affinity of ^{86}Rb for formation of the Mg-ouabain-E_2P[Rb] complex is >10-fold higher than for Na [89]. The difference in apparent affinity for the complexes in Fig. 5 is sufficient for allowing exchange of Na for Rb(K) at the extracellular surface, but the transition from E_2P[2Na] to E_2[2Rb] has no influence on the protein conformation as judged by the patterns of proteolytic digestion or the fluorescence levels.

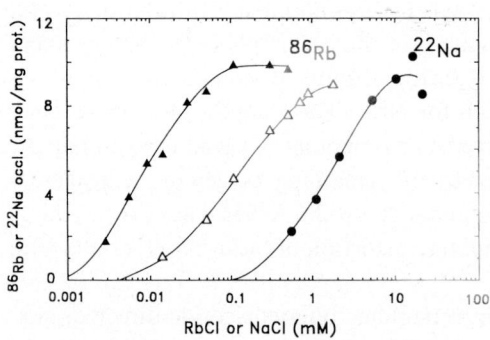

Fig. 5. Occlusion of ouabain complexes with (●) ^{22}Na or (△) ^{86}Rb and unliganded Na,K-ATPase with (▲) ^{86}Rb. Procedure as described before [89] using incubation at increasing concentrations of Na^+ or ^{86}Rb for 15 min at 20°C with 1 mM ouabain, 1 mM $MgCl_2$ and 1 mM P_i-Tris.

4.2. Isolation of the cation occlusion and transport path after tryptic digestion

Using the renal membrane bound Na,K-ATPase, extensive tryptic digestion provides a method for isolating the cation sites and the pathway for cations across the membrane as a structure quite separate from the ATP binding and phosphorylation sites. The intramembrane part of Na,K-ATPase is protected from tryptic digestion by occlusion of either K^+ or Na^+ while ATP binding and phosphorylation are rapidly abolished as the entire cytoplasmic protrusion is digested away. As the fragments remaining in the membrane are the same whether Rb^+ or Na^+ ions are protecting the sites against inactivation, the two ions may alternately bind to the same coordinating groups in the intramembrane part of the enzyme. The sites seem to remain accessible from both membrane surfaces after the extensive cleavage [7,58].

The intramembrane fragments capable of occluding cations in the digested membranes (Table I, Fig. 3B) comprise a 19-kDa fragment plus 8-kDa, 9-kDa and 11-kDa fragments of the α subunit and a 14-kDa fragment of the β subunit. The presence of residues coordinating cations is demonstrated by [^{14}C]DCCD binding to the fragments. Thus K^+-protected labelling was demonstrated in the 19-kDa fragment [7,58] suggesting that Glu^{954} and Glu^{955} are candidates for K^+-binding carboxyls.

4.3. Transport stoichiometry and net charge of Na^+ and K^+ complexes with Na,K-ATPase

The membrane potential can be used as a tool for characterization of the properties and net charge of the major conformations of Na,K-ATPase and their complexes with the cations. The potential difference across the membrane represents large electric fields of up to 500 kV/cm. If the conformational states of the Na,K-pump have different electrical properties with respect to net charge or movable dipoles, the equilibria between conformations should interact with the field. The field may also affect binding of Na^+ or K^+ by altering the concentration of cation near the pump protein or by altering the position of charged groups in the cation sites.

In phospholipid vesicles reconstituted with renal Na,K-pumps, the membrane potential can be controlled using gradients of K^+ or Li^+ and appropriate ionophores. At high or low ATP concentrations, the conformational change E_1P3Na–E_2P3Na is accelerated by voltage, whereas the $E_2[2K]$–$E_1[2K]$ step is independent of potential [97]. At limiting low Na^+ concentrations the imposed potential stimulates Na-transport showing that the electric field has an effect on the intrinsic binding affinity for Na^+. This may mean that Na^+ has to cross part of the electric field in order to reach its sites, thus becoming concentrated by the field. Alternatively, the voltage may cause charges to move or dipoles to reorient in the cation binding area of the protein. Studies of the conformational transitions using fluorescein attached covalently to Lys^{501} [98] also shows that the E_1P–E_2P isomerization is voltage dependent while the E_2K–E_1K step is electrically silent.

The purified membrane bound Na,K-ATPase from kidney adsorbs to planar lipid bilayers and transient currents can be elicited upon release of caged ATP [99,100]. The transient pump currents depend on Mg^{2+} and Na^+, but not K^+, and they are abolished by chymotryptic cleavage. This agrees with the inference that Na^+ translocation and charge movements precede the K^+-transport step. Similar conclusions are reached from studies of presteady state fluxes of Na^+-ions in plasma membrane or reconstituted vesicles [101]. With the assumption that $3Na^+$-ions are transported into and $2K^+$ ions out of the vesicles in a ping-pong reaction, these results mean that the net charge is $+1$ with $3Na^+$ ions bound and zero with $2K^+$ ions bound [98].

5. Structural transitions in the protein related to energy transformation and Na,K-transport

Transduction of the energy from ATP to movement of the cations may involve long-range structural transitions in the protein since ATP binding and phosphorylation takes place in the large cytoplasmic protrusion of the α subunit, while cation sites may be located in intramembrane domains. It is therefore important to establish relationships between the structural changes in the α subunit and ion binding or occlusion to see if the different exposure of bonds to proteolysis reflect the orientation and specificity of the cation sites.

The conformational transitions in Na,K-ATPase are large and easy to detect. In addition to bonds exposed to proteolysis [6], the transition involves tryptophans, sulfhydryl groups, protonizable groups and residues binding FITC [102] and iodoacetamide fluorescein [103], and the conformational changes involve residues in ATP binding and phosphorylation domains and they are transmitted to the extracytoplasmic surface with changes in binding affinity for cations and ouabain, see previous reviews for details [6,92].

5.1. Conformation dependent proteolytic cleavage of Na,K-ATPase

Definition of E_1 and E_2 conformations of the α subunit of Na,K-ATPase involves identification of cleavage points in the protein as well as association of cleavage with different rates of inactivation of Na,K-ATPase and K-phosphatase activities [104,105]. In the E_1 form of Na,K-ATPase the cleavage patterns of the two serine proteases are clearly distinct. Chymotrypsin cleaves at Leu^{266} (C_3), Fig. 3A, and both Na,K-ATPase and K-phosphatase are inactivated in a monoexponential pattern [33,106]. Trypsin cleaves the E_1 form rapidly at Lys^{30} (T_2) and more slowly at Arg^{262} (T_3) to produce the characteristic biphasic pattern of inactivation. Localization of these splits was determined by sequencing N-termini of fragments after isolation on high resolution gel filtration columns [107].

The E_2 form is not cleaved by chymotrypsin, but trypsin cleaves at Arg^{438} (T_1) and subsequently at Lys^{30} (T_2) and tryptic inactivation of E_2K or E_2P forms is linear and associated with cleavage at Arg^{438} (T_1) [104,108]. Inactivation of K-phosphatase is delayed because cleavage of T_1 and T_2 in sequence is required for inactivation of K-phosphatase activity [105].

Thus, transition from E_1 to E_2 consists of an integrated structural change involving protection of bond C_3 or T_3 in the second cytoplasmic domain and exposure of T_1 in the central domain, while the position of T_2 in the N-terminus is altered relative to the central domain (T_1) so that cleavage of T_2 becomes secondary to cleavage of T_1 within the same α subunit in the E_2 form.

5.2. Tryptophan fluorescence and secondary structure changes

Movement of peptide segments from cytoplasmic to hydrophobic environments accompanying E_1–E_2 transitions in Na,K-ATPase may be related to changes in intensity of intrinsic fluorescence in the proteins. Quenching analysis of Na,K-ATPase suggests that the tryptophan fluorescence changes are directly related to change of overall protein conformation [108]. Tryptophan fluorescence is increased 2–3% by transition from E_1 to E_2 in Na,K-ATPase [cf. 109]. This quantitative difference in fluorescence of E_1 and E_2 forms may be understood in terms of different locations of the tryptophans in the protein structures. In Na,K-ATPase, only two of a total of twelve tryptophans in the α subunit are located in predicted transmembrane segments.

Circular dichroism (CD) spectroscopy may detect changes in the ratio between secondary structure elements, but even large shifts in position of α-helices and β-sheets relative to each other do not affect CD spectra. The results of experiments addressing the question whether E_1–E_2 transitions are accompanied by changes in the ratio among secondary structure elements raised some controversy. CD spectroscopy shows that Na,K-ATPase contains a roughly equal mixture of α-helical, β-sheet and random coil structures [110] and changes in CD spectra accompanying exchange of Na^+ for K^+ are interpreted to involve α-helix–β-sheet transition of about 7%. A conversion in the opposite direction of 100 residues from β- to α-helical conformation is suggested by Raman spectroscopy [111]. In apparent contrast to this, another group found that addition of K^+ to a Tris-HCl medium did not cause changes in CD spectra [112]. Infrared spectroscopy in the amide I region shows that Na^+ and K^+ bound forms of renal Na,K-ATPase have almost the same structure [113], but it is not known if the enzyme is active in D_2O. It is not clear to what extent these contradictory results concerning secondary structure transitions are related to the special structure of the membrane bound Na,K-ATPase.

5.3. Cleaved derivatives; cleavage of bond 2 and the regulatory function of the N-terminus

Selective cleavage of bonds in the α subunit of Na,K-ATPase is important for examining structure–function relationships for the protein. The N-terminus of the α subunit is strongly hydrophilic with clusters of alternating positive and negative residues between residue 15 and 60. It is a flexible structure with a strong propensity for α-helix formation and several predicted turns, notably at residues 14–16, 35–36, 49–50 and 70–72. The removal of residues 1–30 by selective tryptic cleavage at Lys^{30} (T_2) is possible because the rate of cleavage of this bond is up to 60-fold higher than the rate of tryptic cleavage at Arg^{262} (T_3) in the second slow phase of inactivation. T_2 cleavage reduces Na,K-ATPase activity by 50–60% [104] with a parallel loss of Na,K-transport [109]. The loss of activity is explained by a poise of equilibria between cation bound (E_1Na–E_2K) and phosphoforms (E_1P–E_2P) in the direction of E_1 forms [114]. These properties of the selectively cleaved derivative suggested that charged residues in the N-terminus engage in salt bridge formation as part of E_1–E_2 transition and that this is important for control of the rate of active Na,K-transport by pumps containing the α1 isoform.

These data suggest that the substantial differences between the amino acid sequences of the N-terminal region in the α1, α2, and α3 isoforms [19] reflect different regulatory functions of these segments. The first 30 residues of the α1 isoform have a high frequency of charges with eight lysins and two arginins. In the α2 isoform there are two negative and three positive charges fewer in this segment than in the α1 isoform suggesting that the strength of the salt bridges formed with other segments may be weaker. The first 11 residues of the α3 isoform are not homologous with the other isoforms, but all isoforms of the α subunit have a stretch of lysins around residue 30. Also the N-terminus of H,K-ATPase [115] possesses a lysine-rich sequence with strong homology with the α1 subunit around T_2. Biphasic cleavage patterns resembling those in Na,K-ATPase have also been demonstrated in H,K-ATPase [116], but so far without identification of cleavage points.

5.4. Effect of C_3 cleavage on E_1P–E_2P transition and cation exchange

The alternating exposure of C_3 (Leu^{266}) or T_3 (Arg^{262}) in the E_1 form and T_1 (Arg^{438}) in the E_2 form reflects that motion within the segment ($M_r = 18\,170$) between these bonds including the phosphorylated residue (Asp^{369}) is an important element in E_1–E_2 transition. This is illustrated by the widely different consequences of selective cleavage of C_3 and T_1 for E_1–E_2 transition and cation exchange.

C_3 cleavage is a selective and particularly efficient tool for examining structure–function relationships of the second cytoplasmic domain. Binding affinities for ADP and ATP are reduced 4–5-fold, while TNP–ATP binds with the same affinity as in native Na,K-ATPase. Nucleotide binding is not affected by K^+ or Rb^+ although

TABLE III
Properties of the C_3-cleaved derivative of Na,K-ATPase

Enzymatic activity, ligand binding or transport capacity	Cleavage C_3 Leu^{266}	Control
Na,K-ATPase	0%	100%
ATP–ADP exchange	4–500%	100%
ADP binding capacity	100%	100%
ADP binding affinity (K_D, μM)	0.075	0.045
Phosphorylation capacity	100%	100%
$\quad E_1P/E_2P$ ratio 100/0	14/84	
Vanadate binding capacity	0%	100%
Rb-binding capacity	100%	100%
Rb-binding affinity (K_D, μM)	9–12	9–12

Compiled from [6] and [33].

cation sites are undamaged. Conversely the cleaved enzyme also binds ^{86}Rb with high affinity and it occludes the cations, but cation binding and occlusion are unaffected by nucleotides, Table III.

Transport studies in reconstituted vesicles show that C_3 cleavage blocks the relatively fast Na–Na or K–K exchange (20–40 s^{-1}) and Na–K exchange (500 s^{-1}), but the slow passive ouabain-sensitive Rb–Rb exchange (1 s^{-1}) and occlusion of K$^+$ or Na$^+$ are only partially affected. C_3 cleavage allows formation of E_1P[3Na], but prevents charge transfer coupled to Na$^+$ translocation in purified Na,K-ATPase after adsorption to a planar lipid bilayer [117]. In combination with the observation of the occlusion of 2Na$^+$ in the E_2P[2Na] these data show that E_2P–E_2P transition represents a charge translocating step and that the single Na$^+$ ion released at the extracellular surface represents this transfer of charge. C_3 cleavage thus interferes with structural changes that alter the capacity of the cation sites for occlusion of Na$^+$ and presumably their orientation.

5.5. Mutagenesis in yeast H-ATPase and Ca-ATPase from sarcoplasmic reticulum

The notion that the segment containing C_3 and T_3 is important for conformational adaptability of the protein is supported by mutations in yeast. Mutations of the genes of the H-ATPase of *Saccharomyces cerevisiae* resulted in a thermo-sensitive mutant (Gly^{254}→Ser) [118]. In *Schizosaccharomyces pombe* the Gly^{268}→Asp substitution is responsible for a mutant phenotype with reduced ATPase and proton pumping activity and vanadate resistance [119]. The mutation may thus produce a higher concentration of E_1 forms and less of the vanadate binding E_2 form during steady state ATP hydrolysis. In interpreting the work on mutations in H-ATPase, the view was forwarded that mutations in the second cytoplasmic loop disrupt an endogenous phosphatase activity [118]. Inactivation of phosphatase activity after the Glu^{233}→Gln mutation (Glu^{233} in H-ATPase is homologous to Glu^{183} in Ca-ATPase) suggested that the defect was in the hydrolytic step of the catalytic cycle. However, the measurements of dephosphorylation rates in Ca-ATPase indicate that the conformational change and the E_1P–E_2P interconversion are blocked in the mutants rather than the hydrolysis of E_2P [96].

In the second cytoplasmic domain between M2 and M3 in Ca-ATPase of sarcoplasmic reticulum, cleavage of bonds at Lys^{234} and Arg^{236} [120] or cleavage between Glu^{231} and Ile^{232} with V8 protease [121] inactivates the enzyme. The cleavage product receives phosphate from ATP, but it is unable to undergo the interconversion between the E_1P and E_2P forms of the phosphoenzyme. Phosphorylation is activated by Ca^{2+}, but occlusion data are not reported. Site-directed mutagenesis of Gly^{233} (for Val, Gln, or Arg) of Ca-ATPase located between the two proteolytic splits gives mutants that are inactive with respect to Ca-ATPase or Ca-transport. Again, ATP and Ca^{2+} can bind and form an ADP-sensitive phosphoenzyme which is deficient in the transition to the ADP-insensitive form. In this mutant the apparent affinity in the backward reaction with P_i is reduced ($K_{1/2}$ 250 μM) relative to the wild type ($K_{1/2}$ 50 μM) suggesting that the E_2P intermediate is destabilized [122]. It is proposed that the region around Gly^{233} is involved in energy transduction between the phosphate site and the cation binding sites in the transmembrane region.

Mutations in another region, the second cytoplasmic loop between M2 and M3 in Ca-ATPase of sarcoplasmic reticulum (Thr^{181}→Ala, Gly^{182}→Ala, and Glu^{183}→Gln) also result in a complete loss of Ca-transport and Ca-ATPase activity associated with a dramatic reduction in the rate of phosphoenzyme turnover [96]. These mutations do not affect the affinity of the enzyme for P_i and therefore resemble the Pro^{312} mutants [123] in that they affect only the E_1P–E_2P conformational change and not the affinities for ATP, Ca^{2+} or P_i.

5.6. Coupling to ion translocation

The sequence around C_3 in the α subunit of Na,K-ATPase is homologous with segments of Ca-ATPase from sarcoplasmic reticulum and H-ATPase from yeast and plants. Selective proteolytic cleavage of Ca-ATPase and site-directed mutagenesis of both pumps stabilizes E_1P and interferes with structural transitions and cation translocation in the same manner as observed after C_3 cleavage in Na,K-ATPase. The observations show that it is a general feature that this region of the cation pump proteins is involved in structural transitions accompanying energy transduction between the phosphate site and the cation sites. In spite of these interesting similarities between conformational transitions in Ca-ATPase in sarcoplasmic reticulum and the α1 subunit of Na,K-ATPase, caution is necessary in drawing parallels between the Ca-transport reaction and Na,K-pumping. A Ca-occluded E_2 form of Ca-ATPase has not been demonstrated and there is no exchange of Ca^{2+} with another cation at the extracellular surface. It has also been difficult to produce evidence for two state models with different orientations of the calcium binding sites relative to the membrane. In the Ca-ATPase, dephosphorylation is followed by reaction with cytoplasmic Ca^{2+}, while in the normal cycle of Na,K-ATPase, dephosphorylation is followed by binding and occlusion of two extracellular K^+ ions to form the relatively stable $E_2[2K]$ form that must react with ATP for transition to the E_1 form.

The data discussed above provide evidence for the existence of two conformations with different capacities and orientations of cation sites. In the E_1 form of Na,K-ATPase, the exposure of $C_3(T_3)$ to cleavage reflects that the cation sites of the phosphoprotein are in an inward oriented conformation with a capacity for occlusion of $3Na^+$ ions. The E_2 form with exposed T_1 and protected $C_3(T_3)$ occludes either $2Na^+$ or $2Rb^+(K^+)$ in the phosphoform or $2Rb^+(K^+)$ in the unliganded enzyme.

The selective C_3-cleavage experiments shows that the E_1P–E_2P isomerization is coupled to the major charge carrying step in the reaction cycle, the deocclusion of one Na^+ ion during transition from $E_1P[3Na]$ to $E_2P[2Na]$ with release of the Na^+ ion to the extracellular side. The implication of two Na-occluded forms does not require more than one set of sites. The change in conformation of the protein that opens and closes barriers may also alter the position of a given set of cation binding sites. Extensive proteolysis of Na,K-ATPase [58] or mutagenesis experiments in Ca-ATPase [96] have shown that carboxyl groups in transmembrane segments contribute to formation of Na^+ and Ca^{2+} binding sites. It is possible that the conformational transition involves rotation or tilting of the M4 transmembrane helix altering the number of cation coordinating groups and their orientation as previously proposed by Tanford [124].

Acknowledgements

Work in the author's laboratory is supported by the Danish Medical Research and Natural Sciences Council and Novo's Foundation.

References

1. Jørgensen, P.L. (1974 and 1988) Methods Enzymol. 32, 277–294 and 156, 29–143.
2. Jørgensen, P.L. (1975) Quart. Rev. Biophys. 7, 239–274.
3. Jørgensen, P.L. and Andersen, J.P. (1986) Biochemistry 25, 2889–2897.
4. Kyte, J. (1981) Nature (London) 292, 201–204.
5. Jørgensen, P.L. (1982) Biochim. Biophys. Acta 694, 27–68.
6. Jørgensen, P.L. and Andersen, J.P. (1988) J. Membr. Biol. 103, 95–120.
7. Karlish, S.J., Goldshleger, R. and Stein, W.D. (1990) Proc. Natl. Acad. Sci. U.S.A. 87, 4566–4570.
8. Hebert, H., Jørgensen, P.L., Skriver, E. and Maunsbach, A.B. (1982) Biochim. Biophys. Acta 689, 571–574.

9. Hebert, H., Skriver, E. and Maunsbach, A.B. (1985) FEBS Lett. 187, 182–187.
10. Mohraz, M., Simpson, M.V. and Smith, P.R. (1987) J. Cell. Biol. 105, 1–8.
11. Ovchinnikov, Y.A., Demin, V.V., Barnakov, A.N., Kuzin, A.P., Lunev, A.V., Modyanov, N.N. and Dzhandzhugazyan, K.N. (1985) FEBS Lett. 190, 73–76.
12. Jørgensen, P.L. (1980) Physiol. Rev. 60, 864–917.
13. Jørgensen, P.L. (1986) Kidney Int. 29, 10–20.
14. Shull, G.E., Schwartz, A. and Lingrel, J.E. (1985) Nature (London) 316, 691–695.
15. Ovchinnikov, Y.A., Modyanov, N.N., Broude, N.E., Petrukhin, K.E., Grishin, A.V., Arzamazova, N.M., Aldanova, N.A., Monastyrskaya, G.S. and Sverdlov, E.D. (1986) FEBS Lett. 201, 237–245.
16. Shull, G.E., Lane, L.K. and Lingrel, J.B. (1986) Nature (London) 321, 429–431.
17. Lingrel, J.B., Orlowski, J., Shull, M.M. and Price, E.M. (1990) Prog. Nucleic Acid Res. Mol. Biol. 38, 37–89.
18. Sweadner, K. (1989) Biochim. Biophys. Acta 988, 185–220.
19. Shull, G.E., Greeb, J. and Lingrel, J.B. (1986) Biochemistry 25, 8125–8132.
20. Gehring, K. (1990) J. Membr. Biol. 115, 109–121.
21. Takeyasu, K., Lemas, M.V. and Fambrough, D.M. (1990) Am. J. Physiol. 259, C619–C630.
22. Horisberger, J.D., Lemas, V., Kraehenbühl, J.P. and Rossier, B.C. (1991) Annu. Rev. Physiol. 53, 565–584.
23. Cornelius, F. (1991) Biochim. Biophys. Acta 1071, 1–66.
24. Pedemonte, C.H. and Kaplan, J.H. (1990) Am. J. Physiol. 258, C1–C23.
25. Glynn, I.M. and Karlish, S.J.D. (1990) Annu. Rev. Biochem. 59, 171–205.
26. Kaplan, J.H. and DeWeer, P. (Eds.) (1991) The Sodium Pump: Structure, Mechanism, and Regulation, Rockefeller University Press, New York.
27. Kaplan, J.H. and DeWeer, P. (Eds.) (1991) The Sodium Pump: Recent Developments, Rockefeller University Press, New York.
28. Brotherus, J.R., Møller, J.V. and Jørgensen, P.L. (1981) Biochem. Biophys. Res. Commun. 100, 146–154.
29. Vilsen, B., Andersen, J.P., Petersen, J. and Jørgensen, P.L. (1987) J. Biol. Chem. 62, 10511–10517.
30. Deguchi, N., Jørgensen, P.L. and Maunsbach, A.B. (1977) J. Cell. Biol. 74, 619–634.
31. Skriver, E., Maunsbach, A.B. and Jørgensen, P.L. (1981) FEBS Lett. 131, 219–222.
32. Hebert, H., Jørgensen, P.L., Skriver, E. and Maunsbach, A.B. (1982) Biochim. Biophys. Acta 689, 571–574.
33. Jørgensen, P.L., Skriver, E., Hebert, H. and Maunsbach, A.B. (1982) Ann N.Y. Acad. Sci. 402, 203–219.
34. Mohraz, M., Simpson, M.V. and Smith, P.R. (1987) J. Cell Biol. 105, 1–8.
35. Ovchinnikov, Y.A., Demin, V.V., Barnakov, A.N., Kuzin, A.P., Lunev, A.V., Modyanov, N.N. and Dzhandzhugazyan, K.N. (1985) FEBS Lett. 190, 73–76.
36. Zamphigi, G., Kyte, J. and Freytag, W. (1984) J. Cell Biol. 98, 1851-1864.
37. Dux, L., Taylor, K.A., Ting-Beall, H.P. and Martonosi, A. (1985) J. Biol. Chem. 260, 11730–11743.
38. Rabon, E., Wilke, M., Sachs, G. and Zampighi, G. (1986) J. Biol. Chem. 261, 1434–1439.
39. Hebert, H., Skriver, E. and Maunsbach, A.B. (1985) FEBS Lett. 187, 182–187.
40. Castellani, L., Hardwicke, P.M.D. and Vibert, P. (1985) J. Mol. Biol. 185, 579–594.
41. Taylor, K.A., Dux, L. and Martonosi, A. (1986) J. Mol. Biol. 187, 417–427.
42. Dux, L., Pikula, S., Mullner, N. and Martonosi, A. (1987) J. Biol. Chem. 262, 6439–6442.
43. Taylor, K.A., Mullner, N., Pikula, S., Dux, L., Peracchia, C., Varga, S. and Martonosi, A. (1988) J. Biol. Chem. 263, 5287–5294.
44. Stokes, D.L. and Green, N.M. (1990) J. Mol. Biol. 213, 529–538.
45. Stokes, D.L. and Green, N.M. (1990) Biophys. J. 57, 1–14.
46. Molitoris, B.A., Chan, L.K., Shapiro, J.I., Conger, J.D. and Falk, S.A. (1989) J. Membr. Biol. 107, 119–127.

47. Wilson, P.D. and Hreniuk, D. (1987) J. Cell Biol. 105, 176a.
48. Nelson, W.J. and Veshnock, P.J. (1987) Nature (London) 328, 533–535.
49. Nagafuchi, A., Shirayoshi, Y., Okazaki, K., Yasuda, K. and Takeichi, M. (1987) Nature (London) 329, 340–343.
50. Nelson, W.J., Shore, E.M., Wang, A.Z. and Hammerton, R.W. (1990) J. Cell Biol. 110, 349–357.
51. Gloor, S.H., Antonicek, H., Sweadner, K.J., Pagliusi, S., Frank, R., Moos, M. and Schachner, M. (1990) J. Cell Biol. 110, 165–174.
52. Jørgensen, P.L. (1975) Biochim. Biophys. Acta 401, 399–415.
53. Jørgensen, P.L. and Petersen, J. (1985) Biochim. Biophys. Acta 821, 319–333.
54. Jørgensen, P.L. and Collins, J.H. (1986) Biochim. Biophys. Acta 860, 570–576.
55. Bercovici, T. and Gitler, C. (1978) Biochemistry 17, 1484–1489.
56. Brunner, J. (1989) Methods Enzymol. 172, 628–687.
57. Modyanov, N., Lutsenko, S., Chertova, E. and Efremov, R. (1991) In: The Sodium Pump: Structure, Mechanism, and Regulation (Kaplan, J.H. and DeWeer, P., Eds.), pp. 99–115, Rockefeller University Press, New York.
58. Karlish, S.J.D., Goldshleger, R., Tal, D.M. and Stein, W.D. (1991) In: The Sodium Pump: Structure, Mechanism, and Regulation (Kaplan, J.H. and DeWeer, P., Eds.), pp. 129–141, Rockefeller University Press, New York.
59. Jørgensen, P.L., Karlish, S.J.D. and Gitler, C. (1982) J. Biol. Chem. 257, 7435–7441.
60. Jørgensen, P.L. and Brunner, J. (1983) Biochim. Biophys. Acta 735, 291–296.
61. Brandl, C.J. and Deber, C.M. (1986) Proc. Natl. Acad. Sci. U.S.A. 83, 917–921.
62. Brotherus, J., Jost, P.C., Griffith, O.H., Keana, J.F.W. and Hokin, L.E. (1980) Proc. Natl. Acad. Sci. U.S.A. 77, 272–276.
63. Matthews, I., Colyer, J., Mata, A.M., Green, N.M., Sharma, R.P., Lee, A.G. and East, J.M. (1989) Biochim. Biophys. Res. Commun. 161, 683–688.
64. Kirley, T.L. (1989) J. Biol. Chem. 264, 7185–7192.
65. Ohta, T., Yoshida, M., Nagano, K., Hirano, H. and Kawamura, M. (1986) FEBS Lett. 204, 297–301.
66. Esmann, M. (1982) Biochim. Biophys. Acta 688, 251–259.
67. Dzhandzhugazyan, K.N. and Jørgensen, P.L. (1985) Biochim. Biophys. Acta 817, 165–173.
68. Chin, G.J. (1985) Biochemistry 24, 5943–5947.
69. Zibirre, R., Hippler-Feldtmann, G., Kühne, J., Poronnik, P., Warnecke, G. and Koch, G. (1987) J. Biol. Chem. 262, 4349–4354.
70. Bairoch, A. (1992) Nucleic Acids Res. Vol. 20, 2013–2018.
71. Martin-Vasallo, P., Dackowski, W., Emanuel, J.R. and Levenson, R. (1989) J. Biol. Chem. 264, 4613–4618.
72. Mercer, R.W., Schneider, J.W., Savitz, A., Emanuel, J.R., Benz, E.J. and Levenson, R. (1986) Mol. Cell. Biol. 6, 3884–3890.
73. Gloor, S. (1989) Nucleic Acids Res. 17, 10117.
74. Good, P.J., Richter, K. and Dawid, I.B. (1990) Proc. Natl. Acad. Sci. U.S.A. 87, 9088–9092.
75. Canfield, V.A., Okamoto, C.T., Chow, D., Dorfman, J., Gros, P., Forte, J.G. and Levenson, R. (1990) J. Biol. Chem. 265, 19878–19884.
76. Mildvan, A.S. (1989) FASEB J. 3, 1705–1714.
77. Walker, J.E., Saraste, M., Ronswick, M.J. and Gay, N.J. (1982) EMBO J. 1, 945–951.
78. Fry, D.C., Kuby, S.A. and Mildvan, A.S. (1986) Proc. Natl. Acad. Sci. U.S.A. 83, 907–911.
79. Moczydlowski, E.G. and Fortes, P.A.G. (1981) J. Biol. Chem. 256, 2357–2366.
80. Skou, J.C. and Esmann, M. (1981) Biochim. Biophys. Acta 647, 232–240.
81. Stewart, J.M.M., Jørgensen, P.L. and Grisham, C.M. (1989) Biochemistry 28, 4695–4701.
82. Ohtsubo, M., Noguchi, S., Takeda, K., Morohashi, M. and Kawamura, M. (1990) Biochim. Biophys. Acta 1021, 157–160.
83. Ovchinnikov, Y.A., Dzhandzugazyan, K.N., Lutsenko, S.V., Mustayev, A.A. and Modyanov, N.N.

(1987) FEBS Lett. 217, 111–116.
84. Hinz, H.R. and Kirley, T.L. (1990) J. Biol. Chem. 265, 10260–10265.
85. Ellis-Davies, G.C. and Kaplan, J.H. (1990) J. Biol. Chem. 265, 20570–20576.
86. Scheiner-Bobis, G., Buxbaum, E. and Schoner, W. (1988) Prog. Clin. Biol. Res. 268A, 219–226.
87. Maruyama, K., Clarke, D.M., Fujii, J., Inesi, G., Loo, T.W. and MacLennan, D.H. (1989) J. Biol. Chem. 264, 13038–13042.
88. Serrano, R. and Portillo, F. (1990) Biochim. Biophys. Acta 1018, 195–199.
89. Jørgensen, P.L. (1991) In: The Sodium Pump: Structure, Mechanism, and Regulation (Kaplan, J.H. and DeWeer, P., Eds.), pp. 189–200, Rockefeller University Press, New York.
90. Yoda, A. and Yoda, S. (1987) J. Biol. Chem. 262, 110–115.
91. Lee, J.A. and Fortes, P.A.G. (1985) In: The Sodium Pump (Glynn, I.M. and Ellory, C., Eds.), pp. 277–282, Company of Biologists, Cambridge.
92. Glynn, I.M. (1985) In: The Enzymes of Biological Membranes (Martonosi, A.N., Ed.), Vol. 3, pp. 35–114, Plenum, New York.
93. De Meis, L., Martins, O.B. and Alves, E.W. (1980) Biochemistry 19, 4252–4261.
94. Grell, E. and Oberbaumer, I. (1977) Mol. Biol. Biochem. Biophys. 24, 371–379.
95. Boyer, P.D. (1988) Trends Biochem. Sci. 13, 5–7.
96. Clarke, D.M., Loo, T.W. and MacLennan, D.H. (1990) J. Biol. Chem. 265, 6262–6267.
97. Shani-Sekler, M., Goldschleger, R., Tal, D.M. and Karlish, S.J.D. (1988) J. Biol. Chem. 263, 19331–19341.
98. Rephaeli, A., Richards, D.E. and Karlish, S.J.D. (1986) J. Biol. Chem. 261, 6248–6254.
99. Apell, H.-J., Borlinghaus, R. and Lauger, P. (1987) J. Membr. Biol. 97, 179–191.
100. Nagel, G., Fendler, F., Grell, E. and Bamberg, E. (1987) Biochim. Biophys. Acta 901, 239–249.
101. Forbush, B. III (1984) Anal. Biochem. 140, 159–163.
102. Karlish, S.J.D. (1980) J. Bioenerg. Biomemb. 12, 111–136.
103. Kapakos, J.G. and Steinberg, M. (1986) J. Biol. Chem. 261, 2084–2089.
104. Jørgensen, P.L. (1975) Biochim. Biophys. Acta 401, 399–415.
105. Jørgensen, P.L. (1977) Biochim. Biophys. Acta 466, 97–108.
106. Jørgensen, P.L. and Petersen, J. (1985) Biochim. Biophys. Acta 821, 319–333.
107. Jørgensen, P.L. and Collins, J.H. (1986) Biochim. Biophys. Acta 860, 570–576.
108. Tyson, P. and Steinberg, M. (1987) J. Biol. Chem. 262, 4644–4648.
109. Karlish, S.J.D. and Pick, U. (1981) J. Physiol. 312, 505–529.
110. Gresalfi, T.V. and Wallace, B.A. (1984) J. Biol. Chem. 259, 2622–2628.
111. Nabiev, I.R., Dzandzhugazyan, K.N., Efrenov, R.G. and Modyanov, N.N. (1988) FEBS Lett. 236, 235–239.
112. Hastings, D.F., Reynolds, J.A. and Tanford, C. (1986) Biochim. Biophys. Acta, 860, 566–569.
113. Chetverin, A.B. and Brazhnikov, E.V. (1985) J. Biol. Chem. 260, 7817–7819.
114. Jørgensen, P.L. and Karlish, S.J.D. (1980) Biochim. Biophys. Acta 597, 305–317.
115. Shull, G.E. and Lingrel, J.B. (1986) J. Biol. Chem. 261, 16788–16791.
116. Saccomani, G., Dailey, D.W. and Sachs, G. (1979) J. Biol. Chem. 254, 2821–2827.
117. Apell, H.-J., Borlinghaus, R. and Lauger, P. (1987) J. Membr. Biol. 97, 179–191.
118. Portillo, F. and Serrano, R. (1988) EMBO J. 7, 1793–1798.
119. Ghislain, M., Schlesser, A. and Goffeau, A. (1987) J. Biol. Chem. 262, 17549–17555.
120. Imamura, Y. and Kawakita, M. (1989) J. Biochem. (Tokyo) 105, 775–781.
121. le Maire, M., Lund, S., Viel, A., Champeil, P. and Møller, J.V. (1990) J. Biol. Chem. 265, 1111–1123.
122. Andersen, J.P., Vilsen, B., Leberer, E. and MacLennan, D.H. (1989) J. Biol. Chem. 264, 21018–21023.
123. Vilsen, B., Andersen, J.P., Clarke, D.M. and MacLennan, D.H. (1989) J. Biol. Chem. 264, 21024–21030.
124. Tanford, C. (1982) Proc. Natl. Acad. Sci. U.S.A. 79, 2881–2884.

CHAPTER 2

Structure and function of gastric H,K-ATPase

TOM J.F. VAN UEM and JAN JOEP H.H.M. DE PONT

Department of Biochemistry, University of Nijmegen, 6500 HB Nijmegen, The Netherlands

1. Introduction

The secretion of gastric acid by the parietal cell in the gastric mucosa is a complex process that is regulated by a variety of stimuli, such as histamine, gastrin and acetylcholine [1–3]. The final link in this process is formed by the gastric, Mg^{2+}-dependent, H^+-transporting, and K^+-stimulated adenosinetriphosphatase (E.C. 3.6.1.36) (H,K-ATPase), which delivers H^+ to the lumen in exchange for K^+ by an ATP-driven catalytic cycle, thereby generating a pH gradient of more than six units across the secretory membrane [4–6]. Stimulation of the parietal cell leads to large morphological changes and results both in a translocation of H,K-ATPase from the intracellular tubulovesicular membrane system to the apical secretory membrane, and in the activation or insertion, of K^+ and Cl^- pathways in the secretory membrane of the cell [7–9]. The ATP-driven H^+/K^+ exchange, together with the K^+ and Cl^- flux from cell to lumen, then results in a net hydrochloric acid secretion.

Gastric H,K-ATPase belongs to the class of phosphoenzyme-forming ATPases or P-type ATPases, which includes the more generally distributed Ca-ATPase and Na,K-ATPase [10,11]. These ATPases are also called E_1/E_2-type ATPases, since these pumps exist in two major conformational states, E_1 and E_2, with different affinities and orientation of cation binding sites. Characteristic for the P-type ATPases is a catalytic α subunit with an apparent molecular mass of 100 kDa that is phosphorylated at an aspartyl residue during ATP hydrolysis, and the inhibition by micromolar concentrations of vanadate. Until recently, Na,K-ATPase was the only enzyme of this class that was also reported to contain a glycosylated β subunit. Recently however, it has become clear that gastric H,K-ATPase contains also, in addition to the catalytic α subunit, a glycosylated β subunit [12]. This makes gastric H,K-ATPase structurally more closely related to Na,K-ATPase than to Ca-ATPase.

A number of reviews dealing with the physiological role of H,K-ATPase in acid secretion [13,14], the catalytic cycle of H,K-ATPase [14,15], the structure of H,K-ATPase [15], functional domains of H,K-ATPase [16], the preparation of gastric H,K-ATPase [17], the ion transport properties of H,K-ATPase preparations of stimulated and non-stimulated parietal cells [7] and H,K-ATPase as a pharmacological target [18] have been published.

In this chapter we will review the recent investigations of the structure of both the α and β subunit, and the function of gastric H,K-ATPase. We will proceed from a brief overview of the tissue distribution to a successive discussion of structure, kinetics, transport properties, lipid dependency, solubilization and reconstitution, and inhibitors of H,K-ATPase that may label functionally important domains of the enzyme.

2. Tissue and cell distribution

Gastric H,K-ATPase has been purified from several species, including frog [4], dog [5], hog [19], rabbit [20], rat [21] and human [22]. The enzyme is unique for the parietal cell in the gastric mucosa as has been demonstrated by immunocytochemical studies with polyclonal and monoclonal antibodies directed against the catalytic α subunit and the recently identified β subunit of the enzyme [23–25]. A K^+-dependent ATP-driven proton pump, however, similar to gastric H,K-ATPase, appears also to be located in the rabbit distal colon [26,27] and the rabbit outer medullary collecting duct [28–30]. The enzyme and transport activities that have been purified and studied in these studies were all ouabain insensitive. In rat and guinea pig distal colon an ouabain-sensitive ATP-driven H^+–K^+ exchange mechanism has also been demonstrated that is not identical with either Na,K-ATPase or gastric H,K-ATPase [31,32]. The fact that these colonic and kidney H,K-ATPases are just similar and not identical to gastric H,K-ATPase has recently been confirmed by Northern blot analysis of mRNA from a number of rat tissues with specific α- and β-subunit probes, demonstrating that both subunits are only expressed in the stomach [12,25].

3. Structure

3.1. The catalytic α subunit

In the case of H,K-ATPase it appears that the catalytic α subunit is a single protein with an apparent molecular mass of around 95 kDa on reducing acrylamide gels. The homogeneity of the 95-kDa peptide however is subject of dispute. Partial resistance of the 95-kDa peptide against tryptic digestion [33] and the generation of three groups of peptides upon isoelectric focusing of the 95-kDa peptide [24] favored heterogeneity of this subunit. Contrary to the results of Saccomani et al. [33], Helmich-De Jong et al. [34] found a nearly complete trypsin sensitivity of the catalytic subunit and therefore suggested homogeneity of it. Sedimentation studies, N-terminal amino acid analysis and alkylation with citraconic anhydride [35] did not reveal heterogeneity of the catalytic subunit either. The existence however of multiple forms of the catalytic subunit was suggested by immunoblotting of pig gastric microsomes, yielding signals at $M_r = 88$ kDa and $M_r = 94$ kDa [36], and by cell-free translation of rabbit gastric mucosal messenger RNA, resulting in immunoprecipitable polypeptides with apparent molecular masses between 88 and 94 kDa [37]. In the latter study it was concluded that co- or posttranslational modifications and proteolytic activity could not be the cause for the observed heterogeneity. Contrary to Na,K-ATPase, however, which has three isoforms of the catalytic α subunit [38,39], so far only a single α subunit of H,K-ATPase has been identified by molecular cloning [40-42].

The catalytic subunit of rat, pig, and human H,K-ATPase has a molecular mass of 114 kDa in each case and comprises 1 034 or 1 035 amino acid residues [40–42]. The predicted porcine amino acid sequence corresponds to the N-terminal 17 amino acid sequence identified by Lane et al. [43]. The interspecies homology is 98%, whereas the overall amino acid homology with the sheep kidney Na,K-ATPase catalytic subunit is 62% [44]. Moreover, the H,K-ATPase gene also shows a great similarity to the Na,K-ATPase gene in exon–intron organization [42,45], indicating that these two ATPases are derived from a common ancestral ATPase. The regions of greatest homology between the H,K-ATPase and Na,K-ATPase catalytic subunit appear around the phosphorylation site at Asp^{386} (numbering refers to the porcine sequence) [46], the putative ATP-binding region that includes the pyridoxal phosphate reactive site Lys^{497} [47,48] and the fluorescein-5-isothiocyanate (FITC) reactive site Lys^{518} [49,50], as well as a region between residue 157 and 300, which has been proposed as part of an energy transduction region [40,44].

Information about the putative folding of the H,K-ATPase catalytic subunit through the membrane has been obtained by the combined use of hydropathy analysis according to the criteria of Kyte and Doolittle [51], identification of sites sensitive to chemical modification [46,48,50,52–55], and localization of epitopes of monoclonal antibodies [56]. The model of the H,K-ATPase catalytic subunit (Fig. 1) which has emerged from these studies shows ten transmembrane segments and contains cytosolic N- and C-termini [53]. This secondary structure of the catalytic subunit is probably a common feature of the catalytic subunits of P-type ATPases, since evidence supporting a ten α-helical model with cytosolic N- and C-termini has also been published recently for both Ca-ATPase of the sarcoplasmic reticulum and Na,K-ATPase [57–59].

Hydropathy analysis predicted that there are four major transmembrane domains (M_1–M_4) prior to the phosphorylation site at Asp^{386}. The existence of these four transmembrane segments in the N-terminal half of the catalytic subunit is generally accepted for all P-type ATPases. The four transmembrane sequences are followed by a large cytosolic loop that contains the phosphorylation site Asp^{386}, the pyridoxal

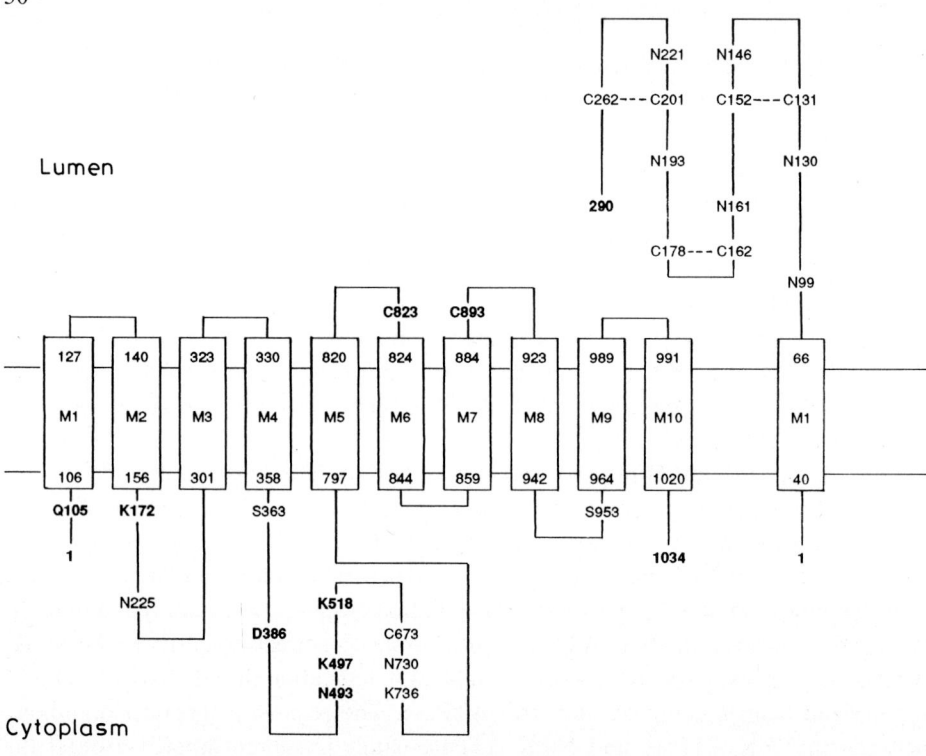

Fig. 1. Model for the possible transmembranous organization of the porcine H,K-ATPase αβ-protomer, based on sequence information [40,41]. The α subunit is modeled with ten transmembrane segments (M_1–M_{10}) and shows the following identified amino acid residues (bold): D386, the phosphorylation site [46]; K518, the FITC-reactive site [50]; K497, the pyridoxal phosphate-reactive site [48]; C823 and C893, the omeprazole-reactive sites [52]; Q105 to K172, the mDAZIP-labelled (an analog of SCH 28080) fragment [55]; and N493, the carbohydrate containing residue [62]. Potential N-glycosylation sites at N225 and N730 [41], potential sites of phosphorylation by cAMP-dependent protein kinase at S363 and S953 [41], and the amino acid residues C673 and K736, corresponding to FSBA-reactive residues in Na,K-ATPase [60], are also shown. The β subunit shows six potential N-glycosylation sites and six conserved cysteine residues that form the three disulfide bonds of the Na,K-ATPase β subunit [83,84].

phosphate site Lys^{497} and the FITC site Lys^{518}. These residues are suggested to be located at or near the catalytic ATP-binding site. Besides the latter two sites the predicted cytosolic loop contains also Cys^{673} and Lys^{736}, whose corresponding amino acid residues in Na,K-ATPase can be modified by the ATP-site directed probe fluorosulfonylbenzoyl adenosine (FSBA) [60]. Although it is generally presumed that this loop has a cytosolic disposition, the possibility that part of the cytosolic loop between M_4 and M_5 is membrane inserted has recently been proposed after demonstrating that one of three potential N-linked glycosylation sites at residues 225, 493 and 730, namely Asn^{493}, contains N-linked carbohydrate [61,62]. Electron microscopy of inside-out-oriented gastric vesicles indicated that the majority of this oligosaccharide moiety is at the extracellular side [62]. Two extra membrane-spanning segments between M_4 and M_5, however, leading to an extracellular location of the oligosaccharide moiety at Asn^{493}, do not fit with the observed solubility of proteolytic fragments of H,K-ATPase that contain the phosphorylation site, the FITC-binding site, and the carbohydrate moiety at Asn^{493} [61,63]. Hall et al. [62] attempted to explain the luminal disposition of Asn^{493} by interaction of part of the cytosolic domain between M_4 and M_5 with the hydrophobic membrane-spanning peptides. This hypothesis, however, does not fit well with the presence of the pyridoxal phosphate and FITC-reactive sites at residues 497 and 518 that would leave insufficient distance for a membrane-spanning segment, comprising about 20 amino acid residues. A cytosolic assignment of the glycosylated amino acid residue, however, which is contradictory to the current models of glycosylation of membrane-bound proteins, might be possible since it recently has been demonstrated that the Na,K-ATPase catalytic subunit does contain N-linked carbohydrate moieties located at the cytosolic face of the cell membrane [64].

The transmembrane organization after the large cytosolic domain is uncertain and hydrophobicity plots predicted up to six additional transmembrane segments [40,41,65]. At the end of the large cytosolic domain there is a hydrophobic domain that is long enough to contain two transmembrane segments. On the basis of the recently suggested binding sites Cys^{823} and Cys^{893} of the luminal-acting inhibitor omeprazole [52], it is very probable that there are indeed two transmembrane passages, M_5 and M_6, in this region, i.e., from residue 797–820 and from residue 824–844 [40]. Hydrophobicity analysis revealed that another four putative transmembrane segments, M_7–M_{10}, probably span between residues 859 and 884, 923 and 942, 964 and 989 and 991 and 1 020 [40,65]. This ten α-helical model for the structure of the H,K-ATPase catalytic subunit would account for a luminal localization of the omeprazole-labelled residues Cys^{823} and Cys^{893} [52] and for the cytoplasmic C-terminus [53]. Moreover, this model would also fit with a cytosolic disposition of two potential sites of phosphorylation by cAMP-dependent protein kinase at residues 363 and 953 in case these sites are functional regulatory sites [41].

3.2. The β subunit

Until recently, the possibility that H,K-ATPase consists not only of a catalytic α subunit but also of other subunits was not examined. This was mainly due to the fact that SDS-PAGE of purified gastric H,K-ATPase preparations principally gave one protein band with an apparent molecular mass of about 100 kDa, which was reported to comprise 75% or more of the total amount of protein [6,66,67]. This mass is lower than the mass deduced from its cloned cDNA [40], but may be due to the higher electrophoretic mobility of membrane-bound proteins, as consequence of having relatively high contents of hydrophobic amino acid residues [68].

One of the first observations suggesting that the functional unit of H,K-ATPase is not a monomer of the catalytic subunit, but instead a heterodimer of the catalytic subunit and another protein, was made with radiation inactivation studies of frozen H,K-ATPase [69]. It was demonstrated that the Mg-ATP-dependent formation of a beta-aspartyl phosphate exhibited a target size of 133–147 kDa and that the target size of the protein migrating on SDS-polyacrylamide gels at 94 kDa was 92–143 kDa. This was clearly larger than a monomer and clearly smaller than a dimer of the catalytic subunit.

Further indications for an additional subunit were provided by a crosslinking analysis of $C_{12}E_8$ solubilized H,K-ATPase, which exhibited ATPase and phosphatase activities, and ligand affinities comparable to the native enzyme [70]. Glutaraldehyde treatment of soluble protein fractions resolved on a linear glycerol gradient revealed no active fraction enriched in monomeric (M_r = 94 kDa) H,K-ATPase. Instead, K^+-ATPase activity was only obtained in fractions enriched in particles of M_r = 175 kDa. This size also suggested that the functional H,K-ATPase unit is a heterodimer of a catalytic subunit and an additional subunit, since the apparent molecular mass of 175 kDa is probably too small to be a homodimer of the catalytic subunit.

Several groups then independently provided evidence for the existence of an extensively glycosylated β subunit of gastric H,K-ATPase that has an apparent molecular mass of 60–90 kDa. Okamoto et al. [71] used wheat germ agglutinin affinity column chromatography and immunoprecipitation procedures to demonstrate that the catalytic subunit of porcine H,K-ATPase is tightly associated with a glycoprotein of 60–80 kDa. This noncovalent association was confirmed by co-purification of the 60–80 kDa transmembrane glycoprotein when performing a purification of H,K-ATPase by a modified Jørgensen SDS-extraction procedure [72]. This co-purification of a glycoprotein was also observed before by Nandi et al. [73]. Deglycosylation of this 60–80 kDa glycoprotein by endoglycosidase F revealed that it had five asparagine-linked oligosaccharide chains and an apparent core molecular mass of 32 kDa, which was approximately the same size as that of the Na,K-ATPase β subunit. As was in the case of Na,K-ATPase, densitometric scans of SDS-gels of purified gastric H,K-ATPase revealed an approximate stoichiometric ratio of 1 : 1 for the catalytic α subunit/60–80 kDa glycoprotein.

Hall et al. [62] identified in a separate study the same glycoprotein in H,K-ATPase vesicles isolated from porcine gastric mucosa. A stoichiometric ratio of 1.2 : 1.0 was found for the deglycosylated protein (35 kDa)/catalytic 94-kDa protein. Furthermore, compelling evidence that this glycoprotein is the H,K-ATPase β subunit was provided by N-terminal sequence analysis of three protease V8-obtained peptides of the 35-kDa core protein. These peptides showed 30% and 45% homology with the Na,K-ATPase β1 and β2 subunit, respectively.

With the use of oligonucleotide probes based on the amino acid sequences of these protease V8-obtained peptides and of cyanogen bromide fragments of the porcine H,K-ATPase β subunit, cDNA clones for the rat [12,25] and rabbit [74] H,K-ATPase β subunit were then isolated.

Parallel to these studies it was reported that, besides the H,K-ATPase α subunit [75], a 60–90 kDa glycoprotein is one of the main parietal cell antigens associated with autoimmune gastritis and pernicious anemia [76]. This 60–90 kDa antigen was shown to be a conserved molecule, comprising a 34-kDa core protein extensively glycosylated with N-linked oligosaccharides. By analogy with the approach of cDNA cloning of the rat and the rabbit H,K-ATPase β subunit, the cDNA encoding the porcine H,K-ATPase β subunit was then isolated [77].

The β subunits from rat [12,25], pig [77] and rabbit [74] have a molecular weight of about 33 kDa and comprise 290–294 amino acid residues. With 82% interspecies homology, the β subunit is not as well conserved as the α subunit, having 98% interspecies homology. Also the homology with the Na,K-ATPase β1 subunit [78–80] and β2 subunit [81,82], being 33% and 42%, respectively, is much lower than the homology between the H,K-ATPase and Na,K-ATPase α subunits. The hydropathy profile predicts that the H,K-ATPase β subunit contains a short polar cytoplasmic N-terminus followed by a single membrane-spanning segment and an extensive extracellular region including six or seven potential N-glycosylation sites. Two of these sites (residues 161 and 193) are conserved relative to the predicted N-linked glycosylation sites of the Na,K-ATPase β1 and β2 subunit, which have three and eight to nine potential sites, respectively. It is likely that most of the potential glycosylation sites have carbohydrate attached, since four or five deglycosylated products were observed during enzymatic deglycosylation of the porcine H,K-ATPase β subunit [71,76]. There are six cysteine residues in the presumed extracellular domain that are highly conserved between the H,K-ATPase β subunit and the Na,K-ATPase β1 and β2 subunits. These cysteine residues in the Na,K-ATPase β1 subunit have been demonstrated to form three disulfide linkages [83,84]. Hence, it can be assumed that the H,K-ATPase β subunit also contains three disulfide bonds that are similar to those of the Na,K-ATPase β1 subunit. In summary, these structural similarities strongly indicate that besides the H,K-ATPase and Na,K-ATPase α subunits, the H,K-ATPase and Na,K-ATPase β subunits are also derived from a common ancestral ATPase.

The identification of the H,K-ATPase β subunit raises important questions regarding the functional significance of this subunit. By analogy with the function of the Na,K-ATPase β subunit that has been identified up until now [85], a possible function of the H,K-ATPase β subunit could be a stabilization of the α subunit during biogenesis, subsequently leading to a functional enzyme in the plasma membrane. The highly conserved C-terminus of the H,K-ATPase β subunit (95% for the last 20 amino acid residues) could play a role in exerting this function as has been demonstrated for Na,K-ATPase recently. Expression of deletion mutants of the Na,K-ATPase β subunit revealed that deletion of as few as 11 amino acid residues from the C-terminus of the β subunit results in a molecule that does not appear to assemble with the Na,K-ATPase α subunit [86]. However, transfection of polarized Madin Darby canine kidney (MDCK) cells with a cDNA encoding the rat H,K- ATPase α subunit resulted in the appearance of the apparently complete α subunit in the apical membrane of the stably transfected cells [87]. Nevertheless, it is possible that the H,K-ATPase β subunit is not necessary for plasma membrane insertion of the H,K-ATPase α subunit, since the stabilizing role of the H,K-ATPase β subunit might be taken over by the endogenous Na,K-ATPase β subunit of these cells. Whether the protein that is expressed in these cells is functional or not is not known. One important conclusion that can be drawn from these transfection experiments is that putative determinants involved in targeting H,K-ATPase and Na,K-ATPase to their proper membrane destinations are not located on their respective β subunits.

3.3. Molecular organization

The recent identification of the H,K-ATPase β subunit and the demonstration that the ratio for the H,K-ATPase catalytic α subunit/β subunit is about 1:1 [62,71] might bring new insight into the structural and functional assembly of H,K-ATPase *in situ*.

Earlier work using radiation inactivation analysis of frozen H,K-ATPase has revealed that the target size for the K^+-ATPase activity ranges from 232 to 270 kDa [69,88]. In lyophilized samples, however, a target size of 444 kDa was found for this enzyme activity [89]. In all of these irradiation studies a smaller target size was measured for the K^+-*p*-nitrophenylphosphatase (K^+-*p*NPPase) activity. For the phosphoenzyme formation and for the destruction of the catalytic subunit Rabon et al. [69] measured target sizes of 133–147 kDa and 92–143 kDa, respectively. With the use of reconstituted H,K-ATPase these authors obtained for the passive Rb^+ exchange and for H^+ transport target sizes of 233 and 388 kDa, respectively.

A glutaraldehyde crosslinking analysis of $C_{12}E_8$ solubilized H,K-ATPase revealed that only particles of $M_r = 175$ kDa exhibited K^+-ATPase activity [70]. This size suggests that the functional H,K-ATPase unit is a heterodimer of the catalytic α subunit and the β subunit, since the molecular mass of 175 kDa is probably too small to be a homodimer of the catalytic subunit.

Soumarmon et al. [90,91] determined in an octylglycoside solubilized H,K-ATPase preparation by gel filtration and glycerol gradient centrifugation molecular masses between 390 and 420 kDa for the holomeric enzyme.

Thus the picture that emerges at this point is that in the membrane probably a diprotomeric $(\alpha\beta)_2$ structure (α is 114 kDa, β is 60–80 kDa) is responsible for the active transport process. On the other hand the minimal structure for K^+-ATPase activity, K^+-*p*NPPase activity, passive transport and phosphorylation is either an αβ protomer or another larger oligomeric structure.

3.4. Conformations of H,K-ATPase

The widely accepted model, of how the class of P-type ATPases transports ions across the membrane, assumes the existence of two enzyme conformations called E_1 and E_2, which have different affinities for the ions to be transported. Generally, E_1 represents the conformation that has high affinity for ions at the cytoplasmic side and E_2 is the conformation with high-affinity ion-binding sites at the extracellular side. In the case of H,K-ATPase the existence of these two conformations has been demonstrated by several indirect and direct approaches.

The existence of the E_1, and E_2 states of the phosphorylated protein, i.e., the high- and low-energy phosphoenzyme intermediate, has been demonstrated by the ATP–ADP exchange reaction [92,93] and by the ^{18}O exchange between inorganic phosphate and water [94].

Indirect evidence for the existence of different conformations of H,K-ATPase has been gained by site-selective reagents. For example, Schrijen et al. [67,95] demonstrated that Mg^{2+} increased exposure of an essential arginine residue near the ATP-binding site and Mg^{2+} caused an increase in the number of reactive sulfhydryl groups on the enzyme.

Direct proof of different conformational states of H,K-ATPase has been provided by limited trypsin digestion of H,K-ATPase as the primary cleavage sites were shown to be conformationally dependent [34]. In the ATP-dependent E_1 conformation proteolytic digestion gave peptide fragments of $M_r = 67$ and 35 kDa, whereas in the K^+-dependent E_2 conformation fragments of $M_r = 42$ and 56 kDa were obtained. Both the 67- and 42-kDa peptide fragments contained the residue Asp^{386} that is phosphorylated during ATP hydrolysis. The existence of a putative third conformation was demonstrated by tryptic digestion in the presence of Mg^{2+}, revealing a digestion pattern unlike ATP or K^+. Recently it has been demonstrated that the tryptic-cleavage sites for the E_1, pattern are at Lys^{47} and probably at Lys^{668}, while the tryptic-cleavage site for the E_2 pattern is at Arg^{455} [56].

With the aid of three different fluorescent reporter groups attempts to observe conformational changes directly within or near the nucleotide domain have been reported. Jackson et al. [49] demonstrated that covalent labelling of H,K-ATPase with FITC inhibited the K^+-ATPase activity while largely sparing the associated K^+-pNPPase activity. The inhibition of the former activity was prevented by ATP suggesting that the ligand binds at or near to the ATP-binding site. Saturation of the high-affinity luminal K^+ site or binding of vanadate in the presence of Mg^{2+} induced a low fluorescence state that was defined as the E_2 conformation. Rabon et al. [96] have recently shown that Na^+ ions reversed the K^+-induced fluorescence quench and somewhat enhanced fluorescence in the absence of K^+ ions, demonstrating that under that condition about 70% of H,K-ATPase resides in the E_1 conformation. The apparent affinity of the Na^+ ions for the high fluorescence E_1 conformation increased upon elevation of pH, which is consistent with competition between protons and Na^+ ions.

Conformational changes within or near the ATP-binding site of H,K-ATPase have also been demonstrated with the reversible fluorescent probes TNP–ATP [97,98] and eosin [99]. TNP–ATP binding to H,K-ATPase enhanced the fluorescence five-fold and was suggested to be due to binding to a hydrophobic region of the protein. K^+ ions caused a rapid fluorescence quench, while Mg^{2+} rapidly and completely reversed the K^+ quench and then caused a slow fluorescence quench. This fluorescence quench induced by K^+ and Mg^{2+} was caused by a change in the enhancement factor of bound fluorophore and was not due to a change in the amount of bound fluorophore. These cofactor-induced fluorescence quenches could be explained by assuming that there are three different conformational states, whose relationship to E_1 and E_2 was not established.

Contrary to the observations with TNP–ATP, however, Mg^{2+} enhanced eosin fluorescence, whereas K^+ ions decreased the fluorescence enhancement of Mg^{2+} and caused a fluorescence quench in the absence of added ions [99]. The fluorescence enhancement caused by Mg^{2+} was explained by an increase in the number of eosin-binding sites. Faller [98] on the other hand, has challenged this explanation and argued that only an increase in enhancement factor (i.e., movement of the fluorophore to a more hydrophobic region of the protein) can explain the Mg^{2+}-induced fluorescence increase.

Recently Rabon et al. [100] reported on a new conformational probe of H,K-ATPase. This fluorescent quinoline derivative MDPQ was shown to be a reversible luminal K^+-site inhibitor of both K^+-ATPase and K^+-pNPPase activity. High-affinity MgATP binding induced a conformational change with fluorophore movement into a more hydrophobic environment.

The conformational changes which have been described so far are probably all relatively small local changes in the structure of H,K-ATPase. This has been confirmed by Mitchell et al. [101] who demonstrated by Fourier transform infrared spectroscopy that a gross change in the protein secondary structure does not occur upon a conformational change from E_1 to E_2. Circular dichroism measurements, however [102,103], indicated an increase in α-helical structure upon addition of ATP to H,K-ATPase in the presence of Mg^{2+} and K^+.

4. Kinetics of H,K-ATPase

4.1. Overall reaction

Gastric H,K-ATPase catalyses the 1:1 exchange of H^+ for K^+ upon hydrolysis of ATP, which results in vivo in a pH difference of more than six between the cytoplasm and the lumen [4–6]. Several reaction schemes have been developed to account for the hydrolysis of ATP and the transport of H^+ and K^+ [14,104–107]. The basis for all these schemes is the Albers–Post scheme originally postulated for the mechanism of action of Na,K-ATPase [108,109].

The scheme in Fig. 2 describes the various reactions of H,K-ATPase. In this

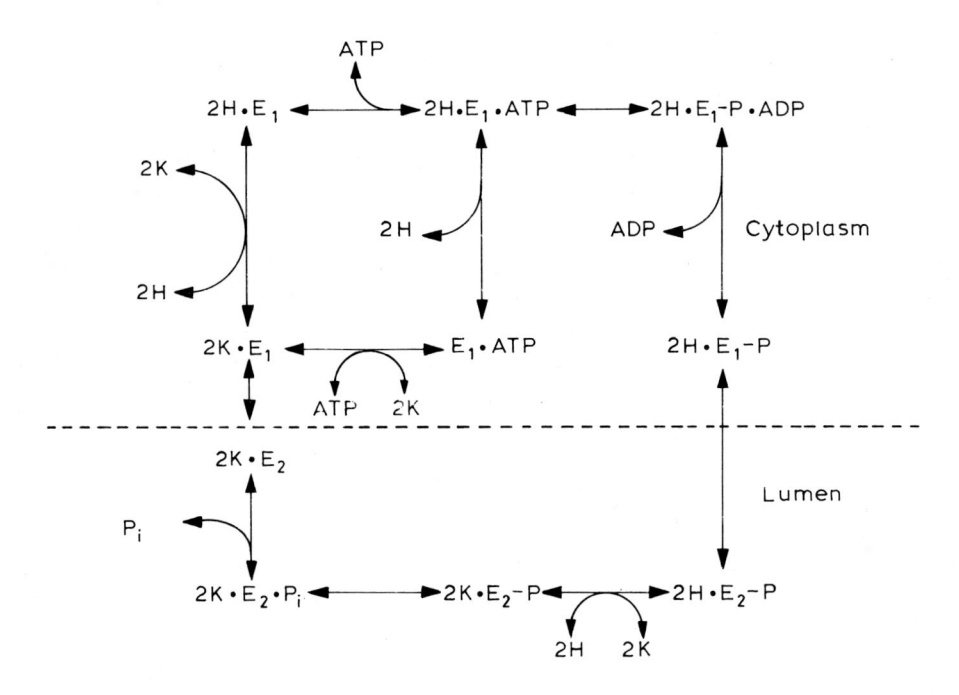

Fig. 2. E_1–E_2 reaction cycle of H,K-ATPase, accounting for the transport of two H^+ and two K^+ ions per molecule of hydrolysed ATP. The ATPase reaction proceeds from $2H \cdot E_1 \cdot ATP$ through $2H \cdot E_1$–P and $2H \cdot E_2$–P to $2K \cdot E_1$. Details of the reaction cycle are described in the text.

scheme, which requires several steps related to binding of substrate and cationic ligands, E_1 and E_2 represent conformations of the enzyme with high-affinity cation-binding sites facing the cytoplasm and the lumen, respectively. Phosphoenzyme forms are represented as E–P in the case of covalent bound phosphate, and $E \cdot P$ in the case of bound phosphate. The existence of the latter form has been established by measurements of ^{18}O exchange between P_i and H_2O [94]. The scheme accounts for the sequential transport of two H^+ ions from the cytoplasm to the lumen and for the transport of two K^+ ions in the opposite direction [110–112].

The reaction starts with the binding of ATP to the H^+-liganded form of the enzyme, $2H \cdot E_1$. In the presence of K^+, this binding is to the K^+-liganded enzyme form, $2K \cdot E_1$ or $2K \cdot E_2$, or to an occluded form between these two forms. The existence of such an occluded form has not yet been demonstrated, but its detection with filtration or column techniques similar to those used previously to measure occluded transported cations for Na,K-ATPase [113] will be very difficult, because of the rapid dissociation of K^+ from the enzyme [96]. Subsequent binding of Mg^{2+} to $2H \cdot E_1$ then leads to phosphorylation at an aspartyl residue [46,114]. The major phosphoenzyme then formed is a K^+-sensitive intermediate ($2H \cdot E_2$–P), whereas a minor part (20%) exists as an ADP-sensitive intermediate ($2H \cdot E_1$–P) [92,93]. With the conformational change from $2H \cdot E_1$–P to $2H \cdot E_2$–P the transport of two protons from the cytoplasm to the lumen is very probable [115]. The next step is the displacement at the luminal side of two H^+ ions and the subsequent binding of two K^+ ions, which then leads to a rapid biphasic dephosphorylation of the phosphoenzyme $2K \cdot E_2$–P to $2K \cdot E_2$ [104,105]. Finally, the conformational change from $2K \cdot E_2$ to $2K \cdot E_1$, which reflects the transport of two K^+ ions from the luminal to the cytoplasmic side, completes the transport cycle. This step in the reaction cycle of H,K-ATPase has been studied in both native and reconstituted gastric vesicles as a K^+–K^+ exchange reaction, using ^{86}Rb in lieu of K^+ as a tracer [116–118]. Contrary to Na,K-ATPase [119,120], the rate of K^+–K^+ exchange by H,K-ATPase is not stimulated by ATP or ADP, and is nearly equivalent to the overall rate of hydrolysis. This also indicates, as mentioned before, that a rate-limiting K^+ occlusion as defined for Na,K-ATPase [121] is not very likely to occur.

4.2. Phosphorylation from ATP

In the presence of Mg^{2+} gastric H,K-ATPase can be phosphorylated by ATP at aspartyl residue 386 [46,114]. The mixed-anhydride that is formed, is acid stable and hydroxylamine sensitive [122]. The initial step is the reversible binding of ATP to the protein with a pseudo first-order rate constant, in the absence of K^+ and at pH 7.4, of $1\,400$ min^{-1}. This is followed by a Mg^{2+}-dependent transfer of the terminal phosphate of ATP to the catalytic α subunit with a rate constant of $4\,400$ min^{-1} [114]. Increasing H^+ concentrations on the cytoplasmic side accelerate the phosphorylation rate, whereas increasing K^+ concentrations on this side inhibit the phosphorylation rate [104–106,123]. Normally, the rate of formation of phosphoenzyme greatly exceeds the overall room temperature rate constant of about 200 min^{-1}, but at high K^+/ATP ratios the displacement of K^+ from the cytoplasmic side of the enzyme by ATP becomes rate limiting for overall ATP hydrolysis [104].

It has been established by substitution of Ca^{2+} for Mg^{2+} that, prior to phosphorylation, the divalent cation binds at a cytosolic site with a stoichiometry of about 1 mol per phosphorylation site [124,125]. These experiments also demonstrated that the phosphorylation rate is sensitive to the nature of the divalent cation bound. With Mg^{2+} bound, the phosphorylation rate is about 20 times faster than with Ca^{2+} bound. The divalent cation dissociates after dephosphorylation, suggesting that it is tightly bound to the phosphoenzyme during the reaction cycle. It was also demonstrated that the type of divalent cation that occupies the divalent cation site required for phosphorylation is important for the step $2K \cdot E_2$–P to $2K \cdot E_2 \cdot P$ to $2K \cdot E_2$ [124,125]. With Mg^{2+} bound, the $2K \cdot E_2$–P conformer is K^+-sensitive, whereas with Ca^{2+} bound, the intermediate is K^+-insensitive.

A subject of discussion is the maximal phosphorylation capacity of H,K-ATPase that reaches a maximum of 1.5 nmol mg^{-1} [104,126]. Under the conditions used for the determination of the phosphoenzyme intermediate, the rate of phosphorylation is more than 100 times that of dephosphorylation [104], so that the phosphorylation capacity can be used as an index of the number of functional pumps present. With a protein correction factor of 1.42 [127] this maximal phosphorylation capacity increases to 2.1 nmol mg^{-1}, which corresponds to 0.3 phosphorylation sites per H,K-ATPase $\alpha\beta$-protomer (147 kDa). This value suggests that a large fraction of the enzyme is either inactive or that only a fraction of the sites may be phosphorylated at any one time. Recently, however, phosphorylation capacities of 2.1 and 2.8 nmol mg^{-1} have been measured [125,128]. The latter value is obtained in the presence of the nonionic detergent $C_{12}E_8$ and corresponds to almost 0.6 phosphorylation sites per $\alpha\beta$ protomer, suggesting that each H,K-ATPase α subunit can be phosphorylated at the same time.

4.3. Characteristics of ATP hydrolysis

Gastric H,K-ATPase catalyzes the hydrolysis of ATP in the presence of Mg^{2+} and K^+. This K^+-ATPase activity of the enzyme has a high specificity for ATP. Compared to ATP, the K^+-stimulated hydrolysis of GTP, CTP and ADP occurs at a rate of only 12, 15 and 0.7%, whereas ITP and β/γ-methylene ATP are not hydrolyzed at all [6]. For the stimulation of the ATPase activity various cations can act as K^+ surrogates with the following order of affinity: $Tl^+ > K^+ > Rb^+ > NH_4^+ > Cs^+ \gg Na^+, Li^+$ [6,67,122].

Steady state kinetics demonstrated that K^+ affects the K^+-ATPase activity biphasically [104,106]. Low concentrations of K^+ stimulate ATP hydrolysis with a $K_{0.5}$ between 0.2 and 5 mM, whereas high concentrations of K^+ inhibit ATP hydrolysis with a $K_{0.5}$ of about 20 mM. The $K_{0.5}$ values for both the stimulatory and inhibitory K^+-sites are dependent on the pH and ATP concentration used. Pre-steady state experiments with inside-out-oriented and ion-tight H,K-ATPase vesicles showed that the rapid biphasic dephosphorylation of the phosphoenzyme is induced by luminal K^+, giving evidence that the stimulatory K^+ site faces the lumen [104,105]. The inhibitory K^+ site is shown in intact vesicles to be cytoplasmic and to reduce the rate of enzyme phosphorylation by ATP [104,105]. Lorentzon et al. [123] demonstrated that this inhibitory K^+ site is potential-sensitive, suggesting that there is a charge-carrying step in the K^+ limb of H,K-ATPase.

In the absence of K^+ the enzyme exhibits a basal Mg^{2+}-ATPase activity that can be reduced, but not completely removed, upon further purification of the enzyme by free-flow or zonal electrophoresis [66,89]. Wallmark et al. [104] demonstrated that the rate of spontaneous breakdown of phosphoenzyme corresponded very well to the Mg^{2+}-ATPase activity at low ATP concentrations, implying that this activity was not due to a contaminating Mg^{2+}-ATPase with a reaction path independent of the phosphoenzyme. This conclusion was confirmed by Reenstra et al. [129] in a study on the nonhyperbolic ATP dependence of ATPase activity and phosphoenzyme formation of H,K-ATPase, and by Mendlein and Sachs [130], who observed that the K^+-site inhibitor SCH 28080 inhibits both the K^+-ATPase and the Mg^+-ATPase activity of gastric H,K-ATPase, inhibiting the latter activity by a K^+-like prevention of phosphoenzyme formation. Recently, however, Van der Hijden et al. [131] demonstrated with the ATP-analog *lin*-benzo-ATP that H,K-ATPase also contains a Mg^{2+}-stimulated hydrolytic activity with a reaction path independent of the formation of phosphoenzyme. An answer to the question whether this latter activity is due to another enzyme or due to an intrinsic reaction of H,K-ATPase, comparable to the hydrolysis of *p*-nitrophenylphosphate was not given in this study.

A prominent feature of H,K-ATPase is the nonhyperbolic ATP dependence of the phosphorylation reaction and the ATPase activity. Depending on the pH and K^+ concentration used, two affinities for ATP in the ATPase reaction are observed with $K_{0.5}$ varying from 0.4 to 74 µM and from 50 µM–1.1 mM, respectively [103,104,129,132,133]. Brzezinski et al. [107] explained the nonhyperbolic ATP dependence of the K^+-ATPase reaction by assuming that ATP can react with one nucleotide site on the enzyme in both the E_1 and E_2 state, but with a lower affinity in E_2. On the other hand, Reenstra et al. [129] presented evidence that ATP can also bind to the phosphoenzyme, be it E–P or E·P_i, which then leads, besides the K^+-stimulated breakdown, to an additional breakdown of phosphoenzyme. The existence of a low-affinity, second nucleotide site on the phosphoenzyme (probably E_1–P) was also demonstrated by Helmich-De Jong et al. [93,134] who observed inhibition of the dephosphorylation rate by millimolar nucleotide concentrations in the presence of micromolar Mg^{2+} concentrations. These authors also presented evidence that increasing Mg^{2+} concentrations reduce the inhibitory effect of nucleotides on the dephosphorylation rate, i.e., millimolar Mg^{2+} together with nucleotides enhance the dephosphorylation process.

The kinetically deduced existence of two classes of substrate sites may also account for the molar ratio between ATP analogs and inhibitors on the one hand and phosphoenzyme on the other hand. This ratio has been reported to be 2:1 for the ATP analogs adenylyl imido diphosphate (AMP–PNP) [135] and 2′,3′-*O*-(2,4,6-trinitrophenylcyclohexadienylidine)-ATP (TNP-ATP) [97], and also 2:1 for the ATP-site directed fluorescent inhibitors eosin [99] and FITC [49,50] and the transition-state inhibitor vanadate [126].

4.4. Hydrolysis of p-nitrophenylphosphate

Like Na,K-ATPase, gastric H,K-ATPase also exhibits a *p*-nitrophenylphosphatase (*p*NPPase) activity. This phosphatase activity is dependent on Mg^{2+} and K^+, or one of its congeners with the same order of selectivity as for the ATPase activity, yielding a specific activity of 61–84% of the maximal ATPase activity [4,136,137]. Phosphorylation by *p*NPP has not been demonstrated and transport is also not catalyzed by this substrate. As in the ATPase reaction the effect of K^+ on the pNPPase activity is biphasic. Contrary to the ATPase reaction, however, the K^+ stimulatory site for the phosphatase activity is at the cytoplasmic side of the membrane, and shows a lower affinity than for the ATPase activity, whereas the K^+ inhibitory site is also located at the cytoplasmic side [66,106,138]. For the phosphatase activity an additional luminal K^+ stimulatory site may also be involved [139], as has been demonstrated with the carboxyl activating reagent ethoxy carbonyl-ethoxy-dihydroquinoline (EEDQ) [140]. Therefore the enzyme form for this phosphatase activity may be either $2K \cdot E_1$, $2K \cdot E_2$ or an occluded form between these two. It is disputable, however, whether or not the luminal K^+ stimulatory site for the pNPPase reaction is identical with the K^+ stimulatory site for the ATPase reaction. Experiments with the hydrophobic amine SCH 28080, which inhibits the enzyme competitively with K^+, with either pNPP or ATP as substrates, indicate that the K^+ stimulatory site is identical for both substrates [141,142]. Ray and Nandi on the other hand [139], observed that spermine competes with K^+ for the K^+-pNPPase reaction without affecting the K^+-ATPase activity, suggesting that the K^+ stimulatory sites for both substrates are distinct. A difference between the K^+ stimulatory sites of the pNPPase and ATPase activity was also recently indicated by the different effects a dibenzoyl derivative of Scopadulcic acid, an ingredient of a Paraguayan traditional medicinal herb, has on both activities [143]. The catalytic sites for pNPP and ATP seem to be essentially different as has been demonstrated by a noncompetitive inhibition of the pNPPase activity by ATP and ADP [133], and by the observation that FITC-modified enzyme is inhibited with respect to the ATPase but not the pNPPase [49].

4.5. Phosphorylation from inorganic phosphate

In the presence of Mg^{2+} gastric H,K-ATPase can also be phosphorylated by P_i with a $K_{0.5}$ of about $50\,\mu M$ [144,145]. With respect to acid stability, $K^+ \cdot$ sensitivity and maximal phosphorylation level that can be obtained, the phosphoenzyme formed with inorganic phosphate is identical to the phosphoenzyme formed with ATP [145]. The reaction can be considered as the reverse of the dephosphorylation reaction of E_2–P to E_2 plus P_i. This reversibility has also been demonstrated by measurements of ^{18}O exchange between P_i and H_2O [94,146]. One of the characteristic features of phosphorylation by inorganic phosphate is the rate of phosphorylation that is, in the absence of K^+, about two to three orders of magnitude lower compared to phosphorylation by ATP. Without having an effect on the phosphorylation level, low concentrations of K^+ increase the phosphorylation rate ($K_{0.5} = 0.2\,\text{mM}$), whereas at higher concentrations of K^+ ($K_{0.5} = 2.4\,\text{mM}$) the phosphorylation level decreases [145]. The $K_{0.5}$ value for accelerating the phosphorylation rate corresponds rather well to the $K_{0.5}$ values required for the fluorescence quench of FITC-labelled enzyme ($K_{0.5} = 0.28\,\text{mM}$) [49], for the slow phase of the dephosphorylation rate at the luminal site ($K_{0.5} = 0.2\,\text{mM}$) [104], and for the rate of ^{18}O exchange between P_i and H_2O ($K_{0.5} = 0.3\,\text{mM}$) [146], indicating that it is the E_2 conformer that is phosphorylated.

5. Transport by H,K-ATPase

5.1. The H^+–ATP stoichiometry

Gastric H,K-ATPase produces in vivo a pH gradient of 6 pH units. Thermodynamical considerations would therefore preclude a H^+–ATP stoichiometry greater than one [147]. With the use of isolated gastric vesicles or reconstituted gastric H,K-ATPase many reports have been published concerning the correlation between the transport of H^+ and the hydrolysis of ATP. The in vitro values obtained for the H^+–ATP stoichiometry, however, varied from 4 to 1 [6,110–112,147,148]. The reported H^+–ATP stoichiometries of 1 are underestimated because ATP hydrolysis by Mg^{2+}-ATPase activity and K^+-ATPase activity from broken H,K-ATPase vesicles were neglected [147,148]. On the other hand, the measured H^+–ATP stoichiometries of 2 [110–112] are compatible with the pH gradient of 4.5 pH units obtained in gastric H,K-ATPase vesicles [149,150]. The difference between the expected H^+–ATP stoichiometry in vivo and the H^+–ATP stoichiometry measured in vitro therefore suggests that the coupling between transport of H^+ and K^+, and ATP hydrolysis might not be tight.

5.2. Ion selectivity

Various effects of Na^+ have been observed on the kinetics of gastric H,K-ATPase and have suggested the existence of Na^+-binding sites on the enzyme. Na^+ inhibits the phosphorylation rate, whereas it does not accelerate the dephosphorylation rate [104,114]. The inhibition of the phosphorylation rate is increased by an alkaline pH at the cytoplasmic side [106]. The $K_{0.5}$ for ATP at pH 7.0 increases from about $0.1\,\mu\text{M}$ in the absence of Na^+ to $3.0\,\mu\text{M}$ in the presence of 100 mM Na^+, with a concomitant decrease of the steady state phosphorylation level by 18% [151]. At an alkaline pH, Na^+ also inhibits both the K^+-ATPase and K^+-pNPPase activities as well as the transport of H^+ [106,152]. The transport activity is inhibited by Na^+ binding to the cytoplasmic side and is strongest at an alkaline pH and only slight at a neutral or an acidic pH. These observations suggested very strongly a competition between Na^+ ions and H^+ ions, which may be recognized as hydronium ions [153], on cytoplasmically oriented sites on the enzyme. Another indication that Na^+ ions may substitute for protons at the cytoplasmic side was provided by the observation that ^{86}RbCl-loaded gastric vesicles showed an extravesicular Na^+-stimulated ^{86}Rb efflux [116].

This competition between protons and Na^+ ions has been confirmed by Polvani et al. [154] who observed an ATP-dependent ^{22}Na influx into inside-out-oriented H,K-ATPase vesicles at an extravesicular alkaline pH that is stimulated by intravesicular K^+. At an extravesicular neutral pH no ^{22}Na influx was observed. The competition between Na^+ ions and protons on the E_1 conformer is also pointed out by a recent study on the effect of Na^+ ions on the conformational state of FITC-labelled H,K-ATPase [96]. Whereas at pH 6.4 equilibrium titrations showed that Na^+ ions stabilize the E_1 conformation with an apparent dissociation constant of 62 mM, an apparent dissociation constant of 10.2 mM was observed at pH 8.0, consistent with competition between protons and Na^+ ions.

Besides the competition between Na^+ ions and protons at the cytoplasmic side, competition between Na^+ ions and K^+ ions at the luminal side of the enzyme is also observed. Experiments with NaCl-loaded proteoliposomes containing reconstituted H,K-ATPase showed an ATP-dependent proton uptake, requiring internal K^+, after dilution of the proteoliposomes into KCl, but not NaCl, medium [118]. This indicated that a Na^+–K^+ exchange, Na^+ outward and K^+ inward, must have occurred before ATP was added, Na^+ ions thereby interacting with the intravesicular K^+-binding sites. Ditmars et al. [155] measured an ATP-dependent proton uptake in gastric vesicles, loaded with 500 mM NaCl and diluted in Na^+-free medium, suggesting an active transport of Na^+ ions from the luminal to the cytoplasmic side. Furthermore, a strict competition between Na^+ and K^+ ions is also observed with equilibrium titrations of the cations and FITC-labelled H,K-ATPase, showing that Na^+ ions bind to and stabilize the high fluorescence E_1 form of the enzyme while K^+ ions stabilize the low fluorescence E_2 form [96].

5.3. Electrogenicity of ion transport

Gastric H,K-ATPase is an electroneutral ion pump, which exchanges protons for K^+ ions. The electroneutrality of the pump is demonstrated not only in vivo [156], but also in vitro [6]. This indicates that in the consecutive transport cycle of the enzyme either the transporting step of both protons and K^+ ions is electroneutral, or the transporting step of each ion is electrogenic, resulting in an overall electroneutral countertransport. This first possibility would occur if there is binding of protons and K^+ ions to negatively charged binding sites. Such a mechanism has been suggested for Na,K-ATPase where two carboxyl residues should be involved in the cation-binding sites, resulting in an electroneutral transport of two K^+ ions and an electrogenic transport of three Na^+ ions [157]. For H,K-ATPase, however, it has become clear that the transport cycle includes equal and opposite charge movements [115,123]. Lorentzon et al. [123] demonstrated with gastric vesicles that the K^+-transporting step of H,K-ATPase, i.e., $2K \cdot E_2$ to $2K \cdot E_1$ or even E_1 is potential-sensitive and therefore must be charge carrying. The electrogenicity of the H^+-transporting step was proven by Van der Hijden et al. [115]. Upon phosphorylation of H,K-ATPase, which was adsorbed to a black lipid membrane, with a photolabile analog of ATP they observed a transient capacitive charge movement. In the presence of K^+ they observed no pump currents, demonstrating also that the K^+ limb of the transport cycle is electrogenic too.

The electrogenicity of both the H^+- and K^+-transporting steps implicates that the cation-binding sites must be neutral, i.e., do not contain negatively charged amino acid residues. For the H^+-binding site the involvement of a carboxyl residue would actually be very unlikely because of the low luminal pH of about 1, requiring an extremely low pK_a value for the carboxyl group. By using the histidine derivatizing reagent diethyl pyrocarbonate (DEPC), it has been demonstrated that an essential histidine group with a pK_a of about 7 is involved in the binding of protons at the cytoplasmic side [158]. For the luminal K^+-binding site it has been demonstrated with the carboxyl activating reagent N-ethoxycarbonyl-2-ethoxy-1,2-dihydroquinoline (EEDQ) that a carboxyl group is involved in K^+ binding, but this amino acid residue has a pK_a of about 7, showing that it is in a nonaqueous environment, i.e., in a hydrophobic domain [140].

6. Lipid dependency of H,K-ATPase

Gastric H,K-ATPase is embedded in a lipid bilayer consisting of cholesterol and phospholipids. The composition of the phospholipid fraction is little different from the purified Na,K-ATPase membranes [159] and consists of 33% phosphatidylcholine, 22% phosphatidylethanolamine, 11% phosphatidylserine, 8% phosphatidylinositol and 25% sphingomyelin [102,160]. The requirement of phospholipids for enzyme activity has been studied by digestion with specific phospholipases [102,160–162]. Studies with phospholipase A_2 [102,161] demonstrated that the K^+-ATPase activity is much more sensitive to phospholipase digestion than the K^+-pNPPase activity. Besides this inhibitory action on the hydrolytic activities of H,K-ATPase, it was demonstrated that in the case of phospholipase A_2 treatment the ATP-induced conformational changes, as indicated by circular dichroism spectra, were inhibited too [102]. Schrijen et al. [160], however, observed that during phospholipase C treatment the K^+-pNPPase activity is lowered parallel to the K^+-ATPase activity. Moreover, they demonstrated a parallelism between the lowering of the phospholipid content and of the K^+-ATPase activity during treatment with phospholipase C, both in time and in final level.

In studies with specific phospholipases the asymmetry in the composition of the lipid bilayer was also suggested [102,162]. The requirement of specific phospholipids, which are essential for enzyme activity, however, has not been established. For example, Saccomani et al. [102] demonstrated that readdition of various phospholipids, after phospholipase A_2 treatment, results in a restoration of the K^+-ATPase activity. On the other hand, Nandi et al. [161] observed a restoration of the K^+-ATPase activity with addition of phosphatidylcholine and not with phosphatidylethanolamine. Also Schrijen et al. [160] did not observe a specific phospholipid requirement.

7. Solubilization and reconstitution

Although gastric H,K-ATPase can be isolated rather purely by density gradient centrifugation and subsequent free-flow electrophoresis [66] or zonal electrophoresis [89], solubilization experiments have been undertaken to further purify active H,K-ATPase. Solubilization of H,K-ATPase was also carried out in order to identify molecular associations of the active enzyme, and for reconstitution experiments. Soumarmon et al. [90] first reported a nondegradative solubilization of H,K-ATPase by n-octylglucoside. Extracted H,K-ATPase displayed the same enzymatic properties as the native enzyme and in glycerol gradient centrifugation it equilibrated as a single peak corresponding to an apparent molecular mass of 390 kDa. Takaya et al. [163] succeeded in solubilizing H,K-ATPase, which had been purified before by sucrose density gradient centrifugation, with Emulgen, with apparent preservation of the enzyme activity. Further purification of the solubilized enzyme by polyethylene glycol fractionation and Sepharose chromatography steps, however, did not result in an increase of the specific enzyme activity, indicating an inactivation of the enzyme activity during purification. On SDS-polyacrylamide gel the purified, but inactivated, enzyme showed a single band with an apparent molecular mass of 94 kDa. Recently Rabon et al. [70] reported the solubilization of an active H,K-ATPase with the nonionic detergent $C_{12}E_8$. The retention of enzyme activity was shown to be dependent upon pH, K^+ concentration, nucleotide concentration and temperature. Optimal retention of enzyme activity was obtained at pH 6.1 and 2 mM ATP. The high stability of the soluble enzyme ($t_{1/2}$ for ATPase activity of 6 h between 4 and 11°C) enabled the authors to study the subunit associations of functional H,K-ATPase. By means of glutaraldehyde crosslinking analysis of $C_{12}E_8$ solubilized H,K-ATPase it was then demonstrated that functional porcine gastric H,K-ATPase is organized as a structural dimer.

Solubilization of an active H,K-ATPase is also a prerequisite for reconstitution of the enzyme into liposomes. With these H,K-ATPase proteoliposomes it is then possible to study the transport characteristics of pure H,K-ATPase, without the interference of residual protein contamination that is usually present in native vesicular H,K-ATPase preparations. Rabon et al. [118] first reported the reconstitution of cholate or n-octylglucoside solubilized H,K-ATPase into phosphatidylcholine–cholesterol liposomes. The enzyme was reconstituted asymmetrically into the proteoliposomes with 70% of the pump molecules having the cytoplasmic side extravesicular. In the presence of intravesicular K^+, the proteoliposomes exhibited an Mg-ATP-dependent H^+ transport, as monitored by acridine orange fluorescence quenching. Moreover, as seen with native H,K-ATPase vesicles, reconstituted H,K-

ATPase also catalyzed a passive Rb^+–Rb^+ exchange, the rate of which was comparable to the rate of active Rb^+ efflux. This suggested that the K^+-transporting step of H,K-ATPase is not severely limited by a K^+-occluded enzyme form, as was observed for Na,K-ATPase. Skrabanja et al. [164] also described the reconstitution of choleate solubilized H,K-ATPase into phosphatidylcholine–cholesterol liposomes. With the use of a pH electrode to measure the rate of H^+ transport they observed not only an active H^+ transport, which is dependent on intravesicular K^+, but also a passive H^+–K^+ exchange. This passive transport process, which exhibited a maximal rate of 5% of the active transport process, could be inhibited by vanadate and the specific inhibitor omeprazole, giving evidence that it is a function of gastric H,K-ATPase. The same authors demonstrated, by separation of non-incorporated H,K-ATPase from reconstituted H,K-ATPase on a sucrose gradient, that H,K-ATPase transports two protons and two K^+ ions per hydrolyzed ATP [112].

8. *Inhibitors of H,K-ATPase*

The search for specific inhibitors of gastric H,K-ATPase has a dual purpose. First, with the help of suitable inhibitors it is possible to get insight into the molecular mechanisms of H,K-ATPase, and second, a specific inhibitor might be clinically useful for inhibition of gastric acid secretion in anti-ulcer therapy.

There are many ligands and group-specific reagents that have been demonstrated to alter the properties of H,K-ATPase, and which are not clinically useful. For example, there is a variety of chemicals that have been used in studies on structure–function relations of H,K-ATPase and that inhibit the enzyme in vitro by modification of its amino [49,67,158], sulfhydryl [95,165,166] or carboxyl groups [140].

Another approach that has gained attention in recent years in investigating structure–function relationships of gastric H,K-ATPase is the use of monoclonal antibodies that specifically modify a particular reaction step of the enzyme. Asano et al. [167] raised a monoclonal antibody that inhibited K^+-ATPase activity, K^+-*p*NPPase activity and chloride conductance, which had been induced before with the S–S cross-linking reagent Cu^{2+}-*o*-phenanthroline, in gastric vesicles [168,169]. The inhibitory effects of the antibody on both the hydrolytic reactions and the chloride conductance led the authors to the conclusion that the Cl^--channel in gastric vesicles is part of the function of H,K-ATPase. This hypothesis was challenged by Benkouka et al. [170] who described a monoclonal antibody with similar inhibitory effects on the hydrolytic reactions of H,K-ATPase, but which did not inhibit the chloride conductance induced by Cu^{2+}-*o*-phenanthroline. In certain studies monoclonal antibodies against gastric H,K-ATPase have been reported that completely inhibit the K^+-ATPase activity and not the K^+-*p*NPPase activity [171,172]. Both monoclonal antibodies exhibited a pH-dependent inhibitory effect on the ATPase reaction that was maximal at an alkaline pH. The pattern of inhibition of one of these antibodies was uncompetitive with respect to K^+, indicating that it bound to the K^+-bound enzyme ($E_1 \cdot K$) [172]. This antibody recognized the catalytic subunit of the enzyme on Western blot, which made determination of the epitope possible [56].

There are several antisecretory drugs such as fenoctimine [173], nolinium bromide [174], the antidepressant trifluoperazine and the Ca^{2+} antagonists verapamil and TMB-8 (8-(N-N-diethylamino)octyl-3,4,5-trimethoxybenzoate) [175] that interfere with gastric H,K-ATPase. The mechanism of interaction with H,K-ATPase is different for the respective drugs. Fenoctimine has been demonstrated to be nonspecific as it inhibits also other membrane ATPases and inhibition of H,K-ATPase is the result of a fenoctimine-induced disruption of the microsomal membrane [173,176]. Nolinium bromide on the other hand inhibits H,K-ATPase by acting as a K^+-competitive inhibitor, although with low affinity [177,178]. Conflicting results exist about the mechanism of action of the Ca^{2+} antagonists. Several groups have reported that these drugs are K^+-competitive inhibitors of H,K-ATPase [175,179], whereas other groups ascribe the antisecretory activity of these compounds in vitro to a nonspecific, i.e., a protonophoric or detergent-like, action [180,181].

Characteristic for H,K-ATPase and other P-type ATPases is the inhibition by vanadate, a transition-state analog of phosphate [126]. Inhibition by vanadate is thought to occur since vanadium in the $5+$ oxidation state can assume the trigonal-bipyramidal geometry of pentacoordinate phosphorus. It therefore can occupy the phosphate-binding site and prevent the phosphorylation of the enzyme. Besides the phosphorylation reaction, the Mg^{2+}-ATPase, the K^+-ATPase and the proton transport activity, the K^+-pNPPase activity of H,K-ATPase is also inhibited by vanadate. Maximal inhibition of these activities is reached at a vanadate binding stoichiometry of 3 nmol mg^{-1} protein, whereas a maximum of 1.5 nmol mg^{-1} protein acid-stable phosphoenzyme is formed. Faller et al. [126] interpreted these conflicting data by assuming ATP-competitive vanadate binding at a high- and low-affinity catalytic site.

The substituted benzimidazoles constitute a new class of specific and irreversible inhibitors of H,K-ATPase. Omeprazole (5-methoxy-2-[[(4-methoxy-3,5-dimethyl-2-pyridinyl)methyl]sulfinyl]-1H-benzimidazole), as main representative [182], Ro 18–5364 [183], lansoprazole [184], E3810 [185], HOE 731 [186] and BY1023/SK&F 96022 [187] belong to this class of inhibitor. The substituted benzimidazoles were initially reported to inhibit acid secretion both in vivo and in vitro [188]. A correlation between inhibition of acid secretion and blockade of gastric H,K-ATPase was then strongly suggested by Fellenius et al. [189] who observed that H,K-ATPase in vitro is inhibited by the substituted benzimidazole picoprazole, a precursor of the clinically used omeprazole. With omeprazole it was then definitely demonstrated that inhibition of H,K-ATPase is correlated with inhibition of gastric acid secretion [190]. The mechanism of inhibition by omeprazole was investigated and reviewed [18,191].

Characteristic for omeprazole is its low pK_a of approximately 4 that leads to trapping within the acid canaliculus of the parietal cell. The compound is rapidly transformed there to an active and permanent cationic inhibitor of H,K-ATPase [192,193]. The active inhibitor thus formed reacts with essential sulfhydryl groups on the luminal face of the enzyme resulting in inhibition of K^+-ATPase and K^+-pNPPase reactions (IC_{50} = 8 μM at pH 6.0) [194–196]. This acid-induced transformation of omeprazole into an active inhibitor compound makes omeprazole a highly specific drug. Under conditions, in which no pH-gradient is generated, a stoichiometry of 10–20 mol of inhibitor per mol phosphoenzyme is observed at complete inhibition of the enzyme [194,196]. Both in vitro and in vivo studies, however, demonstrate that in the case of acid accumulation, complete inhibition of the K^+-ATPase activity is reached at a stoichiometry of 2 mol of inhibitor per mol phosphoenzyme [197–199]. Recently Mercier et al. [52] have suggested, after sequencing omeprazole-labelled tryptic-digest fragments of H,K-ATPase, that the reactive cysteine residues of the catalytic α subunit are Cys^{823} and Cys^{893}. Morii et al. [200] on the other hand, proposed that omeprazole specifically reacts with Cys^{322}.

An important class of reversible H,K-ATPase inhibitors, exemplified by SCH 28080 (3-(cyanomethyl)-2-methyl-8-(phenylmethoxy)imidazo[1,2a]pyridine) are the substituted imidazo-[1,2a]pyridines. These compounds were initially reported as effective inhibitors of acid secretion [201,202]. Studies in isolated guinea pig fundic mucosa indicated that the antisecretory activity involved direct action on the parietal cell distal to the histamine or cholinergic receptors and at a site beyond the formation of cAMP. Scott and Sundell [203] then demonstrated that SCH 28080 inhibits the K^+-ATPase activity of isolated H,K-ATPase, suggesting that this compound may exert its antisecretory activity via a direct inhibition of H,K-ATPase. Subsequent investigations confirmed this hypothesis and showed that SCH 28080 is competitive with the activating cation K^+ for both the ATPase and the pNPPase reactions, having a K_i between 0.02 and 0.12 μM for the ATPase and 0.28 and 2.4 μM for the pNPPase reaction [130,141,142,204,205]. It was also demonstrated that the potency of SCH 28080, which is a protonatable weak base (pK_a = 5.6), increased at low pH, commensurate with the proportion of SCH 28080 in the protonated form, indicating that the protonated form is the luminally active inhibitory species [142,206]. SCH 28080 binds to both the E_2 and E_2P conformers of H,K-ATPase and with higher affinity to the phosphorylated form; a binding stoichiometry of 2.2 mol of SCH 28080 per mol of catalytic phosphorylation site is observed [130,141,207]. The mechanism by which SCH 28080 inhibits H,K-ATPase suggests that SCH 28080 could be used as a probe for the luminal K^+ transport site. For this reason two photoaffinity derivatives of SCH 28080 have been recently synthesized that show a qualitatively similar interaction with H,K-ATPase to that of SCH 28080 [54,208]. The photoaffinity derivative m-ATIP (8-(3-azidophenylmethoxy)-1,2,3-trimethylimidazo[1,2a]pyridinium iodide), showed a level of incorporation required to produce complete inhibition of ATPase activity of 1.9 times the number of catalytic phosphorylation sites [208]. This stoichiometry corresponded to the stoichiometry for the binding of SCH 28080 to H,K-ATPase [207]. Trypsinisation and sequencing of H,K-ATPase labelled with the light-sensitive derivative mDAZIP (2,3-dimethyl-8-[(4-azidophenyl)methoxy]imidazo[1,2a]pyridine), which, however, exhibited a maximal amount of specific labelling similar to the number of phosphorylation sites, showed that the site of labelling is contained within the region between residues 105 and 172 [55]. This finding indicates that this region of H,K-ATPase is possibly part of the K^+ pathway in the enzyme.

9. Conclusions

Although many studies on the mechanism of gastric H,K-ATPase have led to the reaction scheme of Fig. 2, much more information is required to settle the questions about a possible variable coupling in the ion transport process, a putative regulation of H,K-ATPase by other cations like Na^+, the minimal structure required for transport activity and the phosphorylation state of each catalytic subunit during ATP hydrolysis. Part of the answers to these questions will be given by comparative studies of H,K-ATPase and other P-type ATPases, a strategy which has already been successfully applied to the structural field of the H,K-ATPase α subunit and which may also be very fruitful for the recently identified H,K-ATPase β subunit. Important structural features of H,K-ATPase may also be identified by sequence comparisons with other related ATPases. A very promising approach for unravelling structural and functional problems will be the in vitro expression of H,K-ATPase. In such an expression system the basic structure for transport could be established. Moreover, site-directed mutations to single amino acid residues may contribute to elucidation of the cation-binding sites, for which the first indications are given by the K^+-competitive inhibitor SCH 28080.

References

1. Chew, C.S., Hersey, S.J., Sachs, G. and Berglindh, T. (1980) Am. J. Physiol. 238, G312–320.
2. Chew, C.S. (1986) Am. J. Physiol. 250, G814–G823.
3. Negulescu, P.A. and Machen, T.E. (1988) Am. J. Physiol. 254, C130–C140.
4. Ganser, A.L. and Forte, J.G. (1973) Biochim. Biophys. Acta 307, 169–180.
5. Lee, J., Simpson, G. and Scholes, P. (1974) Biochem. Biophys. Res. Commun. 60, 825–832.
6. Sachs, G., Chang, H.H., Rabon, E., Schackmann, R., Lewin, M. and Saccomani, G. (1976) J. Biol. Chem. 251, 7690–7698.
7. Wolosin, J.M. (1985) Am. J. Physiol. 248, G595–G607.
8. Forte, J.G., Hanzel, D.K., Urushidani, T. and Wolosin, J.M. (1989) Ann. N.Y. Acad. Sci. 574, 145–158.
9. Demarest, J.R. and Loo, D.D.F. (1990) Annu. Rev. Physiol. 52, 307–319.

10. Pedersen, P.L. and Carafoli, E. (1987) Trends Biochem. Sci. 12, 146–150.
11. Pedersen, P.L. and Carafoli, E. (1987) Trends Biochem. Sci. 12, 186–189.
12. Shull, G.E. (1990) J. Biol. Chem. 265, 12123–12126.
13. Forte, J.G., Machen, T.E. and Obrink, K.J. (1980) Annu. Rev. Physiol. 42, 111–126.
14. Sachs, G. (1987) In: Physiology of the Gastrointestinal Tract, (Johnson, L.R., Ed.) pp. 865–881, Raven Press, New York, 2nd ed.
15. Rabon, E.C. and Reuben, M.A. (1990) Annu. Rev. Physiol. 52, 321–344.
16. Sachs, G., Munson, K., Balaji, V.N., Aures-Fischer, D., Hersey, S.J. and Hall, K. (1989) J. Bioenerg. Biomemb. 21, 573–588.
17. Rabon, E.C., Im, W.B. and Sachs, G. (1988) Methods Enzymol. 157, 649–654.
18. Sachs, G., Carlsson, E., Lindberg, P. and Wallmark, B. (1988) Annu. Rev. Pharmacol. Toxicol. 28, 269–284.
19. Saccomani, G., Shah, G., Spenney, J.G. and Sachs, G. (1975) J. Biol. Chem. 250, 4802–4809.
20. Forte, J.G., Ganser, A.L. and Tanisawa, A.S. (1974) Ann. N.Y. Acad. Sci. 242, 255–267.
21. Im, W.B. and Blakeman, D.P. (1982) Biochem. Biophys. Res. Commun. 108, 635–639.
22. Saccomani, G., Chang, H.H., Mihas, A.A., Crago, S. and Sachs, G. (1979) J. Clin. Invest. 64, 627–635.
23. Saccomani, G., Helander, H.F., Crago, S., Chang, H.H., Dailey, D.W. and Sachs, G. (1979) J. Cell. Biol. 83, 271–283.
24. Smolka, A., Helander, H.F. and Sachs, G. (1983) Am. J. Physiol. 245, G589–G596.
25. Canfield, V.A., Okamoto, C.T., Chow, D., Dorfman, J., Gros, P., Forte, J.G. and Levenson, R. (1990) J. Biol. Chem. 265, 19878–19884.
26. Gustin, M.C. and Goodman, D.B.P. (1982) J. Biol. Chem. 257, 9629–9633.
27. Kaunitz, J. and Sachs, G. (1986) J. Biol. Chem. 261, 14005–14010.
28. Doucet, A. and Marsy, S. (1987) Am. J. Physiol. 253, F418–F423.
29. Garg, L.C. and Narang, N. (1988) J. Clin. Invest. 81, 1204–1208.
30. Wingo, C.S. (1989) J. Clin. Invest. 84, 361–365.
31. Perrone, R.D. and McBride, D.E. (1988) Am. J. Physiol. 254, G898–G906.
32. Watanabe, T., Suzuki, T. and Suzuki, Y. (1990) Am. J. Physiol. 258, G506–G511.
33. Saccomani, G., Dailey, D.W. and Sachs, G. (1979) J. Biol. Chem. 254, 2821–2827.
34. Helmich-De Jong, M.L., Van Emst-De Vries, S.E. and De Pont, J.J.H.H.M. (1987) Biochim. Biophys. Acta 905, 358–370.
35. Peters, W.H.M., Fleuren-Jakobs, A.M.M., Schrijen, J.J., De Pont, J.J.H.H.M. and Bonting, S.L. (1982) Biochim. Biophys. Acta 690, 251–260.
36. Smolka, A. and Hughes, F.M. (1989) J. Cell Biol. 107, 125A.
37. Smolka, A., Sachs, G. and Lorentzon, P. (1989) Gastroenterology 97, 873–881.
38. Shull, G.E., Greeb, J. and Lingrel, J.B. (1986) Biochemistry, 25, 8125–8132.
39. Sweadner, K.J. (1989) Biochim. Biophys. Acta 988, 185–220.
40. Shull, G.E. and Lingrel, J.B. (1986) J. Biol. Chem. 261, 16788–16791.
41. Maeda, M., Ishizaki, J. and Futai, M. (1988) Biochem. Biophys. Res. Commun. 157, 203–209.
42. Maeda, M., Oshiman, K., Tamura, S. and Futai, M. (1990) J. Biol. Chem. 265, 9027–9032.
43. Lane, L.K., Kirley, T.L. and Ball, W.J. (1986) Biochem. Biophys. Res. Commun. 138, 185–192.
44. Shull, G.E., Schwartz, A. and Lingrel, J.B. (1985) Nature (London) 316, 691–695.
45. Newman, P.R., Greeb, J., Keeton, T.P., Reyes, A.A. and Shull, G.E. (1990) DNA Cell Biol. 9, 749–762.
46. Walderhaug, M.O., Post, R.L., Saccomani, G., Leonard, R.T. and Briskin, D.P. (1985) J. Biol. Chem. 260, 3852–3859.
47. Maeda, M., Tagaya, M. and Futai, M. (1988) J. Biol. Chem. 263, 3652–3656.
48. Tamura, S., Tagaya, M., Maeda, M. and Futai, M. (1989) J. Biol. Chem. 264, 8580–8584.
49. Jackson, R.J., Mendlein, J. and Sachs, G. (1983) Biochim. Biophys. Acta 731, 9–15.

50. Farley, R.A. and Faller, L.D. (1985) J. Biol. Chem. 260, 3899–3901.
51. Kyte, J. and Doolittle, R.F. (1982) J. Mol. Biol. 157, 105–132.
52. Mercier, F., Besançon, M., Hersey, S.J. and Sachs, G. (1991) FASEB J. 5, A749.
53. Scott, D., Munson, K., Modyanov, N. and Sachs, G. (1991) FASEB J. 5, A748.
54. Munson, K.B. and Sachs, G. (1988) Biochemistry 27, 3932–3938.
55. Munson, K.B., Gutierrez, C., Balaji, V.N., Ramnarayan, K. and Sachs, G. (1991) J. Biol. Chem. 266, 18976–18988.
56. Van Uem, T.J.F., Swarts, H.G.P. and De Pont, J.J.H.H.M. (1991) Biochem. J., 280, 243–248.
57. Clarke, D.M., Loo, T.W. and MacLennan, D.H. (1990) J. Biol. Chem. 265, 17405–17408.
58. Matthews, I., Sharma, R.P., Lee, A.G. and East, J.M. (1990) J. Biol. Chem. 265, 18737–18740.
59. Jørgensen, P.L. (1992) Chapter 1, this volume.
60. Ohta, T., Nagano, K. and Yoshida, M. (1986) Proc. Natl. Acad. Sci. U.S.A. 83, 1–5.
61. Tai, M.M., Im, W.B., Davis, J.P., Blakeman, D.P., Zurcher-Neely, H.A. and Heinrikson, R.L. (1989) Biochemistry 28, 3183–3187.
62. Hall, K., Perez, G., Anderson, D., Gutierrez, C., Munson, K., Hersey, S.J., Kaplan, J.H. and Sachs, G. (1990) Biochemistry 29, 701–706.
63. Saccomani, G. and Mukidjam, E. (1987) Biochim. Biophys. Acta 912, 63–73.
64. Pedemonte, C.H., Sachs, G. and Kaplan, J.H. (1990) Proc. Natl. Acad. Sci. U.S.A. 87, 9789–9793.
65. Nakamoto, R.K., Rao, R. and Slayman, C.W. (1989) Ann. N.Y. Acad. Sci. 574, 165–179.
66. Saccomani, G., Stewart, H.B., Shaw, D., Lewin, M. and Sachs, G. (1977) Biochim. Biophys. Acta 465, 311–330.
67. Schrijen, J.J., Luyben, W.A.H.M., De Pont, J.J.H.H.M. and Bonting, S.L. (1980) Biochim. Biophys. Acta 597, 331–344.
68. De Jong, W.W., Zweers, A. and Cohen, L.A. (1978) Biochim. Biophys. Res. Commun. 82, 532–539.
69. Rabon, E.C., Gunther, R.D., Bassilian, S. and Kempner, E.S. (1988) J. Biol. Chem. 263, 16189–16194.
70. Rabon, E.C., Bassilian, S. and Jakobsen, L.J. (1990) Biochim. Biophys. Acta 1039, 277–289.
71. Okamoto, C.T., Karpilow, J.M., Smolka, A. and Forte, J.G. (1990) Biochim. Biophys. Acta 1037, 360–372.
72. Jørgensen, P.L. (1974) Biochim. Biophys. Acta 356, 36–52.
73. Nandi, J., Meng-Ai, Z. and Ray, T.K. (1987) Biochemistry 26, 4264–4272.
74. Reuben, M.A., Lasater, L.S. and Sachs, G. (1990) Proc. Natl. Acad. Sci. U.S.A. 87, 6767–6771.
75. Karlsson, F.A., Burman, P., Loof, L. and Mårdh, S. (1988) J. Clin. Invest. 81, 475–479.
76. Goldkorn, I., Gleeson, P.A. and Toh, B.-H. (1989) J. Biol. Chem. 264, 18768–18774.
77. Toh, B.-H., Gleeson, P.A., Simpson, R.J., Moritz, R.L., Callaghan, J.M., Goldkorn, I., Jones, C.M., Martinelli, T.M., Mu, F.T., Humphris, D.C., Pettitt, J.M., Mori, Y., Masuda, T., Sobieszczuk, P., Weinstock, J., Mantamadiotis, T. and Baldwin, G.S. (1990) Proc. Natl. Acad. Sci. U.S.A. 87, 6418–6422.
78. Mercer, R.W., Schneider, J.W., Savitz, A., Emanuel, J., Benz, E.J., Jr. and Levenson, R. (1986) Mol. Cell. Biol. 6, 3884–3890.
79. Ovchinnikov, Yu.A., Modyanov, N.N., Broude, N.E., Petrukhin, K.E., Grishin, A.V., Arzamazova, N.M., Aldanova, N.A., Monastyrskaya, G.S. and Sverdlov, E.D. (1986) FEBS Lett. 201, 237–245.
80. Young, R.M., Shull, G.E. and Lingrel, J.B. (1987) J. Biol. Chem. 262, 4905–4910.
81. Gloor, S., Antonicek, H., Sweadner, K.J., Pagliusi, S., Frank, R., Moos, M. and Schachner, M. (1990) J. Cell Biol. 110, 165–174.
82. Martin-Vasallo, P., Dackowski, W., Emanuel, J.R. and Levenson, R. (1989) J. Biol. Chem. 264, 4613–4618.
83. Kirley, T.L. (1989) J. Biol. Chem. 264, 7185–7192.
84. Miller, R.P. and Farley, R.A. (1990) Biochemistry 29, 1524–1532.
85. McDonough, A.A., Geering, K. and Farley, R.A. (1990) FASEB J. 4, 1598–1605.

86. Renaud, K.J. and Fambrough, D.M. (1990) J. Gen. Physiol. 96, 15a.
87. Gottardi, C.J. and Caplan, M.J. (1990) J. Gen. Physiol. 96, 26a.
88. Saccomani, G., Sachs, G., Cuppoletti, J. and Jung, C.Y. (1981) J. Biol. Chem. 256, 7727–7729.
89. Schrijen, J.J., Van Groningen-Luyben, W.A.H.M., Nauta, H., De Pont, J.J.H.H.M. and Bonting, S.L. (1983) Biochim. Biophys. Acta 731, 329–337.
90. Soumarmon, A., Grelac, F. and Lewin, M.J.M. (1983) Biochim. Biophys. Acta 732, 579–585.
91. Soumarmon, A., Robert, J.C. and Lewin, M.J. (1986) Biochim. Biophys. Acta 860, 109–117.
92. Rabon, E., Sachs, G., Mårdh, S. and Wallmark, B. (1982) Biochim. Biophys. Acta 688, 515–524.
93. Helmich-De Jong, M.L., Van Emst-De Vries, S.E., De Pont, J.J.H.H.M., Schuurmans Stekhoven, F.M.A.H. and Bonting, S.L. (1985) Biochim. Biophys. Acta 821, 377–383.
94. Faller, L.D. and Elgavish, G.A. (1984) Biochemistry 23, 6584–6590.
95. Schrijen, J.J., Van Groningen-Luyben, W.A.H.M., De Pont, J.J.H.H.M. and Bonting, S.L. (1981) Biochim. Biophys. Acta 640, 473–486.
96. Rabon, E.C., Bassilian, S., Sachs, G. and Karlish, S.J.D. (1990) J. Biol. Chem. 265, 19594–19599.
97. Faller, L.D. (1989) Biochemistry 28, 6771–6778.
98. Faller, L.D. (1990) Biochemistry 29, 3179–3186.
99. Helmich-De Jong, M.L., Van Duynhoven, J.P.M., Schuurmans Stekhoven, F.M.A.H. and De Pont, J.J.H.H.M. (1986) Biochim. Biophys. Acta 858, 254–262.
100. Rabon, E.C., Sachs, G., Bassilian, S., Leach, C. and Keeling, D. (1991) J. Biol. Chem. 266, 12395–12401.
101. Mitchell, R.C., Haris, P.I., Fallowfield, C., Keeling, D.J. and Chapman, D. (1988) Biochim. Biophys. Acta 941, 31–38.
102. Saccomani, G., Chang, H.H., Spisni, A., Helander, H.F., Spitzer, H.L. and Sachs, G. (1979) J. Supramol. Struct. 11, 429–444.
103. Sachs, G., Berglindh, T., Rabon, E., Stewart, H.B., Barcellona, M.L., Wallmark, B. and Saccomani, G. (1980) Ann. N.Y. Acad. Sci. 341, 312–334.
104. Wallmark, B., Stewart, H.B., Rabon, E., Saccomani, G. and Sachs, G. (1980) J. Biol. Chem. 255, 5313–5319.
105. Stewart, B., Wallmark, B. and Sachs, G. (1981) J. Biol. Chem. 256, 2682–2690.
106. Ljungström, M., Vega, F.V. and Mårdh, S. (1984) Biochim. Biophys. Acta 769, 220–230.
107. Brzezinski, P., Malmström, B.G., Lorentzon, P. and Wallmark, B. (1988) Biochim. Biophys. Acta 942, 215–219.
108. Post, R.L., Sen, A.K. and Rosenthal, A.S. (1965) J. Biol. Chem. 240, 1437–1445.
109. Albers, R.W., Fahn, S. and Koval, G.J. (1963) Proc. Natl. Acad. Sci. U.S.A. 50, 474–481.
110. Rabon, E.C., McFall, T.L. and Sachs, G. (1982) J. Biol. Chem. 257, 6296–6299.
111. Skrabanja, A.T.P., De Pont, J.J.H.H.M. and Bonting, S.L. (1984) Biochim. Biophys. Acta 774, 91–95.
112. Skrabanja, A.T.P., Van der Hijden, H.T.W.M. and De Pont, J.J.H.H.M. (1987) Biochim. Biophys. Acta 903, 434–440.
113. Glynn, I.M., Richards, D.E. and Beaugé, L.A. (1988) Methods Enzymol. 156, 281–287.
114. Wallmark, B. and Mårdh, S. (1979) J. Biol. Chem. 254, 11899–11902.
115. Van der Hijden, H.T.W.M., Grell, E., De Pont, J.J.H.H.M. and Bamberg, E. (1990) J. Membr. Biol. 114, 245–256.
116. Schackmann, R., Schwartz, A., Saccomani, G. and Sachs, G. (1977) J. Membr. Biol. 32, 361–381.
117. Soumarmon, A., Rangachari, P.K. and Lewin, M.J. (1984) J. Biol. Chem. 259, 11861–11867.
118. Rabon, E., Gunther, R.D., Soumarmon, A., Bassilian, S., Lewin, M. and Sachs, G. (1985) J. Biol. Chem. 260, 10200–10207.
119. Karlish, S.J.D. and Stein, W.D. (1982) J. Physiol. 328, 295–316.
120. Karlish, S.J.D. and Stein, W.D. (1982) J. Physiol. 328, 317–331.
121. Glynn, I.M. and Karlish, S.J.D. (1975) Annu. Rev. Physiol. 37, 13–55.

122. Ray, T.K. and Forte, J.G. (1976) Biochim. Biophys. Acta 443, 451–467.
123. Lorentzon, P., Sachs, G. and Wallmark, B. (1988) J. Biol. Chem. 263, 10705–10710.
124. Mendlein, J. and Sachs, G. (1989) J. Biol. Chem. 264, 18512–18519.
125. Mendlein, J., Ditmars, M.L. and Sachs, G. (1990) J. Biol. Chem. 265, 15590–15598.
126. Faller, L.D., Rabon, E. and Sachs, G. (1983) Biochemistry 22, 4676–4685.
127. Peters, W.H.M., Fleuren-Jakobs, A.M.M., Kamps, K.M.P., De Pont, J.J.H.H.M. and Bonting, S.L. (1982) Anal. Biochem. 124, 349–352.
128. Swarts, H.G.P., Van Uem, T.J.F., Hoving, S., Fransen, J.A.M. and De Pont, J.J.H.H.M. (1991) Biochim. Biophys. Acta 1070, 283–292.
129. Reenstra, W.W., Bettencourt, J.D. and Forte, J.G. (1988) J. Biol. Chem. 263, 19618–19625.
130. Mendlein, J. and Sachs, G. (1990) J. Biol. Chem. 265, 5030–5036.
131. Van der Hijden, H.T.W.M., Kramer-Schmitt, S., Grell, E. and De Pont, J.J.H.H.M. (1990) Biochem. J. 267, 565–572.
132. Schrijen, J.J. (1981) Ph.D. Dissertation, University of Nijmegen, The Netherlands.
133. Ljungström, M. and Mårdh, S. (1985) J. Biol. Chem. 260, 5440–5444.
134. Helmich-De Jong, M.L., Van Emst-De Vries, S.E., Swarts, H.G.P., Schuurmans Stekhoven, F.M.A.H. and De Pont, J.J.H.H.M. (1986) Biochim. Biophys. Acta 860, 641–649.
135. Van de Ven, F.J.M., Schrijen, J.J., De Pont, J.J.H.H.M. and Bonting, S.L. (1981) Biochim. Biophys. Acta 640, 487–499.
136. Forte, J.G., Ganser, A.L. and Tanisawa, A.S. (1974) Ann. N.Y. Acad. Sci. 242, 255–267.
137. Chang, H., Saccomani, G., Rabon, E., Schackmann, R. and Sachs, G. (1977) Biochim. Biophys. Acta 464, 313–327.
138. Hersey, S.J., Steiner, L., Matheravidathu, S. and Sachs, G. (1988) Am. J. Physiol. 254, G856–863.
139. Ray, T.K. and Nandi, J. (1986) Biochem. J. 233, 231–238.
140. Saccomani, G., Barcellona, M.L. and Sachs, G. (1981) J. Biol. Chem. 256, 12405–12410.
141. Wallmark, B., Briving, C., Fryklund, J., Munson, K., Jackson, R., Mendlein, J., Rabon, E. and Sachs, G. (1987) J. Biol. Chem. 262, 2077–2084.
142. Keeling, D.J., Laing, S.M. and Senn-Bilfinger, J. (1988) Biochem. Pharmacol. 37, 2231–2236.
143. Asano, S., Mizutani, M., Hayashi, T., Morita, N. and Takeguchi, N. (1990) J. Biol. Chem. 265, 22167–22173.
144. Jackson, R.J. and Saccomani, G. (1984) Biophys. J. 45, 83a.
145. Van der Hijden, H.T.W.M., Koster, H.P.G., Swarts, H.G.P. and De Pont, J.J.H.H.M. (1991) Biochim. Biophys. Acta 1061, 141–148.
146. Faller, L.D. and Diaz, R.A. (1989) Biochemistry 28, 6908–6914.
147. Reenstra, W.W. and Forte, J.G. (1981) J. Membr. Biol. 61, 55–60.
148. Smith, G.S. and Scholes, P.B. (1982) Biochim. Biophys. Acta 688, 803–807.
149. Lee, H.C. and Forte, J.G. (1978) Biochim. Biophys. Acta 508, 339–356.
150. Rabon, E., Chang, H. and Sachs, G. (1978) Biochemistry 17, 3345–3353.
151. De Jong, M.L. (1986) Ph.D. Dissertation, University of Nijmegen, The Netherlands.
152. Ray, T.K. and Nandi, J. (1985) FEBS Lett. 185, 24–28.
153. Boyer, P.D. (1988) Trends. Biochem. Sci. 12, 5–7.
154. Polvani, C., Sachs, G. and Blostein, R. (1989) J. Biol. Chem. 264, 17854–17859.
155. Ditmars, M., Mendlein, J. and Sachs, G. (1989) FASEB J. 3, A873.
156. Hersey, S.J., Sachs, G. and Kasbakar, D.K. (1985) Am. J. Physiol. 248, G246–G250.
157. De Weer, P., Gadsby, D.C. and Rakowski, R.F. (1988) Annu. Rev. Physiol. 50, 225–241.
158. Saccomani, G., Barcellona, M.L., Rabon, E. and Sachs, G. (1980) In: Hydrogen Ion Transport in Epithelia (Schulz, I., Sachs, G., Forte, J.G. and Ullrich, K.J. Eds.) pp. 175–183, Elsevier, Amsterdam.
159. Schuurmans Stekhoven, F. and Bonting, S.L. (1981) Physiol. Rev. 61, 1–76.
160. Schrijen, J.J., Omachi, A., Van Groningen-Luyben, W.A.H.M., De Pont, J.J.H.H.M. and Bonting, S.L. (1981) Biochim. Biophys. Acta 649, 1–12.

161. Nandi, J., Wright, M.V. and Ray, T.K. (1983) Biochemistry, 22, 5814–5821.
162. Olaisson, H., Mårdh, S. and Arvidson, G. (1985) J. Biol. Chem. 260, 11262–11267.
163. Takaya, J., Omori, K., Taketani, S., Kobayashi, Y. and Tashiro, Y. (1987) J. Biochem. (Tokyo) 102, 903–911.
164. Skrabanja, A.T.P., Asty, P., Soumarmon, A., De Pont, J.J.H.H.M. and Lewin, M.J.M. (1986) Biochim. Biophys. Acta 860, 131–136.
165. Forte, J.G., Poulter, J.L., Dykstra, R., Rivas, J. and Lee, H.C. (1981) Biochim. Biophys. Acta 644, 257–265.
166. Nandi, J., Meng-Ai, Z. and Ray, T.K. (1983) Biochem. J. 213, 587–594.
167. Asano, S., Inoie, M. and Takeguchi, N. (1987) J. Biol. Chem. 262, 13263–13268.
168. Takeguchi, N., Joshima, R., Inoue, Y., Kashiwagura, T. and Morii, M. (1983) J. Biol. Chem. 258, 3094–3098.
169. Takeguchi, N. and Yamazaki, Y. (1986) J. Biol. Chem. 261, 2560–2566.
170. Benkouka, F., Péranzi, G., Robert, J.C., Lewin, M.J.M. and Soumarmon, A. (1989) Biochim. Biophys. Acta 987, 205–211.
171. Asano, S., Tabuchi, Y. and Takeguchi, N. (1989) J. Biochem. (Tokyo) 106, 1074–1079.
172. Van Uem, T.J.F., Peters, W.H.M. and De Pont, J.J.H.H.M. (1990) Biochim. Biophys. Acta 1023, 56–62.
173. Reenstra, W.W., Shortridge, B. and Forte, J.G. (1985) Biochem. Pharmacol. 34, 2331–2334.
174. Goldenberg, M.M. and Moore, R.B. (1980) Arch. Int. Pharmacodyn. Ther. 247, 163–176.
175. Im, W.B., Blakeman, D.P., Mendlein, J. and Sachs, G. (1984) Biochim. Biophys. Acta 770, 65–72.
176. Nagaya, H., Satoh, H. and Maki, Y. (1987) Biochem. Pharmacol. 36, 513–519.
177. Nandi, J., Wright, M.V. and Ray, T.K. (1983) Gastroenterology 85, 938–945.
178. Nandi, J. and Ray, T.K. (1987) Biochem. J. 241, 175–181.
179. Nandi, J., King, R.L., Kaplan, D.S. and Levine, R.A. (1990) J. Pharmacol. Exp. Ther. 252, 1102–1107.
180. Herling, A.W. and Ljungström, M. (1988) Eur. J. Pharmacol. 156, 341–350.
181. Beil, W., Bersibaev, R.J., Hannemann, H. and Sewing, K.-F. (1990) Pharmacology 40, 8–20.
182. Wallmark, B., Jaresten, B.-M., Larsson, H., Ryberg, B., Brändström, A. and Fellenius, E. (1983) Am. J. Physiol. 245, G64–G71.
183. Sigrist-Nelson, K., Krasso, A., Müller, R.K.M. and Fischli, A.E. (1987) Eur. J. Biochem. 166, 453–459.
184. Satoh, H., Inatomi, N., Nagaya, H., Inada, I., Nohara, A., Nakamura, N. and Maki, Y. (1989) J. Pharmacol. Exp. Ther. 248, 806–815.
185. Morii, M., Takata, H., Fujisaki, H. and Takeguchi, N. (1990) Biochem. Pharmacol. 39, 661–667.
186. Herling, A.W., Becht, M., Lang, H.-J., Scheunemann, K.-H., Weidmann, K., Scholl, T. and Rippel, R. (1990) Biochem. Pharmacol. 40, 1809–1814.
187. Simon, W.A., Keeling, D.J., Laing, S.M., Fallowfield, C. and Taylor, A.G. (1990) Biochem. Pharmacol. 39, 1799–1806.
188. Olbe, L., Berglindh, T., Elander, B., Helander, H., Fellenius, E., Sjöstrand, S.-E., Sundell, G. and Wallmark, B. (1979) Scand. J. Gastroenterol. 14 (suppl. 55), 131–135.
189. Fellenius, E., Berglindh, T., Sachs, G., Olbe, L., Elander, B., Sjöstrand, S.-E. and Wallmark, B. (1981) Nature (London) 290, 159–161.
190. Wallmark, B., Larsson, H. and Humble, L. (1985) J. Biol. Chem. 260, 13681–13684.
191. Wallmark, B. (1989) Scand. J. Gastroenterol. 24 (suppl. 166), 12–18.
192. Wallmark, B., Brändström, A. and Larsson, H. (1984) Biochim. Biophys. Acta 778, 549–558.
193. Lindberg, P., Nordberg, P., Alminger, T., Brändström, A. and Wallmark, B. (1986) J. Med. Chem. 29, 1327–1329.
194. Lorentzon, P., Eklundh, B., Brändström, A. and Wallmark, B. (1985) Biochim. Biophys. Acta 817, 25–32.

195. Im, W.B., Sih, J.C., Blakeman, D.P. and McGrath, J.P. (1985) J. Biol. Chem. 260, 4591–4597.
196. Keeling, D.J., Fallowfield, C., Milliner, K.J., Tingley, S.K., Ife, R.J. and Underwood, A.H. (1985) Biochem. Pharmacol. 34, 2967–2973.
197. Lorentzon, P., Jackson, R., Wallmark, B. and Sachs, G. (1987) Biochim. Biophys. Acta 897, 41–51.
198. Keeling, D.J., Fallowfield, C. and Underwood, A.H. (1987) Biochem. Pharmacol. 36, 339–344.
199. Fryklund, J., Gedda, K. and Wallmark, B. (1988) Biochem. Pharmacol. 37, 2543–2549.
200. Morii, M., Takata, H. and Takeguchi, N. (1990) Biochem. Biophys. Res. Commun. 167, 754–760.
201. Ene, M.D., Khan-Daneshmend, T. and Roberts, C.J. (1982) Br. J. Pharmacol. 76, 389–391.
202. Long, J.F., Chiu, P.J.S., Derelanko, M.J. and Steinberg, M. (1983) J. Pharmacol. Exp. Ther. 226, 114–120.
203. Scott, C.K. and Sundell, E. (1985) Eur. J. Pharmacol. 112, 268–270.
204. Beil, W., Hackbarth, I. and Sewing, K.-F. (1986) Br. J. Pharmacol. 88, 19–23.
205. Scott, C.K., Sundell, E. and Castrovilly, L. (1987) Biochem. Pharmacol. 36, 97–104.
206. Briving, C., Andersson, B.M., Nordberg, P. and Wallmark, B. (1988) Biochim. Biophys. Acta 946, 185–192.
207. Keeling, D.J., Taylor, A.G. and Schudt, C. (1989) J. Biol. Chem. 264, 5545–5551.
208. Keeling, D.J., Fallowfield, C., Lawrie, K.M.W., Saunders, D., Richardson, S. and Ife, R.J. (1989) J. Biol. Chem. 264, 5552–5558.

CHAPTER 3

The Ca^{2+} transport ATPases of sarco(endo)-plasmic reticulum and plasma membranes

ANTHONY MARTONOSI

Department of Biochemistry and Molecular Biology, State University of New York Health Science Center, Syracuse, NY 13210, U.S.A.

1. Introduction

Most living cells, including muscle, maintain the cytoplasmic Ca^{2+} concentration at submicromolar levels, against steep gradients of $[Ca^{2+}]$, both at the cell surface and across the endoplasmic reticulum membrane [1–7]. In the muscle cell two membrane systems are primarily involved in this function: the sarcoplasmic reticulum and the surface membrane.

The Ca^{2+} transport ATPase of sarcoplasmic reticulum is an intrinsic membrane protein of 110 kDa [8–11] that controls the distribution of intracellular Ca^{2+} by ATP-dependent translocation of Ca^{2+} ions from the cytoplasm into the lumen of the sarcoplasmic reticulum [12–16].

The total Ca^{2+} content of the muscle cell is regulated by Ca^{2+} fluxes through the sarcolemma, in which Ca^{2+} channels [17–19], Na^+–Ca^{2+} exchange systems [20–27] and ATP-dependent Ca^{2+} pumps [2–5,28–34] participate.

The Ca^{2+} transport ATPase of the surface membrane is a Ca^{2+}-calmodulin-dependent enzyme of approximately 138-kDa mass that is structurally distinct from the sarcoplasmic reticulum Ca^{2+}-ATPase, but shares with it some similarities in the mechanism of Ca^{2+} translocation [2,3,34]. In both enzymes the Ca^{2+}-dependent phosphorylation of an aspartyl-carboxyl-group by ATP leads to the formation of an acyl phosphate intermediate that provides the coupling between ATP hydrolysis and Ca^{2+} translocation.

Both enzymes have been continuously in the limelight of interest, as attested by thousands of reports and dozens of recent review articles on their structure and mechanism of action. Less is known about the regulation of the activity of the two enzymes under physiological conditions [2,3,35–38] and about the control of their rate of synthesis and degradation that adjusts their concentration to the physiological requirements [39–41].

This chapter will summarize recent developments on the structure of the Ca^{2+}-ATPase of the sarcoplasmic reticulum with occasional references to the Ca^{2+}-ATPases in the plasma membranes and endoplasmic reticulum of non-muscle cells.

2. The classification of Ca^{2+}-ATPase isoenzymes

The determination of the amino acid sequences of the sarcoplasmic reticulum Ca^{2+}-ATPase [42] and of the closely related Na^+,K^+-ATPase [43,44] have opened a new era in the analysis of ion transport mechanisms. Since 1985, several large families of structurally related ion transport enzymes were discovered [3,34,45–50] that are the products of different genes. Within each family several isoenzymes may be produced from a single gene-product by alternative splicing (Table I).

Our discussion here will concentrate on the various forms of the Ca^{2+} transport ATPases that occur in the sarcoplasmic reticulum of muscle cells of diverse fiber types and in the endoplasmic reticulum of nonmuscle cells (SERCA). The structure of these enzymes will be compared with the Ca^{2+} transport ATPases of surface membranes (PMCA) [3,29–32,34] and with other ATP-dependent ion pumps that transport Na^+, K^+, and H^+ [46,50–52].

2.1. The Ca^{2+} transport ATPases of sarco(endo)plasmic reticulum (SERCA)

The sarco(endo)plasmic reticulum Ca^{2+}-ATPases of mammalian tissues can be divided structurally into three main groups (SERCA1–3) representing the products of different genes (Table I) [8,9,11,53–57].

2.1.1. SERCA1
The SERCA1 gene produces two isoforms of the Ca^{2+}-ATPase, that are derived by alternative splicing of the primary gene-product (Table I).

SERCA1a denotes the Ca^{2+}-ATPase of adult fast-twitch skeletal muscle with glycine at its C-terminus in the rabbit [53,58], and alanine at the C-terminus in the chicken [59,60]. The C-terminus of the lobster enzyme is apparently blocked [59].

SERCA1b is the alternatively spliced neonatal form of SERCA1, in which the glycine at the C-terminus is replaced by the alternative sequence Asp–Pro–Glu–Asp–Glu–Arg–Arg–Lys [8,9].

The gene encoding SERCA1 is on human chromosome 16 [61]; a selective defect in its expression is the cause of Brody's disease [62–66].

2.1.2. SERCA2
The SERCA2 gene also produces at least two isoforms that are tissue specific.

SERCA2a is the principal form of the Ca^{2+}-ATPase in adult slow-twitch skeletal and cardiac muscles and in neonatal skeletal muscles [8,9,42,53,54,67]. It is also expressed at much lower levels in nonmuscle cells [57]. Its C-terminus is Pro–Ala–Ile–Leu–Glu (Table I).

SERCA2b is an alternatively spliced product of the same gene. It is located primarily in non-muscle tissues and in smooth muscles, where it serves as the major intracellular Ca^{2+} pump. SERCA2b is characterized by a long C-terminal extension ending in Trp–Ser (Table I) [54–56]. The gene for both forms of SERCA2 is located on human chromosome 12; its expression is not affected in Brody's disease, where the cellular concentration of the SERCA1 gene product is severely reduced [64]. Therefore the two major forms of sarcoplasmic reticulum Ca^{2+}-ATPases are independently regulated.

2.1.3. SERCA3

This gene is broadly distributed in skeletal muscle, heart, uterus, and in a variety of non-muscle cells. The mRNA levels are particularly high in intestine, lung and spleen, whereas they are very low in liver, testes, kidney and pancreas. In the muscle tissue SERCA3 may be confined primarily to non-muscle cells (vascular smooth muscle, endothelial cells, etc.). The C-terminus of SERCA3 is Asp–Gly–Lys–Lys–Asp–Leu–Lys (Table I); it may serve as a sorting signal for retention of the enzyme in the endoplasmic reticulum [57].

2.1.4. SERCA-type Ca^{2+}-ATPases from non-mammalian cells (SERCAMED)

Sequences of SERCA-type Ca^{2+}-ATPases were also obtained from *Plasmodium yoelii* [68], *Artemia* [69] and *Drosophila* [70]. These enzymes are similar in size to the SERCA1- and SERCA2a-type Ca^{2+}-ATPases from mammalian muscles, but based on their N- and C-terminal sequences they represent a distinct group. In spite of the wide philogenetic variations between them they all share a common N-terminal sequence (MED) that differs from mammalian enzymes. None of the corresponding proteins were isolated and characterized.

The molecular weights of all SERCA-type Ca^{2+} transport ATPases are in the range of 100–110 kDa. Their N-terminal sequences are similar: Met–Glu–X(Ala, Asn, Glu, Asp)–X (Ala, Gly, Ile). The Met–Glu–X–X sequence serves as a signal for the acetylation of N-terminal methionine both in soluble and in membrane proteins [71,72].

2.2. The plasma membrane Ca^{2+} transport ATPases (PMCA)

There are at least five distinct isoforms of plasma membrane Ca^{2+}-ATPases in mammalian tissues that differ in distribution and C-terminal sequences [34]. The molecular weight of these enzymes is in the range of 127 300–134 683 (Table I) and they all contain calmodulin-binding domains [3], in contrast to the much smaller ($\simeq$110 kDa) SERCA enzymes that are calmodulin independent.

TABLE I
Classification of sarco(endo)plasmic reticulum (SERCA) and surface membrane (PMCA) Ca^{2+}-ATPases

Cellular location	Tissue	Species	# of amino acids	MW (Da)	Sequence N-terminal	Sequence C-terminal	References
Sarco(endo)plasmic reticulum							
SERCA1a	Adult fast-twitch skeletal muscle	Rabbit, Chicken	994 994	109 361 109 375	MEAA MENA	EG EA	53 60
SERCA1b	Neonatal fast-twitch skeletal muscle	Rabbit	1001	110 331	MEAA	EDPEDERRK	8, 53
SERCA2a	Slow-twitch skeletal, cardiac and smooth muscle	Rabbit	997	109 529	MENA	EPAILE	8, 42, 53, 54, 67
SERCA2b	Non-muscle and smooth muscle	Rat	1043	114 759	MENA	EPGKECAQPATKPSCSLSACTDGI STPFVLLIMPLVVWVYSTDTNFSDMFWS	55
SERCA2b	Non-muscle, smooth muscle	Human	1042	115 000	MENA	EPGKECVQPATKSCSFSACTDGIS WPFVLLIMPLVIWVYSTDTNFSDMFWS	54
SERCA2b	Smooth muscle	Rabbit	1042	115 000	MENA	EPGKECVQPAPQSCSLWACTEGVS WPFVLLIVPLVMWVYSTDTNFSDLLWS	56
SERCA2b	Smooth muscle	Pig	1042	115 000	MENA	GKECVQPATKSCSFSACTDGISWP FVLLIMPLVIWVYSTDTNFSDMFWS	295, 434
SERCA3	Both muscle and non-muscle tissues	Rat	999	109 223	MEEA	PLSGRQWGVVLQMSLPVILLDEA LKYLSRHHVDEKKDLK	57
SERCAMED	Adult head	*Drosophila*	1002	109 540	MEDG	LLDETLKFVARK IADVPDVVDRM	70
SERCAMED	Cysts and nauplii	*Artemia*	1003	–	MEDA	LLDEVLKFVARKYTDEFSFIK	69
SERCAMED		*Plasmodium yoelii*	1115	126 717	MEDI	IIDEIIKFYAKK QLNKELGYGQKL KTQ	68

TABLE I
(continued)

Plasma membrane						
rPMCA1a	Brain	Rat	1176	129 500	MGDMA	QHHDVTNVSTPTHVVFSSSTASTP VGYPSGECIS	30
hPMCA1a	Teratoma	Human	1220	134 683	MGDMA	NNNAVNSGIHLTIEMNKSATSSSP GSPLHSLETSL	31
rPMCA2	Brain	Rat	1198	132 605	MGDMT	AIDSGINLTTDTSKSATSSSPGSPIHSLETSL	30
rPMCA3	Brain and skeletal muscle	Rat	1159	127 300	MGDMA	QLHDVTNLSTPTHVTLSAAKPTSA AGNPSGESIP	32
hPMCA4a	Erythrocyte	Human	1170	129 400	MTNPS	NQSGQSVP	33
hPMCA4b	Erythrocyte	Human	1205	133 930	MTNPS	LQSLETSV	34

For details, see text.

2.2.1. rPMCA1 and rPMCA2
The cDNAs for these two isoforms of the plasma membrane Ca^{2+}-ATPases were isolated from rat brain. Although PMCA1 and PMCA2 exhibit 82% amino acid identity, they are encoded on different genes, as shown by the significant differences between them near the C-terminus and between residues 296–368 in the hydrophilic region [30]. The 3′ nontranslated sequence of PMCA1 may encode a C-terminus similar to that of PMCA2 in an alternative reading frame. Therefore alternative splicing may produce additional diversity among PMCA1 and PMCA2 classes of Ca^{2+}-ATPases. PMCA1 is expressed in all tissues, while PMCA2 is found primarily in brain and heart.

2.2.2. rPMCA3
This isoform of the plasma membrane Ca^{2+}-ATPase occurs predominantly in brain and skeletal muscle [32]. Its amino acid sequence shows 81 and 85% identity with PMCA1 and PMCA2, respectively, but it is clearly the product of a different gene. The PMCA3 gene also has the potential to express, by alternative splicing, variant isoforms of the enzyme with distinct calmodulin binding sites and C-terminal sequences. The Ca^{2+} transport ATPase of skeletal muscle sarcolemma [73] probably belongs to this group.

2.2.3. rPMCA4
An mRNA of $\simeq 8.8$ kb was detected in rat uterus and stomach [32], which is different from the mRNAs of PMCA1–3 and may represent a distinct member of the PMCA series. Its amino acid sequence is not yet available.

2.2.4. hPMCA
cDNAs coding for two distinct human plasma membrane Ca^{2+} pumps were isolated and sequenced from human teratoma [31] and from human small intestinal mucosa libraries [33]. The teratoma enzyme [31] is similar to rPMCA1 and it was designated as hPMCA1 [34]. Its N-terminal sequence is MGDMA. This enzyme is distinct from the erythrocyte Ca^{2+} pump. The second hPMCA isoform [33] differs from all three isoforms of PMCA identified in rat or human tissues and was designated as hPMCA4 [34]. Its N-terminal sequence is MTPNPS and it may correspond to the Ca^{2+}-ATPase of human erythrocyte membrane [3,6,28,29,34,74].

3. *The deduced amino acid sequences of the fast-twitch and slow-twitch isoforms of the sarcoplasmic reticulum Ca^{2+}-ATPases*

The first Ca^{2+}-ATPase clones were isolated by probing cDNA libraries with radiolabeled synthetic oligonucleotides [42] that represented an established amino acid sequence ((Trp–) Phe–Met–Tyr–Ala) in the fast-twitch skeletal muscle

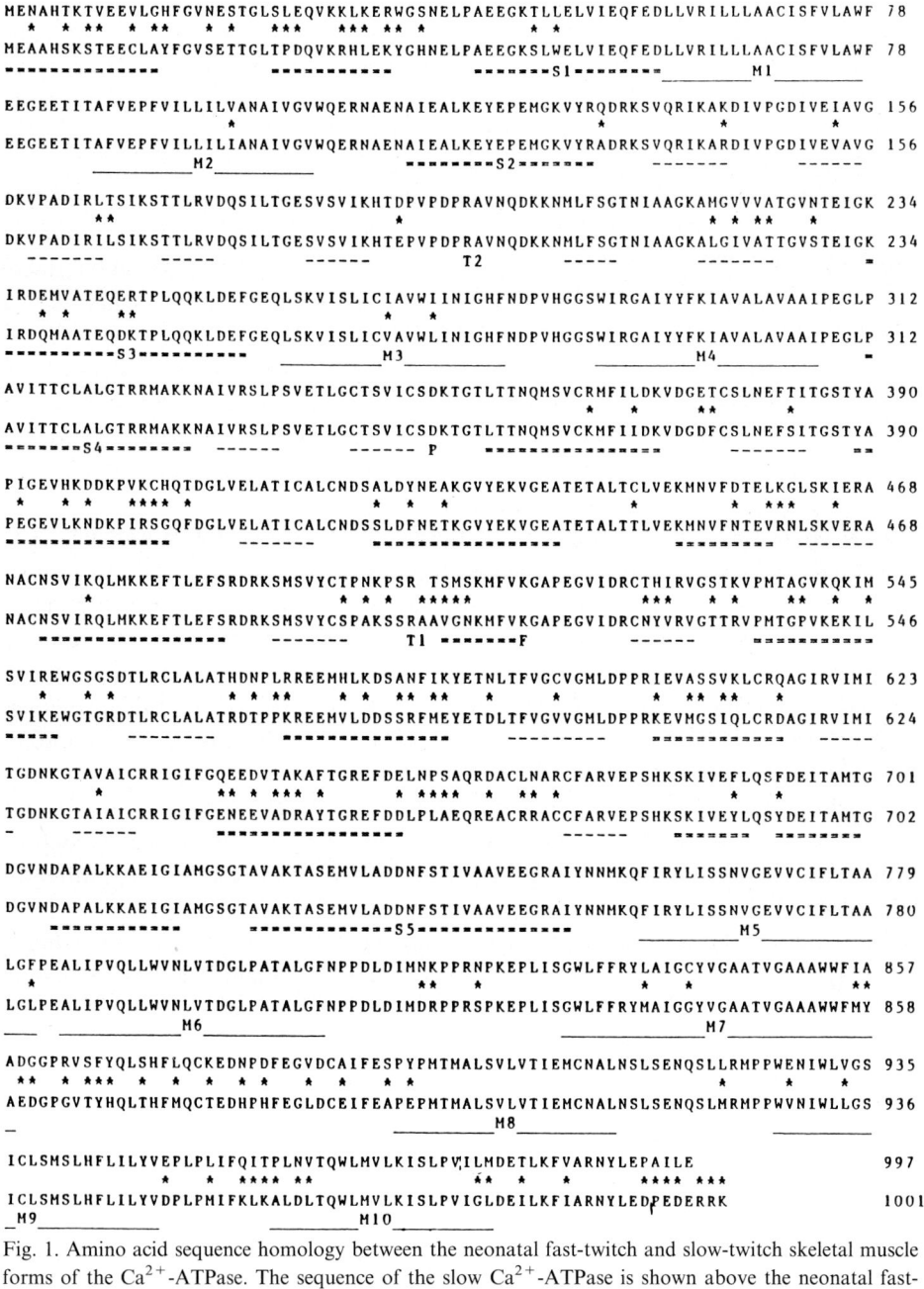

```
MENAHTKTVEEVLGHFGVNESTGLSLEQVKKLKERWGSNELPAEEGKTLLELVIEQFEDLLVRILLLAACISFVLAWF  78
   *   *  **   * *       ***    *** **  *     *
MEAAHSKSTEECLAYFGVSETTGLTPDQVKRHLEKYGHNELPAEEGKSLWELVIEQFEDLLVRILLLAACISFVLAWF  78
---------------          ------------       --------S1---------_____M1_____

EEGEETITAFVEPFVILLILVANAIVGVWQERNAENAIEALKEYEPEMGKVYRQDRKSVQRIKAKDIVPGDIVEIAVG  156
                    *                                *        *     *    *
EEGEETITAFVEPFVILLIANAIVGVWQERNAENAIEALKEYEPEMGKVYRADRKSVQRIKARDIVPGDIVEAVG    156
_____M2_____              ---------S2--------         -------      ------

DKVPADIRLTSIKSTTLRVDQSILTGESVSVIKHTDPVPDPRAVNQDKKNMLFSGTNIAAGKAMGVVVATGVNTEIGK  234
         **                  *                              * * **    *
DKVPADIRILSIKSTTLRVDQSILTGESVSVIKHTEPVPDPRAVNQDKKNMLFSGTNIAAGKALGIVATTGVSTEIGK  234
-------         -----        ------     T2      -----        -------          =

IRDEMVATEQERTPLQQKLDEFGEQLSKVISLICIAVWIINIGHFNDPVHGGSWIRGAIYYFKIAVALAVAAIPEGLP  312
   * *    **                   *   *
IRDQMAATEQDKTPLQQKLDEFGEQLSKVISLICVAVWLINIGHFNDPVHGGSWIRGAIYYFKIAVALAVAAIPEGLP  312
-----------S3-----------  _____M3_____         _____M4_____  -

AVITTCLALGTRRMAKKNAIVRSLPSVETLGCTSVICSDKTGLTTTNQMSVCRMFILDKVDGETCSLNEFTITGSTYA  390
                                                  *  *   **       *
AVITTCLALGTRRMAKKNAIVRSLPSVETLGCTSVICSDKTGLTTTNQMSVCKMFIIDKVDGDFCSLNEFSITGSTYA  390
-------S4---------   ------        ------ P    ================         -------  ==

PIGEVHKDDKPVKCHQTDGLVELATICALCNDSALDYNEAKGVYEKVGEATETALTCLVEKMNVFDTELKGLSKIERA  468
 * *  *   ****   *              *  **          *           *   ***   *
PEGEVLKNDKPIRSGQFDGLVELATICALCNDSSLDFNETKGVYEKVGEATETALTTLVEKMNVFNTEVRNLSKVERA  468
---------------  ---------        ----------------        =============  -------

NACNSVIKQLMKKEFTLEFSRDRKSMSVYCTPNKPSR TSMSKMFVKGAPEGVIDRCTHIRVGSTKVPMTAGVKQKIM  545
      *                      * *  *****             ***  *  *  ** *  *  *
NACNSVIRQLMKKEFTLEFSRDRKSMSVYCSPAKSSRAAVGNKMFVKGAPEGVIDRCNYVRVGTTRVPMTGPVKEKIL  546
----------------          -------     T1 --------F             ------  =========

SVIREWGSGSDTLRCLALATHDNPLRREEMHLKDSANFIKYETNLTFVGCVGMLDPPRIEVASSVKLCRQAGIRVIMI  623
   *   **  *         *        *   **  **  *                * **  **
SVIKEWGTGRDTLRCLALATRDTPPKREEMVLDDSSRFMEYETDLTFVGVVGMLDPPRKEVMGSIQLCRDAGIRVIMI  624
-----   --------       ==============     ---------     ==============    -----

TGDNKGTAVAICRRIGIFGQEEDVTAKAFTGREFDELNPSAQRDACLNARCFARVEPSHKSKIVEFLQSFDEITAMTG  701
       *         ** * ***  *       *  ****  *   ** *                *     *
TGDNKGTAIAICRRIGIFGENEEVADRAYTGREFDDLPLAEQREACRRACCFARVEPSHKSKIVEYLQSYDEITAMTG  702
-  -------       ===================       ------     =======   ========

DGVNDAPALKKAEIGIAMGSGTAVAKTASEMVLADDNFSTIVAAVEEGRAIYNNMKQFIRYLISSNVGEVVCIFLTAA  779

DGVNDAPALKKAEIGIAMGSGTAVAKTASEMVLADDNFSTIVAAVEEGRAIYNNMKQFIRYLISSNVGEVVCIFLTAA  780
-------------       =============S5==============          _____M5_____

LGFPEALIPVQLLWVNLVTDGLPATALGFNPPDLDIMNKPPRNPKEPLISGWLFFRYLAIGCYVGAATVGAAAWWFIA  857
  *                                **              **  *    *               **
LGLPEALIPVQLLWVNLVTDGLPATALGFNPPDLDIMDRPPRSPKEPLISGWLFFRYMAIGGYVGAATVGAAAWWFMY  858
____        _____M6_____                    _____M7_____

ADGGPRVSFYQLSHFLQCKEDNPDFEGVDCAIFESPYPMTHALSVLVTIEMCNALNSLSENQSLLRMPPWENIWLVGS  935
 * ***  *   *     *    *    *    *                            *  *       *
AEDGPGVTYHQLTHFMQCTEDHPHFEGLDCEIFEAPEPMTHALSVLVTIEMCNALNSLSENQSLMRMPPWVNIWLLGS  936
-              _____M8_____

ICLSMSLHFLILYVEPLPLIFQITPLNVTQWLMVLKISLPVILMDETLKFVARNYLEPAILE                  997
    *   **** **             **  *   *    **** ***
ICLSMSLHFLILYVDPLPMIFKLKALDLTQWLMVLKISLPVIGLDEILKFIARNYLEDPEDERRK              1001
_M9_____          _____M10_____
```

Fig. 1. Amino acid sequence homology between the neonatal fast-twitch and slow-twitch skeletal muscle forms of the Ca^{2+}-ATPase. The sequence of the slow Ca^{2+}-ATPase is shown above the neonatal fast-twitch form, with nonhomologous amino acids indicated by asterisks. The sequence of the slow ATPase is shifted to the right by one residue at residue 505 to allow realignment after the difference in sequence length. M1–M10, membrane spanning regions; S1–S5, stalk sectors; T1, T2, major tryptic cleavage sites; P, phosphorylation site; F, FITC binding site. (= =) regions of predicted α helix; (- - - -) regions of predicted β strand. From Brandl et al. [8].

sarcoplasmic reticulum Ca^{2+} transport ATPase [75]. Two sets of crosshybridizing clones (pFA and PCA) were obtained, which encoded for two distinct forms of the Ca^{2+}-ATPase corresponding to the isoenzymes found in neonatal fast-twitch (SERCA1b) and in slow-twitch skeletal muscles (SERCA2a), respectively [8,9,53,76].

The cDNA clone for the neonatal rabbit fast-twitch skeletal muscle Ca^{2+}-ATPase encodes for 1 001 amino acids giving a product with an estimated molecular weight of 110 331 Da [8]. The clone for the Ca^{2+}-ATPase of slow-twitch skeletal muscle sarcoplasmic reticulum (S-Ca^{2+}-ATPase) encoded for 997 amino acids with a relative molecular mass (M_r) of 109 529 kDa [42].

The structures predicted for the fast and slow Ca^{2+}-ATPase (Fig. 1) are 84% identical [8]. There are 164 differences in the amino acid sequences between the two isoenzymes, 66 of which are conservative replacements, involving substitution of serine for threonine, aspartic for glutamic, lysine for arginine, or interchanges between aromatic or hydrophobic amino acids [8].

The deduced amino acid sequence of the fast-twitch Ca^{2+}-ATPase of neonatal muscle generally agrees with the sequence earlier established for the Ca^{2+}-ATPase of adult rabbits, using amino acid sequencing techniques [75,77–85]. The only major difference between the two isoenzymes is that the highly charged C-terminal sequence of neonatal fast-twitch Ca^{2+}-ATPase, Asp–Pro–Glu–Asp–Glu–Arg–Arg–Lys, is replaced in the adult fast-twitch Ca^{2+}-ATPase by Gly. Since the rabbit genome contains only one gene encoding the fast-twitch Ca^{2+}-ATPase isoforms, Brandl et al. [8,53] suggested that the adult and neonatal forms of the Ca^{2+}-ATPase arise as a result of alternative splicing of the same primary transcript. Allelic variation is not likely to contribute to this difference because C-terminal glycine has been independently reported for the adult rabbit fast-twitch F-Ca^{2+}-ATPase by three different laboratories [59,78,85] from different countries.

The N-terminal methionine is posttranslationally acetylated [86]. Similar N-terminal sequences were found in several other proteins that contain N-acetylmethionine as the N-terminal amino acid, suggesting that the signal for acetylation may be the Met–Asp or Met–Glu sequence [71].

4. The predicted topology of the Ca^{2+}-ATPases

4.1. The Ca^{2+}-ATPase of the sarcoplasmic reticulum

Combining structural and biochemical information, MacLennan and his colleagues [8,11,42,45,48,87] constructed a hypothetical model of the tertiary structure of Ca^{2+}-ATPase that has interesting mechanistic implications (Fig. 2). The structure was divided into three major parts, designated as the cytoplasmic headpiece, the stalk domain and the transmembrane domain; each was assigned distinct functional

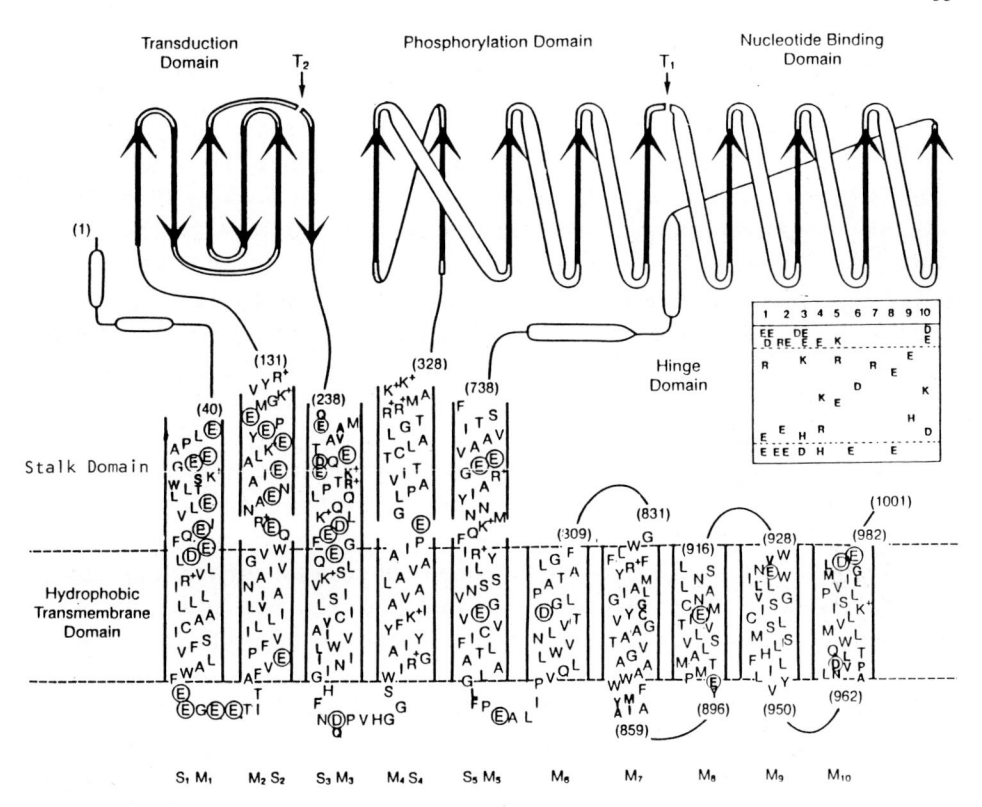

Fig. 2. The hypothetical structure of Ca^{2+}-ATPase. The structure consists of three major cytoplasmic domains, a pentahelical stalk region, and an intramembranous domain with ten, presumably helical, transmembrane segments. T_1 and T_2 mark the tryptic cleavage sites. Inset: charged amino acids in and near the transmembrane region that may contribute to a Ca^{2+}-channel. Adapted from Brandl et al. [8].

and structural roles. Only short loops were assumed to be exposed on the luminal side of the membrane.

4.1.1. The cytoplasmic headpiece

More than half of the total mass of the ATPase molecule is exposed on the cytoplasmic surface of the membrane, forming the 40-Å × 60-Å particles seen by negative staining electron microscopy [88–93].

The headpiece contains five subdomains: the N-terminal region (residues 1–40), the transduction or B domain (residues 131–238), the phosphorylation domain (residues 328–505), the nucleotide binding domain (residues 505–680) and the hinge domain (residues 681–738). The evidence supporting each of these assignments will be discussed in turn.

4.1.1.1. The phosphorylation and nucleotide binding domains. The phosphorylation domain contains the phosphate acceptor Asp351, that is phosphorylated by ATP during Ca^{2+} transport [45,80,87,94,95]. Site-specific mutagenesis of Asp351 and Lys352 inhibits the phosphorylation of the enzyme by ATP [96,97].

The nucleotide binding domain contains the binding sites for fluorescein-5′-isothiocyanate (FITC) on Lys515 [98], the site of reaction of adenosine triphosphopyridoxal on Lys684 [99,100], and the sequences which are homologous with the sites of reaction of 5′-(p-fluorosulfonyl)-benzoyl-adenosine [101] and γ[4-(N-2-dichloroethyl-N-methylamino)] benzylamide ATP (ClrATP) [102] in the Na^+,K^+-ATPase. The role of this domain in the binding of substrates is further supported by mutagenesis of Asp601, Pro603, Gly626, Asp627 and Asp707 that inhibits the formation of phosphoenzyme intermediate from ATP or inorganic phosphate [97,103]. Gly626 and Asp627 are in a highly conserved region of the molecule extending from Arg616 to Lys629.

Vanadate-catalyzed photocleavage of Ca^{2+}-ATPase occurs in the phosphorylation domain in the absence of Ca^{2+} and in the nucleotide binding domain in the presence of Ca^{2+} [104,105]; therefore vanadate bound at the substrate binding site may be brought, by Ca^{2+}-induced changes in the conformation of the enzyme, into reaction with either one of these segments of the molecule. The close proximity of the phosphorylation and the nucleotide binding domains to each other in the native structure of the Ca^{2+}-ATPase is also supported by intramolecular crosslinking of the two segments with glutaraldehyde [106–108] and with the carbodiimide adduct of ATP [109,110].

The phosphorylation and nucleotide binding domains are predicted to contain alternating α helices and β strands that fold into a parallel β sheet. The proposed location of the bound nucleotides is at the C-termini of the parallel β strands [49]. This brings the α and β phosphates of ATP close to highly conserved loops (K515GAPE and T701GDGVND) and allows the γ phosphate to extend toward the phosphorylation domain (TKD351). The adenine ring would be placed near T625GD and the ribose next to D601PPR [103]. Indeed, mutagenesis of amino acids in these regions of the structure interferes with substrate binding and phosphorylation [103].

The Ca^{2+}-ATPase of adult rabbit muscle is cleaved by trypsin at the T_1 site into two major fragments (A and B), followed by a second cleavage of the A fragment at the T_2 cleavage site into A_1 and A_2 [111,112]. The T_1 and T_2 sites are located at arginine residues 505 and 198, respectively (Fig. 2). The binding site for fluorescein isothiocyanate (FITC), an inhibitor of ATP binding to the Ca^{2+}-ATPase [98,113–117] is lysine 515, which is only 10 residues away from the T_1 cleavage site. The accessibility of residues 505 and 515 to trypsin and FITC, respectively, suggests that both sites are freely exposed to the cytoplasm. The T_1, T_2, phosphate acceptor and FITC binding sites are in similar locations in the sequences of the fast and slow Ca^{2+}-ATPase isoenzymes.

The T_1 cleavage site is at the boundary between the phosphorylation and nucleotide binding domains, while the T_2 site demarcates the phosphorylation from the transduction or B domain. The separation into subdomains is presumed to enhance the flexibility of the structure required for the hinge-bending and domain shifting motions that may accompany Ca^{2+} transport.

The T_2 cleavage site is blocked by vanadate in a Ca^{2+}-free medium either by direct interaction with Arg198 or by inducing the E_2-V conformation of the Ca^{2+}-ATPase that may interfere with access to the T_2 site [118–121].

4.1.1.2. The transduction or B domain. The transduction or B domain (residues 131–238) lies between two extended helical hairpin loops that traverse the membrane (M_2S_2 and M_3S_3); it consists of seven β strands separated by short loops that are presumed to fold into an antiparallel β sandwich [8,9,42]. The B domain was originally thought to participate in the coupling of ATP hydrolysis to Ca^{2+} translocation, based on the observations of Scott and Shamoo [122], that cleavage of the Ca^{2+}-ATPase by trypsin at the T_2 cleavage site (Arg198) inhibited Ca^{2+} transport without inhibition of ATP hydrolysis. However, these observations were not confirmed [123]. The inhibition of Ca^{2+} transport by trypsin is apparently due to an increase in the passive Ca^{2+} permeability of the membrane that is independent of the digestion of Ca^{2+}-ATPase and may indicate an effect of trypsin on the junctional Ca^{2+} channels [123–125]. In fact the Ca^{2+} activation of ATP hydrolysis persists after T_2 cleavage, indicating that the coupling between the active site and the Ca^{2+} binding site is fully maintained. Therefore the originally suggested 'transduction' function of the B domain is without experimental support. However, mutagenesis of Thr181, Gly182, Glu183 and Gly233 interferes with the conformational changes associated with the transition between the E_1P and E_2P intermediates [126,127], supporting a conformational role of the B domain.

4.1.1.3. The hinge domain. The proposed hinge region between residues 680 and 738 is one of the most highly conserved segments of the Ca^{2+}-ATPase. By analogy with kinases, MacLennan and his colleagues proposed that the hinge domain interacts with the nucleotide binding and phosphorylation domains transmitting the hinge-bending motions of these domains to the Ca^{2+} transport sites [8,45,49].

4.1.1.4. The stalk region. The pentahelical stalk connects the headpiece to the membrane. In the early models of the enzyme [8,42] the 18 glutamic acid and three aspartic acid residues in the stalk helices (S1–S5) were assumed to form the high-affinity binding site for Ca^{2+}, at the entrance to the putative Ca^{2+} transport channel (Fig. 2). The low-affinity Ca^{2+} binding sites were tentatively assigned to the cluster of 4 glutamic acid residues located in a loop between transmembrane helices M_1 and M_2 on the luminal side of the membrane (Fig. 2). This assignment is supported by resonance X-ray diffraction measurements on oriented multilayers of sarcoplasmic reticulum [128] that revealed a high affinity lanthanide binding site $\simeq 12$ Å outside the polar headgroups of phospholipids, in a region of the molecule that corresponds to the putative stalk helices.

However, mutagenesis of the acidic amino acids in the stalk region or in the luminal loop had no effect on ATP-dependent Ca^{2+} transport [129], while mutagenesis of Glu309, Glu771, Thr799, Asn796, Asp800 and Glu908 in the transmembrane segments inhibited the Ca^{2+}-dependent reactions [129–131]. These observations would suggest that the high-affinity Ca^{2+} binding site of Ca^{2+}-ATPase is located in the transmembrane helices M_4, M_5, M_6 and M_8, in agreement with energy transfer data [132]. The helices of the stalk are amphipathic with a high density of negative charges at the interfaces between stalk helices S1–S3 and the corresponding membrane spanning helices M_1–M_3. A concentration of helix breaking amino acids at the interfaces of S2, S4 and S5 and M_2, M_4 and M_5 may impart the flexibility needed for the structural changes involved in the translocation of calcium.

4.1.2. The transmembrane domain
The intramembranous part of the molecule may consist of 8–10 hydrophobic transmembrane helices (M_1–M_{10}) that anchor the Ca^{2+}-ATPase to the lipid bilayer and form the transmembrane channel for the passage of Ca^{2+} (Fig. 2).

Hydropathy plots [133] of the slow and fast Ca^{2+}-ATPase isoenzymes are nearly identical and provide unambiguous prediction of four of the proposed transmembrane segments (M_1, M_2, M_3 and M_4) [8,11]. Similar hydropathy plots were also obtained for other closely related cation transporting ATPases [31,46,47,134].

The assignment of the six additional transmembrane helices (M_5–M_{10}) is less conclusive. Since both the N-terminal and the C-terminal segments of the Ca^{2+}-ATPase molecule are exposed on the cytoplasmic surface of the membrane [112,135–138], the number of transmembrane sequences must be even, probably eight or ten; odd numbered transmembrane domains (seven or nine) appear to be excluded [138,139].

The intramembranous domain of Ca^{2+}-ATPase contains $\simeq 1/3$ of the mass of the ATPase molecule based on electron microscopy of Ca^{2+}-ATPase crystals [90,91] and X-ray diffraction analysis of oriented multilayers of sarcoplasmic reticulum [140]. Although in speculative models developed from these reconstructed structures the intramembranous domain was pictured as containing ten transmembrane helices [141,142], at the resolution attainable so far, several alternative transmembrane arrangements would be equally possible.

The large intrinsic birefringence of the sarcoplasmic reticulum [143] and the polarized attenuated total reflectance FTIR spectroscopy data obtained on oriented films of sarcoplasmic reticulum [144] indicate that a sizeable portion of the secondary structural elements are arranged perpendicularly to the plane of the membrane in a manner reminiscent to the structure of bacteriorhodopsin [145–148].

4.2. The predicted domains of the plasma membrane Ca^{2+}-ATPase

Although the sequence identity averaged over the whole length of the molecule is generally low among different P-type ion transport ATPases, the conserved sequences around the phosphate acceptor aspartyl group and in the ATP binding domain are well preserved [30,32,46]. Structure predictions based on the hydropathy plots and on the primary sequence suggest that the overall topology of the domain structure of the PMCAs is very similar to the Ca^{2+}-ATPase of sarcoplasmic reticulum [34].

A significant functional and structural feature of the plasma membrane Ca^{2+} pumps is the presence of the calmodulin-binding subdomains A and B near the C-terminus (Fig. 3), that imparts calmodulin sensitivity on the Ca^{2+} transport and ATP hydrolysis [3]. Adjacent to the calmodulin-binding region are two acidic segments (AC) and the P(S) region containing a serine residue that is susceptible to phosphorylation by cAMP-dependent protein kinase [34]. A unique feature of the plasma membrane Ca^{2+} pump is its activation by acidic phospholipids that are presumed to

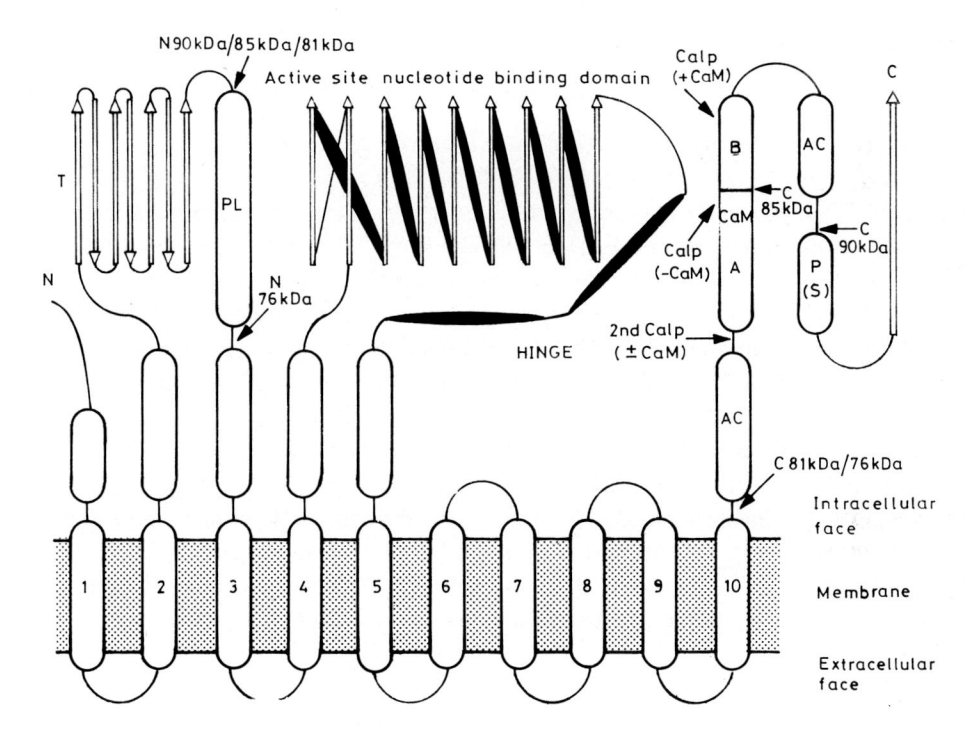

Fig. 3. Proposed model for the overall topology of PMCAs. A planar representation of the PMCA is shown, including the putative transmembrane topology (TM_1 to TM_{10} and the assignment of important domains. Open rods and black cigar-shaped bars correspond to putative α helices, and arrows denote β sheet secondary structural elements. The N-terminal (N 90 kDa/85 kDa/81 kDa, N 76 kDa) and C-terminal location (C 90 kDa, C 85 kDa, C 81 kDa/76 kDa) of the tryptic cleavage sites leading to the production of major proteolytic fragments is also indicated, as are the sites of calpain attack in the presence (Calp (+ CaM)) and absence (Calp (−CaM)) of calmodulin. The site of secondary, calmodulin-independent, calpain attack is labeled 2nd Calp(±CaM). AC, acidic regions flanking the calmodulin-binding domain; C, C-terminus; CaM, calmodulin-binding domain consisting of subdomains A and B; N, N-terminus; T, transduction domain; P(S), region containing the serine residue susceptible to phosphorylation by the cAMP-dependent protein kinase; PL, phospholipid-sensitive region. From Strehler [34].

interact with a phospholipid-binding domain close to the phosphorylation site (PL). The plasma membrane Ca^{2+} pump is specifically activated by the protease calpain, that cleaves the enzyme at three locations near the calmodulin-binding site [34].

There are indications for the existence of two ATP binding sites both in the sarcoplasmic reticulum [149] and in the plasma membrane Ca^{2+}-ATPases [30]. These two ATP binding sites may be located at two TGD containing conserved sequences that are present in the ATP binding domain of PMCA1 and 2, $\simeq 90$ residues apart [30,34].

As in the Ca^{2+}-ATPase of sarcoplasmic reticulum, the predicted number of membrane spanning sequences in PMCA1 and PMCA2 is even, with both N- and C-terminus located on the cytoplasmic side, but their actual number is uncertain and may be 10 or less [30].

5. Reconstruction of Ca^{2+}-ATPase structure by electron microscopy

During ATP-dependent Ca^{2+} transport the Ca^{2+}-ATPase alternates between two distinct conformations, E_1 and E_2 [12,150]. The E_1 conformation of the enzyme can be stabilized by saturation of the high-affinity Ca^{2+} binding sites with Ca^{2+} or lanthanides [119]. Removal of Ca^{2+} from the enzyme with the chelating agent ethyleneglycol bis-(β-aminoethyl ether)-N,N,N',N'-tetraacetic acid (EGTA) and the binding of inorganic phosphate or vanadate stabilize the E_2 conformation [115,116,151].

The Ca^{2+}-ATPase has been crystallized in both conformations [119,152–155]. The two crystal forms are quite different [10,88–93,156–161], suggesting significant differences between the interactions of Ca^{2+}-ATPase in the E_1 and E_2 conformations. Since the E_1–E_2-transition does not involve changes in the circular dichroism spectrum of the Ca^{2+}-ATPase [162], the structural differences between the two states presumably arise by hinge-like or sliding motions of domains rather than by a rearrangement of the secondary structure of the protein.

5.1. The vanadate-induced E_2-type crystals

Sarcoplasmic reticulum vesicles suspended in a medium of 0.1 M KCl, 10 mM imidazole pH 7.4, 5 mM $MgCl_2$, 0.5 mM EGTA, and 5 mM Na_3VO_4, for 1–3 days at 2°C develop crystalline arrays of Ca^{2+}-ATPase molecules on the surface of 60–90% of the vesicles [152–155,163–171]. Vesicles with extensive crystalline arrays usually acquire an elongated tubular shape. The negatively stained vanadate-induced Ca^{2+}-ATPase crystals consist of right-handed helical chains of Ca^{2+}-ATPase dimers [88,90,91]. Digital Fourier transforms of images of the negatively stained flattened tubules [88] yielded unit cell dimensions of $a = 66$ Å, $b = 114$ Å and $\gamma = 78°$. The average structure calculated by Fourier synthesis is consistent with ATPase dimers as structural units. The space group of the vanadate-induced E_2-V crystals is P2.

The dimer chains of Ca^{2+}-ATPase can also be observed by freeze-fracture electron microscopy [119,165,166,172–174], forming regular arrays of oblique parallel ridges on the concave P fracture faces of the membrane, with complementary grooves or furrows on the convex E fracture faces. Resolution of the surface projections of individual Ca^{2+}-ATPase molecules within the crystalline arrays has also been achieved on freeze-dried rotary shadowed preparations of vanadate treated rabbit sarcoplasmic reticulum [163,166,173,175]. The unit cell dimensions derived from these preparations are $a = 6.5$ nm; $b = 10.7$ nm and $\gamma = 85.5°$ [175], in reasonable agreement with earlier estimates on negatively stained preparations [88].

5.1.1. Image reconstruction in three dimensions from negatively stained and frozen hydrated crystals

The three-dimensional reconstruction of the structure of vanadate-induced Ca^{2+}-ATPase crystals preserved in uranyl acetate yielded an image of the cytoplasmic region of the molecule [90]. In the view normal to the membrane plane, each map shows pear-shaped densities arranged in antiparallel strands (Fig. 4) that correspond to the ribbons of Ca^{2+}-ATPase dimers described earlier [88]. The Ca^{2+}-ATPase molecules extend about 60 Å above the surface of the bilayer. The profiles of Ca^{2+}-ATPase molecules are 65 Å long in a direction parallel to the 'a' axis of the crystal and $\simeq 40$ Å wide in the direction of the 'b' axis. An 18-Å wide gap separates the two molecules that make up the Ca^{2+}-ATPase dimers and the only visible cytoplasmic connection between them is a 17-Å thick bridge that crosses the intradimer gap at a height of 42 Å above the surface of the bilayer [90]. There are additional interactions in the lipid bilayer. The gap under the bridge is likely to be accessible in the native membrane to ATP, Ca^{2+} and other solutes of relatively low molecular weight that interact with the Ca^{2+}-ATPase. A small lobe projects from the main body of the Ca^{2+}-ATPase monomers giving the molecule its pear shape. This lobe is centered some 28 Å above the surface of the bilayer, leaving a 16-Å gap between it and the membrane surface. The Ca^{2+}-ATPase molecules are arranged within the dimer ribbons with their long axes about 13° to the 'a' axis of the unit cell, and the dimers are joined head to tail through the lobe.

The cytoplasmic domain of the Ca^{2+}-ATPase of rabbit sarcoplasmic reticulum is very similar to the structure derived from Fourier–Bessel reconstructions of the Ca^{2+}-ATPase tubules of scallop sarcoplasmic reticulum [176].

The cytoplasmic domains reconstructed from negatively stained [90] and from frozen-hydrated samples [91,177] have similar shapes. Both include the protruding lobe and the bridge region that links the Ca^{2+}-ATPase molecules into dimers. The intramembranous peptide domains of the two ATPase molecules which make up a dimer spread apart as they pass through the bilayer toward the luminal side of the membrane, establishing contacts with the Ca^{2+}-ATPase molecules in the neighboring dimer chains. The lateral association of dimer chains into extended crystal lattice is

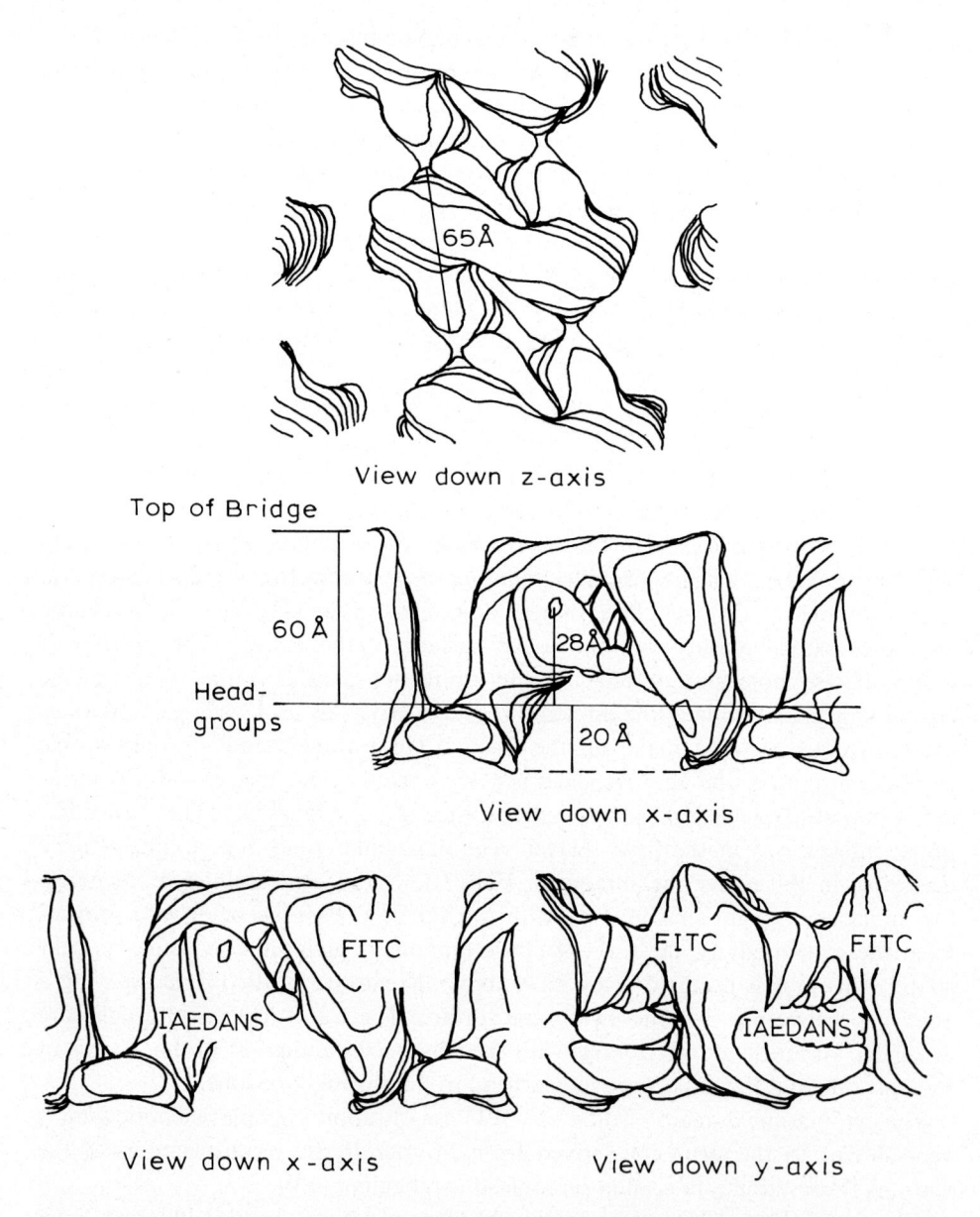

Fig. 4. Tentative allocation of probe binding sites within the three-dimensional structure of Ca^{2+}-ATPase derived from vanadate-induced E_2-type crystals. The top picture is the projection view of the Ca^{2+}-ATPase down the x-axis, revealing the pear-shaped contours of ATPase molecules. The maximum length of the cytoplasmic domain to the tip of the lobe is $\simeq 65$ Å. In the middle and bottom pictures the same structure is viewed down the x-axis, revealing the gap between the bridge and the bilayer surface and the connections between ATPase molecules in neighboring dimer chains. The proposed binding sites for IAEDANS and FITC are indicated. The bottom right picture is the same structure viewed down the y-axis. Adapted from Taylor et al. [90].

determined by these intramembrane contacts. The structural arrangement derived from three-dimensional reconstruction of the frozen-hydrated samples suggests that the Ca^{2+} transport channel cannot be located between the two ATPase molecules constituting the dimer, but it is probably part of each ATPase molecule.

5.2. Crystallization of Ca^{2+}-ATPase by Ca^{2+} and lanthanides in the E_1 state

Sarcoplasmic reticulum vesicles prepared from rabbit skeletal muscle were crystallized in a medium of 0.1 M KCl, 10 mM imidazole (pH 8), and 5 mM $MgCl_2$ by the addition of either $CaCl_2$ (100 μM) or lanthanide ions (1–8 μM) that stabilize the E_1 conformation of the Ca^{2+}-ATPase [119]. After incubation at 2°C for 5–48 hours, crystalline arrays were observed on the surface of about 10–20% of the vesicles in sarcoplasmic reticulum preparations obtained from fast-twitch rabbit skeletal muscles.

CrATP, a suicide inhibitor of Ca^{2+}-ATPase [178], that arrests the enzyme in a Ca^{2+} occluded $E_1 \sim P$ state, also produced E_1-type crystals very similar to those obtained with lanthanides [119]. These observations further support the assignment of the P1-type crystals to the E_1 and $E_1 \sim P$ conformation of the Ca^{2+}-ATPase.

Analysis of the lanthanide-induced crystalline arrays by negative staining (Fig. 5) or freeze-fracture electron microscopy reveals obliquely oriented rows of particles, corresponding to individual Ca^{2+}-ATPase molecules [119]. The unit cell dimensions for the gadolinium-induced Ca^{2+}-ATPase crystals are a = 61.7 Å, b = 54.4 Å and γ = 111°. Similar cell constants were obtained for the crystals induced by lanthanum, praseodymium and calcium. The unit cell dimensions of the E_1 crystals are consistent with a single Ca^{2+}-ATPase monomer per unit cell. The space group of the E_1-type crystals is P1 [119], while that of the E_2 crystals is P2 [88,90].

By proper selection of the experimental conditions a reversible interconversion between the two crystal forms (E_1 and E_2) can be observed [119], that may be related to the structural transitions between the two major conformations of the Ca^{2+}-ATPase.

5.3. Crystallization of Ca^{2+}-ATPase in detergent-solubilized sarcoplasmic reticulum

Further advance toward a high-resolution structure of Ca^{2+}-ATPase requires three-dimensional crystals of sufficient size and quality for X-ray diffraction analysis [179]. A prerequisite for the formation of three-dimensional crystals is the solubilization of the enzyme from its membrane environment by detergents [180,181]. Since the detergent-solubilized Ca^{2+}-ATPase is notoriously unstable, the first task was to find conditions that preserve the ATPase activity of solubilized enzyme for several months.

By systematically testing several hundred conditions, Pikula et al. [182] found that the Ca^{2+}-modulated ATPase activity was preserved for several months at 2°C under

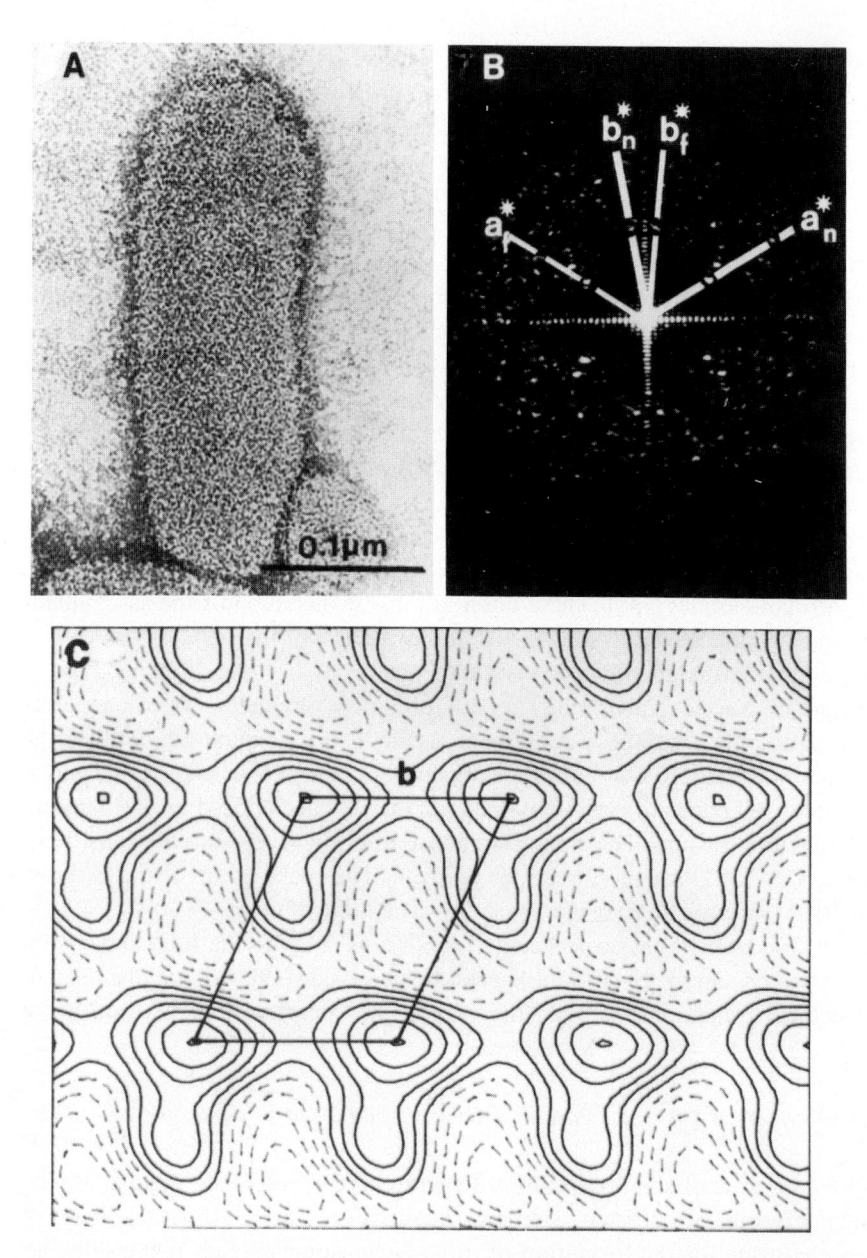

Fig. 5. Image and optical diffraction pattern of praseodymium-induced crystals. (A). Crystallization was induced with 8 μM PrCl$_3$. Doublet tracks so prominent in vanadate-induced crystals are not evident in crystals induced with lanthanides. This results in an approximate halving of the b-axis of the unit cell. Magnification × 222 000. (B) The image of the superimposed top and bottom lattices of the flattened cylinder give rise to two separate diffraction patterns. (C) Projection map of praseodymium-induced crystals. Map scale: 0.55 mm per Å. From Dux et al. [119].

nitrogen in a crystallization medium of 0.1 M KCl, 10 mM K-MOPS, pH 6.0, 3 mM MgCl$_2$, 3 mM NaN$_3$, 5 mM dithiothreitol, 25 IU/ml Trasylol, 2 μg/ml 1,6-di-tert-butyl-*p*-cresol, 20 mM CaCl$_2$, 20% glycerol, 2 mg/ml sarcoplasmic reticulum protein, and 4–8 mg/ml of the appropriate detergent, such as C$_{12}$E$_8$, Brij 36T, Brij 56, or Brij 96 [182,183]. Under these conditions the sarcoplasmic reticulum membranes readily dissolved upon addition of the detergent and formed a clear solution. Negative staining electron microscopy revealed individual Ca^{2+}-ATPase particles in various stages of aggregation and the absence of intact vesicles. After incubation for 6–10 days under nitrogen, ordered crystalline arrays began to appear that increased in number and size during the next several weeks [156,182]. The average diameter of the crystalline aggregates was about 2000–10000 Å, containing an estimated 10^5–10^6 molecules of the Ca^{2+}-ATPase.

The Ca^{2+}-ATPase microcrystals formed in detergent-solubilized sarcoplasmic reticulum in the presence of 20 mM Ca^{2+} and 20% glycerol contain highly ordered crystalline sheets of Ca^{2+}-ATPase molecules, that associate into multilamellar stacks (Fig. 6) consisting frequently of more than 100 layers [156,183–185].

Two distinct patterns of repeats were observed by electron microscopy of sectioned, negatively stained, frozen-hydrated, or freeze-fractured specimens of Ca^{2+}-ATPase crystals that represent different projections of the same structure [156,183,186].

In the first view, layers of densities are seen that repeat at $\simeq$ 103–147 Å in sectioned specimens, at 130–170 Å in negatively stained material, and at 170–180 Å in images of frozen-hydrated crystals [156]. These layered structures represent side-views of stacked multilamellar arrays of ATPase molecules (Fig. 6). The $\simeq$ 40-Å thick stain-excluding core of the lamellae contains a lipid-detergent phase into which the hydrophobic tail portions of the ATPase molecules are inserted symmetrically on both sides [156]. The periodicity of the lamellae is defined by contacts between the hydrophilic headgroups of the ATPase molecules. The 170-Å spacing of the layers is consistent with the dimensions of the ATPase molecules defined by the analysis of the E$_2$-V-type crystals [90,91], suggesting minimal interdigitation between the cytoplasmic domains of Ca^{2+}-ATPase molecules that interact from adjacent lamellae [156,186]. The interactions between the exposed headgroups are responsible for the association of lamellae into three-dimensional structures. The high Ca^{2+} concentration ($\simeq$ 20 mM), low temperature ($\simeq$ 2°C and the low pH ($\simeq$ 6.0) required for crystallization presumably promote these interactions.

High glycerol concentration and low temperature reduce stacking presumably by interfering with the interactions between the hydrophilic headgroups of Ca^{2+}-ATPase molecules in adjacent lamellae, while not affecting or promoting the ordering of ATPase molecules within the individual sheets [184,185].

In the second view of the three-dimensional crystals the projected image normal to the plane of the lamellae (Fig. 7) shows ordered arrays of 40–50-Å-diameter particles, that represent the cytoplasmic domains of ATPase molecules [156]. The crystals

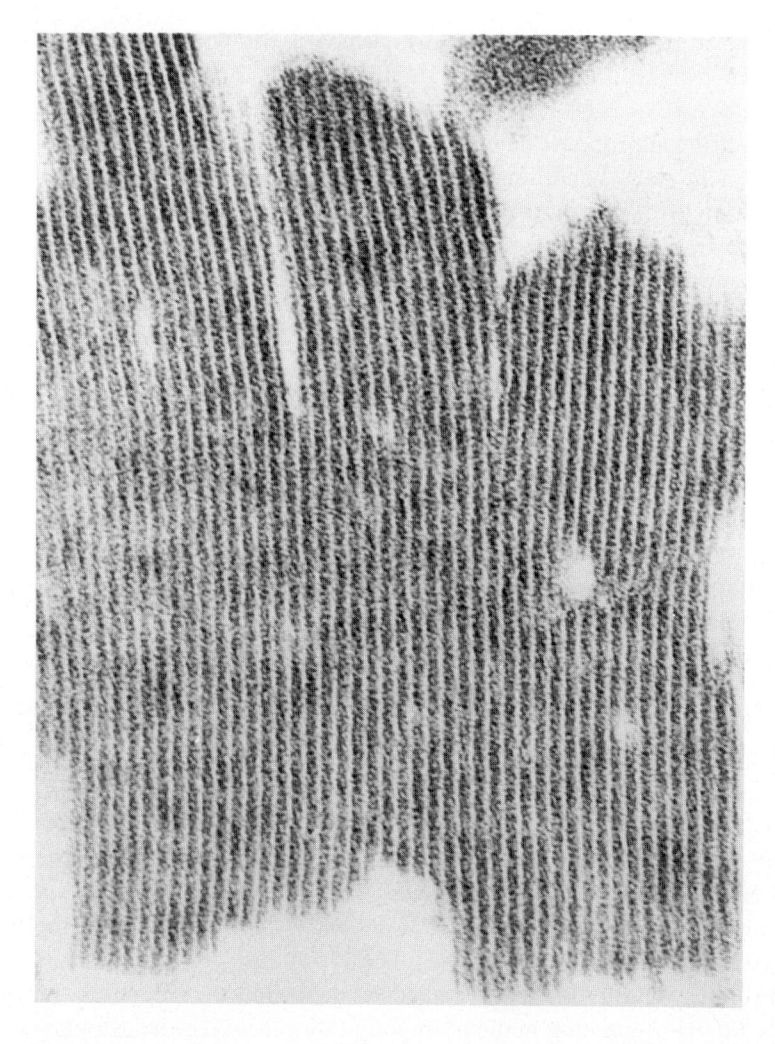

Fig. 6. Electron microscopy of Ca^{2+}-ATPase crystals in thin sections. Sarcoplasmic reticulum (2 mg of protein/ml) was solubilized in the standard crystallization medium with $C_{12}E_8$ (2 mg/mg protein) and incubated under nitrogen at 2°C for 15 days. The crystalline sediment was embedded in Epon–Araldite mixture and processed for electron microscopy. Depending on conditions during fixation, embedding, sectioning and viewing, the observed periodicities in different specimens varied between 103 and 147 Å. Magnification, × 207 000. From Taylor et al. [156].

diffracted to 7.2 Å in X-ray powder patterns and to 4.1 Å in electron diffraction [141,186].

Taylor et al. [156] suggested that the crystals belong to the two-sided plane group C12, in which there are four ATPase molecules per unit cell of 9 113 Å^2, with ATPase dimers related by a two-fold rotational axis within the membrane plane parallel to the b cell axis. While the arrangement of ATPase molecules was highly ordered within

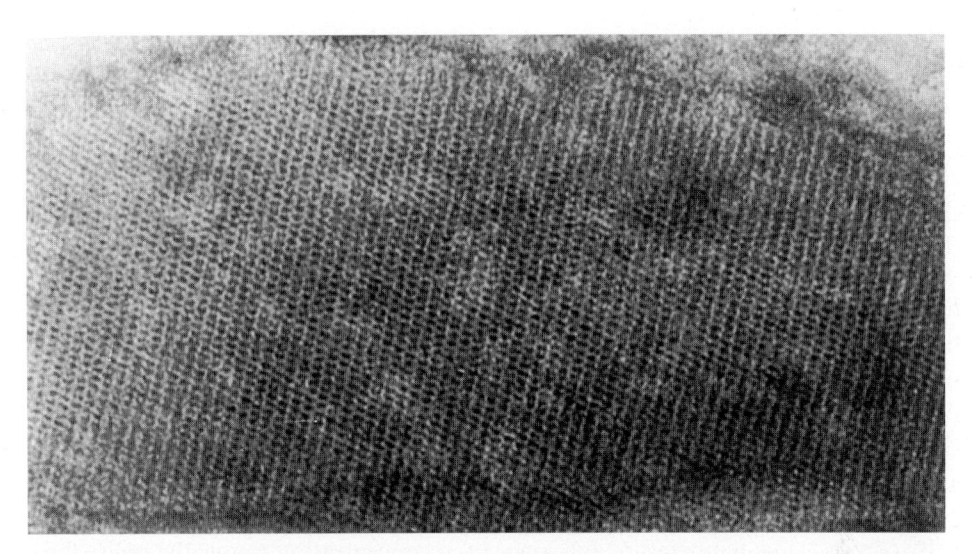

Fig. 7. Projection view of negatively stained Ca^{2+}-ATPase crystals in sarcoplasmic reticulum solubilized with $C_{12}E_8$ (2 mg/mg protein) in the standard crystallization medium. The prominent large spacing is the half-period of the 'a' cell dimension. Striations oblique to this direction are the (1,1) and the (−1,1) periodicities. Magnification, × 308 000. From Taylor et al. [156].

each sheet, there was a slight rotational misalignment between the successive layers in the stacks, which prevented the separation of the projections of individual layers [156]. Stokes and Green [186] confirmed the C12 symmetry, but they found ordered stacking in the third dimension, leading to the conclusion that the crystals belonged to the three-sided space group C2. The four ATPase molecules occupy ≃35% of the unit cell, leaving the remainder of space for lipids and detergents.

The projection map of the unstained frozen-hydrated three-dimensional crystals was compared with the map of negatively stained crystals [141,142] to define the densities associated with the intramembranous domains. The ten transmembrane helices were predicted to be arranged in two crescent shaped rows that provide intramembranous contacts between ATPase molecules. The scheme is entirely speculative and with few exceptions there is no clear relationship between the density contours seen in the map and the proposed transmembrane helices.

6. X-ray and neutron diffraction analysis of the Ca^{2+}-ATPase of sarcoplasmic reticulum

Blasie and his colleagues have determined the separate profile structures of the lipid bilayer and of the Ca^{2+} transport ATPase molecule within the sarcoplasmic reticulum membrane to 11 Å resolution by a combination of X-ray and neutron diffraction techniques [128,140,187–199].

In oriented, partially dehydrated multilayers, under conditions suitable for X-ray diffraction studies, the sarcoplasmic reticulum vesicles retain much of their ATP energized Ca^{2+} transport activity [200–202]. The Ca^{2+} transport can be initiated by flash-photolysis of P^3-1(2-nitro)phenyl-ethyladenosine-5'-triphosphate, 'caged ATP' [203–208]. The flash-photolysis of caged ATP rapidly releases ATP and effectively synchronizes the Ca^{2+} transport cycle of the ensemble of Ca^{2+}-ATPase molecules [190–192,201,209].

The changes in the profile structure of the sarcoplasmic reticulum during Ca^{2+} transport imply that about 8% of the total mass of the Ca^{2+}-ATPase is redistributed from the extravesicular surface to the membrane bilayer region and to the intravesicular surface within 200–500 ms after the flash-photolysis of caged ATP [140,191]. During the next 5 s there was no further change in the profile structure, although the Ca^{2+}-ATPase completed several cycles of Ca^{2+} transport. This may be explained by assuming that the $E_1 \sim P$ form of the enzyme is the dominant intermediate during the steady state.

By contrast, low temperature and low Mg^{2+} concentration causes the redistribution of as much as 15% of the Ca^{2+}-ATPase mass from the lipid hydrocarbon region into the cytoplasm [140,196,198]. These effects of temperature and Mg^{2+} concentration on the structure of the lipid phase and on the transmembrane disposition of Ca^{2+}-ATPase are manifested in a slower rate of E_1P formation and a longer lifetime of E_1P at near zero °C temperatures and at low Mg^{2+} concentrations [196,198].

7. Site specific mutagenesis of sacroplasmic reticulum Ca^{2+}-ATPase

7.1. The search for the Ca^{2+} binding site

7.1.1. Mutation of amino acids in the stalk sector
The predicted location of the high-affinity Ca^{2+} binding sites in stalk helices 1, 2, 3 and 5 [8,9,42] was tested by mutation of glutamate-, aspartate-, glutamine- and asparagine-residues in this region of the molecule [129]. Mutations of glutamate 55, 56, 58, 109, 113, 117, 121, 123, 125, 192, 243, 244, 255, 258; aspartate 59, 196, 245, 254; and glutamine 108 or 259 produced no significant changes in ATP-dependent Ca^{2+} transport, ruling out the stalk sector as the site of high-affinity Ca^{2+} binding. Even mutations of clusters of amino acids such as DELPAEE45 → AALPAA, EEGEE83 → QQGQQ, EEGEE83 → AAGAA, ERNAE113 → ARNAA, EE749 → QQ or NN756 → AA left the Ca^{2+} transport activity entirely unaffected.

The luminal loop (residues 79–87) between transmembrane helices M_1 and M_2 was proposed by MacLennan et al. [42] to serve as the low-affinity Ca^{2+} binding site. This proposal was tested by mutation of EEGEE83 into either QQGQQ or AAGAA. Neither of these mutations affected either Ca^{2+} transport or Ca^{2+} affinity. Therefore the Ca^{2+} binding sites are not likely to involve the region of the molecule between residues 40–131 (S_1 M_1 M_2 S_2), 238–262 (S_3) and 738–759 (S_5).

Proline 195 was suggested by Gangola and Shamoo [210] as a component of the hypothetical Ca^{2+} binding site. Mutation of Pro195 to alanine and the mutation of several acidic residues in the same area [129] had no effect on the Ca^{2+} transport, making this proposition also unlikely.

7.1.2. The probable location of Ca^{2+} binding sites in the transmembrane domain
The negative outcome of the observations described above directed attention to the charged residues in the transmembrane helices M_3–M_6, M_8, and M_{10} as possible components of the Ca^{2+} channel [130,131]. Mutations of Glu309 → Gln or Asp, Glu771 → Gln or Asp, Asn796 → Asp or Ala, Thr799 → Ala, Asp800 → Glu or Asn and Glu908 → Ala or Asp completely inhibited Ca^{2+} transport and the Ca^{2+}-dependent phosphorylation of the Ca^{2+}-ATPase by ^{32}P-ATP. The phosphorylation of the enzyme by Pi in the absence of Ca^{2+} was not affected, in fact, the mutant enzyme was phosphorylated by Pi even at Ca^{2+} concentrations that completely blocked the phosphorylation of the wild enzyme. ATP competitively inhibited the phosphorylation of the enzyme by Pi with $K_I \simeq 5$–$10\,\mu M$, indicating that the ATP binding site remained intact. These observations strongly suggest that Glu309 (helix M_4), Glu771 (helix M_5), Asn796 (helix M_6), Thr799 (helix M_6), Asp800 (helix M_6) and Glu908 (helix M_8) provide oxygen ligands to the high-affinity Ca^{2+} binding site.

The six amino acids implicated in Ca^{2+} binding – Glu309, Glu771, Thr799, Asn796, Asp800 and Glu908 – are all located near the middle of the proposed transmembrane domain of Ca^{2+}-ATPase, distributed in transmembrane helices M_4–M_6 and M_8 (Fig. 2). This location is consistent with the location of the high-affinity lanthanide binding site derived from fluorescence data [132].

7.2. Mutations in the putative catalytic site

The assignment of the catalytic site in the cytoplasmic domain between residues 328 and 738 initially rested largely on the identification of aspartate 351 as the phosphate acceptor in the formation of aspartyl β-phosphate intermediate [42,80,211] and on lysine 515 as the site of reaction of fluorescein-5′-isothiocyanate in competition with ATP [98,212]. Other conserved regions of the enzyme that may be involved in ATP binding are in the region of Asp601, Pro603, Gly626, Asp627, Asp703 and Asp707 [49,134].

7.2.1. Mutations around Asp351
Mutations of Asp351 to Glu, Asn, Ser, Thr, His or Ala and of Lys352 to Arg, Glu or Gln completely inhibit Ca^{2+} transport and enzyme phosphorylation by ATP in the presence of Ca^{2+} [96,97]. Inhibition of Ca^{2+} transport was also caused by interchange of Asp351–Lys352 to Lys351–Asp352 [97], supporting the essential role

of these two amino acids in the formation of the phosphorylated enzyme intermediate.

The functional consequences of mutations in other amino acids within the conserved sequence around Asp351,

Ile–Cys–Ser–Asp351–Lys352–Thr–Gly–Thr–Leu–Thr357

is dependent on the type of substitution. The mutants Ile348 → Ser, Thr355 → Ala, Leu356 → Thr and Thr357 → Ala lost Ca^{2+} uptake, together with phosphoenzyme formation from either ATP in the presence of Ca^{2+} or from Pi in the absence of Ca^{2+}. The inhibition of Ca^{2+} transport in these mutations is probably due to disruption of the structure of the phosphorylation site [97]. The mutants Ser350 → Leu, and Thr353 → Ala or Ser had reduced Ca^{2+} uptake without impairment of Ca^{2+}-dependent phosphorylation by ATP. Therefore the inhibition of Ca^{2+} transport in these cases is due to impairment of some step after E~P formation.

7.2.2. The mutations around Lys515

The K515-G-A-P-E519 sequence is highly conserved in most cation transport ATPases [3,34,46,47,213,214]. Mutation of Lys515 to Arg, Ala, Gln or Glu yielded enzymes with 60, 30, 25 and 5% of the Ca^{2+} transport rate of the wild-type Ca^{2+}-ATPase measured at 10^{-5} M free Ca^{2+} concentration [96,97]. While the reaction of Lys515 with fluorescein-5′-isothiocyanate inhibited only the ATP-dependent Ca^{2+} transport but left the Ca^{2+} transport energized by acetylphosphate largely unaffected [212], the Ca^{2+} transport activity of the Lys515 → Ala mutant was reduced to $\simeq 30\%$ of the control rate both with ATP or acetylphosphate as substrates [97]. The phosphorylation of the enzyme by ATP + Ca^{2+} or by Pi in the absence of Ca^{2+} was unaffected after changing Lys515 to alanine. The Ca^{2+} transport rate was slightly reduced in the Gly516 → Ala mutant but remained unaffected by mutation of Ala517 → Val, Pro518 → Ala and Glu519 → Ala or Gln.

These observations indicate that the ATP binding was not grossly altered by mutations affecting Lys515 and remained essentially unaffected by mutations of the neighboring residues in spite of the high degree of conservation of the sequence. These observations suggest a less direct involvement of Lys515 in ATP binding than previously assumed.

7.2.3. The role of sequences 601–604 in ATP binding and Ca^{2+} transport

In the conserved region Asp–Pro–Pro–Arg604, mutation of Pro602 → Leu or Arg604 → Met caused only modest inhibition of Ca^{2+} transport [103]. The Ca^{2+} transport was completely blocked, however, after mutation of Asp601 to Asn or Glu, and of Pro603 to Leu or Gly, using either ATP or acetylphosphate as substrate. The mechanism of inhibition appears to depend on the type of substitution. In the Asp601 → Glu and Pro603 → Gly mutants the Ca^{2+} transport was inhibited due to inhibition of E~P formation. In the Asp601 → Asn and Pro603 → Leu mutants the formation of E~P from either ATP or Pi remained unaffected, in fact there was an increase in the Ca^{2+} affinity of the enzyme in the phosphorylation reaction. In these mutants the inhibition of Ca^{2+} transport was due to inhibition of the transition from the E_1P to the E_2P conformation of the Ca^{2+}-ATPase [103].

7.2.4. Mutations in the R616–K629 region of the Ca^{2+}-ATPase (Thr625, Gly626, Asp627)

Mutations of Gly626 for Ala or Pro and of Asp627 to Ala or Asn completely abolished both Ca^{2+} transport and the phosphorylation of the enzyme by ATP or Pi [97,103]. It is plausible to attribute this inhibition to a requirement for Gly626 and Asp627 in ATP binding. Mutation of Asp627 to glutamate also produced significant inhibition of Ca^{2+} transport but in this mutant the phosphorylation of the enzyme by ATP or Pi and the Ca^{2+} dependence of phosphorylation remained unaffected. The mutations in the Thr–Gly–Asp627 region suggest a tight relationship between the structure of the ATP binding site and the folding of Ca^{2+}-ATPase required for its insertion into the membrane and for the conformational changes associated with Ca^{2+} translocation.

7.2.5. Mutations in the 701–707 region

Substitution of Asn or Glu for Asp707 inhibited Ca^{2+} transport both with ATP or acetylphosphate as energy donors; the formation of $E \sim P$ from ATP in the presence of 0.1 mM Ca^{2+} or from Pi in the absence of Ca^{2+} was also inhibited [103]. This data is consistent with a requirement for Asp707 in the phosphorylation of the enzyme by ATP, acetylphosphate and inorganic phosphate.

7.2.6. Mutations of Lys712

The ATP analogue 5'-fluorosulfonyl benzoyl adenosine (FSBA) inhibits the Na^+,K^+-ATPase by interaction with a lysine residue analogous in position to Lys712 in the Ca^{2+}-ATPase [101]. Mutation of Lys–Lys713 to Met–Met caused only slight reduction of Ca^{2+} transport ($\simeq 13\%$) and had no effect on the phosphorylation of the enzyme by ATP or Pi [97]. Therefore the Lys–Lys713 sequence may be close enough to the ATP binding site to permit steric interference by FSBA with ATP binding, but Lys712 is probably not a contact residue for the bound ATP.

7.2.7. The structure of the ATP binding site

The mutation data on the conserved sequences of the putative ATP binding site is summarized in Fig. 8 [103].

The phosphate acceptor Asp351 is positioned adjacent to the γ phosphate of ATP, while the loop containing Lys515 is close to the α and β phosphoryl groups. The loops containing Asp627, Asp701 and Asp707 influence the formation of the phosphorylated enzyme intermediate, participate in the conformational changes that occur after enzyme phosphorylation and establish communication between the catalytic site and the Ca^{2+} binding site of the transmembrane Ca^{2+} channel.

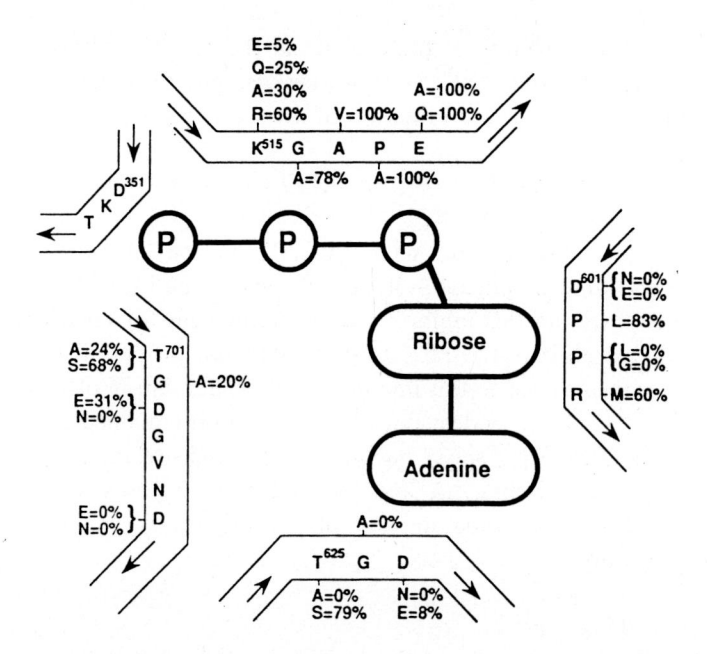

Fig. 8. Mutagenesis of the predicted ATP binding site. ATP is shown in proximity to amino acids in four loops predicted to form the ATP binding site in the nucleotide binding domain [49,134] and a fifth loop representing the phosphorylation site at Asp351 [97]. Mutations and the corresponding Ca^{2+} transport activity of the mutants relative to wild-type are indicated. From Clarke et al. [103].

7.3. The β strand sector. Conformational change mutants

The antiparallel β strand structure between residues 131 and 238 in the cytoplasmic portion of Ca^{2+}-ATPase was originally designated as 'transduction' domain; the name suggested its possible role in the conformational coupling between the nucleotide binding and phosphorylation sites exposed to the cytoplasm and the Ca^{2+} channel located at some distance from each other in the lipid bilayer [8,42]. The site specific mutagenesis of conserved amino acids in the β strand sector of the molecule provides support for its proposed function in conformational transitions [103,126,127,215].

Mutation of glycine 233 to Val, Glu or Arg inhibited Ca^{2+} transport, without inhibition of the phosphorylation of the enzyme by ATP in the presence of 0.1 mM Ca^{2+} [126]. In fact, the Ca^{2+} affinity of the Gly233 → Val or Glu mutants was greater ($K_{Ca} \simeq 0.1\,\mu M$) than that of the wild type enzyme ($K_{Ca} \simeq 0.3\,\mu M$). The phosphorylation of the mutant enzymes by Pi in a Ca^{2+}-free medium was reduced, presumably due to a decrease in their affinity for Pi ($K_{Pi} \simeq 200-300\,\mu M$) compared with wild-type enzyme ($K_{Pi} \simeq 50\,\mu M$).

In the mutants the phosphoenzyme formed from ATP was ADP-sensitive, even under conditions (low K^+ concentration and alkaline pH) that caused the accumulation of large amounts of ADP-insensitive E_2P in the wild-type enzyme. Furthermore, the decomposition of $E \sim P$ in the presence of EGTA proceeded much more slowly in the mutant than in the wild-type enzyme. Since the $E \sim P$ decomposition in EGTA-containing solutions requires the conversion of ADP-sensitive E_1P into the ADP-insensitive E_2P, these observations suggest that the mutation of Gly233 into either valine or glutamine stabilizes the E_1 and E_1P states and interferes with the $E_1P \to E_2P$ transition that is required for the translocation of Ca^{2+}. Mutants of this type in which the binding sites for Ca^{2+} and ATP remain functional and the inhibition of Ca^{2+} transport is caused by impairment of the $E_1P \to E_2P$ transition are designated as conformational change mutants [126,215].

The mutation of Thr181, Gly182, or Glu183 to alanine, or of Glu183 to glutamine also completely inhibited the ATP or acetylphosphate-dependent Ca^{2+} transport, without effect on the phosphorylation of the enzyme by ATP in the presence of Ca^{2+} or by Pi in the absence of Ca^{2+} [127]. The phosphoenzyme formed from ATP retained its ADP-sensitivity at low K^+ concentration and alkaline pH, but its rate of decomposition was much slower than that of the wild-type enzyme in the presence of EGTA. These observations implicate the 181–183 region in the conformational changes related to Ca^{2+} translocation.

7.4. The transmembrane segments of the Ca^{2+}-ATPase

The pattern of insertion of intrinsic membrane proteins into the lipid bilayer during or immediately after their biosynthesis [216,217] affects their three-dimensional structure. For this reason, reliable identification of the transmembrane segments should be the starting point of structure prediction. Unfortunately the techniques available for this purpose [133,218-221] have only been tested so far against the structures of bacteriorhodopsin [220], and the photosynthetic reaction center of *Rhodopseudomonas viridis* [222–225] and *Rhodobacter spheroides* R-26 [226–228] and may not be entirely reliable for the prediction of transmembrane segments in ion transport ATPases [229–233]. The reason for this is that the two basic criteria of predictive algorithms for transmembrane sequences – the $\simeq 22$ amino acid length and the high level of hydrophobicity or amphipathic character – may not be necessarily valid for ion channel proteins. The hydrophilic pore of these proteins could be lined by short hydrophilic helices forming contact with an outer ring of amphipathic helices, rather than with the bilayer [229]. Furthermore, some hydrophobic regions that would fit the criteria for transmembrane sequences may be located in folded regions of the proteins outside the bilayer domain [234]. As a result of these uncertainties, the number of transmembrane sequences proposed for ion transport ATPases varies between four and ten in spite of considerable similarities in their hydrophobicity profiles.

Reliable information about the transmembrane topology of ion transport ATPases can be obtained only by a combination of predictions based on amino acid sequence by site-specific mutagenesis, immunochemical techniques, *in situ* proteolysis, vectorial labeling and labeling with hydrophobic reagents. Most of this data is still not available.

8. *In situ proteolysis of Ca^{2+}-ATPase*

The fragmentation of Ca^{2+}-ATPase by proteolytic enzymes [42,85,235,236] and by vanadate-catalyzed photocleavage [104,105] occurs at well defined and conformationally sensitive cleavage sites that delineate functional domains within the Ca^{2+}-ATPase. The functional changes that follow the cleavage of the polypeptide chain provide useful hints about the role of various domains in the mechanism of Ca^{2+} transport.

8.1. *Hydrolysis of Ca^{2+}-ATPase by trypsin*

The successive cleavage of the Ca^{2+}-ATPase by trypsin at the T_1, T_2, T_{3a}, T_{3b} and T_4 cleavage sites produces a series of well defined cleavage fragments (A, B, A_1, A_2, A_{1a}, A_{1b} and C).

8.1.1. *The T_1 cleavage*
The primary cleavage of the rabbit fast-twitch Ca^{2+}-ATPase isoenzyme by trypsin occurs at the T_1 cleavage site between Arg505 and Ala506, yielding two large fragments (Fig. 9). The apparent molecular weight of fragment A is about 57 kDa, while that of fragment B is $\simeq 52$ kDa [111,237–247]. The A fragment originates from the N-terminal half of the Ca^{2+}-ATPase and contains the phosphate acceptor aspartyl 351 residue [72,85,135]. The B fragment is derived from the C-terminal half of the enzyme and it is assumed to contain much of the substrate binding site [42,85]. The A and B fragments are held together in the membrane by secondary forces [248,249], with preservation of ATPase activity and Ca^{2+} transport. The ATPase vesicles are still readily phosphorylated by ^{32}P-ATP after cleavage into the A and B fragments and on polyacrylamide gel electrophoresis the covalently bound radioactivity was associated with Asp351 in the A fragment [42,80,94,95,237,246].

The rate of tryptic cleavage at the T_1 site is rapid, whether the enzyme is stabilized in the E_1 state by Ca^{2+} or in the E_2V state by EGTA and vanadate [118–120]. Vanadate protected both fragments from further hydrolysis [118], and enhanced the fluorescence of the fluorescein-5′-isothiocyanate covalently attached to lysine 515 in the B fragment of the cleaved ATPase. Therefore the A and B fragments respond with concerted conformational changes to vanadate binding at the substrate site within the A + B complex, even when there is no covalent linkage between them.

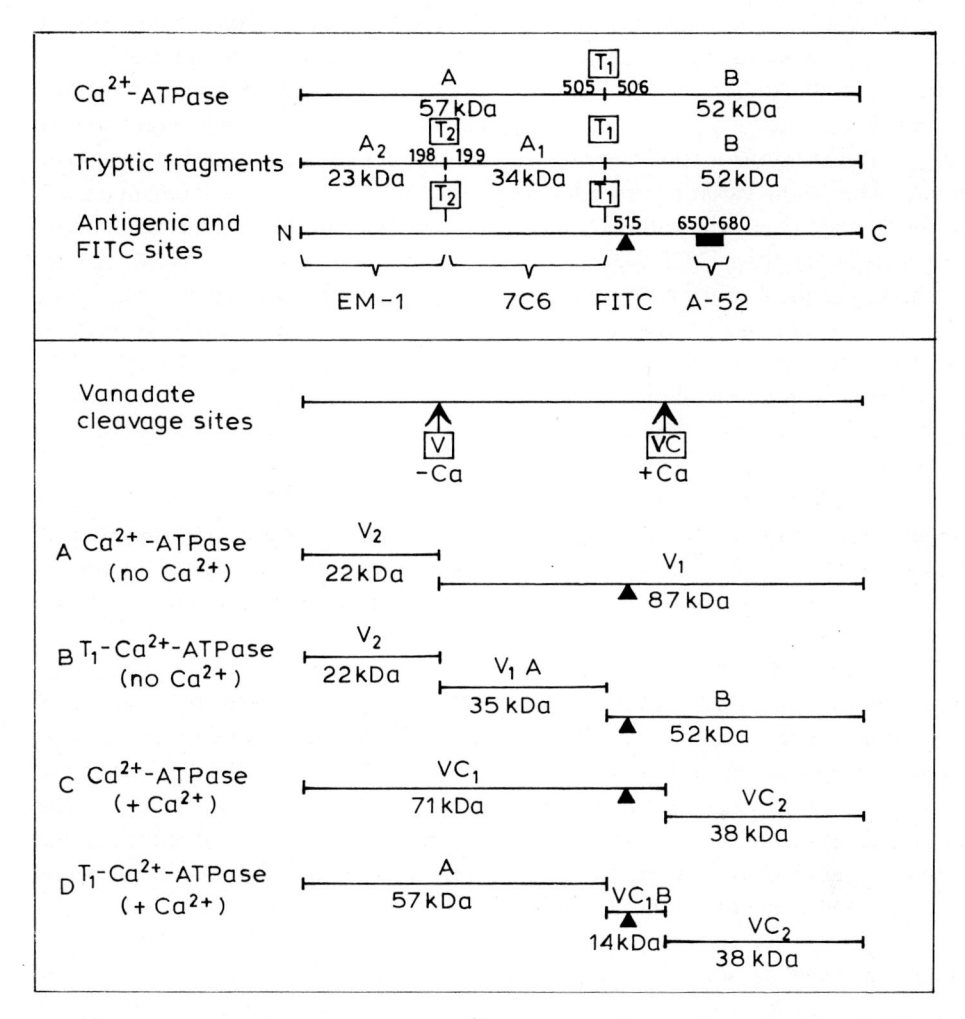

Fig. 9. The topography of cleavage sites in the Ca^{2+}-ATPase. A schematic illustration is provided for the distribution of tryptic and photolytic cleavage sites in the Ca^{2+}-ATPase, together with the designation, approximate size and antibody specificity of the cleavage fragments. T_1-Ca^{2+}-ATPase, Ca^{2+}-ATPase cleaved by trypsin at the T_1 site. From Vegh et al. [104].

8.1.2. The T_2 cleavage

The secondary tryptic cleavage site of the rabbit sarcoplasmic reticulum Ca^{2+}-ATPase is in the A fragment between Arg198 and Ala199, yielding an A_1 fragment of $\simeq 34$ kDa and an A_2 fragment of $\simeq 23$ kDa (Fig. 9) [42,111,237–240,242,246]. The A_2 fragment originates from the N-terminus of the Ca^{2+}-ATPase, while the A_1 fragment contains the phosphate acceptor Asp351 residue [42,85]. A similar pattern of phosphorylation was observed in cardiac sarcoplasmic reticulum of the dog [250].

The T_2 cleavage is highly sensitive to the conformation of the Ca^{2+}-ATPase and proceeds at a rapid rate only in the E_1 state stabilized by Ca^{2+} or lanthanides, while it is completely blocked in the presence of EGTA and vanadate which stabilize the E_2V state [118–120]. Addition of Ca^{2+} or Pr^{3+} relieves the inhibition caused by vanadate and EGTA, as shown by the rapid appearance of the A_1 and A_2 fragments at Ca^{2+} or Ln^{3+} concentrations in slight excess over EGTA. Other lanthanides gave similar results with slight variations in effectiveness [119]. The effects of lanthanides are probably due to the stabilization of the E_1 conformation, although other mechanisms may also be operative [251].

The T_2 site also became protected from tryptic hydrolysis after phosphorylation of the native or solubilized sarcoplasmic reticulum vesicles with inorganic phosphate in a calcium free medium in the presence of dimethylsulfoxide or glycerol [121,252]. Under these conditions the Ca^{2+}-ATPase is converted into a covalent E_2–P intermediate, that is analogous in conformation to the E_2V intermediate formed in the presence of vanadate. In contrast to this, the T_2 site in the stable phosphorylated $Ca_2E_1 \sim P$ intermediate generated by the reaction of the Ca^{2+}-ATPase with chromium-ATP in the presence of Ca^{2+} [178,253] was fully exposed to trypsin, just as it was in the nonphosphorylated Ca_2E_1 form. Therefore the phosphorylated intermediates show the same sensitivity to trypsin at the T_2 site as the corresponding nonphosphorylated enzyme forms.

The hydrophobic photoactivatable reagent trifluoromethyl[^{125}I]-iodophenyl diazirine (TID) gave 14–19% greater labeling of the Ca^{2+}-ATPase in the E_2 and E_2V states than in the Ca_2E_1 or $Ca_2E_1 \sim P$ states [252]. Much of this difference was associated with increased labeling of the A_1 region of the Ca^{2+}-ATPase, while the A_2 and B fragments did not show preferential labeling in the E_2 state. Whether the increased labeling in the E_2 conformation is due to greater 'immersion' of the enzyme into the lipid phase or to greater reactivity of the hydrophobic cytoplasmic domains due to a rearrangement of the structure remains to be determined.

8.1.3. The cleavage of Ca^{2+}-ATPase by trypsin at the T_3 and T_4 sites
During continued digestion of the Ca^{2+}-ATPase the A_1 fragment is further cleaved at several distinct sites [121,235,236,252,254,255]. The cleavage at the T_{3a} site (Lys218–Ala219) produces the A_{1a} subfragment, while cleavage at the two adjacent T_{3b} sites (Lys234–Ile235 or Arg236–Asp237) produces the still shorter A_{1b} subfragment [236]. These two subfragments differ from A_1 by the removal of short polypeptide segments from the N-terminal end of the A_1 fragment. As a result of these changes the decomposition of the A_2 fragment is accelerated [254], suggesting that the region of the molecule between Arg198 and Lys234 plays some role in the stabilization of the conformation of the A_2 region of the Ca^{2+}-ATPase.

Parallel with these changes there is also cleavage in the B fragment at the T_4 site (Lys728–Thr729) releasing the C fragment. The C-terminal sequence of the C fragment is identical with that of the intact Ca^{2+}-ATPase, suggesting that it represents the C-terminal half of the B fragment [236]. Andersen et al. [252] proposed the existence of another cleavage site at the 819–825 region, but its precise location was not defined.

During digestion at 35°C in the presence of Ca^{2+} the formation of the A_{1a} and A_{1b} fragments was accompanied by loss of ATPase activity, but significant $E \sim P$ formation was preserved in the A_{1a} and A_{1b} fragments [235]. The $E \sim P$ formed in the A_{1b} + B complex was ADP sensitive (E_1P), even in the absence of KCl, suggesting that its conversion into the ADP-insensitive E_2P was impaired. This effect may account for the inhibition of ATPase activity after cleavage of the T_{3a} and T_{3b} sites [123].

8.2. The effect of other proteolytic enzymes on the Ca^{2+}-ATPase

8.2.1. Chymotrypsin
The initial cleavage of Ca^{2+}-ATPase by chymotrypsin yields two fragments – CA and CB – with molecular weights of 52 and 62 kDa in Laemmli gels and 64 kDa and 60 kDa in Weber–Osborn gels [256]. The two chymotrypsin cleavage sites are presumably identical with the T_1 and T_2 tryptic sites. The CB fragment does not accumulate in proportion to CA, presumably because CB is degraded at a faster rate during continued digestion. Two of the CB subfragments (CB_1 and CB_2) were water soluble and were released into the medium.

8.2.2. Thermolysin
Thermolysin acts simultaneously at several sites on the Ca^{2+}-ATPase without accumulation of large fragments; this property proved useful in the sequence analysis of the Ca^{2+}-ATPase [78,79,82,83], and in the isolation of SH-group-containing peptides [257]. Small fragments also accumulate after treatment of sarcoplasmic reticulum with subtilisin [256].

8.2.3. Staphylococcal V8 protease
The V8 protease selectively cleaves the Ca^{2+}-ATPase at glutamate and aspartate peptide bonds [258]. In the presence of Ca^{2+} two high molecular weight bands – p85 and p95 – accumulated. The 85 kDa fragment arises by cleavage at Ile232 (site $V8_1$) and contains the intact C-terminal end of the molecule. In the absence of Ca^{2+} a large number of fragments appeared within a few minutes, with only a 29 kDa fragment showing slight accumulation during continued digestion.

8.3. Vanadate-catalyzed photocleavage of the Ca^{2+}-ATPase

Vanadate-catalyzed photocleavage of the Ca^{2+}-ATPase was observed after illumination of sarcoplasmic reticulum vesicles or the purified Ca^{2+}-ATPase with ultraviolet light in the presence of 1 mM monovanadate or decavanadate [104]. Two sites of photocleavage were identified depending on the Ca^{2+} concentration of the medium (Fig. 9). When the $[Ca^{2+}]$ was maintained below 10 nM with EGTA, the vanadate-catalyzed photocleavage occurred near the T_2 tryptic cleavage site yielding fragments of 87 kDa (V_1), and 22 kDa (V_2); in the presence of 2–20 mM Ca^{2+} polypeptides of 71 kDa (VC_1) and 38 kDa (VC_2) were obtained as the principal cleavage products [104]. These observations indicate that the site of photocleavage is determined by the conformation of the Ca^{2+}-ATPase.

The two cleavage sites were located in the structure of the Ca^{2+}-ATPase after partial proteolysis of the FITC-labeled enzyme by SDS-polyacrylamide gel electrophoresis and labeling with anti-ATPase antibodies of defined specificity [104]. The V_2 fragment (22 kDa) obtained in the absence of Ca^{2+} contains the N-terminus of the intact Ca^{2+}-ATPase and it is similar in size to the A_2 fragment (23 kDa) produced by trypsin. The V_1 fragment (87 kDa) contains the regions of the Ca^{2+}-ATPase corresponding to A_1 + B (Fig. 9). The VC_1 fragment (71 kDa) formed in the presence of Ca^{2+} contained the A region extending to the N-terminus, together with a short segment of B including Lys515 that is labeled with FITC. The VC_2 fragment (38 kDa) contains the C-terminal, $1/3$ of the intact Ca^{2+}-ATPase (Fig. 9).

The photocleavage is accompanied by loss of ATPase activity and inhibition of the crystallization of Ca^{2+}-ATPase; both processes are accelerated by Ca^{2+} [104,105]. The faster loss of ATPase activity during illumination in the presence of calcium is consistent with simultaneous binding of vanadate and calcium to the Ca^{2+}-ATPase [259–261].

The target of vanadate-catalyzed photolysis is presumably an amino acid near the catalytic site of the Ca^{2+}-ATPase. The vanadate-catalyzed photocleavage at the V cleavage site in the absence of Ca^{2+} is $\simeq 500$ amino acids away from the C cleavage site which is attacked in the presence of Ca^{2+}. Both sites are probably adjacent in the native structure to the catalytic site of the Ca^{2+}-ATPase.

9. Monoclonal and polyclonal anti-ATPase antibodies

A wide selection of monoclonal and polyclonal anti-Ca^{2+}-ATPase antibodies have become available in recent years. Studies with these antibodies defined the localization of Ca^{2+}-ATPase in the sarcoplasmic reticulum of developing and mature skeletal muscles [60,262–270] and established a pattern of cross reactivity with various Ca^{2+}-ATPase isoenzymes in the sarco(endo)plasmic reticulum [270–286] and in the plasma membrane [284,287–290] of skeletal, cardiac and smooth muscles. Antibodies have also proved useful in the quantitation of Ca^{2+}-ATPase, both in muscles of diverse fiber types [291–294] and in COS-1 cells transfected with Ca^{2+}-ATPase cDNA [97,103,126,127,129,215].

With the arrival of new information on the amino acid sequence of various Ca^{2+}-ATPase isoforms [8,9,11,32,53–57,295] it became possible to identify the epitopes for the various antibodies, and to relate their positions to the three-dimensional structure of Ca^{2+}-ATPase emerging from crystallographic studies [90,91,141,142,156,

157,160,161,182,186] from site-specific mutagenesis [11] and from covalent labeling with substrates, substrate analogs and fluorescent probes [104,132,160,213, 214,296,297].

9.1. Antibodies reacting with the N- and C-terminal regions of the Ca^{2+}-ATPase

The Ca^{2+}-ATPase is synthesized on membrane bound polysomes without an N-terminal signal sequence [86,298–300]. The location of its N-terminus on the cytoplasmic surface of the membrane was first established by the reaction of cysteine 12 with nonpenetrating SH-group reagents added to the cytoplasmic side [135]. Further support for this arrangement was provided by Matthews et al. [137,301] using polyclonal antibodies produced against peptides representing the N-terminal 1–11 sequence (MEAAHSKTEEC) and the C-terminal 985–994 sequence (CKFIAR-NYLEG) of the Ca^{2+}-ATPase. Both antibodies recognized the Ca^{2+}-ATPase in ELISA tests using either intact sarcoplasmic reticulum vesicles or purified ATPase preparations and reacted to the same extent in competitive ELISA assays with intact and SDS disrupted sarcoplasmic reticulum. These observations clearly support models in which the number of transmembrane segments is even and both N- and C-termini are exposed on the cytoplasmic surface. The actual number of membrane crossing segments remains undefined.

9.2. Distribution of epitopes in the cytoplasmic domain of Ca^{2+}-ATPase

The approximate location of the epitopes for more than 40 monoclonal anti-ATPase antibodies has been mapped to various regions within the cytoplasmic domain of the Ca^{2+}-ATPase [285,302–304]. All antibodies were found to bind with high affinity to denatured Ca^{2+}-ATPase, but the binding to the native enzyme showed significant differences depending on the location of antigenic sites within the ATPase molecule.

Antibody A52 with its epitope at residues 657–672 [129,139,274,275] inhibited the vanadate-induced crystallization of Ca^{2+}-ATPase and decreased the stability of preformed Ca^{2+}-ATPase crystals [285]. The vanadate-induced crystals arise by the association of the ATPase monomers into dimers (type A interaction), the dimers into dimer chains (type B interaction), and the dimer chains into 2-dimensional arrays (type C interaction). It is suggested that antibody A52 interferes with type B interactions, preventing the formation of dimer chains, without exerting major effect on the concentration of Ca^{2+}-ATPase dimers in the membrane. The simplest interpretation of the destabilization of Ca^{2+}-ATPase crystals by mAb A52 is that binding of the antibody to its antigenic site physically blocks the interaction between ATPase molecules [285]. Considering the large bulk of the antibody, such interference is not unexpected, yet only a few of the antibodies that bind to the Ca^{2+}-ATPase in native sarcoplasmic reticulum interfered with crystallization.

Some of the antibodies directed against the B fragment of the Ca^{2+}-ATPase caused moderate inhibition of ATPase activity and Ca^{2+} transport; the inhibition usually did not exceed 50%, even at high antibody : Ca^{2+}-ATPase ratios where the antigenic sites are expected to be fully saturated [285,302,304].

The powerful inhibition of Ca^{2+} transport without inhibition of ATPase activity seen with polyclonal anti-ATPase sera in earlier studies [305] was probably due to complement dependent lesion of the membrane that permitted the leakage of accumulated calcium [306–308]. The scarcity of inhibitory antibodies may imply that the active site of the Ca^{2+}-ATPase is either inaccessible to antibodies or poorly antigenic, perhaps due to a unique secondary structure.

According to Mata et al. [309], solubilization in $C_{12}E_8$ solution unmasked the inhibitory effect of several antibodies (B/3D6, Y/1F4, Y/2EG, Y/3G6, B/4H3, A/4H3 and I/2H7) on the Ca^{2+}-ATPase. They suggested that these antibodies bind to protein–protein contact sites opened by the dissociation of ATPase oligomers, thus causing inhibition. Alternatively, the binding of antibody to the solubilized ATPase may promote its folding into a conformation that is unfavorable for enzymatic activity.

9.3. Antibodies reacting with the putative luminal domain of the Ca^{2+}-ATPase

The small luminal loop in the proposed structure of Ca^{2+}-ATPase [11] between transmembrane helices 7 and 8 (residues 870–890) was suggested to contain the epitope for mAb A20 [139], and for an antipeptide antibody directed against the 877–888 sequence (TEDHPHFEGLDC) of the Ca^{2+}-ATPase [138].

The A20 antibody did not bind significantly to native SR vesicles, but solubilization of the membrane with $C_{12}E_8$ or permeabilization of the vesicles by EGTA exposed its epitope and increased the binding more than 20-fold [139]. By contrast, the A52 antibody reacted freely with the native sarcoplasmic reticulum, while the A25 antibody did not react either in the native or in the $C_{12}E_8$ solubilized or permeabilized preparations, and required denaturation of Ca^{2+}-ATPase for reaction. Clarke et al. [139] concluded that the epitope for A52 is freely exposed on the cytoplasmic surface, while the epitope for A20 was assigned to the luminal surface, where it became accessible to cytoplasmic antibodies only after solubilization or permeabilization of the membrane. The epitope for A25 is assumed to be on the cytoplasmic surface in a folded structure and becomes accessible only after denaturation.

Essentially identical conclusions arose from the studies of Matthews et al. [138]. An anti-peptide antibody directed against the cytoplasmically exposed C-terminal region of the Ca^{2+}-ATPase (985–994) reacted freely in native sarcoplasmic reticulum, in agreement with earlier observations [137], while the antibody directed against the putative luminal loop (877–888) reacted strongly only after solubilization of sarcoplasmic reticulum with $C_{12}E_8$. Purified ATPase preparations reacted freely with both antibodies under both conditions. A 30-kDa protease-resistant fragment obtained after digestion with proteinase K contained the epitopes both for the antibodies directed against the C-terminus and the 877–888 region [138]. This was rationalized by the largely transmembranous arrangement of the 720–994 region of the Ca^{2+}-ATPase that would protect it from proteinase K action.

10. Covalent modification of side-chain groups in the Ca^{2+}-ATPase

10.1. Sulfhydryl groups

The slow and fast isoenzymes of Ca^{2+}-ATPase contain 26 and 24 cysteine residues, respectively [8]. Of these, 22 cysteine residues are in the same position of the primary sequence of the two isoenzymes and most of the substitutions, except one, are conservative (valine, serine, glycine). The one exception is cysteine 674 in the fast isoenzyme of Ca^{2+}-ATPase, which is replaced by Arg673 in the slow isoenzyme.

Of the $\simeq 20$ residues that react with N-ethylmaleimide in the non-reduced denatured Ca^{2+}-ATPase at least 15 are available for reaction with various SH reagents in the native enzyme [75,239,310]. These residues are all exposed on the cytoplasmic surface. After reaction of these SH groups with Hg-phenyl azoferritin, tightly packed ferritin particles can be seen by electron microscopy only on the outer surface of the sarcoplasmic reticulum vesicles [143,311–314]. Even after the vesicles were ruptured by sonication, aging, or exposure to distilled water, alkaline solutions or oleate, the asymmetric localization of the ferritin particles on the outer surface was preserved [311,313,314].

10.1.1. Identification of cysteine residues that react with N-ethylmaleimide (MalNEt)

Reaction of the Ca^{2+}-ATPase at pH 7.0 with low concentration of N-ethylmaleimide causes first the modification of two SH equivalents (SH_N) that were not required for Ca^{2+} transport. The modification of the next class (SH_D) inhibited Ca^{2+} transport and ATP hydrolysis without interference with the formation of phosphoenzyme intermediate [315,316]. Finally, modification of the least reactive class of SH groups (SH_F) interfered with phosphoenzyme formation [315–317].

Surprisingly, the two major radioactive peptides isolated from thermolysine digests of the SH_D labeled membranes were identical to those obtained after SH_N-type labeling, indicating that the same SH groups, identified as Cys344 and Cys364, were involved in both types of reactions [257,318]. Apparently the Ca^{2+} transport activity remains unaffected when either one or the other of these cysteine residues are derivatized, but becomes inhibited when both Cys344 and Cys364 are simultaneously modified by N-ethylmaleimide. Therefore the SH_N- and SH_D-type labeling affects the same region of the molecule, in spite of the rather different functional consequences [318,319]. Cys344 and Cys364 are both close to Asp351 that serves as phosphate acceptor in the formation of the phosphoenzyme intermediate from ATP. Single modification of either SH group by MalNEt may be accommodated by the active site without major kinetic consequences, but simultaneous reaction of MalNEt with both SH groups sufficiently distorts the active site to prevent the decomposition of phosphoenzyme intermediate.

10.1.2. The reaction of iodoacetamide and its N-substituted derivatives with the Ca^{2+}-ATPase

10.1.2.1. Iodoacetamide (IAA) and 5-(2-acetamidoethyl)aminonaphthalene-1-sulfonate (IAEDANS). The IAA binding site was located by reacting sarcoplasmic reticulum vesicles with 150 μM [^{14}C]IAA, followed by cleavage of the labeled enzyme with CNBr and trypsin [320]. The sequence of the radioactive peptide was Ala–Cys*674–Cys–Phe–Ala–Arg with the label exclusively in Cys674. An identical fluorescent peptide was isolated after labeling the Ca^{2+}-ATPase with IAEDANS [320,321], indicating that IAA and IAEDANS react at the same site. The reaction of IAA and IAEDANS with Cys674 is highly specific, leaving Cys675 entirely unmodified.

As modification of the Ca^{2+}-ATPase by IAEDANS has no effect on the ATPase activity, the covalently bound AEDANS is a convenient reporter group of the conformational state of the Ca^{2+}-ATPase and its interactions [160,296,297,322–333].

10.1.2.2. The reaction of 6-(iodoacetamido)fluorescein (IAF) with the Ca^{2+}-ATPase. Reaction of sarcoplasmic reticulum vesicles or the purified Ca^{2+}-ATPase with iodoacetamido-fluorescein at dye:ATPase mole ratios of 2–10 yields 0.5–1 mole covalently bound IAF per mole ATPase without inhibition of Ca^{2+}-stimulated ATP hydrolysis [322,334]. About half of the total label was recovered in Cys674 and 20.8% in Cys670. These observations imply somewhat greater heterogeneity in the labeling by IAF than by IAEDANS. In spite of this, Cys675 was left untouched by IAF.

10.1.3. Modification of Ca^{2+}-ATPase with 7-chloro-4-nitrobenzo-2-oxa-1,3-diazole (NBD-Cl)

Reaction of purified Ca^{2+}-ATPase with 0.3 mM NBD-Cl in the presence of 1 mM AMP-PNP and 1 mM CaCl$_2$ caused inhibition of ATPase activity with the incorporation of ≃15 nmol NBD-Cl per mg protein [335]. The inhibition was attributed to the binding of 7–8 nmol NBD-Cl/mg enzyme protein, corresponding to ≃1 mol NBD-Cl per mol ATPase. The NBD-labeled enzyme was digested with pepsin and several NBD-labeled peptides were isolated [335]. All peptides contained the Gly–X (Cys) sequence that occurs only in one place in the Ca^{2+}-ATPase, i.e., at Gly343–Cys344. Therefore NBD-Cl reacts with the same cysteine 344 residue that is also modified by maleimide derivatives [319]. The NBD modified enzyme had only 5–10% of the ATPase activity of the control ATPase, but the steady state concentration of the phosphoenzyme intermediate was only slightly reduced [335]. The Ca^{2+}

release from the phosphoenzyme in the presence of EGTA was very slow, indicating an impairment of the conformational change associated with the $E_1P \rightarrow E_2P$ transition [335–337]. $C_{12}E_8$ or tetrachlorosalicyl-anilide accelerated 40–50-fold the rate of ATP hydrolysis by the NBD-modified enzyme. Therefore the modification of Cys344 by itself is not fully responsible for the inhibition of ATPase activity, but the state of the enzyme is apparently altered in the modified membrane.

10.1.4. The disulfide of 3'(2')-O-biotinyl-thioinosine triphosphate (biotinyl-S^6-ITP$_2$)
Biotinyl-S^6-ITP$_2$ inactivates the Ca^{2+}-ATPase at two distinct rates. ATP protects only against the slow rate of inactivation [338]. Differential labeling with [^{14}C]iodoacetic acid after affinity labeling with biotinyl-S^6-ITP$_2$ led to the isolation of a weakly radioactive peptide containing the sequence A468–N–A–C471–N–S–V–I–R476. According to the authors, Cys471 is not likely to be part of the active site and its labeling may be due to a conformational change of the enzyme induced by the binding of biotinyl-S^6-ITP$_2$ [338].

10.2. Modification of lysine residues

The slow and fast isoenzymes of Ca^{2+}-ATPase contain 58 and 54 lysine residues, respectively [8], distributed relatively evenly throughout the structure. The C-terminus of the neonatal fast-twitch isoenzyme (SERCA1b) is Arg–Arg–Lys–COOH. Most of the lysine residues occur in similar locations in the slow and fast isoenzymes of the Ca^{2+}-ATPase. In most P-type ion transport ATPases highly conserved sequences containing lysine residues are found next to the phosphate acceptor aspartyl 351 (DKTGTLT), at the reaction site of fluorescein-5'-isothiocyanate on lysine 515 (KGAP), and in the ATP binding domain at Lys712 and Lys713 (KAE(D)IG) [134]. Two of these motifs – DKT and KGAPE – are incorporated in the triphosphate chain binding region of the predicted ATP binding site of the Ca^{2+}-ATPase (Fig. 8; [103]).

Chemical modification studies with fluorescein-5'-isothiocyanate support the proximity of Lys515 to the ATP binding site [98,113–117,212,339]. Fluorescein-5'-isothiocyanate stoichiometrically reacts with the Ca^{2+}-ATPase in intact or solubilized sarcoplasmic reticulum at a mildly alkaline pH, causing inhibition of ATPase activity, ATP-dependent Ca^{2+} transport, and the phosphorylation of the Ca^{2+}-ATPase by ATP; the Ca^{2+} uptake energized by acetylphosphate, carbamylphosphate or *p*-nitrophenyl phosphate is only partially inhibited [113,114,212,339]. The reaction of Ca^{2+}-ATPase with FITC is competitively inhibited by ATP, AMPPNP, TNP-ATP, and less effectively by ADP or ITP; the concentrations of the various nucleotides required for protection are consistent with their affinities for the ATP binding site of the Ca^{2+}-ATPase [114,212,340].

The participation of other lysine residues in ATP hydrolysis is suggested by the inhibition of various partial reactions of Ca^{2+} transport after modification of Ca^{2+}-

ATPase with pyridoxal-5'-phosphate [341–344], adenosine-5'-triphosphopyridoxal [99], 2,4,6-trinitrobenzene sulfonate [344–346], fluorescamine [347,348], methylbenzimidate [349], acetic anhydride and maleic anhydride [344], and o-phthalaldehyde [350].

The inhibition of ATPase activity by adenosine triphosphopyridoxal (AP$_3$PL) in the presence of Ca^{2+} is due to stoichiometric binding of AP$_3$PL to Lys684 in the ATP binding site of the Ca^{2+}-ATPase [99,100]. After labeling the enzyme with [^{3}H]AP$_3$PL the radioactive AP$_3$PL was attached only to Lys684, and Lys686 remained unlabeled. Lysine 684 is only 10 residues away from cysteine 674, which reacts selectively with IAEDANS [257]. In the absence of Ca^{2+} the selectivity of AP$_3$PL toward Lys684 was less pronounced and Lys492 also participated about equally in the reaction [100]. These observations suggest that Lys684 and Lys492 are both near the active site and the spatial arrangements of these groups is changed when Ca^{2+} is bound to the high-affinity Ca^{2+} site [100].

10.3. Modification of arginine residues

The slow and fast isoenzymes of Ca^{2+}-ATPase contain 42 and 50 arginine residues, respectively. The C-terminal sequence of the neonatal fast-twitch isoenzyme is Arg–Arg–Lys. There are only four arginine residues in the putative transmembrane helices, which are probably located near the cytoplasmic or luminal surface of the membrane. The remaining arginine residues are distributed in the cytoplasmic domains.

The guanidino groups of arginine are likely components of the binding site of enzymes acting on phosphate compounds. Pai et al. [351] located six arginine residues in the vicinity of the active site of adenylate kinase and the involvement of guanidino groups in the binding of decavanadate to phosphorylase was demonstrated by Soman et al. [352]. Other enzymes of phosphate metabolism with arginine in their active sites are: Na$^+$,K$^+$,-ATPase [353], F$_1$-ATPase [354] and several glycolytic enzymes [355].

The inhibition of Ca^{2+}-ATPase by 2,3-butanedione [356,357] and by phenylglyoxal [357,358] suggest the involvement of arginine residues in the Ca^{2+} transport. The inactivation of Ca^{2+}-ATPase by butanedione in the presence of borate approached zero order kinetics at high reagent concentrations ($K_I \simeq 10$ mM), indicating the existence of a weak reversible binding site [356]. Mg-ATP, and less effectively Mg-ADP, protected against the inactivation, while Mg-AMP and Ca^{2+}-acetylphosphate had no effect [356,357]. The modified enzyme lost its ability to bind [^{14}C]-ATP in the concentration range of 1–65 μM ATP, where the control enzyme bound 0.9 mole ATP/1.5 × 10^5 g protein with a $K_D \simeq 20 \mu$M [356]. These observations suggest that 2,3-butanedione modifies arginine residues at the active site that may be involved in the binding of the triphosphate chain of ATP. The loss of ATPase activity after modification with butanedione or phenylglyoxal is accompanied by inhibition of the modified Ca^{2+}-ATPase to form crystalline arrays in the presence of decavanadate [357].

10.4. Modification of histidine

The slow and fast isoenzymes of Ca^{2+}-ATPase contain 13 and 12 histidine residues, respectively [8]. Only seven of these occur in identical positions in the two isoenzymes; these correspond to His51, 190, 278, 284, 682, 871 and 943 in the sequence of the slow Ca^{2+}-ATPase. The stretch of five histidine residues, located in the slow isoenzyme at positions 396, 406, 526, 566 and 576, have no counterparts in the fast Ca^{2+}-ATPase. None of the highly conserved sequences of the Ca^{2+}-ATPase appear to contain histidine. This still leaves the possibility open for the direct or indirect involvement of histidine residues in ATP hydrolysis and Ca^{2+} transport. Such a role is suggested:

(a) by the pH dependence of ATP hydrolysis and Ca^{2+} transport [359–361];

(b) by the inhibition of ATPase activity and Ca^{2+} transport by photooxidation in the presence of Methylene blue or Rose Bengal [359,362–365]; and

(c) by the effect of ethoxyformylation by diethylpyrocarbonate on the ATPase activity [359,366–368].

The rate of phosphoprotein formation in the presence of 5 mM $CaCl_2$ was only slightly affected by mild photooxidation in the presence of Rose Bengal, but the hydrolysis of phosphoenzyme intermediate was inhibited sufficiently to account for the inhibition of ATP hydrolysis [359]. The extent of inhibition was similar whether the turnover of $E \sim P$ was followed after chelation of Ca^{2+} with EGTA, or after the addition of large excess of unlabeled ATP. These observations point to the participation of functionally important histidine residues in the hydrolysis of phosphoprotein intermediate [359].

Similarly, the rate of inhibition of phosphoenzyme formation by diethylpyrocarbonate (DEPC) was much slower than the loss of ATPase activity [368]. Even when the reaction approached completion with more than 90% inhibition of ATP hydrolysis, about 70% of the Ca^{2+}-ATPase could still be phosphorylated by ATP (2.3 nmoles of $E \sim P$/mg protein). The remaining 30% of $E \sim P$ formation and the corresponding ATPase activity was not reactivated by hydroxylamine treatment, suggesting some side reaction with other amino acids, presumably lysine. When the reaction of the DEPC-modified ATPase with ^{32}P-ATP was quenched by histidine buffer (pH 7.8) the ^{32}P-phosphoenzyme was found to be exceptionally stable under the same conditions where the phosphoenzyme formed by the native ATPase underwent rapid hydrolysis [368]. The nearly normal phosphorylation of the DEPC-treated enzyme by ^{32}P-ATP implies that the ATP binding site is not affected by the modification, and the inhibition of ATPase activity is due to inhibition of the hydrolysis of the phosphoenzyme intermediate [368]. This is in contrast to an earlier report by Tenu et al. [367], that attributed the inhibition of ATPase activity by

DEPC to inhibition of ATP binding.

The Ca^{2+} binding sites of the enzyme remained functional after DEPC treatment [368], but the cooperativity of Ca^{2+} binding seen in native Ca^{2+}-ATPase was abolished. The Ca^{2+} binding to the DEPC-modified Ca^{2+}-ATPase could be described in terms of two independent Ca^{2+} binding sites with K_{Ca} of 14 μM and 0.5 μM, respectively. The phosphorylation of DEPC-modified Ca^{2+}-ATPase followed closely the saturation of the site with the higher Ca^{2+} affinity [368].

10.5. Modification of carboxyl groups

The slow and fast isoenzymes of Ca^{2+}-ATPase contain 76 and 80 glutamate and 45 and 52 aspartate residues, respectively [8]. Aspartate 351 serves as phosphate acceptor in the formation of phosphoenzyme intermediate, while Glu309, 771, 908 and Asp800 are probable components of the high-affinity Ca^{2+} binding site [130,131]. Considering that the total number of carboxylic groups is 121 in the slow and 133 in the fast Ca^{2+}-ATPase, the identification of functionally important carboxylic groups will require reagents that interact preferentially with functional sites in the Ca^{2+}-ATPase.

10.5.1. The reaction of Ca^{2+}-ATPase with dicyclohexylcarbodiimide
Carbodiimides readily react in aqueous solutions with protein amino, carboxyl and sulfhydryl groups; slower reactions with tyrosine and serine have also been reported [369,370]. The primary reaction product of carboxyl groups with dicyclohexylcarbodiimide is dicyclohexyl-*O*-acyl isourea [370]. Dicyclohexyl-*O*-acyl isourea is susceptible to nucleophilic attack either by water or by endogenous or exogenous nucleophiles, yielding a complex series of reaction products [369–371].

The Ca^{2+} transport and Ca^{2+}-stimulated ATPase activity of sarcoplasmic reticulum is inhibited by 10–30 nmol dicyclohexylcarbodiimide per mg protein in a Ca^{2+} free medium [372]. A23187 enhanced the sensitivity of the enzyme to DCCD, while Ca^{2+} or Sr^{2+} at micromolar concentrations prevented the inhibition. Since Ca^{2+}-loaded vesicles retained their sensitivity to DCCD in a Ca^{2+}-free medium, the reactivity of the enzyme with DCCD is controlled by the occupancy of the high-affinity Ca^{2+} sites on the cytoplasmic surface of the membrane.

The rate of inactivation by DCCD was highly pH dependent, increasing by nearly an order of magnitude upon lowering the pH from 7.7 to 5.5. The midpoint of the transition was at pH 6.4 and the Hill coefficient was 1.8 [373], in agreement with the pH dependence of Ca^{2+} binding [360,361,374]. The rate of reaction of DCCD with the Ca^{2+}-ATPase was 10^4 times greater than with acetic acid or furylacrylic acid. These observations are consistent with the involvement of a highly reactive carboxyl group in the reaction that is located in a relatively hydrophobic environment [373].

Scofano et al. [375] observed that at low reagent to protein ratios (10 nmole/mg protein), DCCD selectively inhibited the phosphorylation of the enzyme by Pi and the

ATP ⇆ Pi exchange, while the binding of Ca^{2+} to the Ca^{2+}-ATPase and the Ca^{2+}-dependent phosphorylation by ATP remained largely unaffected. The inhibition of Pi incorporation by DCCD was prevented by 10 µM Ca^{2+}. Therefore DCCD modifies residues that are required for the phosphorylation of the enzyme by inorganic phosphate but are not involved in Ca^{2+} binding. The inhibition of E~P formation from Pi occurred with only negligible incorporation of [^{14}C]DCCD into the Ca^{2+}-ATPase, suggesting that the primary O-acylisourea adduct underwent nucleophilic reaction with the elimination of radioactive dicyclohexylurea [375]. As DCCD produced similar inhibition of Pi incorporation with or without exogenous nucleophiles (glycine ethylester or ethylene diamine), the reaction with DCCD may result in intramolecular crosslinking at a location that is not accessible to exogenous nucleophiles [256].

10.5.2. The reaction of Ca^{2+}-ATPase with N-cyclohexyl-N'-(4-dimethyl-amino-α-naphthyl)carbodiimide (NCD-4)

NCD-4 is a nonfluorescent carbodiimide derivative that forms a fluorescent adduct with the Ca^{2+}-ATPase, accompanied by inhibition of ATPase activity and phosphoenzyme formation [376–378]. Ca^{2+} protected the enzyme against the inhibition by NCD-4 and reduced the extent of labeling, suggesting that the reaction may involve the Ca^{2+} binding site. The stoichiometry of the Ca^{2+}-protected labeling was ≃ 2 mole/mol ATPase. The fluorescence emission of the modified Ca^{2+}-ATPase is consistent with the formation of a protein bound N-acylurea adduct in a relatively hydrophobic environment. After tryptic proteolysis of the NCD-4 labeled ATPase the fluorescence was associated with the A_2 band of 24 kDa [376,379].

10.5.3. Reaction of Ca^{2+}-ATPase with the carbodiimide derivative of ATP

The adduct of ATP with 1-ethyl-3-(3 dimethylamino propyl)carbodiimide (ATP-EDC) inhibits the Ca^{2+}-ATPase by converting it into a crosslinked species with an apparent molecular weight of ≃ 170 kDa on SDS-PAGE [109,110]. The inhibition of ATPase activity required the presence of Ca^{2+} at micromolar and Mg^{2+} at millimolar concentrations and its initial rate showed a hyperbolic dependence on the concentration of ATP-EDC with a midpoint at ≃ 10 µM. ATP and ADP protected the enzyme against inactivation. The modified enzyme lost its ability to become phosphorylated by either ATP or Pi, and its affinity for TNP-ADP or AMP-PCP was markedly reduced; there was no change in the Ca^{2+}-dependent increase in the intrinsic tryptophan fluorescence. Tryptic digestion of the ATP-EDC modified Ca^{2+}-ATPase decreased its apparent molecular weight from 170 to ≃ 140 kDa with the release of the A_2 fragment, while the A_1 and B fragments remained crosslinked to each other.

Modification of the ATPase by ATP-EDC occurred without significant incorporation of the nucleotide into the enzyme, suggesting that the inhibition of enzymatic activity is entirely due to intramolecular crosslinking by the carbodiimide moiety positioned near the active site of the Ca^{2+}-ATPase.

A similar mechanism was postulated for the Ca^{2+}-dependent inactivation of Ca^{2+}-ATPase by ATP-imidazolidate [380] that results in intramolecular crosslinking with the formation of a new protein band of 125 kDa. In both cases the reactive carboxyl group was suggested to be the phosphate acceptor Asp351.

11. Spatial relationships between functional sites in the sarcoplasmic reticulum Ca^{2+}-ATPase

The mechanism of the coupling between ATP hydrolysis and Ca^{2+} transport is determined by the spatial relationship of the phosphorylation and ATP binding domains of the Ca^{2+}-ATPase to the Ca^{2+} channel involved in the translocation of calcium. Two alternative coupling mechanisms have been proposed, based on two rather different hypothetical models of the structure of the Ca^{2+}-ATPase. In the *conformational coupling mechanism* the energy transfer between ATP hydrolysis and Ca^{2+} transport involves a mechanical coupling over long distances between separate ATP binding and Ca^{2+} transport sites [381]. In the *'ligand conduction'* mechanism a more direct interaction is postulated between the phosphoryl group of ATP and the transported calcium [13,382–388].

The conformational coupling mechanism is supported by the existence of structurally distinct conformational states of the Ca^{2+}-ATPase that differ in their affinities for ATP, Pi, and Ca^{2+} [14,15]. Furthermore, evidence obtained by chemical modification, site-specific mutagenesis, and intramolecular energy transfer suggests that the site responsible for the binding of ATP is located in the cytoplasmic domain of the Ca^{2+}-ATPase [11,45,49,87], while the high-affinity Ca^{2+} binding sites are presumed to be within the lipid bilayer [11,129–132]. The estimated distance of $\simeq 20$–50 Å between these two sites [160,296,297,389,390] would make conformational change an obligatory element of the coupling process. However, due to the scarcity of independent structural information and uncertainties about the localization and orientation of the probes used for energy transfer measurements, the current estimates of intramolecular distances are only tentative. The conformational changes that are assumed to occur during ATP hydrolysis and Ca^{2+} transport [388,391] appear to involve surprisingly small changes in the secondary structure of the Ca^{2+}-ATPase [162,392–398] or in the intramolecular distances determined by energy transfer [297,332,399] and may actually favor a direct coupling mechanism.

11.1. Intramolecular distances determined by fluorescence energy transfer

The shape and dimensions of the ATPase molecule (Fig. 4) determined by the three-dimensional reconstruction of two-dimensional Ca^{2+}-ATPase crystals [88–91,156, 157] can be compared with the distances between fluorophores bound at specified

TABLE II
Intramolecular distances between functional sites of the Ca^{2+}-ATPase determined by fluorescence energy transfer

Sites	Estimated distances	Reference
$Ca^{2+} \rightarrow Ca^{2+}$		
Li^+-NMR, H^+-NMR, Gd^{3+} EPR	16.1–18.0 Å (max.)	404, 405
$Tb^{3+} \rightarrow Pr^{3+}, Ho^{3+}, Er^{3+}$	8.0–10.8 Å	132, 390, 406, 407, 408, 409
$NTP \rightarrow Ca^{2+}$		
FITC $\rightarrow Nd^{3+}$	17–19 Å (min.)	389
FITC $\rightarrow Co^{3+}$	23–26 Å (min.)	389
FITC $\rightarrow Co^{3+}$	11–19 Å	417
$Tb^{3+} \rightarrow$ FITC	47 Å	390
$Tb^{3+} \rightarrow$ TNP-ATP	35 Å	390
Mg^{2+} subsite		
FITC $\rightarrow Nd^{3+}$	10 Å	400, 401
$Eu^{3+} \rightarrow$ CrATP	10 Å (max.)	407
$Eu^{3+} \rightarrow$ CrATP	22 Å	408
Eosin-Gd^{3+}	less than 12 Å	413
Cys 670 $\rightarrow$ NTP or Ca^{2+}		
IAEDANS–FITC	56 Å	296
IAEDANS–FITC	31 and 52 Å	332
IAEDANS–TNP-AMP	68 Å	296
IAEDANS–Pr^{3+}	16–18 Å	296
Ca^{2+}-site		
NCD quenching by doxylstearates	Ca site is within the bilayer	132, 410
NCD quenching by doxylstearate	Ca site $\simeq$ 20 Å above the surface of the bilayer	411
NCD quenching by FITC-PE	Ca site $\simeq$ 20 Å from lipid water interface	411
NCD $\rightarrow$ TNP-ATP	40–50 Å	132
$Tb^{3+} \rightarrow$ RITC–DPPE	Ca site is 10 Å from phospholipid headgroup	412
FITC site		
FITC $\rightarrow$ RITC–DPPE	34–42 Å (Tb-FITC distance by triangulation, 35–43 Å)	412, 414
FITC $\rightarrow$ RITC-PE	FITC site 60–80 Å from glycerol backbone region	414
FITC $\rightarrow$ ANS	51 Å	415
FITC $\rightarrow$ DPH	60 Å	415

(continued)

TABLE II
(continued)

Protein tryptophan		
Tryptophan → TNP-ATP	>30 Å	399
Tryptophan → IAEDANS	>30 Å	399
Intermolecular energy transfer		
ε-ADP → TNP-ADP	44 Å (distance between ATPase molecules)	403

Abbreviations: EPR, electron paramagnetic resonance; FITC, fluorescein-5'-isothiocyanate; IAEDANS, N-iodoacetyl-N'-(5-sulfo-1-naphthyl)ethylenediamine; NCD, fluorescent N-cyclohexyl-N'-(4-dimethyl-amino-α-naphthyl)carbodiimide; RITC, rhodamine-5'-isothiocyanate; DPPE, dipalmitoylphosphatidyl-ethanolamine; PE, egg phosphatidyl-ethanolamine; ANS, 8-anilino-1-naphthalene sulfonate; DPH, diphenylhexatriene; ε-ADP, 1,N^6-ethanoadenosine-5'-diphosphate; TNP-ADP, 2'[3']-O-(2,4,6-trinitrophe-nyl)adenosine-5'-diphosphate.

locations within the ATPase molecule determined by energy transfer (Table II).

Most of these distances are subject to considerable uncertainty due to the restricted mobility of the various fluorophores, as indicated by their relatively high polarization. Furthermore, the binding of lanthanides is not confined to the high-affinity Ca^{2+} binding sites but may include the Mg^{2+} binding subsite of the nucleotide domain [128,251,400,401], and other low-affinity divalent cation binding sites that are not directly involved in Ca^{2+} translocation [400,402]. The calculated intramolecular distances may also be distorted by intermolecular energy transfer between ATPase molecules [328,403] due to the high concentration of Ca^{2+}-ATPase in the membrane [10]. For example, the distance of 40–50 Å derived from energy transfer between two nucleotide site probes, such as fluorescein-5'-isothiocyanate and eosin-5'-isothiocyanate [328] or TNP-ADP and 1,N^6-ethenoadenosine-5'-diphosphate [403] probably represents the distance between the catalytic sites of adjacent ATPase monomers in an ATPase dimer. Therefore the energy transfer 'distances' can only be used as approximate indicators of the relative positions of the various sites within the ATPase molecule.

11.1.1. The location of the high-affinity Ca^{2+} binding site
There is good agreement that the two high-affinity Ca^{2+} sites are within ≃10 Å of each other (Table II) [132,390,404–409]. Their localization within the bilayer is supported by the observation [130,131] that site-specific mutagenesis of several amino acids within the putative transmembrane helices interferes with Ca^{2+} binding and with the Ca^{2+}-dependent phosphorylation of the enzyme by ATP, but has no effect on the Ca^{2+}-independent phosphorylation by inorganic phosphate.

Independent evidence for the localization of the Ca^{2+} binding site within the bilayer was obtained by labeling the Ca^{2+} binding site with the fluorescent N-

cyclohexyl-N'-(4-dimethylamino-α-naphthyl)carbodiimide (NCD-4) and measuring the quenching of NCD fluorescence by a series of doxylstearates (NS), in which the nitroxide group was located at different positions along the fatty acid chain [132,410]. After correction for the differences in the partitioning of the various derivatives between the lipid and water phase, the order of quenching efficiency was found to be 16-NS > 12-NS > 10-NS > 7-NS > 5-NS. The most efficient quenching of NCD fluorescence was obtained when the nitroxide group was on carbon 16 of the fatty acid chain, indicating that the NCD bound to the Ca^{2+}-ATPase was located in the membrane spanning domain near the middle of the bilayer [132]. The opposite order of quenching efficiency was obtained for the intrinsic tryptophan fluorescence (5-NS > 7-NS > 10-NS > 12-NS > 16-NS), suggesting that most of the tryptophan residues are near the surface of the bilayer, in agreement with the predicted structure of Ca^{2+}-ATPase [8].

Interestingly, the fluorescence of covalently bound NCD was also effectively quenched by water soluble quenchers, such as I or acrylamide and the accessibility of the aqueous quenchers was altered by the binding of Ca^{2+} to the Ca^{2+}-ATPase [132]. The simplest explanation of these observations is that the Ca^{2+} binding sites are located in a water-filled channel near the middle of the bilayer. Energy transfer between the bound NCD and TNP-ATP attached to the ATP binding site gave a distance of 40–50 Å between the two sites; this is consistent with distances obtained by energy transfer between the $Tb^{3+} \rightarrow$ TNP-ATP and $Tb^{3+} \rightarrow$ FITC pairs [390,406], where Tb^{3+} serves as a label for the Ca^{2+} binding site and TNP-ATP or FITC for the ATP binding site.

Munkonge et al. [411] obtained distances of $\simeq 20$ Å for the quenching of the fluorescence of NCD-Ca^{2+}-ATPase either by doxylstearate (5-NS), or by FITC-labeled phosphatidylethanolamine (FITC-PE) incorporated into the bilayer; although the authors concluded that the NCD label was located $\simeq 20$ Å *above* the surface of the bilayer, the data actually support the conclusion of Scott [132] placing the NCD $\simeq 20$ Å below the surface of the bilayer. The energy transfer distance of $\simeq 10$ Å for the Tb^{3+} RITC–DPPE pair [412] is also consistent with Tb^{3+} binding in the acyl chain region of the bilayer.

11.1.2. The ATP binding site
The estimated distances between lanthanides bound at various cation binding sites on the Ca^{2+}-ATPase, and FITC, TNP-ATP, Cr-ATP or eosin bound at the nucleotide binding site range between 10 and 47 Å [389,390,400,401,407–409,413] (Table II).

The longer distances (35–47 Å) are usually attributed to the distance between the high-affinity Ca^{2+} site and the ATP binding site [389–390]. This would be consistent with the distance of 40–50 Å obtained for the NCD–TNP-ATP pair [132]. Other evidence that places the ATP binding site high above the surface of the bilayer is the distance of 34–42 Å for the FITC–RITC–DPPE pair [412], the distance of 60–80 Å for the FITC–RITC-PE pair [414], the distance of 51 Å for the FITC–anilinonaphthalenesulfonate (ANS) pair [415] and the distance of 60 Å for the FITC–diphenylhexatriene (DPH) pair [415]. The distance of $\simeq 30$ Å obtained from the energy transfer between protein tryptophan located near the bilayer surface and TNP-ATP bound at the ATP binding site [399] is also consistent with the location of the ATP binding site in the main cytoplasmic domain.

The short distances of $\simeq 10$ Å for the FITC–Nd^{3+} [401] and Eu^{3+}–CrATP [407] pairs may be related to binding of lanthanides to a Mg^{2+} binding subsite of the Mg^{2+}-ATP binding pocket [251,388,400,401,409,416]. Although the existence of a Mg^{2+} binding subsite is well documented, the experimental distinction between the high-affinity Ca^{2+}/Ln^{3+} sites and other cation binding sites is difficult to achieve in the setting of energy transfer measurements. The <12 Å distance for the eosin–Gd^{3+} pair was obtained at a Gd^{3+} concentration of 1 mM [413] and almost certainly involves nonspecific cation binding sites.

The intermediate distances of 10–22 Å between the Eu^{3+} binding site and the catalytic site derived from the quenching of Eu^{3+} luminescence by Cr-ATP [407–409] do not fit readily into either interpretation, suggesting that the position of the Cr atom in the bound Cr-ATP may differ from that of fluorescein or trinitrophenol in the bound FITC or TNP-ATP. Similar uncertainty exists with respect to the 11–19-Å distance reported recently by Cuenda et al. [417] for the FITC–Co pair. These observations raise the possibility that the probes used for marking the putative ATP binding site do not refer to the exact position of ATP in the binding pocket. In fact the inhibition of ATPase activity by FITC can be overcome by high ATP concentration [117], suggesting that FITC merely decreases the affinity of ATP binding, but does not occupy the ATP binding site. Similarly, the trinitrophenyl moiety of TNP-ATP is almost certainly outside the ATP binding pocket. The structure stabilized by Cr-ATP [178] may be different from the structure of native Ca^{2+}-ATPase or the ATPase labeled with FITC or TNP-ATP. Considering the $\simeq 18$-Å length of the ATP molecule, the wide range of estimated distances between the Ca^{2+} and ATP binding sites can accommodate the structural requirements of either direct or indirect coupling mechanisms and more accurate data are needed to refine the choice between them.

11.1.3. The use of IAEDANS as reference point for distance measurements
The large ($\simeq 56$–68-Å) distance between AEDANS bound at cysteine 674 and FITC or TNP-ATP bound at the nucleotide site [296] is of some interest because the AEDANS site serves as a point of reference for other distance measurements (Table II). The only comparable distance in the cytoplasmic domain of the Ca^{2+}-ATPase is the $\simeq 60$-Å length of the pear shaped profile of the cytoplasmic mass shown in Fig. 4. We propose that the binding site for IAEDANS is near the tip of the lobe at a position marked IAEDANS in Fig. 4, while the site labeled by FITC or TNP-ATP is located at the opposite end of the cytoplasmic domain near the bridge at a site

marked FITC in Fig. 4. Such a location of the FITC site is consistent with the proposed distance of $\simeq 47$ Å between FITC bound at lysine 515 and Tb^{3+} bound at the high-affinity Ca^{2+} site [390]. To match the 16–18-Å distance between the AEDANS site and the Ca^{2+} binding site [296] the AEDANS site must be positioned on the lower surface of the lobe close to the cytoplasmic surface of the bilayer (position IAEDANS in Fig. 4).

The intramolecular distances measured at room temperature with the AEDANS–FITC pair were similar in the Ca_2E_1 and E_2V states [297]. Ca^{2+} and lanthanides are expected to stabilize the E_1 conformation of the Ca^{2+}-ATPase, since they induce a similar crystal form of Ca^{2+}-ATPase [119,157] and have similar effects on the tryptophan fluorescence [151] and on the trypsin sensitivity of Ca^{2+}-ATPase [119,120]. It is also likely that the vanadate-stabilized E_2V state is similar to the $E_2 \sim P$ state stabilized by Pi [418]. Therefore the absence of significant difference in the resonance energy transfer distances between the two states implies that the structural differences between the two conformations at sites recorded by currently available probes, fall within the considerable error of resonance energy transfer measurements. Even if these distances would vary by as much as 5 Å the difference between the two conformations could not be established reliably.

11.2. Thermal fluctuations in the structure of the Ca^{2+}-ATPase

Since Ca^{2+} is transferred from one side of the membrane to the other side in association with the Ca^{2+}-ATPase, thermal fluctuation of critical regions of the Ca^{2+}-ATPase influenced in specific ways through the phosphorylation of the enzyme by ATP may play a role in Ca^{2+} translocation. Similar ideas have been proposed some time ago by Huxley [419] in relationship to crossbridge movements during muscle contraction and by Welch and others on the role of protein fluctuations in enzyme action [420–430].

Temperature dependent changes in the normalized energy transfer efficiency [431,432] between site specific donor and acceptor fluorophores indeed reveal structural fluctuations in the Ca^{2+}-ATPase (Fig. 10) [297]. The temperature dependence of normalized energy transfer efficiency varied depending on the location of the donor–acceptor pairs in the ATPase molecule, suggesting the existence of flexibility gradients within the ATPase structure. For example, the normalized energy transfer efficiency between AEDANS and FITC increased by 13 and 20% in the E_1 and E_2V states, respectively, upon raising the temperature from 10° to 38°C [297]. As expected from the conformational coupling hypothesis, considerable thermal fluctuation of the structure was also evident in the energy transfer between FITC and Nd^{3+}, representing the region of the molecule connecting the ATP binding site (FITC) to the Ca^{2+} channel (Nd^{3+}). Some of this flexibility may be related to the relatively large distances between these donor–acceptor pairs. By contrast, the region monitored by the AEDANS–Pr^{3+} pair appears to be rigid, judged from the insensitivity of the

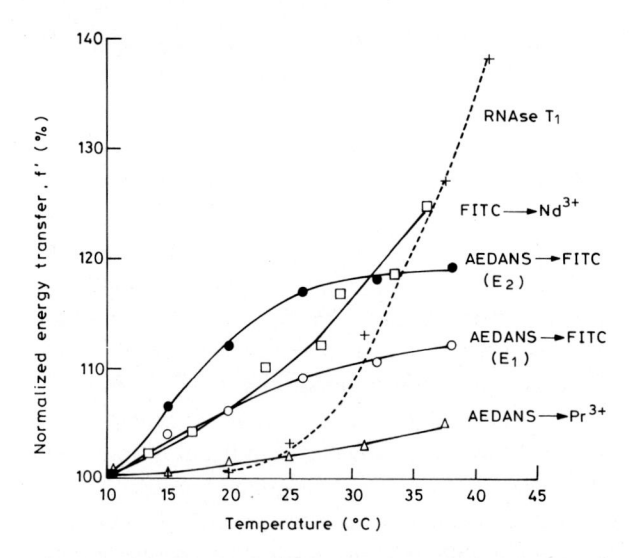

Fig. 10. Temperature dependence of normalized energy transfer, f', for different donor–acceptor systems. The normalized energy transfer, f', was calculated from the average energy transfer efficiencies $<E>$, determined from steady state intensity data. The f' value for data measured at 10°C was taken as 100%. Symbols: (○), energy transfer from AEDANS to FITC in the E_1 state (medium: 0.1 M KCl, 30 mM Tris-MOPS (pH 7.0), 5 mM $MgCl_2$ + 0.1 mM $CaCl_2$); (●), energy transfer from AEDANS to FITC in the E_2V state (medium: 0.1 M KCl, 30 mM Tris-MOPS (pH 7.0), 5 mM $MgCl_2$ + 0.1 mM EGTA, 0.5 mM Na_3VO_4); (□), energy transfer from FITC to Nd^{3+} (1 μM); (△), energy transfer from AEDANS to Pr^{3+} (1 μM); (+), data published by Somogyi et al. [431] on the soluble enzyme RNAse T_1, using tryptophan as donor and pyridoxamine-phosphate as acceptor. In this experiment the f' value determined at 20°C was taken as 100%. From Jona et al. [297].

normalized energy transfer efficiency to temperatures between 10–37°C (Fig. 10). The polarization of fluorescence of the noncovalently attached TNP-AMP sharply increased with increasing temperature, and as a result the energy transfer between AEDANS and TNP-AMP shows a more complex relationship [297].

The increased fluctuations of protein structures at elevated temperatures is reflected in the thermal expansion of proteins [433], that may contribute to the marked temperature dependence of the rate of Ca^{2+} transport.

Acknowledgements

This work was supported by research grants from the NIH (AR 26545), NSF (DMB 8823077; Int. 8617848) and the Muscular Dystrophy Association.

References

1. Martonosi, A. (1983) In: Muscle and Non-Muscle Motility (Stracher, A., Ed.), Vol. 1, pp. 233–357, Academic Press, New York.
2. Carafoli, E. (1987) Annu. Rev. Biochem. 56, 395–433.
3. Carafoli, E. (1991) Physiol. Rev. 71, 129–153.
4. Pedersen, P.L. and Carafoli, E. (1987) Trends Biochem. Sci. 12, 146–150.
5. Pedersen, P.L. and Carafoli, E. (1987) Trends Biochem. Sci. 12, 186–189.
6. Schatzmann, H.J. (1989) Annu. Rev. Physiol. 51, 473–485.
7. Pietrobon, D., Di Virgilio, F. and Pozzan, T. (1990) Eur. J. Biochem. 193, 599–622.
8. Brandl, C.J., Green, N.M., Korczak, B. and MacLennan, D.H. (1986) Cell 44, 597–607.
9. Brandl, C.J., Korczak, B., Green, N.M. and MacLennan, D.H. (1986) In: Muscle Development (Emerson, C., Nadal Ginard, B., Fischan, D. and Siddiqui, M.A.Q., Eds.), pp. 199–212, Alan R. Liss, New York.
10. Martonosi, A., Taylor, K.A., Varga, S., Ting-Beall, H.P. and Dux, L. (1987) In: Electron Microscopy of Proteins (Harris, J.R. and Horne, R.W., Eds.), Vol. 6, Membrane Structures, pp. 255–376, Academic Press, London.
11. MacLennan, D.H. (1990) Biophys. J. 58, 1355–1365.
12. Martonosi, A. and Beeler, T.J. (1983) Skeletal Muscle. In: Handbook of Physiology (Peachey, L.D. and Adrian, R.H., Eds.), pp. 417–485, American Physiological Society, Bethesda.
13. Inesi, G. and de Meis, L. (1985) In: The Enzymes of Biological Membranes (Martonosi, A., Ed.), Vol. 3, 2nd ed., pp. 157–191, Plenum, New York.
14. Jencks, W.P. (1989) Methods Enzymol. 171, 145–164.
15. Jencks, W.P. (1989) J. Biol. Chem. 264, 18855–18858.
16. Inesi, G., Sumbilla, C. and Kirtley, M.E. (1990) Physiol. Rev. 70, 749–760.
17. Catterall, W.A. (1988) Science 242, 50–61.
18. Hess, P. (1988) Can. J. Physiol. Pharmacol. 66, 1218–1223.
19. Glossmann, H. and Striessnig, J. (1988) Vitam. Horm. 44, 155–328.
20. Lauger, P. (1987) Physiol. Rev. 67, 1296–1331.
21. Lauger, P. (1987) J. Membr. Biol. 99, 1–11.
22. Kaczorowski, G.J., Slaughter, R.S., Garcia, M.L. and King, V.F. (1988) Biochem. Soc. Trans. 16, 529–532.
23. Kaczorowski, G.J., Slaughter, R.S., King, V.F. and Garcia, M.L. (1989) Biochim. Biophys. Acta 988, 287–302.
24. Sigel, E., Baur, R., Porzig, H. and Reuter, H. (1988) J. Biol. Chem. 263, 14614–14616.
25. Carafoli, E. (1988) In: Advances in Second Messenger and Phosphoprotein Research (Adelstein, R., Klee, C. and Rodbell, M., Eds.) Vol. 21, pp. 147–155, Raven Press, New York.
26. Blaustein, M.P. (1989) Curr. Top. Membr. Transp. 34, 289–330.
27. Reeves, J.P. (1990) In: Intracellular Calcium Regulation (Bronner, F., Ed.), pp. 305–347, Wiley-Liss, New York.
28. James, P., Zvaritch, E.I., Shakhparonov, M.I., Penniston, J.T. and Carafoli, E. (1987) Biochem. Biophys. Res. Commun. 149, 7–12.
29. James, P., Maeda, M., Fischer, R., Verma, A.K., Krebs, J., Penniston, J.T. and Carafoli, E. (1988) J. Biol. Chem. 263, 2905–2910.
30. Shull, G.E. and Greeb, J. (1988) J. Biol. Chem. 263, 8646–8657.
31. Verma, A.K., Filoteo, A.G., Stanford, D.R., Wieben, E.D., Penniston, J.T., Strehler, E.E., Fischer, R., Heim, R., Vogel, G., Mathews, S., Strehler-Page, M.-A., James, P., Vorherr, T., Krebs, J. and Carafoli, E. (1988) J. Biol. Chem. 263, 14152–14159.
32. Greeb, J. and Shull, G.E. (1989) J. Biol. Chem. 264, 18569–18576.
33. Strehler, E.E., James, P., Fischer, R., Heim, R., Vorherr, T., Filoteo, A.G., Penniston, J.T. and

Carafoli, E. (1990) J. Biol. Chem. 265, 2835–2842.
34. Strehler, E.E. (1991) J. Membr. Biol. 120, 1–15.
35. Martonosi, A., Taylor, K.A., Kracke, G., Dux, L. and Peracchia, C. (1985) In: Regulation and Development of Membrane Transport Processes (Graves, J.S., Ed.), pp. 57–85, John Wiley and Sons, New York.
36. Berridge, M.J. (1988) Proc. R. Soc. London B 234, 359–378.
37. Berridge, M.J. and Galione, A. (1988) FASEB J. 2, 3074–3082.
38. Hawthorne, J.N. (1988) Biochem. Soc. Trans. 16, 657–660.
39. Martonosi, A. (1982) Annu. Rev. Physiol. 44, 337–355.
40. Martonosi, A., Dux, L., Terjung, R.L. and Roufa, D. (1982) Ann. N.Y. Acad. Sci. 402, 485–514.
41. MacLennan, D.H., Zubrzycka–Gaarn, E. and Jorgensen, A.O. (1985) Curr. Top. Membr. Transp. 24, 337–368.
42. MacLennan, D.H., Brandl, C.J., Korczak, B. and Green, N.M. (1985) Nature (London) 316, 696–700.
43. Shull, G.E., Schwartz, A. and Lingrel, J.B. (1985) Nature (London) 316, 691–695.
44. Kawakami, K., Noguchi, S., Noda, M., Takahashi, H., Ohta, T., Kawamura, N., Nojima, H., Nagano, K., Hiroshe, T., Inayama, S., Hayashida, H., Miyata, T. and Numa, S. (1985) Nature (London) 316, 733–736.
45. Green, N.M., Taylor, W.R. and MacLennan, D.H. (1988) In: The Ion Pumps: Structure, Function and Regulation (Stein, W.D., Ed.), pp. 15–24, Alan R. Liss, New York.
46. Shull, G.E., Young, R.M., Greeb, J. and Lingrel, J.B. (1988) In: The Na^+,K^+-Pump. Part A: Molecular Aspects, Progress in Clinical and Biological Research (Skou, J.C., Norby, N.G., Maunsbach, A.B. and Esmann, M., Eds.), Vol. 258A, pp. 3–18, Alan R. Liss, New York.
47. Jorgensen, P.L. (1988) In: The Na^+, K^+-Pump. Part A: Molecular Aspects (Skou, J.C., Norby, J.G., Maunsbach, A.B. and Esmann, M., Eds.), pp. 19–38, Alan R. Liss, New York.
48. Green, N.M. and MacLennan, D.H. (1989) Biochem. Soc. Trans. 17, 819–822.
49. Taylor, W.R. and Green, N.M. (1989) Eur. J. Biochem. 179, 241–248.
50. Green, N.M. (1989) Biochem. Soc. Trans. 17, 970–972.
51. Shull, G.E. and Lingrel, J.B. (1986) J. Biol. Chem. 261, 16788–16791.
52. Maeda, M., Ishizaki, J. and Futai, M. (1988) Biochem. Biophys. Res. Commun. 157, 203–209.
53. Brandl, C.J., de Leon, S., Martin, S. and MacLennan, D.H. (1987) J. Biol. Chem. 262, 3768–3774.
54. Lytton, J. and MacLennan, D.H. (1988) J. Biol. Chem. 263, 15024–15031.
55. Gunteski-Humblin, A.-M., Greeb, J. and Shull, G.E. (1988) J. Biol. Chem. 263, 15032–15040.
56. Lytton, J., Zarain-Herzberg, A., Periasamy, M. and MacLennan, D.H. (1989) J. Biol. Chem. 264, 7059–7065.
57. Burk, S.E., Lytton, J., MacLennan, D.H. and Shull, G.E. (1989) J. Biol. Chem. 264, 18561–18568.
58. Korczak, B., Zarain-Herzberg, A., Brandl, C.J., Ingles, C.J., Green, N.M. and MacLennan, D.H. (1988) J. Biol. Chem. 263, 4813–4819.
59. Ohnoki, S. and Martonosi, A. (1980) Comp. Biochem. Physiol. B 65, 181–189.
60. Karin, N.J., Kaprielian, Z. and Fambrough, D.M. (1989) Mol. Cell. Biol. 9, 1978–1986.
61. MacLennan, D.H., Brandl, C.J., Champaneria, S., Holland, P.C., Powers, V.E. and Willard, H.F. (1987) Somat. Cell Mol. Genet. 13, 341–346.
62. Brody, I.A. (1964) Neurology 14, 1091–1100.
63. Brody, I.A. (1969) N. Engl. J. Med. 281, 187–192.
64. Karpati, G., Charuk, J., Carpenter, S., Jablecki, C. and Holland, P. (1986) Ann. Neurol. 20, 38–49.
65. Danon, M.J., Karpati, G., Charuk, J. and Holland, P. (1988) Neurology 38, 812–815.
66. Martonosi, A. (1989) Biochim. Biophys. Acta 991, 155–242.
67. Lompre, A.-M., de la Bastie, D., Boheler, K.R. and Schwartz, K. (1989) FEBS Lett. 249, 35–41.
68. Murakami, K., Tanabe, K. and Takada, S. (1990) J. Cell Sci. 97, 487–495.
69. Palmero, I. and Sastre, L. (1989) J. Mol. Biol. 210, 737–748.

70. Magyar, A. and Varadi, A. (1990) Biochem. Biophys. Res. Commun. 173, 872–877.
71. Tong, S.W. (1977) Biochem. Biophys. Res. Commun. 74, 1242–1248.
72. Tong, S.W. (1980) Arch. Biochem. Biophys. 203, 780–791.
73. Michalak, M., Famulski, K. and Carafoli, E. (1984) J. Biol. Chem. 259, 15540–15547.
74. Enyedi, A., Vorherr, T., James, P., McCormack, D.J., Filoteo, A.G., Carafoli, E. and Penniston, J.T. (1989) J. Biol. Chem. 264, 12313–12321.
75. Green, N.M. and Toms, E.J. (1985) Biochem. J. 231, 425–429.
76. Brandl, C.J., Fliegel, L. and MacLennan, D.H. (1988) Methods Enzymol. 157, 289–302.
77. Allen, G. (1977) In: Membrane Proteins (Nicholls, P., Møller, J.., Jorgensen, P.L. and Moody, A.J., Eds.), Proc. FEBS 11th, Copenhagen, 1977, pp. 159–168, Pergamon, New York.
78. Allen, G. (1980) Biochem. J. 187, 545–563.
79. Allen, G. (1980) Biochem. J. 187, 565–575.
80. Allen, G. and Green. N.M. (1976) FEBS Lett. 63, 188–192.
81. Allen, G. and Green, N.M. (1978) Biochem. J. 173, 393–402.
82. Allen, G., Bottomley, R.C. and Trinnaman, B.J. (1980) Biochem. J. 187, 577–589.
83. Allen, G., Trinnaman, B.J. and Green, N.M. (1980) Biochem. J. 187, 591–616.
84. Klip, A. and MacLennan, D.H. (1978) In: Frontiers in Biological Energetics (Dutton, L., Leigh, J. and Scarpa, A., Eds.), Vol. II, pp. 1137–1147, Academic, New York.
85. Klip, A., Reithmeier, R.A.F. and MacLennan, D.H. (1980) J. Biol. Chem. 255, 6562–6568.
86. Reithmeier, R.A.F., de Leon, S. and MacLennan, D.H. (1980) J. Biol. Chem. 255, 11839–11846.
87. Green, N.M., Taylor, W.R., Brandl, C., Korczak, B. and MacLennan, D.H. (1986) In: Calcium and the Cell, CIBA Fdn. Symp. 122, pp. 93–107, Wiley, New York.
88. Taylor, K.A., Dux, L. and Martonosi, A. (1984) J. Mol. Biol. 174, 193–204.
89. Taylor, K.A., Dux, L. and Martonosi, A. (1984) Biophys. J. 45, 145a.
90. Taylor, K.A., Dux, L. and Martonosi, A. (1986) J. Mol. Biol. 187, 417–427.
91. Taylor, K.A., Ho, M.H. and Martonosi, A. (1986) Ann. N.Y. Acad. Sci. 483, 31–43.
92. Martonosi, A., Varga, S., Taylor, K.A., Ho, M.H. and Dux, L. (1987) In: Perspectives of Biological Energy Transduction (Mukohata, Y., Morales, M.F. and Fleischer, S., Eds.), pp. 127–138, Academic Press, Tokyo.
93. Martonosi, A., Dux, L., Taylor, K.A., Csermely, P., Mullner, N., Pikula, S., Papp, S., Varga, S., Jona, I. and Keresztes, T. (1987) Acta Biochim. Biophys. Hung. 22, 263–276.
94. Green, N.M., Allen, G., Hebdon, G.M. and Thorley-Lawson, D.A. (1977) In: Calcium-Binding Proteins and Calcium Function (Wasserman, R.H., Corradino, R.A., Carafoli, E., Kretsinger, R.H., MacLennan, D.H. and Siegel, F.L., Eds.), pp. 164–172, North-Holland, New York.
95. Green, N.M., Allen, G. and Hebdon, G.M. (1980) Ann. N.Y. Acad. Sci. 358, 149–158.
96. Maruyama, K. and MacLennan, D.H. (1988) Proc. Natl. Acad. Sci. U.S.A. 85, 3314–3318.
97. Maruyama, K., Clarke, D.M., Fujii, J., Inesi, G., Loo, T.W. and MacLennan, D.H. (1989) J. Biol. Chem. 264, 13038–13042.
98. Mitchinson, C., Wilderspin, A.F., Trinnaman, B.J. and Green, N.M. (1982) FEBS Lett. 146, 87–92.
99. Yamamoto, H., Tagaya, M., Fukui, T. and Kawakita, M. (1988) J. Biochem. 103, 452–457.
100. Yamamoto, H., Imamura, Y., Tagaya, M., Fukui, T. and Kawakita, M. (1989) J. Biochem. (Tokyo) 106, 1121–1125.
101. Ohta, T., Nagano, K. and Yoshida, M. (1986) Proc. Natl. Acad. Sci. U.S.A. 83, 2071–2075.
102. Ovchinnikov, Y.A., Dzhandzugazyan, K.N., Lutsenko, S.V., Mustayev, A.A. and Modyanov, N.N. (1987) FEBS Lett. 217, 111–116.
103. Clarke, D.M., Loo, T.W. and MacLennan, D.H. (1990) J. Biol. Chem. 265, 22223–22227.
104. Vegh, M., Molnar, E. and Martonosi, A. (1990) Biochim. Biophys. Acta 1023, 168–183.
105. Molnar, E., Varga, S. and Martonosi, A. (1991) Biochim. Biophys. Acta 1068, 17–26.
106. McIntosh, D.B. and Ross, D.C. (1985) Biochemistry 24, 1244–1251.
107. Ross, D.C. and McIntosh, D.B. (1987) J. Biol. Chem. 262, 2042–2049.

108. Ross, D.C. and McIntosh, D.B. (1987) J. Biol. Chem. 262, 12977–12983.
109. Murphy, A.J. (1990) FEBS Lett. 263, 175–177.
110. Murphy, A.J. (1990) Biochemistry 29, 11236–11242.
111. Migala, A., Agostini, B. and Hasselbach, W. (1973) Z. Naturforsch. Teil C 28, 178–182.
112. MacLennan, D.H. and Reithmeier, R.A.F. (1982) In: Membranes and Transport (Martonosi, A., Ed.), Vol. 1, pp. 567–571, Plenum, New York.
113. Pick, U. (1981) FEBS Lett. 123, 131–136.
114. Pick, U. (1981) Eur. J. Biochem. 121, 187–195.
115. Pick, U. (1982) J. Biol. Chem. 257, 6111–6119.
116. Pick, U. and Karlish, S.J.D. (1982) J. Biol. Chem. 257, 6120–6126.
117. Champeil, P., Riollet, S., Orlowski, S., Guillain, F., Seebregts, C.J. and McIntosh, D.B. (1988) J. Biol. Chem. 263, 12288–12294.
118. Dux, L. and Martonosi, A. (1983) J. Biol. Chem. 258, 10111–10115.
119. Dux, L., Taylor, K.A., Ting-Beall, H.P. and Martonosi, A. (1985) J. Biol. Chem. 260, 11730–11743.
120. Dux, L., Papp, S. and Martonosi, A. (1985) J. Biol. Chem. 260, 13454–13458.
121. Andersen, J.P. and Jorgensen, P.L. (1985) J. Membr. Biol. 88, 187–198.
122. Scott, T.L. and Shamoo, A.E. (1982) J. Membr. Biol. 64, 137–144.
123. Torok, K., Trinnaman, B.J. and Green, N.M. (1988) Eur. J. Biochem. 173, 361–367.
124. Shoshan-Barmatz, V., Ouziel, N. and Chipman, D.M. (1987) J. Biol. Chem. 262, 11559–11564.
125. Shoshan-Barmatz, V. and Zarka, A. (1988) J. Biol. Chem. 263, 16772–16779.
126. Andersen, J.P., Vilsen, B., Leberer, E. and MacLennan, D.H. (1989) J. Biol. Chem. 264, 21018–21023.
127. Clarke, D.M., Loo, T.W. and MacLennan, D.H. (1990) J. Biol. Chem. 265, 14088–14092.
128. Asturias, F.J. and Blasie, J.K. (1991) Biophys. J. 59, 488–502.
129. Clarke, D.M., Maruyama, K., Loo, T.W., Leberer, E., Inesi, G. and MacLennan, D.H. (1989) J. Biol. Chem. 264, 11246–11251.
130. Clarke, D.M., Loo, T.W., Inesi, G. and MacLennan, D.H. (1989) Nature (London) 339, 476–478.
131. Clarke, D.M., Loo, T.W. and MacLennan, D.H. (1990) J. Biol. Chem. 265, 6262–6267.
132. Scott, T.L. (1988) Mol. Cell. Biochem. 82, 51–54.
133. Eisenberg, D. (1984) Annu. Rev. Biochem. 53, 595–623.
134. Serrano, R. (1989) Annu. Rev. Plant Physiol. & Plant Mol. Biol. 40, 61–94.
135. Reithmeier, R.A.F. and MacLennan, D.H. (1981) J. Biol. Chem. 256, 5957–5960.
136. MacLennan, D.H. and Reithmeier, R.A.F. (1985) In: Structure and Function of Sarcoplasmic Reticulum, Symposium on Structure and Function of Sarcoplasmic Reticulum, Kobe, Japan, November 1–4, 1982 (Fleischer, S. and Tonomura, Y., Eds.), pp. 91–100, Academic Press, Orlando.
137. Matthews, I., Colyer, J., Mata, A.M., Green, N.M., Sharma, R.P., Lee, A.G. and East, J.M. (1989) Biochem. Biophys. Res. Commun. 161, 683–688.
138. Matthews, I., Sharma, R.P., Lee, A.G. and East, J.M. (1990) J. Biol. Chem. 265, 18737–18740.
139. Clarke, D.M., Loo, T.W. and MacLennan, D.H. (1990) J. Biol. Chem. 265, 17405–17408.
140. Blasie, J.K., Pascolini, D., Asturias, F., Herbette, L.G., Pierce, D.H. and Scarpa, A. (1990) Biophys. J. 58, 687–693.
141. Stokes, D.L. and Green, N.M. (1990) Mol. Biol. 213, 529–538.
142. Stokes, D.L. and Green, N.M. (1990) Biochem. Soc. Trans. 18, 841–843.
143. Stromer, M. and Hasselbach, W. (1976) Z. Naturforsch. Teil C 31, 703–707.
144. Buchet, R., Varga, S., Seidler, N.W., Molnar, E. and Martonosi, A. (1991) Biochim. Biophys. Acta 1068, 201–216.
145. Henderson, R. and Unwin, P.N.T. (1975) Nature (London) 257, 28–32.
146. Ovchinnikov, Y.A. (1979) Eur. J. Biochem. 94, 321–336.
147. Henderson, R., Baldwin, J.M., Ceska, T.A., Zemlin, F., Beckmann, E. and Downing, K.H. (1990) J. Mol. Biol. 213, 899–929.

148. Ceska, T.A. and Henderson, R. (1990) J. Mol. Biol. 213, 539–560.
149. Coll, R.J. and Murphy, A.J. (1991) Biochemistry 30, 1456–1461.
150. Inesi, G. (1985) Annu. Rev. Physiol. 47, 573–601.
151. Jona, I. and Martonosi, A. (1986) Biochem. J. 234, 363–371.
152. Dux, L. and Martonosi, A. (1983) J. Biol. Chem. 258, 2599–2603.
153. Dux, L. and Martonosi, A. (1983) J. Biol. Chem. 258, 11896–11902.
154. Dux, L. and Martonosi, A. (1983) J. Biol. Chem. 258, 11903–11907.
155. Dux, L. and Martonosi, A. (1984) Eur. J. Biochem. 141, 43–49.
156. Taylor, K.A., Mullner, N., Pikula, S., Dux, L., Peracchia, C., Varga, S. and Martonosi, A. (1988) J. Biol. Chem. 263, 5287–5294.
157. Taylor, K.A., Dux, L., Varga, S., Ting-Beall, H.P. and Martonosi, A. (1988) Methods Enzymol. 157, 271–289.
158. Martonosi, A., Dux, L., Taylor, K.A., Ting-Beall, H.P., Varga, S., Csermely, P., Mullner, N., Papp, S. and Jona, I. (1987) In: Proteins and Excitable Membranes (Fambrough, D. and Hille, B., Eds), pp. 257–286, Wiley, New York.
159. Martonosi, A., Dux, L., Taylor, K.A. and Pikula, S. (1988) In: Ion Pumps: Structure, Function and Regulation (Stein, W., Ed.), pp. 7–14, Alan R. Liss, New York.
160. Martonosi, A.N., Jona, I., Molnar, E., Seidler, N.W., Buchet, R. and Varga, S. (1990) FEBS Lett. 268, 365–370.
161. Martonosi, A., Taylor, K.A. and Pikula, S. (1990) In: Crystallization of Membrane Proteins (Michel, H., Ed.), pp. 167–182, CRC Press, Boca Raton, FL.
162. Csermely, P., Katopis, C., Wallace, B.A. and Martonosi, A. (1987) Biochem. J. 241, 663–669.
163. Buhle, E.L., Jr., Knox, B.E., Serpersu, E. and Aebi, U. (1983) J. Ultrastruct. Res. 85, 186–203.
164. Aebi, U., Fowler, W.E., Buhle, L.E. Jr. and Smith, P.R. (1984) J. Ultrastruct. Res. 88, 143–176.
165. Scales, D.J. and Highsmith, S.R. (1984) Z. Naturforsch. Teil C 39, 177–179.
166. Maurer, A. and Fleischer, S. (1984) J. Bioenerg. Biomemb. 16, 491–505.
167. Highsmith, S., Barker, D. and Scales, D. (1985) Biochim. Biophys. Acta 817, 123–133.
168. Misra, M. and Malhotra, S.K. (1985) Biosci. Rep. 5, 551–558.
169. Misra, M. and Malhotra, S.K. (1986) Biosci. Rep. 6, 1065–1070.
170. Maurer, A., Tu, A.T. and Volpe, P. (1987) FEBS Lett. 224, 89–96.
171. Jorge-Garcia, I., Bigelow, D.J., Inesi, G. and Wade, J.B. (1988) Arch. Biochem. Biophys. 265, 82–90.
172. Peracchia, C., Dux, L. and Martonosi, A. (1984) J. Muscle Res. Cell Motil. 5, 431–442.
173. Franzini-Armstrong, C. and Ferguson, D.G. (1985) Biophys. J. 48, 607–615.
174. Ting-Beall, H.P., Dux, L. and Martonosi, A.N. (1986) Biophys. J. 49, 570a.
175. Ferguson, D.G., Franzini-Armstrong, C., Castellani, L., Hardwicke, P.M.D. and Kenney, L.J. (1985) Biophys. J. 48, 597–605.
176. Castellani, L., Hardwicke, P.M.D. and Vibert, P. (1985) J. Mol. Biol. 185, 579–594.
177. Ho, M.H., Taylor, K.A. and Martonosi, A. (1986) Biophys. J. 49, 570a.
178. Serpersu, E.H., Kirch, U. and Schoner, W. (1982) Eur. J. Biochem. 122, 347–354.
179. McPherson, A. (1990) In: Crystallization of Membrane Proteins (Michel, H., Ed.), pp. 1–51, CRC Press, Boca Raton, FL.
180. Zulauf, M. (1990) In: Crystallization of Membrane Proteins (Michel, H., Ed.), pp. 53–72, CRC Press, Boca Raton, FL.
181. Welte, W. and Wacker, T. (1990) In: Crystallization of Membrane Proteins (Michel, H., Ed.), pp. 107–123, CRC Press, Boca Raton, FL.
182. Pikula, S., Mullner, N., Dux, L. and Martonosi, A. (1988) J. Biol. Chem. 263, 5277–5286.
183. Dux, L., Pikula, S., Mullner, N. and Martonosi, A. (1987) J. Biol. Chem. 262, 6439–6442.
184. Varga, S. and Martonosi, A. (1991) Biophys. J. 59, 337a.
185. Varga, S., Taylor, K.A. and Martonosi, A. (1991) Biochim. Biophys. Acta 1070, 374–386.
186. Stokes, D.L. and Green, N.M. (1990) Biophys. J. 57, 1–14.

187. Herbette, L., De Foor, P., Fleischer, S., Pascolini, D., Scarpa, A. and Blasie, J.K. (1985) Biochim. Biophys. Acta 817, 103–122.
188. Herbette, L. (1986) In: Sarcoplasmic Reticulum in Muscle Physiology, (Entman, M.L. and Van Winkle, W.B., Eds.), Vol. 1, pp. 127–152, CRC Press, Boca Raton, FL.
189. Herbette, L.G. (1986) In: The Heart and Cardiovascular System, Scientific Foundations (Fozzard, H.A., Jennings, R.B., Haber, E., Katz, A.M. and Morgan, H.E., Eds.), Vol. 1, pp. 263–288, Raven Press, New York.
190. Blasie, J.K., Herbette, L., Pierce, D., Pascolini, D., Scarpa A. and Fleischer, S. (1982) Ann. N.Y. Acad. Sci. 402, 478–484.
191. Blasie, J.K., Pachence, J.M. and Herbette, L. (1984) In: Basic Life Sciences. Neutrons in Biology (Schoenborn, B.P., Ed.), Vol. 27, pp. 201–210, Plenum, New York.
192. Blasie, J.K., Herbette, L.G., Pascolini, D., Skita, V., Pierce, D.H. and Scarpa, A. (1985) Biophys. J. 48, 9–18.
193. Blasie, J.K., Herbette, L., Pierce, D.H., Pascolini, D., Skita, V., Scarpa, A. and Fleischer, S. (1985) In: Structure and Function of Sarcoplasmic Reticulum, Symposium on Structure and Function of Sarcoplasmic Reticulum, Kobe, Japan, November 1–4, 1982 (Fleischer, S. and Tonomura, Y., Eds.), pp. 51–62, Academic Press, Orlando.
194. Blasie, J.K., Herbette, L. and Pachence, J. (1985) J. Membr. Biol. 86, 1–7.
195. Blasie, J.K., Pascolini, D., Herbette, L., Pierce, D., Itshak, F., Skita, V. and Scarpa, A. (1986) Biophys. J. 49, 110–111.
196. Pascolini, D. and Blasie, J.K. (1988) Biophys. J. 54, 669–678.
197. Pascolini, D., Herbette, L.G., Skita, V., Asturias, F., Scarpa, A. and Blasie, J.K. (1988) Biophys. J. 54, 679–688.
198. Asturias, F.J. and Blasie, J.K. (1989) Biophys. J. 55, 739–753.
199. Asturias, F.J., Pascolini, D. and Blasie, J.K. (1990) Biophys. J. 58, 205–217.
200. Herbette, L., Marquardt, J., Scarpa, A. and Blasie, J.K. (1977) Biophys. J. 20, 245–272.
201. Pierce, D.H., Scarpa, A., Topp, M.R. and Blasie, J.K. (1983) Biochemistry 22, 5254–5261.
202. Pierce, D.H., Scarpa, A., Trentham, D.R., Topp, M.R. and Blasie, J.K. (1983) Biophys. J. 44, 365–373.
203. McCray, J.A., Herbette, L., Kihara, T. and Trentham, D.R. (1980) Proc. Natl. Acad. Sci. U.S.A. 77, 7237–7241.
204. Goldman, Y.E., Hibberd, M.G., McCray, J.A. and Trentham, D.R. (1982) Nature (London) 300, 701–705.
205. Hibberd, M.G., Goldman, Y.E. and Trentham, D.R. (1984) Curr. Top. Cell Regul. 24, 357–364.
206. Walker, J.W., Reid, G.P. and Trentham, D.R. (1989) Methods Enzymol. 172, 288–301.
207. Kaplan, J.H. and Somlyo, A.P. (1989) Trends Neurosci. 12, 54–59.
208. Homsher, E. and Millar, N.C. (1990) Annu. Rev. Physiol. 52, 875–896.
209. Herbette, L. and Blasie, J.K. (1980) In: Calcium-Binding Proteins: Structure and Function (Siegel, F.L., Carafoli, E., Kretsinger, R.H., MacLennan, D.H. and Wasserman, R.H., Eds.), pp. 115–120, Elsevier, Amsterdam.
210. Gangola, P. and Shamoo, A.E. (1986) J. Biol. Chem. 261, 8601–8603.
211. Degani, C. and Boyer, P.D. (1973) J. Biol. Chem. 248, 8222–8226.
212. Pick, U. and Bassilian, S. (1981) FEBS Lett. 123, 127–130.
213. Inesi, G. and Kirtley, M.E. (1990) J. Membr. Biol. 116, 1–8.
214. Kirtley, M.E., Sumbilla, C. and Inesi, G. (1990) Intracell. Calcium Regul. 249–270.
215. Vilsen, B., Andersen, J.P., Clarke, D.M. and MacLennan, D.H. (1989) J. Biol. Chem. 264, 21024–21030.
216. Singer, S.J. (1990) Annu. Rev. Cell Biol. 6, 247–296.
217. Singer, S.J. and Yaffe, M.P. (1990) Trends Biochem. Sci. 15, 369–373.
218. Kyte, J. and Doolittle, R.F. (1982) J. Mol. Biol. 157, 105–129.

219. Eisenberg, D., Schwarz, E., Komasomy, M. and Wall, R. (1984) J. Mol. Biol. 179, 125–142.
220. Engelman, D.M., Steitz, T.A. and Goldman, A. (1986) Annu. Rev. Biophys. Chem. 15, 321–353.
221. Fasman, G. (Ed.) (1989) Prediction of Protein Structure and the Principles of Protein Conformation, Plenum Press, New York.
222. Deisenhofer, J., Epp, O., Miki, K., Huber, R. and Michel, H. (1984) J. Mol. Biol. 180, 385–398.
223. Deisenhofer, J., Michel, H. and Huber, R. (1985) Trends Biochem. Sci. 10, 243–248.
224. Deisenhofer, J., Epp, O., Miki, K., Huber, R. and Michel, H. (1985) Nature (London) 318, 618–624.
225. Deisenhofer, J. and Michel, H. (1989) Angew. Chem. 28, 829–968.
226. Allen, J.P., Feher, G., Yeates, T.O., Komiya, H. and Rees, D.C. (1987) Proc. Natl. Acad. Sci. U.S.A. 84, 5730–5734.
227. Rees, D.C., Komiya, H., Yeates, T.O., Allen, J.P. and Feher, G. (1989) Annu. Rev. Biochem. 58, 607–633.
228. Rees, D.C., De Antonio, L. and Eisenberg, D. (1989) Science 245, 510–513.
229. Lodish, H.M. (1988) Trends Biochem. Sci. 13, 332–334.
230. McCrea, P.D., Engelman, D.M. and Popot, J.-L. (1988) Trends Biochem. Sci. 13, 289–290.
231. Jennings, M.L. (1989) Annu. Rev. Biochem. 58, 999–1027.
232. Fasman, G.D. and Gilbert, W.A. (1990) Trends Biochem. Sci. 15, 89–92.
233. Jahnig, F. (1990) Trends Biochem. Sci. 15, 93–95.
234. Ovchinnikov, Y.A. (1987) Trends Biochem. Sci. 12, 434–438.
235. Imamura, Y. and Kawakita, M. (1986) J. Biochem. (Tokyo) 100, 133–141.
236. Imamura, Y. and Kawakita, M. (1989) J. Biochem. 105, 775–781.
237. Thorley-Lawson, D.A. and Green, N.M. (1973) Eur. J. Biochem. 40, 403–413.
238. Thorley-Lawson, D.A. and Green, N.M. (1975) Eur. J. Biochem. 59, 193–200.
239. Thorley-Lawson, D.A. and Green, N.M. (1977) Biochem. J. 167, 739–748.
240. Stewart, P.S. and MacLennan, D.H. (1974) J. Biol. Chem. 249, 985–993.
241. Green, N.M., Hardwicke, P.M.D. and Thorley-Lawson, D.A. (1974) Biochem. Soc. Symp. 39, 111–113.
242. Inesi, G. and Scales, D. (1974) Biochemistry 13, 3298–3306.
243. Sarzala, M.G., Zubrzycka, E. and Drabikowski, W. (1974) In: Calcium Binding Proteins (Drabikowski, W., Strzelecka-Golaszewska, H. and Carafoli, E., Eds.), pp. 315–346, PWN, Polish Sci. Publ., Warsaw.
244. Louis, C.F., Buonaffina, R. and Binks, B. (1974) Arch. Biochem. Biophys. 161, 83–92.
245. Louis, C.F. and Irving, I. (1974) Biochim. Biophys. Acta 365, 193–204.
246. Stewart, P.S., MacLennan, D.H. and Shamoo, A.E. (1976) J. Biol. Chem. 251, 712-719.
247. MacLennan, D.H., Khanna, V.K. and Stewart, P.S. (1976) J. Biol. Chem. 251, 7271–7274.
248. Rizzolo, L.J. and Tanford, C. (1978) Biochemistry 17, 4044–4048.
249. Rizzolo, L.J. and Tanford, C. (1978) Biochemistry 17, 4049–4055.
250. Kirchberger, M.A., Borchman, D. and Kasinathan, C. (1986) Biochemistry 25, 5484–5492.
251. Girardet, J.-L., Dupont, Y. and Lacapere, J.-J. (1989) Eur. J. Biochem. 184, 131–140.
252. Andersen, J.P., Vilsen, B., Collins, J.H. and Jorgensen, P.L. (1986) J. Membr. Biol. 93, 85–92.
253. Vilsen, B. and Andersen, J.P. (1986) Biochim. Biophys. Acta 855, 429–431.
254. Saito, K., Imamura, Y. and Kawakita, M. (1984) J. Biochem. 95, 1297–1304.
255. Imamura, Y., Saito, K. and Kawakita, M. (1984) J. Biochem. (Tokyo) 95, 1305–1313.
256. de Ancos, J.G. and Inesi, G. (1988) Biochemistry 27, 1793–1803.
257. Saito-Nakatsuka, K., Yamashita, T., Kubota, I. and Kawakita, M. (1987) J. Biochem. (Tokyo) 101, 365–376.
258. Le Maire, M., Lund, S., Viel, A., Champeil, P. and Moller, J.V. (1990) J. Biol. Chem. 265, 1111–1123.
259. Markus, S., Priel, Z. and Chipman, D.M. (1986) Biochim. Biophys. Acta 874, 128–135.
260. Markus, S., Priel, Z. and Chipman, D.M. (1989) Biochemistry 28, 793–799.

261. Vanderkooi, J.M., Papp, S., Pikula, S. and Martonosi, A. (1988) Biochim. Biophys. Acta 957, 230–236.
262. Jorgensen, A.O., Kalnins, V. and MacLennan, D.H. (1979) J. Cell Biol. 80, 372–384.
263. Jorgensen, A.O., Shen, A.C.Y., MacLennan, D.H. and Tokuyasu, K.T. (1982) J. Cell Biol. 92, 409–416.
264. Jorgensen, A.O., Shen, A.C.-Y., Daly, P. and MacLennan, D.H. (1982) J. Cell Biol. 93, 883–892.
265. Jorgensen, A.O., Arnold, W., Pepper, D.R., Kahl, S.D., Mandel, F. and Campbell, K.P. (1988) Cell Motil. Cytoskel. 9, 164–174.
266. Dulhunty, A.F., Banyard, M.R.C. and Medveczky, C.J. (1987) J. Membr. Biol. 99, 79–92.
267. Kaprielian, Z. and Fambrough, D.M. (1987) Develop. Biol. 124, 490–503.
268. Maier, A., Leberer, E. and Pette, D. (1988) Histochemistry 88, 273–276.
269. Ohlendieck, K. and Ryan, N.M. (1989) Biochem. Soc. Trans. 17, 673–674.
270. Krenacs, T., Molnar, E., Dobo, E. and Dux, L. (1989) Histochem. J. 21, 145–155.
271. De Foor, P.H., Levitsky, D., Biryukova, T. and Fleischer, S. (1980) Arch. Biochem. Biophys. 200, 196–205.
272. Damiani, E., Betto, R., Salvatori, S., Volpe, P., Salviati, G. and Margreth, A. (1981) Biochem. J. 197, 245–248.
273. Damiani, E., Margreth, A., Furlan, A., Dahms, A.S., Arnn, J. and Sabbadini, R.A. (1987) J. Cell Biol. 104, 461–472.
274. Zubrzycka-Gaarn, E., MacDonald, G., Phillips, L., Jorgensen, A.O. and MacLennan, D.H. (1984) J. Bioenerg. Biomembr. 16, 441–464.
275. Zubrzycka-Gaarn, E., Phillips, L. and MacLennan, D.H. (1984) In: Epithelial Calcium and Phosphate Transport. Molecular and Cellular Aspects, 2nd International Workshop on Calcium and Phosphate Across Biomembranes, Vienna, March 5–7, 1984, Progress in Clinical and Biological Research (Bronner, F. and Peterlik, M., Eds.), Vol. 168, pp. 19–23, Alan R. Liss, New York.
276. Wuytack, F., Raeymaekers, L., Verbist, J., Hartweg, D., Gietzen, K. and Casteels, R. (1987) Arch. Int. Physiol. Biochim. 95, 3.
277. Wuytack, F., Kanmura, Y., Eggermont, J.A., Raeymaekers, L., Verbist, J., Hartweg, D., Gietzen, K. and Casteels, R. (1989) Biochem. J. 257, 117–123.
278. Wuytack, F., Eggermont, J.A., Raeymaekers, L., Plessers, L. and Casteels, R. (1989) Biochem. J. 264, 765–769.
279. Levitsky, D.O., Syrbu, S.I., Cherepakhin, V., Rokhlin, O.V. and Popovich, M.I. (1986) Biol. Memb. 3, 101–107.
280. Levitsky, D.O., Syrbu, S.I., Cherepakhin, V.V. and Rokhlin, O.V. (1987) Eur. J. Biochem. 164, 477–484.
281. Levitsky, D.O., Syrbu, S.I., Cherepakhin, V.V. and Rokhlin, O.V. (1987) Biomed. Biochim. Acta 46, S382–S387.
282. Levitsky, D.O., Benevolensky, D.S., Ikemoto, N., Syrbu, S.I. and Watras, J. (1989) J. Mol. Cell. Cardiol. 12 (Suppl. 1), 55–58.
283. Grover, A.K., Boonstra, J., Garfield, R.E. and Campbell, K.P. (1988) Biochem. Arch. 4, 169–179.
284. Sarkadi, B., Enyedi, A., Penniston, J.T., Verma, K.A., Dux, L., Molnar, E. and Gardos, G. (1988) Biochim. Biophys. Acta 939, 40–46.
285. Molnar, E., Seidler, N.W., Jona, I. and Martonosi, A.N. (1990) Biochim. Biophys. Acta 1023, 147–167.
286. Dux, L., Green, H.J. and Pette, D. (1990) Eur. J. Biochem. 192, 95–100.
287. Wuytack, F., De Schutter, G., Verbist, J. and Casteels, R. (1983) FEBS Lett. 154, 191–195.
288. Grover, A.K. (1988) J. Biol. Chem. 263, 19510–19512.
289. Caride, A.J., Gorski, J.P. and Penniston, J.T. (1988) Biochem. J. 255, 663–670.
290. Verbist, J., Wuytack, F., Raeymaekers, L. and Casteels, R. (1985) Biochem. J. 231, 737–742.
291. Holland, P.C. and MacLennan, D.H. (1976) J. Biol. Chem. 251, 2030–2036.

292. Ha, D.B., Boland, R. and Martonosi, A. (1979) Biochim. Biophys. Acta 585, 165–187.
293. Betto, R., Damiani, E., Biral, D. and Mussini, I. (1981) J. Immunol. Methods 46, 289–298.
294. Leberer, E. and Pette, D. (1986) Eur. J. Biochem. 156, 489–496.
295. Eggermont, J.A., Wuytack, F., Verbist, J. and Casteels, R. (1990) Biochem. J. 271, 649–653.
296. Squier, T.C., Bigelow, D.J., Garcia de Ancos, J. and Inesi, G. (1987) J. Biol. Chem. 262, 4748–4754.
297. Jona, I., Matko, J. and Martonosi, A. (1990) Biochim. Biophys. Acta 1023, 183–199.
298. Greenway, D.C. and MacLennan, D.H. (1978) Can. J. Biochem. 56, 452–456.
299. Chyn, T.L., Martonosi, A.N., Morimoto, T. and Sabatini, D.D. (1979) Proc. Natl. Acad. Sci. U.S.A. 76, 1241–1245.
300. Mostov, K.E., Defoor, P., Fleischer, S. and Blobel, G. (1981) Nature (London) 292, 87–88.
301. Matthews, I., Colyer, J., Mata, A.M., Lee, A.G., Green, N.M. and East, J.M. (1989) Biochem. Soc. Trans. 17, 708–709.
302. Colyer, J., Mata, A.M., Lee, A.G. and East, J.M. (1989) Biochem. J. 262, 439–447.
303. Molnar, E., Varga, S., Jona, I., Seidler, N.W. and Martonosi, A. (1992) Biochim. Biophys. Acta 1103, 281–295.
304. Tunwell, R.E.A., O'Connor, C.D., Mata, A.M., East, J.M. and Lee, A.G. (1991) Biochim. Biophys. Acta 1073, 585–592.
305. Sumida, M. and Sasaki, S. (1975) J. Biochem. (Tokyo) 78, 757–762.
306. Martonosi, A. and Fortier, F. (1974) Biochem. Biophys. Res. Commun. 60, 382–389.
307. Martonosi, A., Jilka, R.L. and Fortier, F. (1974) In: Membrane Proteins in Transport and Phosphorylation (Azzone, G.F., Klingenberg, E.M., Quagliariello, E. and Siliprandi, N., Eds.), pp. 113–124, North-Holland, Amsterdam.
308. Martonosi, A. (1975) In: Biomembranes – Lipids, Proteins and Receptors, Proc. of a NATO Advanced Study Institute (Burton, R.M. and Packer, L., Eds.), pp. 369–390, Bi-Science Publications Div., Webster Groves, Missouri.
309. Mata, A.M., Michelangeli, F., Lee, A.G. and East, J.M. (1989) Biochem. Soc. Trans. 18, 603.
310. Ikemoto, N. (1982) Annu. Rev. Physiol. 44, 297–317.
311. Hasselbach, W. and Elfvin, L.G. (1967) J. Ultrastruct. Res. 17, 598–622.
312. Hasselbach, W. (1978) Biochim. Biophys. Acta 515, 23–53.
313. Agostini, B. and Hasselbach, W. (1971) J. Submicr. Cytol. 3, 231–238.
314. Agostini, B. and Hasselbach, W. (1971) Quaderni Sclavo Diagn. 7, 406–428.
315. Yamada, S. and Ikemoto, N. (1978) J. Biol. Chem. 253, 6801–6807.
316. Kawakita, M., Yasuoka, K. and Kaziro, Y. (1980) J. Biochem. 87, 609–617.
317. Ikemoto, N., Morgan, J.F. and Yamada, S. (1978) J. Biol. Chem. 253, 8027–8033.
318. Kawakita, M. and Yamashita, T. (1987) J. Biochem. (Tokyo) 102, 103–109.
319. Kawakita, M., Yasuoka-Yabe, K., Saito-Nakatsuka, K., Baba, A. and Yamashita, T. (1988) Methods Enzymol. 157, 251–261.
320. Yamashita, T. and Kawakita, M. (1987) J. Biochem. (Tokyo) 101, 377–385.
321. Suzuki, H., Obara, M., Kuwayama, H. and Kanazawa, T. (1987) J. Biol. Chem. 262, 15448–15456.
322. Vanderkooi, J.M., Ierokomos, A., Nakamura, H. and Martonosi, A. (1977) Biochemistry 16, 1262–1267.
323. Martonosi, A., Nakamura, H., Jilka, R.L. and Vanderkooi, J.M. (1977) In: Biochemistry of Membrane Transport (Semenza, G. and Carafoli, E., Eds.), pp. 401–415, Springer, Berlin.
324. Gingold, M.P., Rigaud, J.L. and Champeil, P. (1981) Biochimie 63, 923–925.
325. Champeil, P., Rigaud, J.L. and Gingold, M.P. (1982) Z. Naturforsch. 37c, 513–516.
326. Watanabe, T. and Inesi, G. (1982) Biochemistry 21, 3254–3259.
327. Yantorno, R.E., Yamamoto, T. and Tonomura, Y. (1983) J. Biochem. (Tokyo) 94, 1137–1145.
328. Papp, S., Pikula, S. and Martonosi, A. (1987) Biophys. J. 51, 205–220.
329. Baba, A., Nakamura, T. and Kawakita, M. (1986) J. Biochem. 100, 1137–1147.
330. Suzuki, H., Obara, M., Kubo, K. and Kanazawa, T. (1989) J. Biol. Chem. 264, 920–927.

331. Obara, M., Suzuki, H. and Kanazawa, T. (1988) J. Biol. Chem. 263, 3690–3697.
332. Birmachu, W., Nisswandt, F.L. and Thomas, D.D. (1989) Biochemistry 28, 3940–3947.
333. Kubo, K., Suzuki, H. and Kanazawa, T. (1990) Biochim. Biophys. Acta 1040, 251–259.
334. Bishop, J.E., Squier, T.C., Bigelow, D.J. and Inesi, G. (1988) Biochemistry 27, 5233–5240.
335. Wakabayashi, S., Imagawa, T. and Shigekawa, M. (1990) J. Biochem. 107, 563–571.
336. Wakabayashi, S., Ogurusu, T. and Shigekawa, M. (1990) Biochemistry 29, 10613–10620.
337. Wakabayashi, S. and Shigekawa, M. (1990) Biochemistry 29, 7309–7318.
338. Kison, R., Meyer, H.E. and Schoner, W. (1989) Eur. J. Biochem. 181, 503–511.
339. Pick, U. and Karlish, S.J.D. (1980) Biochim. Biophys. Acta 626, 255–261.
340. Murphy, A.J. (1988) Biochim. Biophys. Acta 946, 57–65.
341. Murphy, A.J. (1977) Arch. Biochem. Biophys. 180, 114–120.
342. Kawakita, M., Yasuoka, K. and Kaziro, Y. (1979) In: Cation Flux Across Biomembranes (Mukohata, Y. and Packer, L., Eds.), pp. 119–124, Academic, New York.
343. Robinson, J.D. (1984) J. Bioenerg. Biomemb. 16, 195–207.
344. Shoshan-Barmatz, V. (1986) Biochem. J. 240, 509–517.
345. Vale, M.G.P. (1977) Biochim. Biophys. Acta 471, 39–48.
346. Vale, M.G.P. and Carvalho, A.P. (1980) Biochim. Biophys. Acta 601, 620–629.
347. Hidalgo, C. (1980) Biochem. Biophys. Res. Commun. 92, 757–765.
348. Hidalgo, C., Petrucci, D.A. and Vergara, C. (1982) J. Biol. Chem. 257, 208–216.
349. Shoshan-Barmatz, V. (1987) Biochem. J. 243, 165–173.
350. Mitsova, I.Z., Tat'yanenko, L.V., Vasyukova, N.V. and Gvosdev, R.I. (1980) Mol. Biol. 14, 206–210.
351. Pai, E.E., Sachsenheimer, W., Schirmer, R.H. and Schulz, G.E. (1977) J. Mol. Biol. 114, 37–45.
352. Soman, G., Chang, Y.C. and Graves, D.J. (1983) Biochemistry 22, 4994–5000.
353. Pedemonte, C.H. and Kaplan, J.H. (1990) Am. J. Physiol. 258, C1–C23.
354. Satre, M., Lunardi, J., Dianoux, A.-C., Dupuis, A., Issartel, J.P., Klein, G., Pougeois, R. and Vignais, P.V. (1986) Methods Enzymol. 126, 712–732.
355. Riordan, J.F., McElvany, K.D. and Borders, C.L. (1977) Science 195, 884–886.
356. Murphy, A.J. (1976) Biochem. Biophys. Res. Commun. 70, 1048–1054.
357. Varga, S., Csermely, P., Mullner, N., Dux, L. and Martonosi, A. (1987) Biochim. Biophys. Acta 896, 187–195.
358. Bishop, J.E. (1989) Biophys. J. 55, 204a.
359. Coffey, R.L., Lagwinska, E., Oliver, M. and Martonosi, A.N. (1975) Arch. Biochem. Biophys. 170, 37–48.
360. Hill, T.L. and Inesi, G. (1982) Proc. Natl. Acad. Sci. U.S.A. 79, 3978–3982.
361. Inesi, G. and Hill, T.L. (1983) Biophys. J. 44, 271–280.
362. Yu, B.P., Masoro, E.J. and De Martinis, F.D. (1967) Nature (London) 216, 822–824.
363. Yu, B.P., Masoro, E.J. and Bertrand, H.A. (1974) Biochemistry 13, 5083–5087.
364. Martonosi, A., Boland, R. and Halpin, R.A. (1972) Cold Spring Harbor Symp. Quant. Biol. 37, 455–468.
365. Masoro, E.J. and Yu, B.P. (1974) Recent Adv. Stud. Cardiac Struct. Metab. 4, 495–506.
366. Tenu, J.P., Ghelis, C., Leger, D.S., Carrette, J. and Chevallier, J. (1976) J. Biol. Chem. 251, 4322–4329.
367. Tenu, J.P., Ghelis, C., Yon, J. and Chevallier, J. (1976) Biochimie 58, 513–519.
368. Coan, C. and DiCarlo, R. (1990) J. Biol. Chem. 265, 5376–5384.
369. Azzi, A., Casey, R.P. and Nalecz, M.J. (1984) Biochim. Biophys. Acta 768, 209–226.
370. Nalecz, M.J., Casey, R.P. and Azzi, A. (1986) Methods Enzymol. 125, 86–108.
371. Carraway, K.L. and Koshland, D.E., Jr. (1972) Methods Enzymol. 25, 616–623.
372. Pick, U. and Racker, E. (1979) Biochemistry 18, 108–113.
373. Murphy, A.J. (1981) J. Biol. Chem. 256, 12046–12050.
374. Chiesi, M. and Inesi, G. (1980) Biochemistry 19, 2912–2918.

375. Scofano, H.M., Barrabin, H., Lewis, D. and Inesi, G. (1985) Biochemistry 24, 1025–1029.
376. Chadwick, C.C. and Thomas, E.W. (1983) Biochim. Biophys. Acta 730, 201–206.
377. Chadwick, C.C. and Thomas, E.W. (1984) Biochim. Biophys. Acta 769, 291–296.
378. Chadwick, C.C. and Thomas, E.W. (1985) Biochim. Biophys. Acta 827, 419–423.
379. Pick, U. and Weiss, M. (1985) Eur. J. Biochem. 152, 83–89.
380. Bill, E., Gutowski, Z. and Baumert, H.G. (1988) Eur. J. Biochem. 176, 119–124.
381. Tanford, C. (1985) In: Structure and Function of Sarcoplasmic Reticulum (Fleischer, S. and Tonomura, Y., Eds.), pp. 259–275, Academic Press, Orlando.
382. Mitchell, P. (1979) Eur. J. Biochem. 95, 1–20.
383. Mitchell, P. (1985) J. Biochem. 97, 1–18.
384. Mitchell, P. and Koppenol, W.H. (1982) Ann. N.Y. Acad. Sci. 402, 584–601.
385. Scarborough, G.A. (1982) Ann. N.Y. Acad. Sci. 402, 99–115.
386. Scarborough, G.A. (1985) Microbiol. Rev. 49, 214–231.
387. Dupont, Y. (1983) FEBS Lett. 161, 14–20.
388. Dupont, Y., Bennett, N., Pougeois, R. and Lacapere, J.-J. (1985) In: Structure and Function of Sarcoplasmic Reticulum (Fleischer, S. and Tonomura, Y., Eds.), pp. 225–248, Academic Press, Orlando.
389. Highsmith, S. and Murphy, A.J. (1984) J. Biol. Chem. 259, 14651–14656.
390. Scott, T.L. (1985) J. Biol. Chem. 260, 14421–14423.
391. Dupont, Y., Guillain, F. and Lacapere, J.J. (1988) Methods Enzymol. 157, 206–219.
392. Nakamoto, R.K. and Inesi, G. (1986) FEBS Lett. 194, 258–262.
393. Arrondo, J.L.R., Mantsch, H.H., Mullner, N., Pikula, S. and Martonosi, A. (1987) J. Biol. Chem. 262, 9037–9043.
394. Villalain, J., Gomez-Fernandez, J.C., Jackson, M. and Chapman, D. (1989) Biochim. Biophys. Acta 978, 305–312.
395. Buchet, R., Jona, I. and Martonosi, A. (1989) Biochim. Biophys. Acta 983, 167–178.
396. Buchet, R., Carrier, D., Wong, P.T.T., Jona, I. and Martonosi, A. (1990) Biochim. Biophys. Acta 1023, 107–118.
397. Barth, A., Kreutz, W. and Mantele, W. (1990) FEBS Lett. 277, 147–150.
398. Barth, A., Kreutz, W. and Mantele, W. (1991) Biophys. J. 59, 339a.
399. Gryczynski, I., Wiczk, W., Inesi, G., Squier, T. and Lakowicz, J.R. (1989) Biochemistry 28, 3490–3498.
400. Highsmith, S.R. and Head, M.R. (1983) J. Biol. Chem. 258, 6858–6862.
401. Highsmith, S. (1984) Biochem. Biophys. Res. Commun. 124, 183–189.
402. Itoh, N. and Kawakita, M. (1984) J. Biochem. 95, 661–669.
403. White, T.E. and Dewey, T.G. (1987) Memb. Biochem. 7, 67–72.
404. Stephens, E.M. and Grisham, C.M. (1978) Fed. Proc. 37, 1483.
405. Stephens, E.M. and Grisham, C.M. (1979) Biochemistry 18, 4876–4885.
406. Scott, T.L. (1984) J. Biol. Chem. 259, 4035–4037.
407. Herrmann, T.R., Gangola, P. and Shamoo, A E. (1986) Eur. J. Biochem. 158, 555–560.
408. Joshi, N.B. and Shamoo, A.E. (1988) Eur. J. Biochem. 178, 483–487.
409. Herrmann, T.R. and Shamoo, A.E. (1988) Mol. Cell. Biochem. 82, 55–58.
410. Scott, T.L. (1986) Biophys. J. 49, 234a.
411. Munkonge, F., East, J.M. and Lee, A.G. (1989) Biochim. Biophys. Acta 979, 113–120.
412. Teruel, J.A. and Gomez-Fernandez, J.C. (1986) Biochim. Biophys. Acta 863, 178–184.
413. Kotel'nikova, R.A., Tat'yanenko, L.V., Mekler, V.M. and Kotel'nikov, A.I. (1982) Mol. Biol. (Engl. Trans.) 16, 949–954.
414. Gutierrez–Merino, C., Munkonge, F., Mata, A.M., East, J.M., Levinson, B.L., Napier, R.M. and Lee, A.G. (1987) Biochim. Biophys. Acta 897, 207–216.
415. Teruel, J.A. and Gomez-Fernandez, J.C. (1987) Biochem. Int. 14, 409–416.

416. Dupont, Y., Pougeois, R., Ronjat, M. and Verjovsky-Almeida, S. (1985) J. Biol. Chem. 260, 7241–7249.
417. Cuenda, A., Henao, F. and Gutierrez-Merino, C. (1990) Eur. J. Biochem. 194, 663–670.
418. De Meis, L. (1989) Biochim. Biophys. Acta 973, 333–349.
419. Huxley, A.F. (1957) Prog. Biophys. Biophys. Chem. 7, 255–318.
420. Welch, G.R., Somogyi, B. and Damjanovich, S. (1982) Prog. Biophys. Mol. Biol. 39, 109-146.
421. Welch, G.R. (Ed.) (1985) The Fluctuating Enzyme, John Wiley and Sons, New York.
422. Welch, G.R. and Kell, D.B. (1986) In: The Fluctuating Enzyme (Welch, G.R., Ed.), pp. 451–492, John Wiley & Sons, New York.
423. Cooper, A. (1984) Prog. Biophys. Mol. Biol. 44, 181–214.
424. Karplus, M. (1985) Ann. N.Y. Acad. Sci. 439, 107–123.
425. Chothia, C. and Lesk, A.M. (1985) Trends Biochem. Sci. 10, 116–118.
426. Somogyi, B., Welch, G.R. and Damjanovich, S. (1984) Biochim. Biophys. Acta 768, 81–112.
427. Somogyi, B. and Damjanovich, S. (1986) In: The Fluctuating Enzyme (Welch, G.R., Ed.), pp. 341–368, John Wiley & Sons, New York.
428. Ikegami, A. (1986) In: The Fluctuating Enzyme (Welch, G.R., Ed.), pp. 191–226, John Wiley & Sons, New York.
429. Karplus, M., Brunger, A.T., Elber, R. and Kuriyan, J. (1987) Cold Spring Harbor Symp. Quant. Biol. 52, 381–390.
430. Bialek, W. and Onuchic, J.N. (1988) Proc. Natl. Acad. Sci. U.S.A. 85, 5908–5912.
431. Somogyi, B., Matko, J., Papp, S., Hevessy, J., Welch, G.R. and Damjanovich, S. (1984) Biochemistry 23, 3403–3411.
432. O'Hara, P.B., Gorski, K.M. and Rosen, M.A. (1988) Biophys. J. 53, 1007–1013.
433. Frauenfelder, H., Hartmann, H., Karplus, M., Kuntz, D. Jr., Kuriyan, J., Parak, F., Petsko, G.A., Ringe, D., Tilton, R.F. Jr., Connolly, M.L. and Max, N. (1987) Biochemistry 26, 254–261.
434. Eggermont, J.A., Wuytack, F., De Jaegere, S., Nelles, L. and Casteels, R. (1989) Biochem. J. 260, 757–761.

CHAPTER 4

The *Neurospora crassa* plasma membrane H^+-ATPase

GENE A. SCARBOROUGH

Department of Pharmacology, University of North Carolina at Chapel Hill, Chapel Hill, NC 27599, U.S.A.

1. Introduction

Electrophysiological studies carried out in the mid-1960s in the laboratory of Slayman demonstrated that the plasma membrane of *Neurospora crassa* generates and maintains a transmembrane electrical potential difference in excess of 200 mV, interior negative [1]. Subsequent studies in the same laboratory correlated the membrane potential with intracellular ATP levels and extracellular acidification, and from these studies the notion of an electrogenic, proton-translocating ATPase in the fungal plasma membrane emerged [2,3]. Upon the development of a procedure for confidently isolating *Neurospora* plasma membranes [4], it became possible to demonstrate the existence of a plasma membrane ATPase and studies of its biochemical properties ensued [5,6]. Shortly thereafter, in studies with isolated, functionally everted plasma membrane vesicles, it was shown that the ATPase is indeed an electrogenic pump [7] and, in a following study, the identity of the translocated ion as H^+ was clearly established [8]. Concomitantly, the hydrolytic moiety of the H^+-ATPase was identified as a polypeptide with a molecular mass of about 100 kDa that is phosphorylated and dephosphorylated at a rate comparable to the overall rate of ATP hydrolysis catalyzed by the isolated membranes establishing that the catalytic cycle of this proton pump involves a phosphoryl-enzyme intermediate [9]. The subsequent demonstration that the kinetically competent intermediate is an enzyme-bound phosphoryl-aspartate [10] made it clear that except for the fact that the fungal plasma membrane ATPase catalyzes electrogenic proton translocation, it is otherwise quite similar to the Na^+/K^+-, Ca^{2+}- and H^+/K^+-translocating ATPases of animal cell membranes. The more recent availability of amino acid sequence information deduced from the corresponding gene sequences has borne this out, and in addition has included a great many other ATPases in this rather large family of transport enzymes, now conveniently referred to as the P-type ATPases [11]. Since the discovery of the *Neurospora* H$^+$-ATPase, a substantial amount of information about it and the closely related yeast and plant plasma membrane H$^+$-ATPases has accumulated. Summaries of this work can be found in numerous reviews [12–20]. In this chapter, selected information specifically relevant to the structure and dynamics of the *Neurospora* ATPase and other P-type ATPases is first described, allowing the formulation of an approximate structural model for these enzymes. Information available on the catalytic mechanism of the H$^+$-ATPase, and enzymes in general, is then discussed within the framework of this structural model. From these considerations, relatively straightforward schemes for the molecular mechanisms by which the H$^+$-ATPase and the other P-type ATPases might catalyze ATP hydrolysis-driven ion translocation can be visualized.

2. Structural features of the H$^+$-ATPase molecule

2.1. H$^+$-ATPase conformational changes

An important advance in our understanding of the conformational dynamics of the H$^+$-ATPase came with the discovery of the profound effects of several specific H$^+$-ATPase ligands on the susceptibility of the H$^+$-ATPase to degradation by trypsin [21]. To summarize these results, with no ligand present, the H$^+$-ATPase in isolated plasma membrane vesicles is rapidly degraded by trypsin to small undefined fragments. In the presence of the nonhydrolyzable competitive inhibitor, MgADP, the ATPase is rapidly degraded to a ca. 88-kDa form, but further degradation of the ca. 88-kDa form occurs much more slowly, indicating that the ATPase changes its conformation upon binding MgADP. Importantly, in the presence of another nonhydrolyzable competitive inhibitor, Mgβ,γ-methylene ATP, the ATPase is degraded to a different, ca. 95-kDa form, which is also largely resistant to further degradation, suggesting that the enzyme assumes a different conformation when bound to Mgβ,γ-methylene ATP, even though MgADP and Mgβ,γ-methylene ATP are both competitive inhibitors of the ATPase. The difference between the responses seen in the presence of the two substrate analogues was interpreted to indicate that Mgβ,γ-methylene ATP can participate in the formation of the transition state of the enzyme phosphorylation reaction, whereas MgADP cannot, due to the absence of the γ-phosphoryl moiety. Finally, in the presence of Mg^{2+} and vanadate, a transition state analogue of the aspartyl-phosphoryl-enzyme intermediate hydrolysis reaction [22–24], the tryptic degradation pattern is essentially the same as that seen in the presence of Mgβ,γ-methylene ATP. This supports the above interpretation as to the effects of Mgβ,γ-methylene ATP and additionally suggests that the ATPase assumes approximately the same conformation in the transition state of the enzyme phosphorylation reaction as it does in the transition state of the enzyme dephosphorylation reaction, which is a significantly different conformation than the one which it assumes in its unliganded state. The practical importance of these findings is that they made it possible to lock the ATPase in several conformational states corresponding to different stages of the catalytic cycle. These observations also proved to be invaluable in many subsequent studies as an index of the functionality of the H^+-ATPase molecule [25–29]. But even more importantly, these findings are of fundamental significance for mechanistic considerations, because they draw specific attention to the intimate relationship between the major detectable conformational states of the ATPase and the chemical transition states of the reactions it catalyzes, i.e., the enzyme phosphorylation and dephosphorylation reactions. This important aspect of the H^+-ATPase dynamics will be discussed again below.

2.2. The purified ATPase preparation

Progress toward understanding any enzyme usually depends on the ease with which the enzyme can be isolated, the amounts obtainable, and the stability of the final preparation. Two early methods for the isolation of the H^+-ATPase [30,31] served their purpose early on, but it soon became clear that the amounts of enzyme obtained were nowhere near enough. Modifications of our original isolation procedure resulted in our present procedure [32,33] in which intact cells of a cell-wall-less strain of *Neurospora* are simply treated with concanavalin A, solubilized in deoxycholate, and centrifuged, which leads to a pellet that contains essentially all of the ATPase, concanavalin A, and significant amounts of only a few other proteins. Most of the concanavalin A can be removed by treatment of the membranes with alpha-methylmannoside, and the ATPase can then be solubilized with lysolecithin and purified to near homogeneity by glycerol density gradient centrifugation. This procedure is highly reproducible and produces 50–100 mg amounts of the H^+-ATPase, 80–90% free of other proteins and stable indefinitely at $-20°C$, in less than two days. An advantage of the *Neurospora* enzyme over its animal cell counterparts such as the Ca^{2+}-, Na^+/K^+- and H^+/K^+-ATPases is that radiolabeled ATPase molecules can readily be prepared by simply including radioactive protein precursors in the growth medium [25,26]. The availability of large quantities of nearly pure ATPase, radioactive if needed, has greatly facilitated most of our recent studies.

2.3. Subunit composition of the H^+-ATPase

A quantitative reconstitution approach was used to gain information as to the subunit composition of the H^+-ATPase molecule [25]. Proteoliposomes prepared from asolectin and purified, radiolabeled ATPase molecules obtained by a freeze–thaw procedure similar to that of Dufour et al. [34] were shown to catalyze ATP hydrolysis-driven proton translocation, as indicated by the extensive quenching of aminochloromethoxyacridine fluorescence that occurs upon the addition of MgATP to the proteoliposomes, and the reversal of this quenching induced by vanadate or proton conductors. ATP hydrolysis was shown to be tightly coupled to proton translocation by the marked stimulation of ATP hydrolysis by proton conductors. The maximum stimulation by proton conductors was about 3-fold, indicating that at least two-thirds of the hydrolytically active ATPase molecules present in the reconstituted preparation were capable of translocating protons into the liposomes. The fraction of the total population of ATPase molecules in the proteoliposomes that were hydrolytically active was estimated to be at least 91% by the extent of protection of the ATPase against tryptic degradation by vanadate, in the presence of MgATP. Together, these findings indicated that at least 61% of the ATPase molecules in the reconstituted proteoliposomes catalyzed proton translocation. This information placed specific limits on the quantity of any polypeptide in the ATPase preparation other than the 100-kDa hydrolytic moiety that must be present to qualify as a subunit, and a quantitative SDS-PAGE analysis of the radiolabeled ATPase preparation used for the reconstitution ruled out the participation of any such subunits larger than about 2500 Da. Thus, it is unlikely that any polypeptides other than the 100-kDa hydrolytic moiety are involved in the H^+-ATPase transport mechanism.

2.4. The minimum functional unit

Reconstituted H^+-ATPase proteoliposomes were also useful for determining the minimum functional unit required for ATP hydrolysis-driven proton translocation [26]. Sonicated asolectin liposomes were first fractionated on the basis of their size by glycerol density-gradient centrifugation [35] and the small liposomes obtained were used for the reconstitution of purified, radiolabeled H^+-ATPase molecules. The reconstituted proteoliposomes were then subjected to two additional rounds of glycerol density gradient centrifugation, which separated the H^+-ATPase-containing proteoliposomes from the ATPase-free liposomes by virtue of their greater density. A large proportion of the H^+-ATPase molecules in the proteoliposomes thus prepared catalyzed ATP hydrolysis-driven proton translocation, as judged from the specific activity of ATP hydrolysis and the marked stimulation of this activity by proton conductors. Importantly, quantitation of the number of ATPase monomers and liposomes in such preparations by radioactivity determination and counting of negatively stained images in electron photomicrographs indicated ATPase monomer to liposome ratios of one. Because every liposome in the preparations must have had at least one ATPase monomer, this ratio indicated that very few had more than one. Thus, the great majority of active ATPase molecules in the preparations must have been present as proton-translocating monomers. With this information and the fact that no other subunits are involved, the problem of the transport mechanism was reduced to considerations of the events occurring in a single, 100-kDa polypeptide chain. Indications that the ATPase can exist as a dimer [36] may reflect a tendency for self-association of the H^+-ATPase monomers under certain conditions. This could be physiologically important, but does not appear to be relevant in considerations of the transport mechanism.

2.5. Primary structure of the H^+-ATPase

A major step toward understanding the molecular structure of the H^+-ATPase was made by Addison [37] and Hager et al. [38] with the determination of the sequence of the ATPase gene and the deduced amino acid sequence. In addition to solidifying the relatedness of the H^+-ATPase to the other ATPases in the P-type family, this information provided the initial indications of the topography of the H^+-ATPase polypeptide chain, through the use of predictive algorithms. The deduced amino acid sequence was also invaluable in our recent studies of the protein chemistry of the ATPase [39], which paved the way for a direct investigation of the ATPase topography to be described below.

2.6. Secondary structure of the H^+-ATPase

As isolated by the large-scale isolation procedure mentioned above, the purified H^+-ATPase is a hexamer of 100-kDa monomers [40]. Interestingly, after purification of the ATPase in the presence of deoxycholate and lysolecithin, the detergents can be removed by molecular sieve chromatography with essentially no effect on the quaternary structure or stability of the hexamers [27,40]. The resulting preparation contains ATPase hexamers with only 5–10% non-ATPase protein, approximately 12 moles of tightly bound lysolecithin per mole of ATPase monomer, less than 1 mole of deoxycholate/mole of ATPase monomer, and little or no plasma membrane phospholipid [27]. Importantly, the H^+-ATPase monomers in the hexamers are fully functional, indicating that the functional properties of the soluble hexamers are relevant to those of the enzyme in its membrane-bound state [27]. This water-soluble hexameric form of the ATPase proved to be quite useful for studies of the secondary structure of the H^+-ATPase by circular dichroism, since such preparations are virtually free of light scattering and other artifacts normally encountered in optical studies of membrane-bound proteins or even detergent-solubilized preparations. Thus it was possible to ascertain the circular dichroism spectrum of the H^+-ATPase from 184 to 260 nm, from which the secondary structure composition of the ATPase was estimated by the singular value decomposition procedure of Hennessey and Johnson [41]. The results indicated that the H^+-ATPase contains approximately 36% helix, 12% antiparallel beta-sheet, 8% parallel beta-sheet, 11% beta-turn, and 26% irregular structure [27]. Importantly, no detectable changes in the circular dichroism spectrum of the ATPase were found to occur in the presence of ATPase ligands known to induce enzyme conformational changes resembling those that occur during the ATP hydrolytic cycle (see above), suggesting that substantial changes in the secondary structure of the ATPase are not an integral feature of the transport mechanism. Finally, the circular dichroism spectrum of the H^+-ATPase was compared to the corresponding spectra of the Na^+/K^+- and Ca^{2+}-translocating ATPases and shown to be quite similar, indicating in yet another way that the P-type ATPases probably share a great deal of structural and mechanistic similarity.

In another study, ATPase reconstituted into liposomes was analyzed by infrared attenuated total reflection spectroscopy and the secondary-structure elements of the molecule were determined from the spectra obtained by Fourier self-deconvolution [42]. Gratifyingly, essentially identical secondary-structure estimates for the ATPase were obtained by this entirely different approach, suggesting quite strongly that these secondary-structure estimates are reasonably accurate. Thus, any future models for the structure of the H^+-ATPase must take this information into account.

2.7. Protein chemistry of the H^+-ATPase

Integral membrane proteins are notoriously resistant to manipulation by conventional techniques of protein chemistry. Therefore, in order to make possible direct chemical approaches to elucidating a variety of features of the H^+-ATPase molecule, procedures for quantitative dissection of the ATPase molecule had to be developed. These efforts led to a highly effective procedure for fragmentation of the ATPase molecule and purification of virtually all of the numerous peptides produced [39]. The enzyme is first cleaved by trypsin to form a limit digest containing both hydrophobic and hydrophilic peptides, and the hydrophobic and hydrophilic peptides are then separated by extraction with a solution of high ionic strength. The numerous hydrophilic peptides, which are soluble in the high-salt solution, can easily be purified by high-performance liquid chromatography (HPLC). The hydrophobic peptides, which are almost totally insoluble in the high-salt solution, are dissolved in neat trifluoroacetic acid and then partially fractionated by Sephadex LH60 chromatography in chloroform/methanol/trifluoroacetic acid. The recoveries in all of these procedures are greater than 90%. Many of the hydrophobic peptides obtained in this procedure cannot be analyzed by conventional SDS-PAGE methods, so as part of these studies, a dependable procedure for analyzing hydrophobic peptides in the ca. 5- to 20-kDa size range was devised [43]. With the development of the above methodology, direct chemical studies of the H^+-ATPase molecule became feasible.

2.8. Chemical state of the H^+-ATPase cysteines

The H^+-ATPase gene sequence indicates the presence of eight cysteine residues in the molecule [37,38]. In order to ascertain the chemical state of these cysteine residues, direct chemical studies with established cysteine and cystine reagents were carried out [44]. Titrations with the cysteine reagent, dithiobisnitrobenzoate, and the cystine reagent, nitrothiosulfobenzoate, indicated the presence of six free cysteines and one disulfide bridge in the molecule. Quantitative carboxymethylation experiments with radioactive iodoacetate under reducing and nonreducing conditions confirmed this conclusion. ATPase carboxymethylated under both conditions was then cleaved with trypsin and the digests resolved into hydrophilic and hydrophobic fractions as described above. Five of the six labeled free cysteine peptides partitioned into the hydrophilic fraction and were purified and established to contain Cys^{376}, Cys^{409}, Cys^{472}, Cys^{532} and Cys^{545}. The labeled free cysteine in the hydrophobic fraction was identified as either Cys^{840} or Cys^{869}, which in turn identified the other as one of the disulfide bridge cysteines. The other disulfide bridge cysteine was identified as Cys^{148} by purification and N-terminal amino acid sequencing of an additional peptide labeled in the reduced enzyme. Thus, the disulfide bridge in the H^+-ATPase molecule is between Cys^{148} and either Cys^{840} or Cys^{869}.

2.9. Transmembrane topography of the H^+-ATPase

In addition to providing information as to the subunit composition and minimal functional unit of the H^+-ATPase, reconstituted ATPase proteoliposomes have been quite useful in studies of the transmembrane topography of the molecule. First, on the basis of susceptibility to degradation by trypsin and protection against this degradation by the ATPase ligands, MgATP and vanadate, it was established that 85–90% of the ATPase molecules in the proteoliposomes are oriented with their cytoplasmic portion facing outward, whereas the remaining 10–15% are present with the opposite orientation and totally resistant to tryptic cleavage [28,29]. It was also shown in this study that tryptic cleavage in the presence of ligands produces ca. 97-, 95- and 88-kDa degradation products similar to those shown by Mandala and Slayman [45] to result from cleavage of the ATPase at Lys^{24}, Lys^{36} and Arg^{73}, establishing these N-terminal sites on the cytoplasmic side of the membrane. Tryptic cleavage in the absence of ligands leads to the release of numerous H^+-ATPase peptides from the proteoliposomes and, in our first investigation of these peptides [28], one was purified by HPLC and shown by N-terminal amino acid sequence analysis to comprise residues 901–911 at the extreme C-terminus of the molecule. These results established the location of the N- and C-termini of the ATPase molecule on the cytoplasmic side of the membrane. The same conclusion was reached by Mandala and Slayman on the basis of antibody binding and tryptic cleavage experiments with intact and permeabilized cells and presumably inside-out plasma membrane vesicles [46]. Following up on our first study, fourteen additional peptides released from the liposomes were purified by HPLC and identified by N-terminal sequence analysis [29]. The results obtained established residues 70–100, 186–219, 238–256, 441–460, 471–512, 545–559, and 567–663 to be located on the cytoplasmic side of the membrane. Moreover, this information identified several additional flanking sequences, including residues 29–32, 41–69, 220–237, 461–470, 513–544, 560–566 and 916–920, as also likely to be cytoplasmically located by virtue of the fact that they are too short to cross the membrane and return. Although residues 2–20 were originally included in this group of sequences with inferred cytoplasmic locations, it is conceivable that these residues could form a single membrane spanning helix with the N-terminus on the exocytoplasmic surface of the membrane. However, hydropathy analyses suggest that this is probably not the case. These results collectively indicate that residues 21–100, 186–256, 441–663 and 897–920 are located on the cytoplasmic side of the membrane. Subsequent experiments have also localized residues 359–440 on the cytoplasmic side of the membrane [109].

After the development of the above-mentioned methodology for manipulating the previously intractable hydrophobic segments of the H^+-ATPase [39,43], it became possible to extend these studies and define the parts of the ATPase molecule remaining associated with the liposomes after removal of the released peptides [47]. The liposome-bound peptides were separated from the phospholipid and partially separated from each other by Sephadex LH60 chromatography in chloroform/methanol/trifluoroacetic acid and the resulting eluate was analyzed by our SDS-PAGE procedure for analyzing hydrophobic peptides [43]. Three major peptides with approximate M_rs of 7, 7.5 and 21 kDa could be identified by N-terminal sequence analysis as H^+-ATPase peptides beginning at residues 100, 272 and 660, respectively. On the basis of their size, these peptides probably end near residues 173, 355 and 891, respectively. As these peptides were also labeled from the liposomal membrane interior by the lipophilic photolabeling reagent, $[^{125}I]$-trifluoromethyliodophenyldiazirine, and little else was, it was concluded that these three peptides constitute the great majority of the membrane-embedded region of the H^+-ATPase molecule. Additional considerations from these studies localized residues 174–185 and 257–271 on the cytoplasmic side of the membrane.

The results of all of these topography studies are summarized in Fig. 1. The open circles indicate residues of the H^+-ATPase shown in one way or another to be located on the cytoplasmic side of the membrane and the closed circles indicate residues in membrane-embedded segments. The lines in the sequence indicate minor regions with locations as yet not established. Thus, the topographical locations of nearly all of the 919 residues in the molecule have been established. It should be emphasized that the exact points of entry and exit of the polypeptide chain into and out of the membrane are not implied in the model.

2.10. A first-generation model for the tertiary structure of the H^+-ATPase

With this information in hand, we may now consider how the H^+-ATPase polypeptide chain might fold into its functional three-dimensional structure. First, regarding the actual number of membrane-spanning stretches, the available experimental data indicate only that each of the three membrane-embedded peptides must have an even number of and a minimum of two such stretches. However, hydropathy analysis by the method of Mohana Rao and Argos [48] suggests that the second mem-

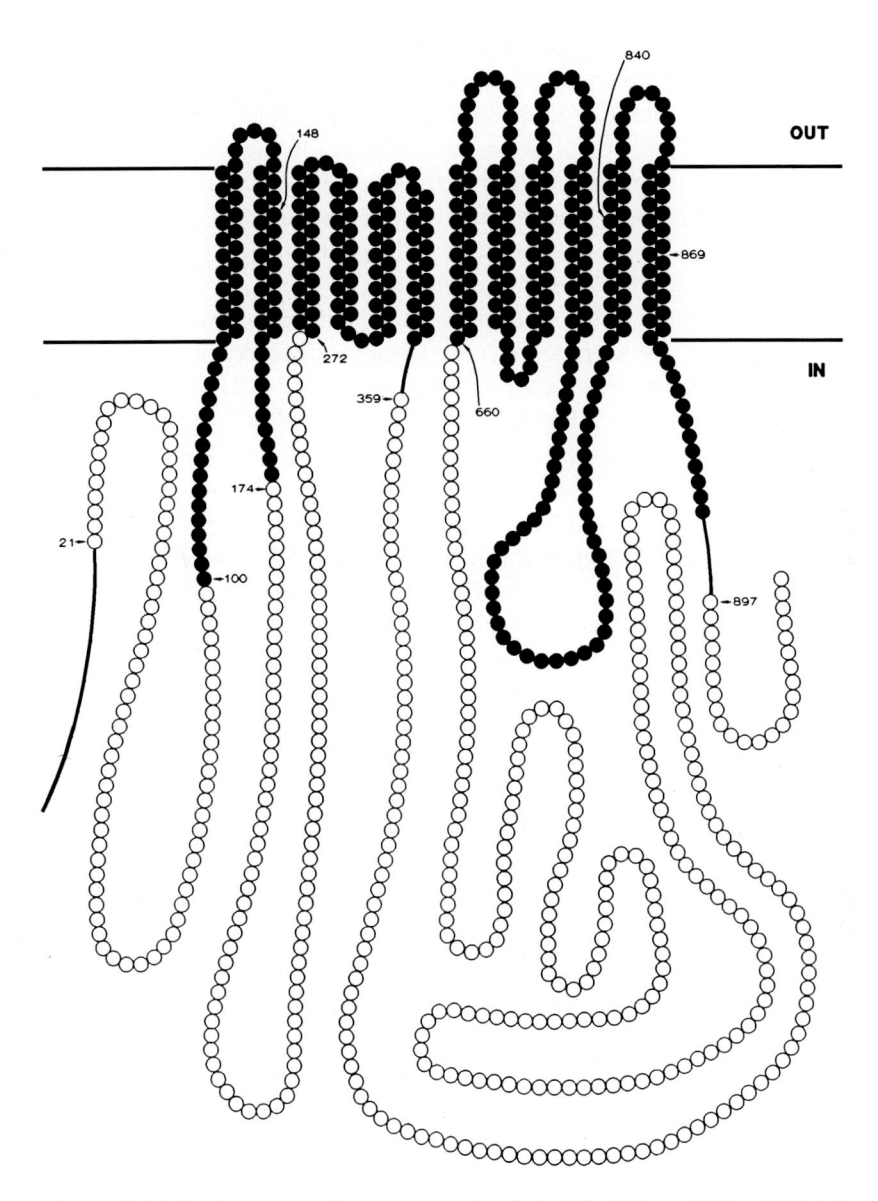

Fig. 1. Model for the transmembrane topography of the H$^+$-ATPase. OUT and IN indicate points of reference outside and inside an intact cell, respectively. See text for additional details.

brane-embedded segment beginning at residue 272 could have four membrane-spanning stretches, and that the third segment beginning at residue 660 could have as many as six [39]. Thus, as indicated in Fig. 1, our current working model for the membrane-embedded region of the ATPase envisions twelve membrane-spanning stretches. If this is correct, it would add the H$^+$-ATPase to the large and rapidly growing family of transport molecules with twelve putative membrane-spanning stretches [49–52]. Although most models for the P-type ATPases do not suggest the existence of this many membrane-spanning stretches, this could be a reflection of the rules for hydropathy analysis, which avoid polar residues in membrane-embedded regions. As it is likely that the membrane-embedded regions of the P-type ATPases contain several ion binding sites (see below), each comprising several charged or otherwise polar residues, it may be that strict hydropathy analyses underestimate the actual number of membrane-embedded stretches.

Except for the two additional membrane-spanning stretches in the second membrane-embedded segment, the proposed membrane-embedded stretches indicated in Fig. 1 are similar to those proposed for the closely related Ca^{2+}-ATPase [53]. In fact, the overall topography proposed in Fig. 1 is quite similar to the Ca^{2+}-ATPase model, lending additional credibility to each.

Although there are probably exceptions [54], it is generally agreed that the membrane-embedded stretches in integral membrane proteins are likely to be present as helices more or less parallel to the membrane normal [55]. This is clearly the case for the two integral membrane proteins of known structure, the photosynthetic reaction center [56] and bacteriorhodopsin [57]. It is therefore reasonable to assume at least tentatively that the transmembrane stretches of the *Neurospora* H^+-ATPase are also folded this way. Four of the proposed helices in the N-terminal segment are connected by short loops, so it is likely that these helices are near one another in the folded molecule. The same holds for at least four of the six proposed helices in the C-terminal membrane-embedded region. Moreover, as described above, there is a disulfide bridge in the H^+-ATPase molecule between Cys^{148} in the N-terminal membrane-embedded region and either Cys^{840} or Cys^{869} in the C-terminal membrane-embedded region [44]. It is therefore probable that many, if not all, of the N- and C-terminal transmembrane helices are clustered together in the folded molecule. Thus, the N-terminal group of membrane-embedded helices in the two-dimensional topographical model of Fig. 1 can be folded over into juxtaposition with the C-terminal group. How the molecule might look after this operation is shown in Fig. 2. The Z-shaped line denotes one of the two possible disulfide bridge configurations.

Although this rendering is highly speculative, several features are worth noting briefly.

First, the transmembrane helices (cylinders) have been grouped into three sets of antiparallel four-helix bundles. Considering the documented marked stability of the antiparallel four-helix bundle [58–62], which is enhanced when helices are closely spaced in a linear sequence or held together by a disulfide bridge, and further enhanced in an hydrophobic milieu [58,62], an antiparallel four-helix bundle configuration for most or all of the transmembrane helices seems entirely feasible. In support of this suggestion, antiparallel helix alignments and four-helix bundle configurations dominate the structures of the membrane-embedded helix sectors of both the photosynthetic reaction center [63] and bacteriorhodopsin [57]. Moreover,

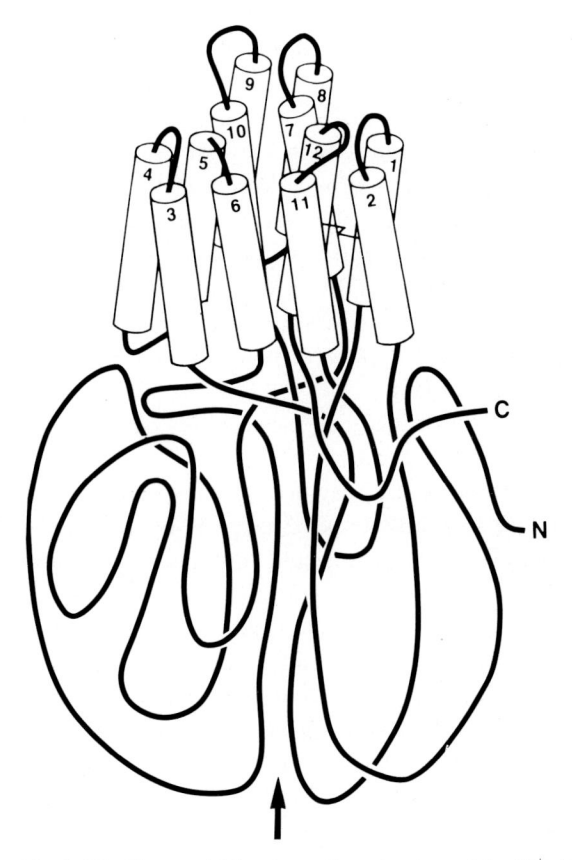

Fig. 2. Working model for the tertiary structure of the H$^+$-ATPase. See text for details.

since four-helix bundles have a marked tendency to interact with each other in highly symmetrical ways [58], the proposed bundles could be present in the molecule as a symmetrical ring, as shown. Of course, many isomers of this structure with different interhelix contacts are equally possible; the helix numbers indicated in the figure are provided only to facilitate tracking the path of the polypeptide chain. Interestingly, a helix arrangement similar to this, but without four-helix bundles, was suggested by Maloney [49] on the basis of an entirely different line of reasoning. It should also be mentioned that the proposed helix arrangement is quite similar to one of the arrangements proposed by Green [64] for the Ca^{2+}-ATPase, if two of the outer helices are removed from the model, as would be necessary if the H$^+$-ATPase has only ten membrane-spanning helices.

Second, our studies of the secondary structure of the H$^+$-ATPase indicate that about 36% of the polypeptide chain is present in a helical configuration [27,42]. If the membrane-embedded sector of the molecule is helical as shown, only 90 or so additional residues in the molecule can be present as helices. Thus, the great majority of the cytoplasmic portion of the molecule must be present as beta-sheet, beta-turn, and other structures, and therefore unlike certain predicted models for the cytoplasmic portions of the Ca^{2+}- and Na^+/K^+-ATPases [65,66].

And third, since virtually all enzymes [67], particularly those that catalyze phosphoryl-transfer reactions [68–74], possess structures with at least two, discrete, relatively rigid structural domains, or lobes, separated by a deep cleft, the cytoplasmic portion of the H^+-ATPase polypeptide chain in the model of Fig. 2 is drawn in such a way as to suggest this situation. The proposed interdomain cleft is indicated by the arrow. No additional structural features of the ATPase molecule are implied in the model. In regard to comparisons with the Ca^{2+}-ATPase, it is of interest to note that the two cytoplasmic domains proposed in Fig. 2 correspond to the C1 and C2 domains in the model of Andersen and Vilsen [53].

Further considerations here do not depend critically on the accuracy of the working model of Fig. 2. Indeed, the interdomain cleft may as well be at the side of the molecule near the surface of the membrane as can be imagined from inspection of the structures proposed by Taylor et al. [75] and Stokes and Green [76] for the Ca^{2+}-ATPase. It is only important to stipulate that the molecule contains at least two domains and a cluster of membrane-embedded helices.

We can now consider possible locations of the active site residues and the proton binding site(s). Suggestions regarding the identity of residues that constitute the nucleotide binding site of the H^+-ATPase can be found in several reviews, [14,17,18] and will not be reiterated here. However, essentially no experimental evidence is available as to the location of the nucleotide and proton binding sites relative to the membrane-embedded sector and the rest of the cytoplasmic sector. Fortunately, a wealth of information is available in this regard for the closely related Ca^{2+}-ATPase of sarcoplasmic reticulum. On the basis of a variety of different experimental approaches, it is now generally agreed that the Ca^{2+} transport sites of the Ca^{2+}-ATPase are located in or near the membrane-embedded helix region. The Ca^{2+}-binding sites of the Ca^{2+}-ATPase have been estimated by Teruel and Gomez-Fernandez [77] to be within 10 Å of the rhodamine group of rhodamine isothiocyanylphosphatidylethanolamine in the membrane. Since it was likely that the fluorophore in these experiments was oriented near the fatty acid chains [78], this estimate places the Ca^{2+}-binding sites in the membrane region. In direct support of this placement, Scott has estimated that the Ca^{2+}-binding sites are close to each other [79] and near the middle of the bilayer [80] and Herrmann et al. [81], Stephens and Grisham [82] and Joshi and Shamoo [83] have provided additional evidence for propinquity of the two Ca^{2+} sites. Moreover, recently reported site-directed mutagenesis and direct biochemical studies also indicate that the Ca^{2+} sites are in the membrane-embedded region of the ATPase molecule [84,85]. Thus, although dissenting views can be found [86,87], the weight of the evidence seems to favor a location for the Ca^{2+} sites in or near the membrane-embedded helix region of the Ca^+-ATPase molecule. A similar situation appears to exist for the Na^+/K^+-ATPase [88]. We can thus visualize the analogous proton binding region of the H^+-ATPase on the surfaces inside the ring of membrane-embedded helices in Fig. 2. It should be mentioned that the H^+-ATPase may have one or two analogous proton binding sites depending on the conditions [89]. On the other hand, it may have no strictly analogous sites if the transported proton(s) arise from the hydrolytic water molecule as has been suggested [90,91].

The location of the nucleotide binding site is more controversial. On the basis of fluorescence energy transfer measurements using fluorescent probes, several investigators have concluded that the nucleotide binding site is quite far removed from the Ca^{2+}-binding sites in the Ca^{2+}-ATPase molecule [77–79,92]. However, other investigators using different approaches have obtained convincing evidence that the ion and nucleotide sites are much closer in this enzyme [81,83,93–95], perhaps with a distance of only 4 Å between the Ca^{2+} sites and the terminal phosphoryl group of ATP [95]. Similar conclusions have been reached for the Na^+/K^+-ATPase [96–98]. The question of the distances between the ion and nucleotide sites is central to the issue of the mechanism by which ATP hydrolysis is coupled to ion translocation, and therefore, a choice between these two contradictory conclusions cannot be made lightly. However, since the fluorescence measurements in question probably overestimate intramolecular distances more often than they underestimate them, it seems more prudent to believe the closer values rather than the longer ones. Thus, we can visualize bound ATP in the nucleotide binding site fairly close to the ion binding sites, i.e., in the cleft (arrow) near the interface between the transmembrane helix cluster and the cytoplasmic domains in Fig. 2.

3. The molecular mechanism of transport

How then might this approximate structure catalyze proton translocation coupled to ATP hydrolysis? Consider the mechanism of ATP hydrolysis first. As proposed by Pauling [99], and later elaborated upon by others [100–102], the essence of enzymatic catalysis is the tendency for enzymes to bind most tightly to the transition state of the reaction catalyzed, thereby stabilizing it and increasing its relative concentration. In keeping with the transition state theory of reaction rates, such an increase in the concentration of the transition state will result in an increase in the rate of the reaction [103]. As mentioned above, enzymes in general comprise two or more relatively rigid domains separated by an interdomain cleft, and for several phosphotransferases, it is quite clear that upon binding of their substrates, these enzymes undergo a so-called hinge-bending conformational change that leads to cleft closure and juxtaposition of the substrates in a relatively non-aqueous environment between the domains [72–74]. As explained elsewhere [90,91], it is likely that the transition state binding affinity is the driving force for this conformational change. As outlined above, conformational changes driven by transition state analogue binding have been directly demonstrated for the *Neurospora* H^+-ATPase [21]. Moreover, as also mentioned above, these apparently substantial conformational changes do not involve measurable changes in the secondary structure of the H^+-ATPase [27], consistent with the hinge-bending maneuver. In view of these considerations, it seems likely that the H^+-ATPase and the other P-type ATPases, which are also phosphotransferases, undergo similar interdomain cleft closures upon binding their respective ligands. Because these enzymes catalyze both an enzyme phosphorylation reaction and an enzyme dephosphorylation reaction, they proceed through two transition states, both of which presumably involve interdomain movements. In the case of the H^+-ATPase, the conformation of the enzyme in each transition state is similar, as reflected by essentially identical tryptic degradation profiles [21]. Thus, in the model of Fig. 2, we can imagine the cleft somewhat more open than shown when the enzyme is taking on substrates or releasing products, and closed approximately as shown in the enzyme phosphorylation and dephosphorylation transition states with the transferred phosphoryl group embedded between the domains in or near the membrane-embedded helix region. The well-known phenomenon of ion 'occlusion' corresponds to cleft closure in the model of Fig. 2.

Viewing the H^+-ATPase reaction cycle in this way clearly de-emphasizes the traditional E_1 and E_2 states of the P-type ATPases, as has been done before [90,91], emphasizing instead established modes of enzyme conformational dynamics. Importantly, in a trenchant ongoing analysis of the Ca^{2+}-ATPase reaction cycle reviewed recently [104], Jencks also questions the traditional E_1E_2 model. In agreement, Andersen and Vilsen have recently suggested that conformational changes directly related to the phosphorylation and dephosphorylation reactions may be more germane to the transport mechanism [53]. In view of the demonstrated interrelationship between the H^+-ATPase conformational changes and transition state analogue binding [21], and the above-mentioned general comments regarding its probable molecular meaning, this would appear to be a suggestion worthy of serious consideration.

Finally, we come to the question of how the events occurring during ATP hydrolysis might be coupled to the translocation of protons or other ions. If the nucleotide binding site is far removed from the ion binding site or sites in the membrane-embedded region as many investigators believe [14,77–80,92,105,106], we are left with a black box to couple the ATP hydrolytic events to the ion-translocation events, and are thus nowhere near an understanding of the molecular mechanism of these enzymes. However, if the nucleotide binding site is within a few angstroms of the ion sites as others suggest [90,91,93–98,107,108], there may be much less left to understand. When the interdomain cleft proposed in Fig. 2 is open during ion and substrate binding and product release, the path from the ion binding sites to the far side of the membrane through the membrane-embedded helices would be blocked. When the interdomain cleft closes during the ATP hydrolytic events, the path from the active site to the far side via the membrane-embedded helices opens.

Additionally, the chemical events related to the phosphoryl-enzyme hydrolysis reaction directly influence the ion binding sites, resulting in decreased affinity for the ions and their facile release to the other side. Suggestions as to how the chemical reactions might influence the ion binding sites for the major P-type ATPases have been described in some detail [90,91] and will not be reiterated here. Finally, when the chemical reactions are completed, the cleft reopens for product release and the path through the helices concomitantly closes. Thus, employing only established modes of protein dynamics and with due attention to the reaction cycle chemistry, feasible molecular mechanisms for the P-type ATPases with experimentally testable features are relatively easy to imagine.

Acknowledgement

This work was supported by United States Public Health Service National Institutes of Health Grant GM24784.

References

1. Slayman, C.L. (1965) J. Gen. Physiol. 49, 69–92.
2. Slayman, C.L., Lu, C.Y.-H. and Shane, L. (1970) Nature (London) 226, 274–276.
3. Slayman, C.L. (1970) Am. Zool. 10, 377–392.
4. Scarborough, G.A. (1975) J. Biol. Chem. 250, 1106–1111.
5. Scarborough, G.A. (1977) Arch. Biochem. Biophys. 180, 384–393.
6. Bowman, B.J. and Slayman, C.W. (1977) J. Biol. Chem. 252, 3357–3363.
7. Scarborough, G.A. (1976) Proc. Natl. Acad. Sci. U.S.A. 73, 1485–1488.
8. Scarborough, G.A. (1980) Biochemistry 19, 2925–2931.
9. Dame, J.B. and Scarborough, G.A. (1980) Biochemistry 19, 2931–2937.
10. Dame, J.B. and Scarborough, G.A. (1981) J. Biol. Chem. 256, 10724–10730.
11. Pedersen, P.L. and Carafoli, E. (1987) TIBS 12, 146–150.
12. Goffeau, A. and Slayman, C.W. (1981) Biochim. Biophys. Acta 639, 197–223.
13. Serrano, R. (1984) Curr. Top. Cell. Regul. 23, 87–126.
14. Serrano, R. (1988) Biochim. Biophys. Acta 947, 1–28.
15. Goffeau, A., Coddington, A. and Schlesser, A. (1989) In: Molecular Biology of the Fission Yeast (Nasim, A., Young, P. and Johnson, B.F., Eds.) pp. 397–429, Academic Press, London.
16. Nakamoto, R.K. and Slayman, C.W. (1989) J. Bioenerg. Biomemb. 21, 621–632.
17. Serrano, R. (1989) Annu. Rev. Plant Physiol. Plant Mol. Biol. 40, 61–94.
18. Rao, R., Nakamoto, R.K. and Slayman, C.W. (1989) In: Ion Transport (Keeling, D. and Benham, C., Eds.) pp. 35–54, Academic Press, London.
19. Goffeau, A. and Green, N.M. (1990) In: Monovalent Cations in Biological Systems (Pasternak, C.A., Ed.) pp. 155–169, CRC Press, Boca Raton.
20. Serrano, R. (1990) In: The Plant Plasma Membrane (Larsson, C. and Moller, I.M., Eds.) pp. 127–153, Springer, Berlin.
21. Addison, R. and Scarborough, G.A. (1982) J. Biol. Chem. 257, 10421–10426.
22. Macara, I.G. (1980) Trends Biochem. Sci. 5, 92–94.

23. Pope, M.T. and Dale, B.W. (1968) Q. Rev. Chem. Soc. 22, 527–548.
24. Cantley, L.C., Jr., Cantley, L.G. and Josephson, L. (1978) J. Biol. Chem. 253, 7361–7368.
25. Scarborough, G.A. and Addison, R. (1984) J. Biol. Chem. 259, 9109–9114.
26. Goormaghtigh, E., Chadwick, C. and Scarborough, G.A. (1986) J. Biol. Chem. 261, 7466–7471.
27. Hennessey, J.P., Jr. and Scarborough, G.A. (1988) J. Biol. Chem. 263, 3123–3130.
28. Hennessey, J.P., Jr. and Scarborough, G.A. (1990) J. Biol. Chem. 265, 532–537.
29. Scarborough, G.A. and Hennessey, J.P., Jr. (1990) J. Biol. Chem. 265, 16145–16149.
30. Addison, R. and Scarborough, G.A. (1981) J. Biol. Chem. 256, 13165–13171.
31. Bowman, B.J., Blasco, F. and Slayman, C.W. (1981) J. Biol. Chem. 256, 12343–12349.
32. Smith, R. and Scarborough, G.A. (1984) Anal. Biochem. 138, 156–163.
33. Scarborough, G.A. (1988) Methods Enzymol. 157, 574–579.
34. Dufour, J.-P., Goffeau, A. and Tsong, T.Y. (1982) J. Biol. Chem. 257, 9365–9371.
35. Goormaghtigh, E. and Scarborough, G.A. (1986) Anal. Biochem. 159, 122–131.
36. Bowman, B.J., Berenski, C.J. and Jung, C.Y. (1985) J. Biol. Chem. 260, 8726–8730.
37. Addison, R. (1986) J. Biol. Chem. 261, 14896–14901.
38. Hager, K.M., Mandala, S.M., Davenport, J.W., Speicher, D.W., Benz, E.J., Jr. and Slayman, C.W. (1986) Proc. Natl. Acad. Sci. U.S.A. 83, 7693–7697.
39. Rao, U.S., Hennessey, J.P., Jr. and Scarborough, G.A. (1988) Anal. Biochem. 173, 251–264.
40. Chadwick, C.C., Goormaghtigh, E. and Scarborough, G.A. (1987) Arch. Biochem. Biophys. 252, 348–356.
41. Hennessey, J.P., Jr. and Johnson, W.C., Jr. (1981) Biochemistry 20, 1085–1094.
42. Goormaghtigh, E., Ruysschaert, J.-M. and Scarborough, G.A. (1988) In: The Ion Pumps: Structure, Function and Regulation, pp. 51–56, Liss, New York.
43. Hennessey, J.P., Jr. and Scarborough, G.A. (1989) Anal. Biochem. 176, 284–289.
44. Rao, U.S. and Scarborough, G.A. (1990) J. Biol. Chem. 265, 7227–7235.
45. Mandala, S.M. and Slayman, C.W. (1988) J. Biol. Chem. 263, 15122–15128.
46. Mandala, S.M. and Slayman, C.W. (1989) J. Biol. Chem. 264, 16276–16281.
47. Rao, U.S., Hennessey, J.P., Jr. and Scarborough, G.A. (1991) J. Biol. Chem. 266, 14740–14746.
48. Mohana Rao, J.K. and Argos, P. (1986) Biochim. Biophys. Acta 869, 197–214.
49. Maloney, P.C. (1989) Philos. Trans. R. Soc. London Ser. B: 326, 437–454.
50. Maloney, P.C., Ambudkar, S.V., Anantharam, V., Sonna, L.A. and Varadhachary, A. (1990) Microbiol. Rev. 54, 1–17.
51. Henderson, P.J.F. (1990) J. Bioenerg. Biomemb. 22, 525–569.
52. Silver, S., Nucifora, G., Chu, L. and Misra, T.K. (1989) TIBS 14, 76–80.
53. Andersen, J.P. and Vilsen, B. (1990) Curr. Opinion Cell Biol. 2, 722–730.
54. Lodish, H.F. (1988) TIBS 13, 332–334.
55. Eisenberg, D. (1984) Ann. Rev. Biochem. 53, 595–623.
56. Deisenhofer, J., Epp, O., Miki, K., Huber, R. and Michel, H. (1985) Nature (London) 318, 618–624.
57. Henderson, R., Baldwin, J.M., Ceska, T.A., Zemlin, F., Beckmann, E. and Downing, K.H. (1990) J. Mol. Biol. 213, 899–929.
58. Sheridan, R.P., Levy, R.M. and Salemme, F.R. (1982) Proc. Natl. Acad. Sci. U.S.A. 79, 4545–4549.
59. DeGrado, W.F., Regan, L. and Ho, S.P. (1987) Cold Spring Harbor Symp. Quant. Biol. 52, 521–526.
60. Regan, L. and DeGrado, W.F. (1988) Science 241, 976–978.
61. Chou, K.-C., Maggiora, G.M., Nemethy, G. and Scheraga, H.A. (1988) Proc. Natl. Acad. Sci. U.S.A. 85, 4295–4299.
62. Hol, W.G.J., Halie, L.M. and Sander, C. (1981) Nature (London) 294, 532–536.
63. Yeates, T.O., Komiya, H., Rees, D.C., Allen, J.P. and Feher, G. (1987) Proc. Natl. Acad. Sci. U.S.A. 84, 6438–6442.
64. Green, N.M. (1989) Nature (London) 339, 424–425.
65. MacLennan, D.H., Brandl, C.J., Korczak, B. and Green, N.M. (1985) Nature (London) 316, 696–700.

66. Ovchinnikov, Yu.A., Monastyrskaya, G.S., Broude, N.E., Ushkaryov, Yu.A., Melkov, A.M., Smirnov, Yu.V., Malyshev, I.V., Allikmets, R.L., Kostina, M.B., Dulubova, I.E., Kiyatkin, N.I., Grishin, A.V., Modyanov, N.N. and Sverdlov, E.D. (1988) FEBS Lett. 233, 87–94.
67. Schulz, G.E. and Schirmer, R.H. (1979) Principles of Protein Structure, Springer, New York.
68. Steitz, T.A., Fletterick, R.J., Anderson, W.F. and Anderson, C.M. (1976) J. Mol. Biol. 104, 197–222.
69. Stuart, D.I., Levine, M., Muirhead, H. and Stammers, D.K. (1979) J. Mol. Biol. 134, 109–142.
70. Sachsenheimer, W. and Schulz, G.E. (1977) J. Mol. Biol. 114, 23–36.
71. Evans, P.R. and Hudson, P.J. (1979) Nature (London) 279, 500–504.
72. Banks, R.D., Blake, C.C.F., Evans, P.R., Haser, R., Rice, D.W., Hardy, G.W., Merrett, M. and Phillips, A.W. (1979) Nature (London) 279, 773–777.
73. Pickover, C.A., McKay, D.B., Engelman, D.M. and Steitz, T.A. (1979) J. Biol. Chem. 254, 11323–11329.
74. Anderson, C.M., Zucker, F.H. and Steitz, T.A. (1979) Science 204, 375–380.
75. Taylor, K.A., Dux, L. and Martonosi, A. (1986) J. Mol. Biol. 187, 417–427.
76. Stokes, D.L. and Green, N.M. (1990) J. Mol. Biol. 213, 529–538.
77. Teruel, J.A. and Gomez-Fernandez, J.C. (1986) Biochim. Biophys. Acta 863, 178–184.
78. Gutierrez-Merino, C., Munkonge, F., Mata, A.M., East, J.M., Levinson, B.L., Napier, R.M. and Lee, A.G. (1987) Biochim. Biophys. Acta 897, 207–216.
79. Scott, T.L. (1985) J. Biol. Chem. 260, 14421–14423.
80. Scott, T.L. (1988) Mol. Cell. Biochem. 82, 51–54.
81. Herrmann, T.R., Gangola, P. and Shamoo, A.E. (1986) Eur. J. Biochem. 158, 555–560.
82. Stephens, E.M. and Grisham, C.M. (1979) Biochemistry 18, 4876–4885.
83. Joshi, N.B. and Shamoo, A.E. (1988) Eur. J. Biochem. 178, 483–487.
84. Clarke, D.M., Loo, T.W., Inesi, G. and MacLennan, D.H. (1989) Nature (London) 339, 476–478.
85. le Maire, M., Lund, S., Viel, A., Champeil, P. and Moller, J.V. (1990) J. Biol. Chem. 265, 1111–1123.
86. Munkonge, F., East, J.M. and Lee, A.G. (1989) Biochim. Biophys. Acta 979, 113–120.
87. Asturias, F.J. and Blasie, J.K. (1991) Biophys. J. 59, 488–502.
88. Karlish, S.J.D., Goldshleger, R. and Stein, W.D. (1990) Proc. Natl. Acad. Sci. U.S.A. 87, 4566–4570.
89. Warncke, J. and Slayman, C.L. (1980) Biochim. Biophys. Acta 591, 224–233.
90. Scarborough, G.A. (1982) Ann. N.Y. Acad. Sci. 402, 99–115.
91. Scarborough, G.A. (1985) Microbiol. Rev. 49, 214–231.
92. Highsmith, S. and Murphy, A.J. (1984) J. Biol. Chem. 259, 14651–14656.
93. Klemens, M.R. and Grisham, C.M. (1988) FEBS Lett. 237, 4–8.
94. Klemens, M.R., Stewart, J.M.MacD., Mahaney, J.E., Kuntzweiler, T.A., Sattler, M.C. and Grisham, C.M. (1988) In: Advances in Biotechnology of Membrane Ion Transport (Jorgensen, P.L. and Verna, R., Eds.), Vol. 51, pp. 107–124, Raven Press, New York.
95. Kuntzweiler, T.A. and Grisham, C.M. (1990) FASEB J. 4, A296.
96. Grisham, C.M., Gupta, R.K., Barnett, R.E. and Mildvan, A.S. (1974) J. Biol. Chem. 249, 6738–6744.
97. Grisham, C.M. and Mildvan, A.S. (1975) J. Supramol. Struct. 3, 304–313.
98. Grisham, C.M. (1981) J. Inorg. Biochem. 14, 45–57.
99. Pauling, L. (1946) Chem. Eng. News 24, 1375–1377.
100. Jencks, W.P. (1966) In: Current Aspects of Biochemical Energetics (Kaplan, N.O. and Kennedy, E.P., Eds.) pp. 273–298, Academic Press, New York.
101. Wolfenden, R. (1969) Nature (London) 223, 704–705.
102. Lienhard, G.E. (1973) Science 180, 149–154.
103. Frost, A.A. and Pearson, R.G. (1961) Kinetics and Mechanism, pp. 77–102, John Wiley and Sons, New York, 2nd ed.
104. Jencks, W.P. (1989) J. Biol. Chem. 264, 18855–18858.
105. Brandl, C.J., Green, N.M., Korczak, B. and MacLennan, D.H. (1986) Cell 44, 597–607.
106. Squier, T.C., Bigelow, D.J., de Ancos, J.G. and Inesi, G. (1987) J. Biol. Chem. 262, 4748–4754.

107. Mitchell, P. (1979) Eur. J. Biochem. 95, 1–20.
108. Dupont, Y. (1983) FEBS Lett. 161, 14–20.
109. Rao, U.S., Bauzon, D.D. and Scarborough, G.A. (1992) Biochim. Biophys. Acta, in press.

CHAPTER 5

The Enzymes II of the phosphoenolpyruvate-dependent carbohydrate transport systems

J.S. LOLKEMA and G.T. ROBILLARD

The BIOSON Research Institute, University of Groningen, 9747 AG Groningen, The Netherlands

1. Introduction

1.1. PTS carbohydrate specificity

The Enzymes II (E-IIs) of the phosphoenolpyruvate (P-enolpyruvate)-dependent phosphotransferase system (PTS) are carbohydrate transporters found only in prokaryotes. They not only transport hexoses and hexitols, but also pentitols and disaccharides. The PTS substrates are listed in Table I. The abbreviations used (as superscripts) throughout the text for these substrates are as follows: Bgl, β-glucoside; Cel, cellobiose; Fru, fructose; Glc, glucose; Gut, glucitol; Lac, lactose; Man, mannose; Mtl, mannitol; Nag, N-acetylglucosamine; Scr, sucrose; Sor, sorbose; Xtl, xylitol.

1.2. PTS components

Carbohydrate transport occurs at the expense of P-enolpyruvate, concomitant with phosphorylation. The entire process is characterized by a number of phospho-enzyme intermediates. Textbooks usually outline these reactions and the associated phospho-enzyme intermediates as shown in Fig. 1.

However, observations early on indicated that the situation was more complicated because some systems, such as the mannitol PTS in *E. coli* lacked an E-III. Saier Jr. et al. [2] proposed that they might consist of an E-II with a covalently bound E-III, since the molecular weight of the E-II in these systems was comparable to that of E-II + E-III in the systems where they were found to be separate. A flurry of nucleotide sequence activity in the past several years has more than confirmed this prediction.

TABLE I
List of PTS carbohydrate substrates[a]

	PTS carbohydrate substrates
Triose	dihydroxyacetone
Hexoses	glucose, fructose, mannose, sorbose, trehalose, galactose, N-acetylglucosamine, β-glucoside
Hexitols	mannitol, glucitol, galactitol
Pentitols	xylitol, ribitol
Disaccharides	lactose, sucrose, cellobiose

[a]See Table 1 of [1] for the primary references.

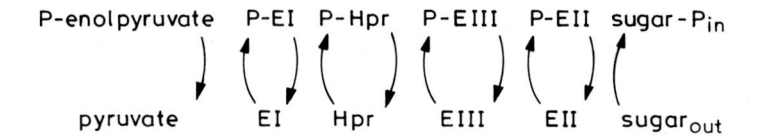

P-enolpyruvate → P-EI → P-Hpr → P-EIII → P-EII → sugar-P_{in}

pyruvate ← EI ← Hpr ← EIII ← EII ← sugar$_{out}$

Fig. 1. Scheme to show the reactions of phospho-enzyme intermediates involved in carbohydrate transport, as usually outlined in textbooks.

Moreover, the data tell us that such fusions have not been restricted to the E-II and E-III structural genes. Shuffling, splicing, fusion, duplication and deletion events have occurred in the various PTS operons during evolution. Consequently, virtually all combinations of the four proteins in Fig. 1 have been identified, E-II/E-III fusions, E-III/HPr fusions and E-III/HPr/E-I fusions. Only a complete fusion, E-II/E-III/HPr/E-I, is still missing. Fig. 2 summarizes the current state of affairs; different portions or domains of the various proteins are indicated by different shadowing [3]. Shuffling has occurred within individual structural genes as well as between structural genes. For instance, hydrophilic regions of E-II which have considerable sequence homology, can be found on the N-terminal or the C-terminal end of large hydrophobic domains as in the case of II^{Bgl} and II^{Scr} versus II^{Glc} in Fig. 2.

1.3. PTS nomenclature

The discovery of this evolutionary activity has reeked havoc with the accepted PTS protein nomenclature. Instead of the time-honored E-I, E-II, E-III and HPr, we must resort to cumbersome phrases such as 'the E-III-like domain of E-II' etc. To alleviate the problem, Saier Jr and Reizer [3] have proposed a new nomenclature (Fig. 2) based on the original IIA/IIB nomenclature of Kundig and Roseman [4]. The old nomenclature is given above the bars representing the permeases in Fig. 2 and the new nomenclature below the bars. H and I still represent HPr and E-I, respectively, but IIA replaces E-III. IIC represents the hydrophobic domain of E-

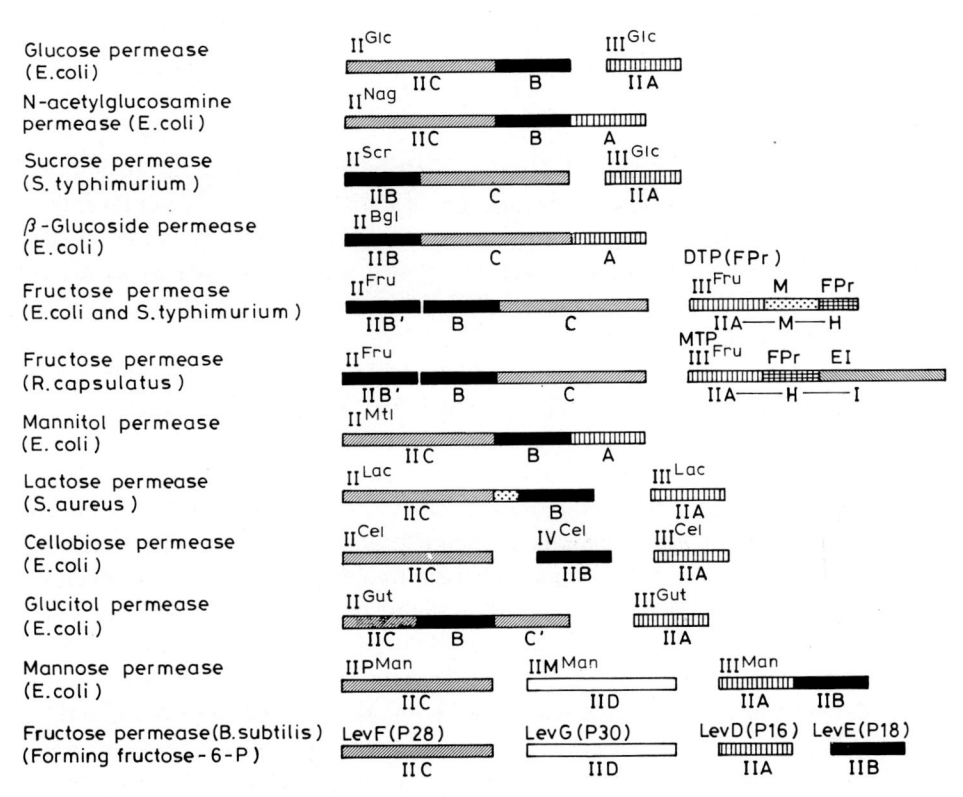

Fig. 2. Schematic representation of various PTS enzymes and their domains (taken from [3]). The different domains are indicated as follows: transmembrane hydrophobic domain (▨, IIC); the E-II domain bearing the first phosphorylation site (▥, IIA); the E-II domain bearing the second phosphorylation site (■, IIB); a transmembrane partially hydrophobic domain of unknown function (□, IID); a non-homologous domain of unknown function (▦); an Hpr-like domain, (▤, H); and an E-I-like domain, (▨, I).

II, and IIB represents the hydrophilic domain. An E-II fused to E-III such as II^{Bgl} thus becomes IICBA^{Bgl} while an E-II separated from an E-III such as II^{Glc} becomes IICB^{Glc}. This nomenclature defines domains on the basis of their function. As additional domains with new functions are discovered, they can also be given a letter code. In this review we will use new nomenclature primarily when referring to domains.

The purpose of this review is to provide an up-to-date picture of the structure and mechanism of E-II, with special reference to the developments of the past few years. We shall only treat older literature when it has a direct bearing on the more recent developments. A review of the older E-II literature can be found in [1].

2. Enzyme II structure

2.1. Sequence homology

The structural genes of seventeen E-IIs and E-IIIs have been sequenced. In addition to the 15 listed in Table 1 of [5], the sequences of E-II^{Fru} of *R. capsulatus* [6] and E-II^{Cel} of *E. coli* [7] have appeared in the last years. The ever-expanding list invites periodic sequence comparisons in search of detailed sequence homology as well as structural and organizational homology [2,5]. The results of these comparisons can be summarized as follows:

(1) With the exception of the *B. subtilis* fructose PTS, the *E. coli* mannose PTS and the *K. pneumoniae* sorbose PTS, all E-II and E-II/E-III pairs cluster in molecular weight range between 62 000 and 68 000 Da.

(2) Sequence homology between all of the proteins is very small; however, if the proteins are divided into families, more homology is evident. Lengeler [8] grouped, into one family, all of the proteins which generate D-glucopryanoside-6-P by vectorial phosphorylation, II^{Scr} II^{Glc} II^{Bgl} and II^{Nag}, and found 30–40% homology between these proteins from enteric bacteria as well as highly conserved consensus sequences. The *E. coli* II^{Mtl} and II^{Fru} family show about 45% similarity in the A domains even though they are members of different fusion proteins in the two systems (see Fig. 2). The C domains have 22% similarity; however, the domains do not occupy similar positions in the structural genes. The C domain in II^{Mtl} is located at the N-terminal end of the enzyme followed by the B domain, while in II^{Fru} it is found at the C-terminal end, preceded by the B domain and another domain of unknown function. The *E. coli* II^{Man}, *K. pneumoniae* II^{Sor} and *B. subtilis* II^{Fru} family differ from the rest in that they each consist of two membrane proteins, a strongly hydrophobic IIP^{Man} and IIA^{Sor} with 61% sequence similarity and a moderately hydrophobic IIM^{Man} and IIM^{Sor} with 63% sequence similarity.

(3) The hydrophobic C domain appears to start with an amphipathic membrane-spanning helix [5].

(4) There is one or more extended hydrophilic stretch in the hydrophobic domain which may be associated with carbohydrate binding.

2.2. Domain structure

Sequence data, in combination with functional studies, reveal several classes of E-IIs as shown in Fig. 3. These classes will most likely change and expand as future studies define additional domains with new functions. We introduce it here, only for the sake of our own convenience, in this chapter. The first class is represented by *E. coli* II^{Cel}. This protein is unique in that, as yet, it is the only representative of its class. It consists only of a single hydrophobic peptide with no hydrophilic domain attached at either end [7,9].

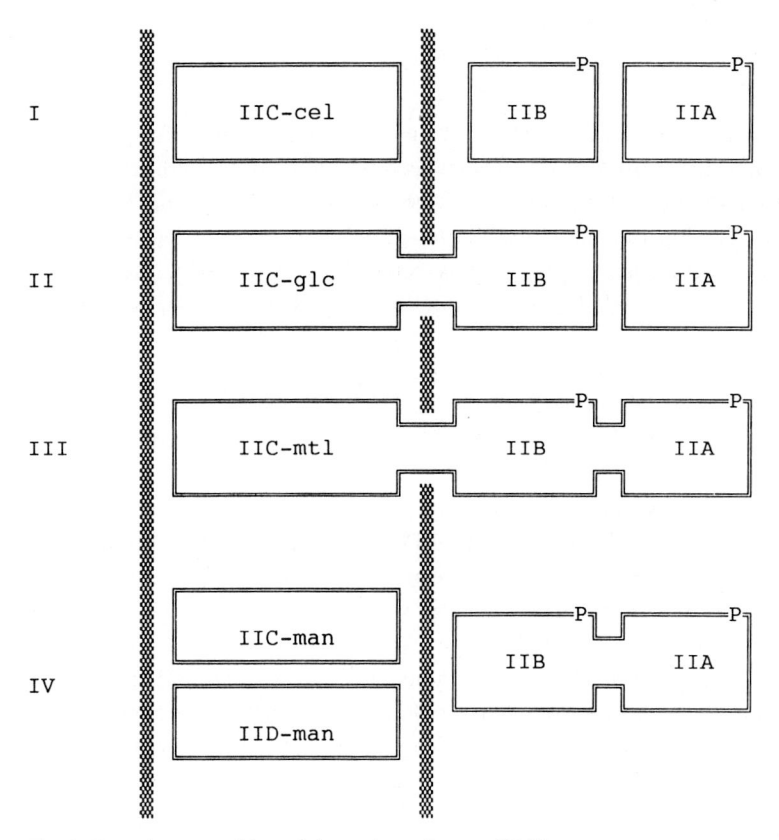

Fig. 3. Domain composition of the various classes of E-II.

The second class, represented by E. coli II^{Glc}, consists of a hydrophobic domain of approximately 360 residues followed by hydrophilic domain of approximately 100 residues [10]. Other representatives of this class are E. coli II^{Gut} [11], S. aureus II^{Lac} [12], E. coli II^{Fru} [13,14] B. subtilis II^{Scr} [15] and II^{Scr} encoded by the plasmid pUR400 [16].

The third class, represented by E. coli II^{Mtl}, consists again of a single hydrophobic domain of approximately 360 residues but with two covalently attached hydrophilic domains, equal, together, in size to the hydrophobic domain [17]. The A domain is proposed to function as a covalently attached E-III. Other representatives of this class include B. subtilis II^{Glc} [18,19], S. mutans II^{Scr} [20], E. coli II^{Bgl} [21,22] and II^{Nag} [23,24].

The last class, represented by E. coli II^{Man}, consists of a membrane domain involving two distinct peptides, one very hydrophobic, and one somewhat less hydrophobic. The domains A and B are not covalently attached to the membrane domains but are separate cytoplasmic proteins [25,107]. Other representatives of this class include B. subtilis II^{Fru} [26] and K. pneumoniae II^{Sor} [8].

2.3. Domain function

2.3.1. The A domain

As shown in Fig. 1, the function of E-III is to transfer the phosphoryl group from P-HPr to E-II. ^{32}P-phosphorylated peptides have been isolated from a number of E-III species and A domains of *S. carnosus* and *S. aureus* III^{Mtl} [27], *E. coli* II^{Mtl} [28], III^{Glc} [29], and III^{Man} [30]. In each case the phosphoryl group is carried on a histidine residue. Enough sequence similarity has been reported between these phosphorylated peptides and other E-IIIs or A domains to confirm that the A domains function as E-IIIs.

The suggestion that the A domain of class III enzymes represented a covalently attached E-III moiety was first tested using the *E. coli* II^{Bgl} and II^{Nag} systems [31,32]. Strong sequence similarities exist between *E. coli* III^{Glc} and the A domain of *E. coli* II^{Bgl} and II^{Nag}, and *K. pneumoniae* II^{Nag}. Transposons were inserted into the A domain region of the *K. pneumoniae* II^{Nag} gene to generate truncated proteins. Strains carrying these plasmids were only able to grow on, transport and phosphorylate *N*-acetylglucosamine when the cells contained III^{Glc}, indicating that III^{Glc} was able to functionally replace the deleted A domain of II^{Nag}. Conversely, the A domain of II^{Nag} was able to substitute for III^{Glc} in II^{Glc}-catalyzed reactions. Finally, antibodies against III^{Glc} inhibited II^{Nag}-dependent phosphorylation of α-methylglucoside and *N*-acetylglucosamine in *in vitro* phosphorylation assays [32]. Similar observations were made by complementing II^{Scr} with II^{Bgl} or II^{Nag}. Sucrose transport and phosphorylation are normally accomplished in conjunction with III^{Glc} since the sucrose PTS lacks a III^{Scr} and plasmid-encoded II^{Scr} lacks an A domain [33]. III^{Glc} deletion mutants carrying the II^{Scr} plasmid were unable to grow on sucrose, but when the cells were also transformed with a plasmid-encoding II^{Nag} which possesses its own A domain, growth on sucrose was restored. II^{Bgl} substituted in a similar manner for III^{Glc} in restoring growth on sucrose.

In contrast to *E. coli*, *B. subtilis* II^{Glc} is a class III enzyme with a covalently attached A domain [19] but its sucrose PTS still lacks its own E-III. Sutrina et al. [18] have recently shown that II^{Glc} complements *B. subtilis* II^{Scr} in sucrose phosphorylation in

Fig. 4. Domain complementation schemes. (A) A domain complementation. The H554A site-directed mutant is inactive in P-enolpyruvate-dependent mannitol phosphorylation because it cannot accept a phosphoryl group from P-Hpr. The measure of A domain activity is its ability to restore mannitol phosphorylation activity to this mutant. A domain activity in the AB subcloned protein can also be measured. (B) B domain complementation. The C384S site-directed mutant is inactive in P-enolpyruvate-dependent mannitol phosphorylation because it cannot pass the phosphoryl group from H554 on its own A domain to mannitol. The measure of B domain activity is its ability to restore mannitol phosphorylation activity to this mutant. B domain activity in the AB subcloned protein can also be measured. (C) C domain complementation. The activity of the C domain is measured by complementation with the purified AB domain.

A

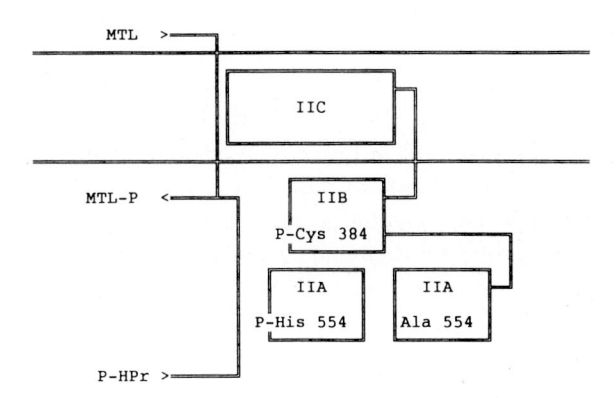

```
MTL       >─┐
────────────┼──────┌─────────┐──────────────
            │      │   IIC   │─┐
            │      └─────────┘ │
────────────┼──────┌─────┐─────┼──────────────
MTL-P     <─┘      │ IIB │     │
                   │P-Cys 384│ │
                   └─────────┘ │
                ┌─────┐ ┌─────┐│
                │ IIA │ │ IIA ││
                │P-His 554│ │Ala 554│
                └─────┘ └─────┘

P-HPr  >────┘
```

B

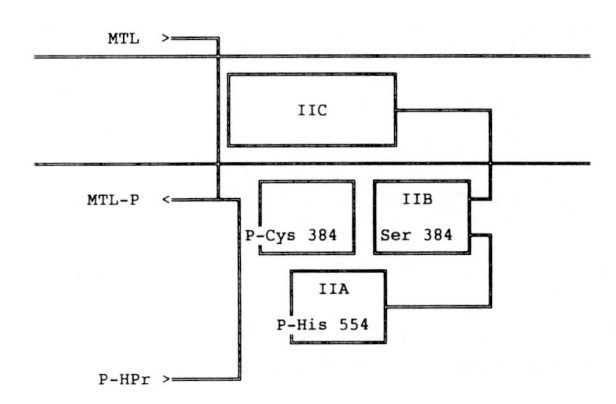

```
MTL       >─┐
────────────┼──────┌─────────┐──────────────
            │      │   IIC   │─┐
            │      └─────────┘ │
────────────┼──────┌─────┐ ┌───┼──┐──────────
MTL-P     <─┘      │     │ │ IIB  │
                   │P-Cys 384│ │Ser 384│
                   └─────┘ └──────┘
                        ┌─────┐
                        │ IIA │
                        │P-His 554│
                        └─────┘

P-HPr  >────────────┘
```

C

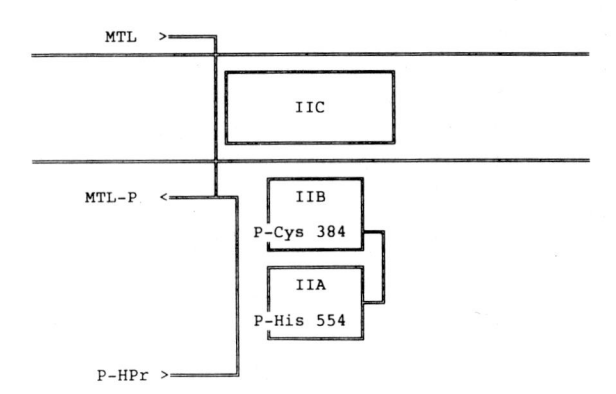

```
MTL       >─┐
────────────┼──────┌─────────┐──────────────
            │      │   IIC   │
            │      └─────────┘
────────────┼──────┌─────┐──────────────
MTL-P     <─┘      │ IIB │
                   │P-Cys 384│
                   └─────┘
                   ┌─────┐
                   │ IIA │
                   │P-His 554│
                   └─────┘

P-HPr  >────┘
```

whole cells and membrane fragments. They have also subcloned the A domain and demonstrated complementation of the A domain alone with II^{Scr}.

Mannitol transport via the PTS in *S. aureus* and *S. carnosus* differs from that in *E. coli* and *S. typhimurium* in that it is dependent on a soluble III^{Mtl}. This protein shows 38% sequence homology with the A domain of *E. coli* II^{Mtl} [27,34]. Grisafi et al. have reported that *E. coli* II^{Mtl} mutants with deletions in the A domain were unable to catalyze P-enolpyruvate-dependent phosphorylation of mannitol in vitro, an E-III-dependent reaction, supporting the E-III-like function of the A domain [35]. Phosphorylation activity was restored using either a C-terminal proteolytic fragment or a partially purified, subcloned fragment starting at residue 377, both of which could both be phosphorylated at H554 [36,37].

We have subcloned the A domain of *E. coli* II^{Mtl} after inserting a restriction site into a region of the structural gene corresponding to a flexible peptide which could function as a linker between domains [38–40]. The purified domain restored 25% of wild-type mannitol phosphorylation activity when used in the mutant complementation assay shown in Fig. 4A. Its E-III-like function was also confirmed by substituting the purified domain for *S. carnosus* III^{Mtl} in an *in vitro* mannitol phosphorylation assay with *S. carnosus* II^{Mtl}. Twenty percent of the original activity was measured when an equal amount of the A domain was substituted for *S. carnosus* III^{Mtl}. Conversely, purified *S. carnosus* III^{Mtl} was able to replace the purified A domain in the *in vitro* mannitol phosphorylation assay with *E. coli* H554A II^{Mtl} shown in Fig. 4A.

2.3.2. The B domain

The existence of a second hydrophilic domain was inferred from the difference in length of the cytoplasmic portions of class II and class III enzymes, from sequence similarities in II^{Glc}, II^{Bgl} and II^{Nag} and from the fact that these homologous regions are found on the C-terminal end of the hydrophobic domain in some cases and on the N-terminal end in others. Experimental evidence for a second hydrophilic domain came again from *E. coli* II^{Mtl} studies. Two phosphoryl groups are transiently incorporated per II^{Mtl} [41]. ^{32}P labelling located both sites in the hydrophilic portion of the enzyme, the H554 site on the A domain mentioned above and C384 [28] which is now considered to be part of the B domain. Grisafi et al. [35] proposed that residues 377–519, encompassing the C384 site, form a separate mannitol phosphorylating domain because deletions in this region eliminated Mtl/Mtl-P exchange activity. Deletions in this region did not affect the mannitol-binding function of the remaining protein.

The function of the B domain has been confirmed by subcloning and preliminary kinetic measurements. We subcloned the AB domain of *E. coli* II^{Mtl}, residues 348–637, after inserting a restriction site at a position corresponding to residue 348. The purified protein restored mannitol phosphorylation activity when measured with the A domain assay in Fig. 4A, and the B domain assay in Fig. 4B [42]. The B domain alone, residues 348–488, was subcloned, over-produced and purified and was also active in the B domain complementation assay (unpublished results). These data confirm that the B domain exists as a separate structural and functional entity.

2.3.3. The A and B domains of E. coli III^{Man}

The *E. coli* III^{Man} is the only case in which the A and B domains are linked together but separate from the hydrophobic domains. They are linked by a characteristic A-P-rich hinge region. The N-terminal A domain is phosphorylated by P-HPr, the C-terminal B domain is phosphorylated by domain A and is the phosphoryl group donor for mannose. It is the only case of a B domain which carries its phosphoryl group on a histidine instead of a cysteine. The domains have been expressed separately with retention of activity again demonstrating the functional independence of these domains [107].

2.3.4. The C domain

The C domain is proposed to be the carbohydrate-binding domain. Experimental proof of this function has been provided only in the case of *E. coli* II^{Mtl}. Mannitol binds with very high affinity; K_ds of 35–150 nM have been reported [30,41]. Deletion analysis revealed that removal of 40% of the C-terminal protein of the native enzyme, leaving only the N-terminal hydrophobic portion intact, did not significantly affect mannitol binding [35]. Lolkema et al. cleaved off domains A and B by trypsin treatment of inside-out vesicles; the high affinity binding remained unaltered [30]. We have also subcloned the C domain, residues 1–347, and expressed it stably in membranes. These membranes show qualitatively the same mannitol binding properties as wild-type enzyme [42].

Wood and Rippon [43] have monitored the binding of mannitol, glucitol and perseitol, a mannitol analogue, to purified *E. coli* II^{Mtl} by following an increase in tryptophan fluorescence which occurs upon binding. There are four tryptophans in II^{Mtl}, all located in the C domain. They also measured binding to the purified C domain, prepared by trypsin cleavage of the intact molecule. Here again, nearly identical binding behavior was observed for the intact protein and the C domain.

2.4. Domain interactions

The complementation experiments in which the A domain of a class III E-II is used as the phosphoryl group donor to the B domain of a second E-II molecule with either the same or different sugar specificity, while both are 'fixed' in a membrane matrix, raises some intriguing issues about the association state of these proteins and the kinetics of their interactions. Do E-IIs form stable homologous complexes in the membranes? If so, is it necessary to postulate the formation of stable heterologous complexes to explain, for example, the phosphorylation of the B domain of *E. coli* II^{Glc} by the A domain of II^{Nag}, or can the data be explained by assuming a transient complex formation between the domains? If stable homologous dimers or tetramers are formed, do the subcloned domains such as the A, AB or B domain of E. coli II^{Mtl} displace inactivated domains to restore activity in the complementation assays in Fig. 4A or B? We will first treat the association state of E-II and then return to this issue of domain interactions.

2.4.1. Association state of E-II

2.4.1.1. E-II^{Mtl}. The majority of association state data has been obtained for *E. coli* II^{Mtl}. The first observation was that there was a linear enzyme concentration dependence for P-enolpyruvate-dependent mannitol phosphorylation catalyzed by II^{Mtl}, but a progressive concentration dependence for the Mtl/Mtl-P exchange reaction, leading to the proposal that a II^{Mtl} dimer was responsible for exchange, while a monomer was responsible for phosphorylation [44]. Similar progressive enzyme concentration dependence was later observed in the phosphorylation reaction [45]. In addition to kinetics, there were also gel filtration and sodium dodecyl sulfate (SDS) electrophoresis data for the occurrence of a II^{Mtl} dimer. ^{32}P-labeled II^{Mtl} could be extracted from the membrane with SDS-containing buffer and appeared as a dimer on polyacrylamide gels [46,47]. The same was observed for *R. sphaeroides* II^{Fru} [48]. Finally, gel filtration of the purified protein in deoxycholate showed a monomeric and dimeric form [49]. The interpretation of the activities of these forms has been controversial. Saier Jr and Jacobson and colleagues link the monomeric form to conditions favoring the phosphorylation reaction and dimeric enzyme to conditions favoring exchange [44,47,50,51]. We, on the other hand, have asserted that a linear enzyme concentration dependence only indicates that association–dissociation does not occur over the concentration range measured; this, by itself, says nothing about the association state of the active species [45]. A reinvestigation of the conditions under which linear and non-linear II^{Mtl} concentration-dependent kinetics are found has resulted in the following observations [52]:

(1) Linear enzyme concentration dependence is observed for both the phosphorylation and exchange reaction over a 1000-fold enzyme concentration range, pM to nM, if the reactions are run in 25 mM Tris, 5 mM Mg^{2+}, pH 7.6. Under these conditions, the enzyme is associated.

(2) Non-linear concentration dependence can be observed in both reactions at low pH and low ionic strength, in the absence of Mg^{2+} (10 mM phosphate, pH 6.3).

(3) The dissociated form is inactive in the exchange reaction, but retained 25% of its maximum activity in the phosphorylation reaction.

(4) An inhibitory complex formed by preincubating Mg^{2+}, phosphate and NaF binds to the enzyme and affects the association/dissociation equilibrium.

To determine the nature of the associated state, gel filtration high-performance liquid chromatography (HPLC) has been performed in a number of detergents with moderate to high critical micelle concentration (CMC) values. II^{Mtl} eluted, in all cases, at a position equivalent to globular proteins with molecular weights ranging from 320 000 to 270 000 Da. The low ionic strength conditions causing dissociation in the kinetic experiments could not be used for gel filtration due to non-specific binding to the column. However, 500 mM phosphate, which also inhibits, causes dissociation of the complex. The association state of the separate A, B and C domains have also been examined. The A, B and AB domains all elute at their monomer positions. The C domain elutes as an associated complex, indicating that it is the hydrophobic domain which is responsible for the association behavior of the intact protein (unpublished results). Whether the associated state of II^{Mtl} and its C domain is a dimer or tetramer has yet to be definitively established.

The molecular weight of II^{Mtl} in *E. coli* membrane vesicles has also been examined by radiation inactivation analysis [53]. The decrease in enzymatic activity as a function of radiation dose was consistent with the molecular weight of a dimer, $140\,000 \pm 20\,000$ Da. Phosphorylation and exchange activity measurements gave the same molecular weight. This technique reports the minimum molecular weight of the catalytically active complex; it suggests that the dimer is the smallest complex capable of catalyzing either phosphorylation or exchange. These data do not agree with the steady-state kinetic analysis which indicates that the dissociated form is also able to catalyze phosphorylation, albeit at a submaximal rate.

2.4.1.2. E-II^{Glc}. Erni [54] characterized purified *S. typhimurium* II^{Glc} in the detergent octyl-polyoxyethylene by analytical equilibrium centrifugation and cross-linking with glutaraldehyde. Both results were consistent with a dimer. The association state of the II^{Glc}/III^{Glc} complex was characterized by immunoprecipitation. Cells were grown on sulfur-limiting medium with [^{35}S]sulfate, harvested and disrupted by freeze-thawing and sonication in 1.5% SDS. The supernatant was precipitated with purified monoclonal IgG against II^{Glc}, fractionated on SDS polyacrylamide gels and quantitated. A ratio of two III^{Glc} per II^{Glc} was found or four moles of III per II^{Glc} dimer.

The association characteristics of purified *E. coli* II^{Glc} were examined using zonal centrifugation through glycerol gradients [55]. Purification in the absence of dithiothreitol (DTT) or dialysis against buffers in the absence of DTT resulted in oxidized, dimeric enzyme, with a different sedimentation velocity than monomeric enzyme. The purified protein kept in the presence of DTT showed a temperature-dependent sedimentation velocity. The enzyme centrifuged at 4°C sedimented as a dimer, while that centrifuged at 15°C sedimented as a monomer suggesting three possible forms of the enzyme, reduced monomer, reduced dimer and oxidized dimer. Intersubunit disulfide formation through Cys 421 resulted from oxidation. The authors concluded that only the monomeric enzyme was active, because DTT was required to restore phosphorylation activity to the oxidized enzyme, and because the oxidized enzyme could not be phosphorylated. Extrapolation from our II^{Mtl} results [28] would now indicate that Cys 421 is the domain B phosphorylation site [54]; consequently, when oxidized, it cannot be phosphorylated. Inactivation results from oxidation, not from dimer formation. This is in keeping with all of the studies published on the sulfhydryl sensitivity of II^{Glc} and II^{Mtl} [40,56–66]. Due to these new insights into the role of the activity-linked cysteine, we must conclude that the above studies are not capable of addressing the issue of the activity of the monomer versus dimer form.

In addition to the experiments listed above, two other results were taken as evidence that II^{Glc} was active as a monomer. Firstly, when α-methylglucoside and glucose-6-P were included in the glycerol gradient during zonal centrifugation, α-methylglucose-6-P was found in all fractions through which monomeric II^{Glc} had passed. Secondly, when the plasmid encoded C421S mutated enzyme was brought to expression in cells containing chromosomally encoded wild-type II^{Glc}, no inhibition of P-enolpyruvate-dependent glucose phosphorylation activity was observed, as would be expected if the dimer were essential for activity. The first result says that the monomer can catalyze exchange in contrast to findings for II^{Mtl}. Since no enzyme concentration-dependent kinetics have been done, it is impossible to comment on the specific activity of the monomer form or to say whether an associated form would be more or less active than the monomer. The second result says that either dimers are not necessary for P-enolpyruvate-dependent phosphorylation or that II^{Glc} heterodimers do not form spontaneously in the membrane.

2.4.2. Kinetics of domain interaction

The II^{Glc} association state data in the previous section are too preliminary to use in a discussion of the nature of the domain interactions in the II^{Nag} or II^{Bgl} complementation of II^{Glc} or II^{Scr} activities [18,31,32,78]. Therefore, we shall confine ourselves to II^{Mtl}.

The A domain complementation assays shown in Fig. 4A has been carried out as a function of the A domain concentration in the presence of enough HPr, E-I and P-enolpyruvate to keep the A domain fully phosphorylated in the steady-state. The kinetics exhibited saturation behavior with respect to the concentration of phospho-A domain. This result indicates that the transfer of the phosphoryl group to the B domain of the H554A-II^{Mtl} is most likely preceded by a formation of a complex. For this to occur, the inactive A domain carrying the H554A mutation might swing out of the way to allow for binding of the active A domain to the active B domain. On the other hand, there could be a complex between the active and mutant A domains and a phosphoryl group transfer across the subunit interface as shown in Fig. 4A [39].

Heterologous dimers have been demonstrated kinetically in the case of Mtl/Mtl-P exchange [40]. Detergent solubilized H554A-II^{Mtl} and C384S-II^{Mtl} were examined for exchange activity. The H554A enzyme catalyzed exchange at wild-type rates, whereas the C384S enzyme was inactive. This is in keeping with the expectation that C384 is the phosphoryl group donor–acceptor site to Mtl/Mtl-P. When exchange kinetics were done at a fixed concentration of the H554A enzyme and increasing concentrations of the C384S enzyme, a C384S enzyme concentration-dependent stimulation of up to 1.8 times the H554A rate was observed. This supports the proposal that association at least to a dimer is necessary for exchange activity in the II^{Mtl} system. In this view, the dimer is maximally active if there is at least one Cys 384 residue per dimer. Such a state is achieved by complementing each H554A subunit with a C348S subunit. A maximum two-fold stimulation would be expected, all other things being equal.

Stable heterologous complexes are not necessary to explain the limited P-enolpyruvate-dependent mannitol phosphorylation kinetic data now available from domain complementation assays; transient complexes between domains are sufficient. The challenge remains, however, to visualize how a subcloned A or B domain would be able to transiently associate with an A or B domain on an E-II^{Mtl} dimer or tetramer, at rates high enough to be comparable with wild-type enzyme.

3. Binding studies

3.1. General considerations

The physiologically relevant function of the E-IIs is vectorial phosphorylation, i.e., transport with concomitant phosphorylation of the sugar. This reaction requires a phosphoryl group donor, for instance, P-HPr, and it may even be argued that the active species is phosphorylated E-II,

$$\text{Sugar}_{out} + \text{II-P} <=> \text{Sugar-P}_{in} + \text{II}. \tag{1}$$

The relevant state of the enzyme with respect to the sugar appears to be the phosphorylated enzyme, which raises the question whether or not the unphosphorylated E-IIs are reactive towards the sugar substrates at all. In fact they are. E-IIs, in the absence of their respective phosphoryl group donors, bind the sugars with high affinity. The relevance of this binding to the overall mechanism is that it allows the study of certain properties of the translocation mechanism without interference of the otherwise dominating phosphorylation reaction. Binding studies potentially give information on important items such as the location of the binding site, the orientation of the binding site, translocation steps and cooperativity between binding sites.

Binding to transport proteins may be of particular interest, since binding not only assays the affinities of the binding site on the transporter protein but also the translocation equilibria [67]. In terms of enzyme catalysis, a transport protein transforms a substrate, a molecule located at one side of the membrane, into a product, the same molecule at the other side of the membrane, without chemical modification. Substrate must bind to a particular conformation of the enzyme with the binding sites accessible only from, for example, the outside. Similarly, the release of the product has to occur from a conformation which opens the binding site to the inside only; this implies at least one transition step between the two types of conformations (see Fig.

```
                    K_f
      E_per  <====>  E_cyt
        ▲              ▲
        ‖              ‖
K_D^per ‖              ‖ K_D^cyt
        ▼              ▼
      E_per:S <====> E_cyt:S
                    K_r
```

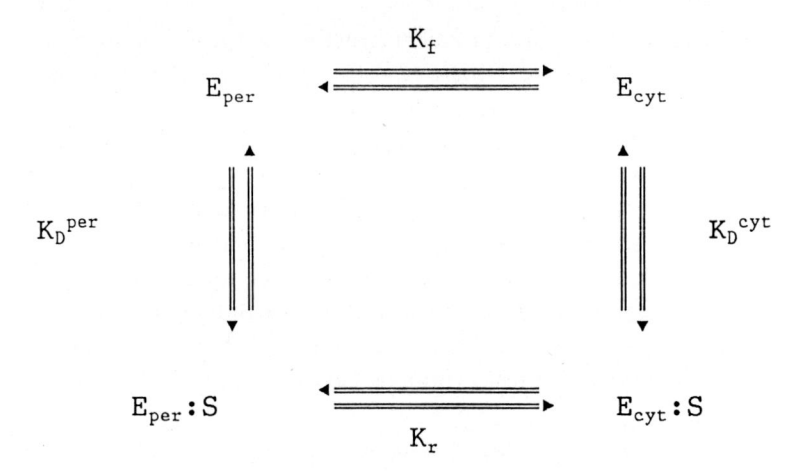

Fig. 5. A simple kinetic representation of a transport reaction catalyzed by a bacterial transport protein. E_{cyt}, and E_{per} denote those conformations of the enzyme with the binding site facing the cytoplasm and periplasm, respectively.

5). Since no chemical reaction is involved, measurements of binding of substrates and products to the enzyme under equilibrium conditions can be easily performed and are adequately described by the following equilibria,

$$S + E_{cyt} \underset{}{\overset{K_d^{cyt}}{<=>}} S:E_{cyt} \underset{}{\overset{K_r}{<=>}} S:E_{per} \underset{}{\overset{K_d^{per}}{<=>}} E_{per} + S. \qquad (2)$$

The overall dissociation constant obtained comprises the intrinsic dissociation constants for the ligands and the equilibrium constant for the conformational transition [68],

$$K_d = \frac{K_d^{cyt} + K_r K_d^{per}}{1 + K_r}. \qquad (3)$$

Note that in equilibria (2) the subscripts 'per' and 'cyt' are omitted where substrate S is concerned. This is obvious when the binding is measured to a solubilized transport protein, but also in the case where the enzyme is embedded in the membrane of closed vesicular structures, internal and external substrate will have equal concentrations at equilibrium (see Fig. 5). Consequently, the binding is independent of the orientation of the enzyme in the membrane.

In the case of E-IIs, we do not deal with a transport protein as described in Fig. 5. Clearly, the most important difference is that the substrate *is* chemically modified. Therefore, we do not know whether or not a state E_{cyt}:S is in contact with the internal water phase, if it exists at all. On the other hand, since transport is part of the overall function of E-IIs, translocations as described in Fig. 5 *may* very well be part of the overall mechanism. These questions have been approached by binding studies to the unphosphorylated E-IIs.

3.2. Equilibrium binding to E-II

E-IIs in the unphosphorylated state bind their substrates with extremely high affinity. Reported affinity constants are in the sub μM range. The *R. sphaeroides* II^{Fru} embedded in cytoplasmic membranes permeabilized with the detergent deoxycholate binds its substrate fructose with an affinity constant K_d = 300 nM [1]. Purified *E. coli* II^{Mtl} solubilized in the detergent decylpolyethylene glycol (decyl-PEG) binds mannitol with K_d = 100 nM [41]. Approximately the same value was reported when cytoplasmic membranes containing II^{Mtl} were solubilized directly in decyl-PEG. However, when the detergent was omitted in the latter case the affinity was even higher, K_d = 35 nM [30]. Experiments in our laboratory indicate an affinity constant for the *E. coli* II^{Glc} in the same order of magnitude (unpublished results).

Equilibrium binding to II^{Mtl} has been studied in more detail. The availability of purified enzyme in large amounts allows for the assessment of the binding site stoichiometry [41]. The high affinity binding with K_d = 100 nM amounted to one binding site per two molecules of enzyme. However, at higher mannitol concentrations a second site became apparent with an estimated affinity constant K_d = 10 μM. The data indicate that each II^{Mtl} monomer possesses a mannitol binding site and that dimer formation is accompanied with functional interaction between these binding sites.

Chemical modifications like alkylation with (*N*-ethylmaleimide (NEM) or oxidation with diamide that inhibit the phosphorylation activity of the enzyme did not seem to have any significant effect on the high affinity binding site when the enzyme was solubilized in the detergent decyl-PEG [69,41]. However, in the intact membrane these treatments reduced the affinity by a factor of 2–3. The reduction of the affinity was exclusively due to modification of the cysteine residue at position 384 in the B domain [69]. Apparently, the detergent effects the interaction between the B and C domains.

3.3. Orientation of the binding site

Early indications that unphosphorylated E-IIs might bind there sugar substrates with high affinity came from the strong inhibition by the sugar observed in the sugar/sugar-P exchange reaction catalyzed by E-IIs (see section 5). It was argued that the mechanism of this inhibition could be competition between sugar and sugar-P for the binding site on the unphosphorylated enzyme [1]. Binding of sugar would prevent the binding of the sugar-P and subsequent phosphorylation of the enzyme. This would imply that the sugar binds to the product leaving site on the enzyme which faces the cytoplasm. The sugar binding site on the unphosphorylated

TABLE II
Outline of the experiment to determine the orientation of a single binding site on unphosphorylated II^{Mtl}

Orientation of site	RSO vesicles	ISO vesicles
Cytoplasmic	−	+
Periplasmic	+	−
Dynamic	+	+
Experimental	+	+

The first column indicates the possible orientations. 'Dynamic' indicates that the site is not fixed at one side of the membrane. Whether or not binding would be measured is indicates by the + and − signs, respectively. The last line in the table gives the results from the actual experiment [30]. RSO, right-side-out; ISO, inside-out.

enzymes would be oriented towards the cytoplasm. More direct evidence for the latter hypothesis was sought by measuring mannitol binding to II^{Mtl} embedded in cytoplasmic membrane vesicles with either an inside-out (ISO) or right-side-out (RSO) orientation [30]. The cytoplasmic domains of II^{Mtl} are at the interior face in the case of the RSO vesicles and at the exterior face in the case of the ISO vesicles. The basic set-up of the experiments is outlined in Table II. With the binding site fixed at the cytoplasmic side of the membrane, binding would only be detected to the ISO membranes and not to the RSO membranes. This would be the other way around with the site fixed at the periplasmic side of the membrane. A third option is the one discussed in section 3.1; the site is not fixed at either site of the membrane but can change its orientation spontaneously as described in Fig. 5. Binding will be observed both to the RSO and ISO membrane vesicles. The experiments agreed with the prediction of the latter option. Scatchard analysis of the binding to both RSO and ISO membranes demonstrated high affinity bindings with about equal K_ds. The total number of sites extrapolated from the high affinity binding curve was not significantly different from the total number of sites found after solubilization of the membranes in the detergent decyl-PEG, indicating that all high affinity sites were accessible in both types of intact membrane structures.

Unphosphorylated II^{Mtl} functioning according to Fig. 5 catalyzes facilitated diffusion of mannitol across the membrane. The same process has been reported for purified II^{Mtl} reconstituted in proteoliposomes [70]. The relevance of this activity in terms of transport of mannitol into the bacterial cell is probably low, but it may have important implications for the mechanism by which E-IIs catalyze vectorial phosphorylation. It would indicate that the transmembrane C domain of II^{Mtl} is a mannitol translocating unit which is somehow coupled to the kinase activity of the cytoplasmic domains. We propose that the inwardly oriented binding site which is in contact with the internal water phase (E$_{cyt}$:Mtl, see Fig. 5) is the site from where mannitol is phosphorylated when transport is coupled to phosphorylation. Mechan-

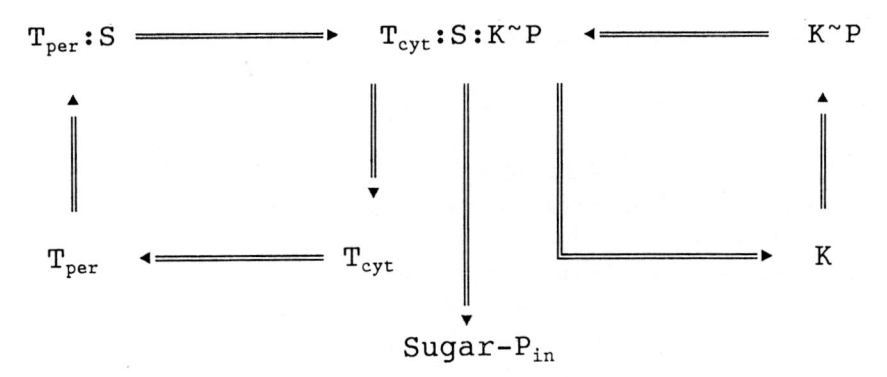

$$T_{per}:S \rightleftharpoons T_{cyt}:S:K{\sim}P \rightleftharpoons K{\sim}P$$

$$T_{per} \rightleftharpoons T_{cyt}$$

$$\text{Sugar-}P_{in} \qquad K$$

Fig. 6. Vectorial phosphorylation by a mechanism in which translocation and phosphorylation of the sugar are two distinct steps. The product binding site of the translocator T (domain C of II^{Mtl}) would be the substrate binding site of the kinase K (domains A and B). Since both the left-hand cycle and the right-hand cycle are catalyzed by the same enzyme they will very likely be kinetically dependent. Note that the kinetic cycle on the left-hand side of the figure is identical to Fig. 5.

istically, transport and phosphorylation of the sugar would be two separate steps as is schematically shown in Fig. 6. This issue will be discussed further in section 4.

3.4. Kinetics of binding

Conformational transitions of the translocator domain of II^{Mtl} have been detected by following the binding events in time [30,71]. The technique used was flow dialysis [72] which has a low time resolution. Our system has been optimized to a response time of $t_{1/2} = 10$ s. Nevertheless, interesting data could be collected with this technique. The time course of the binding of mannitol to RSO membranes in the backward and forward direction was too fast to be measured either at room temperature or at 4°C. Surprisingly, the binding to, and release from, the ISO membranes was slow enough to follow. At room temperature it took about 1 min for the binding to equilibrate after the addition of mannitol to the vesicles. The release of bound mannitol, measured as the rate of exchange of bound [^{3}H]mannitol with excess [^{1}H]mannitol was even slower with a half time of about 1 min. By itself, this asymmetry in the time course of the backward and forward processes indicated that they were enzyme-catalyzed processes and that passive diffusion over the membrane was not significantly involved. Within the framework of the interpretation of the data in the former paragraph, that unphosphorylated II^{Mtl} catalyzes facilitated diffusion according to Fig. 5 (replace S with Mtl) the data was taken as evidence that, at equilibrium, bound mannitol accumulated in state E_{per}:Mtl and therefore was bound to the periplasmic side of the membrane. The rapid binding to and dissociation from the RSO membranes would reflect the equilibrium between E_{per} and E_{per}:Mtl. The slow exchange between mannitol bound to ISO membranes and excess unlabelled mannitol would demonstrate the transition between E_{per}:Mtl and

E_{cyt}:Mtl followed by dissociation of mannitol at the cytoplasmic side (E_{cyt}:Mtl → E_{cyt}). The reverse pathway would be measured with the binding to the ISO membranes. The small rate constants for the latter two processes indicate that turnover through the cycle as depicted in Fig. 5 is very slow and has to be accelerated considerably when transport is coupled to phosphorylation.

The binding kinetics to the ISO membranes were studied in more detail by lowering the temperature to 4°C [71]. This slowed down the transitions, making it possible to measure them more accurately. In fact, two time phases could be discriminated in the binding event. An initial rapid phase that relaxed within a few minutes was followed by a much slower phase that required more than an hour to equilibrate. It was shown that the slow phase was caused by a fraction of the sites that initially were in a state not accessible to the substrate. These empty sites would be slowly recruited to a state where binding could take place. The most straight-forward interpretation of this recruitment of sites would be translocation of a fraction of unloaded sites from the internal to the external phase of the ISO membrane (E_{per} → E_{cyt}). The faster phase was interpreted as a combination of two steps: (1) the slow association step to the cytoplasmic facing binding site (E_{cyt} → E_{cyt}:Mtl); and (2) a conformational change of E_{cyt}:Mtl to a state with a higher affinity, presumably E_{per}:Mtl. Therefore, both phases appear to measure conformational changes of the translocator domain.

The interpretation of much of the binding data given so far is based upon the assumption that the high affinity binding sites represent a population of independent sites. In the unphosphorylated II^{Mtl} these sites would open up either to the periplasmic or cytoplasmic side of the membrane independently of each other. The assumption ignores the evidence that the enzyme is, in fact, multimeric and that the data

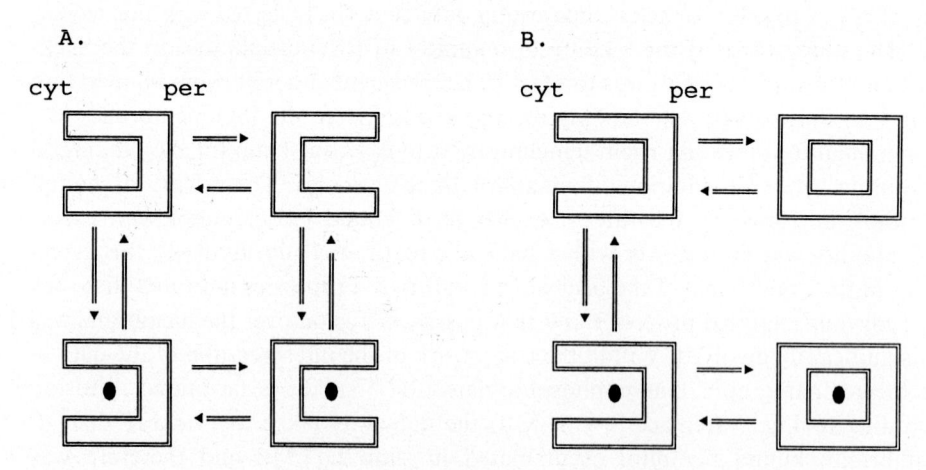

Fig. 7. Two possible interpretations of the transitions of the translocator domain detected by the kinetics of binding to ISO membranes at 4°C. (A) Translocation. The states not in contact with the external medium are the states in contact with the internal medium. (B) Occlusion. The binding sites can be in a state where they are not accessible from either side of the membrane. The spots represent the substrate molecule. cyt and per represent the cytoplasmic and periplasmic side of the enzyme, respectively.

suggest interaction between the binding sites on the monomeric units (see above). To what extent this will affect the present interpretations will have to follow from future experimentation. In particular, these studies will focus on the identification of the two 'hidden' states detected in the binding kinetics to the ISO membranes at 4°C described in the last paragraph. They were interpreted to be those states of the enzyme with the binding sites opened up to the internal phase of the vesicle (E_{per} and E_{per}:Mtl). However, in a more complex situation, it could be possible that they actually are occluded states of the enzyme (see Fig. 7).

4. The coupling between transport and phosphorylation

4.1. General considerations

Two types of coupling should be considered when discussing the activities of E-IIs. Firstly, since E-IIs catalyze 'active' transport, i.e., they are capable of accumulating their substrates inside the cell, the coupling between the free energy donating reaction and the free energy consuming reaction should be considered. Secondly, the coupling between the transport and the phosphorylation of the sugar should be considered.

Accumulation of the sugar moiety catalyzed by E-IIs is achieved by coupling the thermodynamic 'down-hill' hydrolysis of the phosphoryl group donor, for example P-HPr, to the 'up-hill' phosphorylation of the sugar. Strictly taken, experimentally uncoupling of the 'driving' and 'driven' part of the overall reaction is not possible due to the cross-over of the phosphoryl group between the two processes. The best representatives of the two parts in this particular case may be hydrolysis of the phosphoryl group donor catalyzed by E-IIs in the absence of the sugar and transport of the sugar catalyzed by E-IIs in the absence of the phosphoryl group donor, respectively. Both reactions are slow compared to the turnover rates of the enzyme. The latter has been discussed in section 3, whereas the former is related to the stability of the phosphorylated E-IIs. E-II-catalyzed hydrolysis of the phosphoryl group donor proceeds through hydrolysis of the phosphorylated E-IIs. Both stability studies [73] and direct measurements of the hydrolysis rate [74] indicate that E-II-Ps are quite stable. Therefore, from an energetic point of view, the degree of coupling between the 'down-hill' and 'up-hill' part of the reactions catalyzed by E-IIs is high; the slip is small.

The coupling between transport and phosphorylation of the sugar is more interesting from a mechanistic point of view. Two types of activities may be discriminated when transport and phosphorylation of the sugar would not be strictly coupled events:

(1) E-IIs might be capable of phosphorylating free sugar located at the cytoplasmic side of the membrane (phosphorylation without transport).

(2) E-IIs in the phosphorylated state might be capable of catalyzing facilitated diffusion of the sugar across the membrane (transport without phosphorylation). Detection of these activities as such is often difficult because conditions are such that the coupled reaction (transport plus phosphorylation) is also catalyzed. Nevertheless, experiments have been successfully set up in such a way that the enzyme is more or less tricked. This has often led to very unphysiological conditions, but it should then be borne in mind that these experiments were set up to unravel the kinetic mechanism of E-IIs, which requires the detection of all possible kinetic pathways. The experiments will be discussed in more detail in the following paragraphs. Finally, a model in which these activities have been integrated will be discussed. This model gives an explanation for the coupling between transport and phosphorylation of the carbohydrate during vectorial phosphorylation.

4.2. Phosphorylation of free cytoplasmic carbohydrates

The difficulty in demonstrating E-II-catalyzed phosphorylation of free cytoplasmic sugar is that a situation has to be created in which there is free substrate at the cytoplasmic side of the membrane but not at the periplasmic side. A pool of substrate at both sides of the membrane would make it very difficult, if not impossible, to identify the origin of the sugar moiety of the sugar-P. This problem has been solved elegantly by using disaccharides which were accumulated by non-PTS carriers and subsequently hydrolyzed to internal free hexoses by carbohydrate hydrolases. Thompson et al. [75] compared the yields of growth on lactose of wild-type *S. lactis* with a mutant lacking the mannose PTS and glucokinase. The mutant did not grow on externally supplied glucose and its cell yield, per mole of lactose used, was 50% that of wild-type cells. This is in keeping with its inability to phosphorylate the glucose generated internally by the hydrolysis of lactose. Internal phosphorylation was demonstrated in wild-type *S. lactis* cells using lactose derivatives with [^{14}C]2-deoxyglucose (2-DG) as the aglycon substituent of the disaccharide. The derivative 2-D-lactose was taken up by the lactose-PTS yielding internal 2-DG by the action of β-galactosidase. The 2-DG was recovered as 2-DG-6P under conditions where there was endogenous *N*-acetylmannosamine present to prevent efflux and re-entry of 2-DG via the mannose-PTS. Since 2-DG is not a substrate for glucokinase and since 2-DG-6P is not formed during uptake of 2-D-lactose in cells lacking II^{Man}, the 2-DG-6P must be formed by the phosphorylation of internal 2-DG via II^{Man}.

Similar observations have been reported for *E. coli*. Cells growing on lactose-minimal medium do not expel glucose into the medium indicating that the glucose formed from the disaccharide is phosphorylated and metabolized. It is not excreted and taken back up by II^{Glc}, however, because unlabelled α-methylglucoside added to the medium does not compete for the phosphorylation of the glucose derived from lactose. Cells lacking II^{Glc} or carrying the mutated enzyme, II^{Glc}-C421S, do release glucose into the medium. Together, these observations indicate that the internally formed glucose is phosphorylated without first exiting the cell [56].

Kinetic measurements on II^{Mtl} reconstituted in proteoliposomes are also consistent with the phosphorylation without transport. II^{Mtl} reconstituted by the detergent dialysis method into proteoliposomes assumes a random orientation; the cytoplasmic domains face inward for 50% and outward for 50%. Those facing inward catalyze transport of external mannitol to the interior when E-I, HPr and P-enolpyruvate are included on the inside. Those facing outward convert external mannitol to external Mtl-P when HPr, E-I and P-enolpyruvate are included in the external medium. Comparison of the rates showed that the rate of external phosphorylation in this system was higher than the rate of transport. If transport and phosphorylation were obligatorily coupled, the rate of phosphorylation would not exceed the rate of transport [70].

4.3. Facilitated diffusion catalyzed by E-II

4.3.1. Diffusion in uptake studies

Evidence is available from both uptake and efflux experiments that facilitated diffusion can be catalyzed by E-II. In 1976, Kornberg and Riordan [76] reported that strains of *E. coli*, devoid of systems for the active transport of galactose, still grew on galactose, but only at high concentrations. The entry of galactose did not require E-I activity and the sugar appeared in the cell as free galactose. They proposed that galactose was taken up either by a facilitated diffusion carrier or by a PTS carrier operating in a facilitated diffusion mode. Evidence in favor of a PTS carrier operating in a facilitated diffusion mode was provided by Postma [77]. *S. typhimurium* strains incapable of growth on galactose and mannose reverted to a Man^+ phenotype with the reappearance of II^{Man} activity. These revertants all regained their ability to grow on galactose even in strains lacking E-I, indicating that galactose transport was catalyzed by II^{Man} operating in the facilitated diffusion mode.

S. typhimurium normally transport trehalose via the galactose permease and are able to grow on this substrate in the complete absence of PTS phosphorylating activity. However, in *S. typhimurium* which lack a functional galactose permease, II^{Man} appears to be able to transport trehalose [78]. There is no evidence that trehalose is phosphorylated in this process, again pointing to II^{Man}-dependent transport in the facilitated diffusion mode.

The strongest *in vivo* evidence for facilitated diffusion comes from II^{Glc} mutants in which transport and phosphorylation of glucose were uncoupled [79,80]. These were *S. typhimurium* HPr and E-I deletion mutants capable of growing on and fermenting glucose. They possess a low affinity glucose transport system as demonstrated by a 1000-fold higher K_m for glucose oxidation than found in wild-type cells. Insertion of a Tn*10* transposon into the II^{Glc} structural gene resulting in a defective II^{Glc} abolished this property. In these uncoupled mutants, glucose appears to enter the cell via II^{Glc} operating in the facilitated diffusion mode, after which it becomes phosphorylated by glucokinase. Selective pressure and growth in a chemostat at low concentrations of glucose resulted in adaptation by decreasing the K_m for glucose apparently by additional mutations in the II^{Glc} gene.

Postma and Stock [81] showed that HPr or E-I mutants were unable to grow on PTS carbohydrates suggesting that transport without phosphorylation did not take place in apparent contradiction with the studies presented above. The explanation may be that facilitated diffusion via PTS carriers is observed only in abnormal situations, carbohydrate being transported by the incorrect PTS carrier (galactose via the mannose carrier) or transport via a mutated carrier. Efflux, which also reflects facilitated diffusion, is more common for PTS carriers.

4.3.2. Diffusion in efflux studies

The history of observations of efflux associated with PTS carriers is nearly as old as PTS itself. Gachelin [82] reported that N-ethylmaleimide inactivation of α-methylglucoside transport and phosphorylation in *E. coli* was accompanied by the appearance of a facilitated diffusion movement of both α-methylglucoside and glucose in both directions, uptake and efflux. His results could not discriminate, however, between one carrier operating in two different modes, active transport for the native carrier and facilitated diffusion for the alkylated carrier, or two distinct carriers. Haguenauer and Kepes [83] went on to show that alkylation of the carrier was not even necessary to achieve efflux; NaF treatment which inhibits P-enolpyruvate synthesis was sufficient; but this study did not address the question of one carrier or two.

That PTS E-IIs were responsible for efflux was first confirmed by monitoring the efflux of [^{3}H]mannitol from an *E. coli* E-I mutant during growth on the substrate galactosyl-[^{3}H]mannitol [84]. The substrate was taken up by the β-galactoside permease and hydrolyzed by β-galactosidase to mannitol and galactose. The cells did not possess the machinery to metabolize free mannitol and excreted it into the medium if they had been induced for II^{Mtl}. However, no efflux occurred without II^{Mtl} suggesting that II^{Mtl} was responsible for the efflux. Similar observations were made using a mutated form of *E. coli* II^{Glc} in which the activity-linked Cys 412 was replaced by a serine [56]. When an *E. coli* strain lacking II^{Glc}, II^{Man} and glucokinase, which grew poorly on lactose-minimal medium, was transformed with a plasmid containing the C412S-II^{Glc} gene, a stimulation of the growth rate was seen, as well as a stimulation in the rate of efflux of labeled glucose derived from lactose.

Efflux has been most thoroughly studied in Gram-positive micro-organisms where it has been characterized as a one of the mechanisms, inducer expulsion, for the regulation of sugar transport activity [85–89]. Lactose and its non-metabolizable analogue, thiomethyl-β-galactoside (TMG) are accumulated in *S. pyrogenes* by a lactose-specific PTS. Cells preloaded with TMG-P were observed to rapidly expel TMG if glucose or mannose was added to the medium, but not if non-PTS substrates were added. The process required PTS uptake activity of these two sugars and suggested that some intermediate in the glucose/mannose metabolism was involved in the expulsion process [85]. The cells reaccumulated the expelled TMG once the glucose and mannose supply were depleted [86]. Similar efflux processes have been reported for TMG-P and xylitol-P in *L. casei* [88,89]. That the PTS E-IIs were involved in these efflux processes is supported by the following data:

(1) *S. lactis* mutants lacking II^{Lac} accumulated TMG via an adenosine triphosphate (ATP)-dependent permease, but glucose was unable to elicit TMG efflux suggesting that, in wild-type cells, II^{Lac} was essential for the expulsion process.

(2) *S. lactis* II^{Man}-deficient strains accumulated glucose derived from lactose to 100 mM; no efflux was observed. The parent strain containing II^{Man} showed a first-order efflux of glucose [75,90].

(3) Inhibitors which blocked TMG uptake via II^{Lac} in *S. pyrogenes* showed the same selectivity in the inhibition of TMG efflux from these cells [87].

(4) In the case of xylitol efflux in *L. casei*, II^{Xtl} is the only route for xylitol transport.

(5) Efflux of mannitol has been demonstrated using proteoliposomes containing only *E. coli* II^{Mtl}. No II^{Mtl}-dependent efflux of glucose from these proteoliposomes was observed. Furthermore, no efflux of mannitol was observed from liposomes lacking II^{Mtl} [70].

Sutrina et al. demonstrated that inducer expulsion proceeds by a facilitated diffusion mechanism [91]. The efflux of accumulated [^{14}C]TMG from *S. pyrogenes* was not influenced by the presence of high concentrations of unlabelled TMG in the external medium; however, no efflux was observed with [^{14}C]TMG in the external medium at concentrations equal to the internal [^{14}C]TMG. The authors argue that the strong temperature dependence and high activation energy of the efflux are characteristic of an enzyme-catalyzed process, rather than the simple opening of a channel.

4.3.3. Regulation of efflux

Dephosphorylation and expulsion serve as a mechanism for elimination of toxic concentrations of non-metabolizable carbohydrates. Only free carbohydrate is expelled from the cell. Since the phosphorylated carbohydrate is the normal product of a PTS-catalyzed transport event and the substrate for the subsequent metabolic reactions, a carbohydrate transported by the PTS will usually not occur in the dephosphorylated form. They are observed, however, if the carbohydrate cannot be metabolized. Gachelin [82] showed that non-metabolizable α-methylglucoside first appeared inside *E. coli* in the phosphorylated form and was subsequently dephosphorylated when it reached high concentrations, presumably by a sugar phosphate phosphatase. In Gram-positive micro-organisms, the dephosphorylation event is regulated. In the case of *S. lactis* and *S. pyrogenes*, high concentrations of accumulated TMG-P are maintained without efflux until the TMG-P-loaded cells are exposed to glucose. A rapid dephosphorylation and subsequent expulsion of TMG-P followed. ATP and the products of glucose metabolism were necessary for the dephosphorylation event [85,86,92]. The proposed mechanism for this regulation is that products of glucose metabolism activate a kinase which phosphorylates Gram-positive HPrs at a regulatory site, Ser46. The P-Ser46-HPr in turn, activates a phosphatase specific for phosphorylated carbohydrates. Such a phosphatase has been purified from *S. lactis* [93]. The regulatory site is lacking in Gram-negative HPrs where the dephosphorylation event appears to be unregulated.

4.4. Coupling in vectorial phosphorylation

The phosphorylation of cytoplasmic sugar and the facilitated diffusion from the cytoplasm to the periplasm are catalyzed by the E-IIs under conditions where they are also active in the vectorial phosphorylation reaction. Therefore, the former two activities should be integral parts of any kinetic scheme representing the mechanism of E-IIs. Such a scheme should explain how vectorial phosphorylation, transport coupled to phosphorylation, is still achieved while the uncoupled pathways are integral parts of the scheme.

The mechanistic coupling of transport and phosphorylation has been investigated by following the fate of mannitol bound to ISO membrane vesicles upon phosphorylation of II^{Mtl} [94]. The study resulted in the model presented in Fig. 8. The binding of mannitol to II^{Mtl} embedded in the membrane of an ISO vesicle has been discussed in the previous section. At equilibrium, bound mannitol is either at the periplasmic-facing binding site or in some occluded state (see Fig. 7). Transfer of mannitol from this state E_{eq} in Fig. 9) to the cytoplasmic volume involves at least part of the physical pathway for transport of mannitol from the outside to the inside of the cell. In the absence of the phosphoryl group donor, P-HPr, this transfer is very slow due to the slow isomerization step (E_{eq}:Mtl → E_{cyt}:Mtl, Fig. 9A). It is clear that in the presence of P-HPr, when vectorial phosphorylation is catalyzed at a much higher rate, this part of the transport route (E-P$_{eq}$:Mtl → E-P$_{cyt}$:Mtl, Fig. 9B) has to be accelerated significantly. Therefore, phosphorylation of II^{Mtl} drastically reduces the activation energy for the translocation step. It was demonstrated that once mannitol had arrived at state E-P$_{cyt}$:Mtl, less than half of the mannitol molecules were phosphorylated by

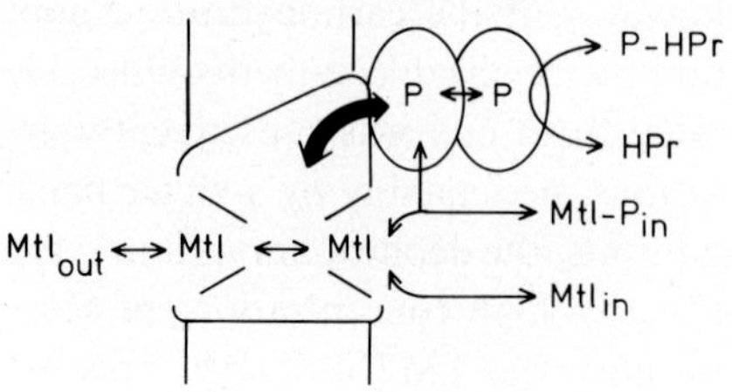

Fig. 8. A model of II^{Mtl} bases on steady-state and pre-steady-state kinetic data. Ps indicate the two phosphorylation sites. CI, CII and NIII refer to domains A, B, and C, respectively.

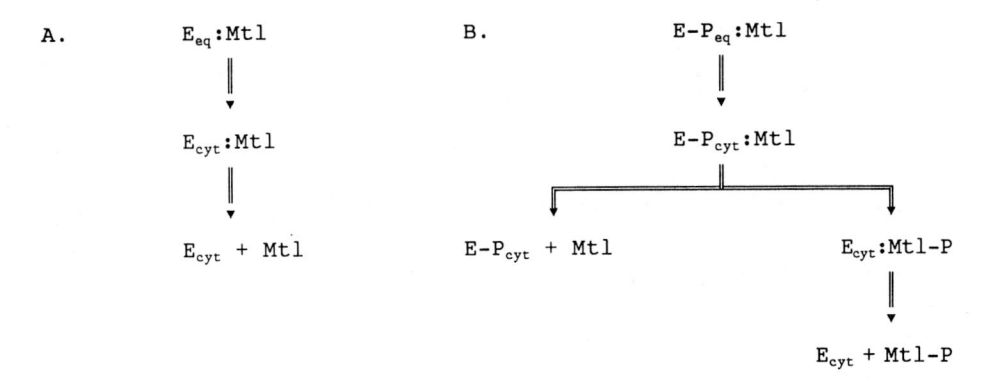

A. E_{eq}:Mtl   B. $E\text{-}P_{eq}$:Mtl
     ‖   ‖
     ▼   ▼
     E_{cyt}:Mtl $E\text{-}P_{cyt}$:Mtl
     ‖   ‖
     ▼   ┌─────────┴─────────┐
     E_{cyt} + Mtl $E\text{-}P_{cyt}$ + Mtl E_{cyt}:Mtl-P
                                                           ‖
                                                           ▼
                                                           E_{cyt} + Mtl-P

Fig. 9. Transfer of mannitol bound to inside-out vesicles to the cytoplasmic volume. (A) Unphosphorylated II^{Mtl}. (B) Phosphorylated II^{Mtl}. It is assumed that the phosphoryl group transfer from the enzyme to the sugar can only take place when the sugar is bound the cytoplasmic-facing binding site, $E\text{-}P_{cyt}$:Mtl (see also Fig. 6).

the enzyme; more than half dissociated as unmodified mannitol into the cytoplasmic volume. These results were placed in the context of the domain structure of II^{Mtl} resulting in the model in Fig. 8. The C domain constitutes a mannitol translocator. Domain A accepts the phosphoryl group from HPr and transfers it to domain B. The latter takes care of the mannitol kinase activity together with the internally oriented binding site on the translocator. In the absence of the phosphoryl group donor, P-HPr, the translocator would be able to catalyze at least part of the translocation of mannitol across the membrane, but at a very low rate. The process is accelerated 2–3 orders of magnitude when domain B is phosphorylated. Therefore, the state of phosphorylation of domain B modulates the activity of domain C. The interaction between the two domains is indicated by the wide arrow in Fig. 8. Within this model, the following events take place when mannitol is added to the outside of the cells. It is assumed that the cells have a high phosphorylation potential, that is, II^{Mtl} will be phosphorylated. Initially, only half of the molecules that are transported into the cell become phosphorylated. The other half are released as free mannitol inside the cell. Mechanistically, the coupling between transport and phosphorylation is less than 50%. External and internal free mannitol will rapidly equilibrate. Once this physiological steady-state is reached, cytoplasmic and periplasmic carbohydrate serve equally well as substrate. As a result, the phenomenological coupling between transport and phosphorylation under these conditions reaches 100%. Every mannitol molecule that is phosphorylated by II^{Mtl} has also been transported by the enzyme, but it may have been in a previous turnover. The uncoupled reactions, cytoplasmic phosphorylation and facilitated diffusion, are readily recognized in the model and, in fact, it is the interplay between these two activities that is responsible for the vectorial phosphorylation of mannitol in the physiological steady-state. II^{Mtl} in the model of Fig. 8 is a facilitated diffusion enzyme with a built-in sugar trap. The most surprising conclusion of our studies is that the coupling between phosphorylation and transport is not at the level of the phosphorylation of the sugar, but at the level of the transport of the sugar. The translocator is activated by phosphorylation of the enzyme.

5. Steady-state kinetics of carbohydrate phosphorylation

5.1. General considerations

The emphasis in kinetic studies of E-IIs has been on the analysis of the rates of phosphorylation of the sugar by the phosphoryl group donor. In the early studies the question was addressed whether phosphorylated E-II would be a catalytic intermediate in the reaction or whether the phosphoryl group would be transferred directly from the donor to the sugar on a ternary complex between the enzyme and its substrates [66,75,95–100]. This matter has been satisfactorily resolved by a number of other techniques in favor of the first option and possible reasons why some systems did not behave according to a 'ping-pong' type of mechanism have been discussed [1].

The most straight-forward interpretation of the phosphorylation reaction would be that exactly the same reaction is catalyzed as during transport of the sugar into the cell where the substrate is offered to the periplasmic side of the membrane (Eq. (1), overall reaction); phosphorylation would measure transport as well. However, this may not be the case: several lines of evidence discussed in the previous sections indicate that the mechanism underlying the phosphorylation reaction could be much more complex. Factors that may complicate the interpretation of the phosphorylation reaction in detergent solutions are:

(1) The phosphoryl group transfer from the phosphoryl group donor to the sugar may proceed through multiple steps. For instance, in the case of II^{Mtl} there is the transfer from P-HPr to domain A, then an internal transfer from domain A to domain B and finally to mannitol.

(2) Binding studies indicate that II^{Mtl} can bind mannitol both at the periplasmic and cytoplasmic side of the membrane.

(3) Demonstration of catalytic activities of E-IIs other than vectorial phosphorylation, like phosphorylation of free substrate at the cytoplasmic side of the membrane and facilitated diffusion catalyzed by E-IIs and phosphorylated E-IIs.

(4) E-IIs solubilized in detergent may exist in a self-associated state [41,44,45,49,50,52,54,101]. In the case of II^{Mtl}, dissociation led to a lower activity in the overall reaction indicating functional interaction between the subunits [52].

More recent kinetic data indeed indicate that the kinetics of sugar phosphorylation is more complicated than originally thought. A kinetic scheme that explains the data may have to include all or some of the items listed above. In the following section two such schemes will be discussed. Firstly, a scheme based upon the kinetic performance of the *R. sphaeroides* II^{Fru}. Secondly, the differences with the kinetics of the *E. coli* II^{Mtl} will be discussed. A more detailed model will be presented that also complies with data other than steady-state kinetics.

5.2. The R. sphaeroides II^{Fru} model

The fructose-specific PTS in *R. sphaeroides* is simpler than the one in *E. coli* or *S. typhimurium* in that it consists of only two proteins. Besides the fructose specific II^{Fru}, a class II enzyme, there is only one cytoplasmic component called soluble factor (SF) [48]. We now know that SF consists of III^{Fru}, HPr and E-I covalently linked [109]. II^{Fru} and SF form a membrane-bound complex whose association–dissociation dynamics is much slower than the turnover of the system. Therefore, the complex is the actual catalytic unit in the overall reaction and P-enolpyruvate is the direct phosphoryl group donor [102],

$$\text{P-enolpyruvate} + \text{Fructose}_{out} \xrightarrow{SF:II^{Fru}} \text{Fructose-1-P}_{in} + \text{Pyruvate}. \qquad (4)$$

Kinetic data for the reactions catalyzed by the complex were available from three types of experiments:
(1) The kinetics of fructose phosphorylation catalyzed by permeabilized membranes [48].
(2) The kinetics of fructose phosphorylation catalyzed by ISO membranes [103].
(3) The kinetics of fructose/fructose-P exchange catalyzed by permeabilized membranes [96].

The major conclusions from these studies were:
(1) The two substrate kinetics of the overall reaction catalyzed by the complex in permeabilized membranes showed classical 'ping-pong' kinetics in accordance with a phosphorylated enzyme intermediate. The affinity constants for fructose and P-enolpyruvate were 8 and $25 \mu M$, respectively.

(2) The kinetics of fructose phosphorylation catalyzed by the complex embedded in the membrane of ISO vesicles was found to be biphasic when plotted in a Lineweaver–Burk plot. The high affinity regime was identical to the kinetics found with the permeabilized membranes and, therefore, was attributed to a fraction of leaky vesicles. The second phase that only became significant at much higher fructose concentrations ($100 \mu M$ to 1 mM) was interpreted as the true fructose phosphorylation catalyzed by intact ISO vesicles. The data suggested that when fructose was offered at the non-physiological side of the membrane, II^{Fru} would still be able to phosphorylate the sugar but did so with a much lower affinity. Remarkably, the maximal rates of fructose phosphorylation catalyzed by the enzyme complex in permeabilized or intact vesicles were identical, indicating merging pathways for the two processes.

(3) The exchange of label between the fructose and fructose-P pools catalyzed by E-II^{Fru} could be satisfactorily described by the partial reaction describing the second step of the overall reaction,

$$\text{Fructose} + \text{II}^{Fru}\text{-P} <=> \text{Fructose-1-P} + \text{II}^{Fru}. \qquad (5)$$

In addition, the data indicated that the strong inhibition ($K_i = 4\,\mu\text{M}$) of the exchange reaction by the unphosphorylated sugar, a phenomenon reported for many E-IIs [73,104–106] was due to competition between fructose and fructose-1-P for the binding site on the enzyme. The important implication of this result was that it would indicate that unphosphorylated II^{Fru} possessed a high affinity for the sugar, which, subsequently, was confirmed by binding studies. Furthermore, it was concluded that

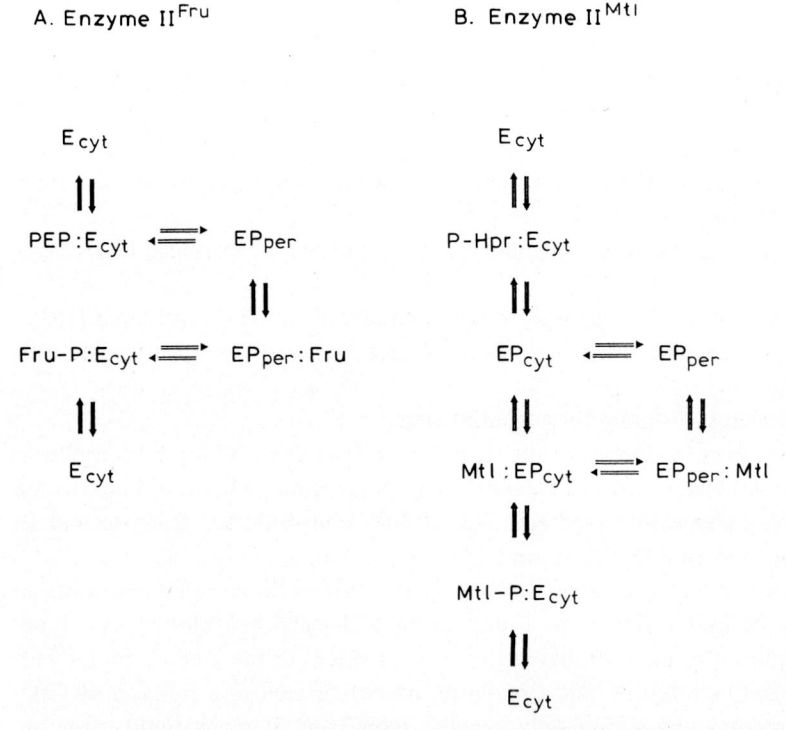

A. Enzyme II^{Fru}

B. Enzyme II^{Mtl}

E_{cyt}
$\Updownarrow$
$PEP:E_{cyt} \rightleftharpoons EP_{per}$
$\Updownarrow$
$Fru-P:E_{cyt} \rightleftharpoons EP_{per}:Fru$
$\Updownarrow$
E_{cyt}

E_{cyt}
$\Updownarrow$
$P-Hpr:E_{cyt}$
$\Updownarrow$
$EP_{cyt} \rightleftharpoons EP_{per}$
$\Updownarrow \qquad \Updownarrow$
$Mtl:EP_{cyt} \rightleftharpoons EP_{per}:Mtl$
$\Updownarrow$
$Mtl-P:E_{cyt}$
$\Updownarrow$
E_{cyt}

Fig. 10. Mechanisms of steady-state kinetics of sugar phosphorylation catalyzed by E-IIs in a non-compartmentalized system. (A) The *R. sphaeroides* II^{Fru} model. The model is based on the kinetic data discussed in the text. Only one kinetic route leads to phosphorylation of fructose. (B) The *E. coli* II^{Mtl} model. The model in Fig. 8 was translated into a kinetic scheme that would describe mannitol phosphorylation catalyzed by II^{Mtl} solubilized in detergent. Two kinetic routes lead to phosphorylation of mannitol. Mannitol can bind either to state EP_{cyt} or EP_{per}. E represents the complex of SF (soluble factor) and II^{Fru} and II^{Mtl} in A and B, respectively. EP represents the phosphorylated states of the E-IIs. Subscripts cyt and per denote the orientation of the sugar binding site to the cytoplasm and periplasm, respectively. PEP, phosphoenolpyruvate.

this high affinity binding site was facing the cytoplasm by virtue of the competition with the product of the phosphotransferase reaction, fructose-1-P.

In an attempt to integrate all this data into a single kinetic scheme describing the mechanism of II^{Fru}, two rather extreme models for the transport step were considered [1]. In one model, the transport step would involve the movement of the substrate through the protein, whereas, in the other model, the transport step would involve the movement of the protein over the substrate. II^{Fru}, according to the first model, would constitute a membrane-spanning channel with fructose binding sites at the two ends. To fit the data the two sites should have alternating low and high affinity depending on the state of phosphorylation of the enzyme. In the unphosphorylated enzyme, the cytoplasmic binding site would have the high affinity and the periplasmic site the low affinity, whereas, in the phosphorylated enzyme, this would be just the other way around. The model was termed 'the two-site affinity-shift model'. The second model under consideration was termed 'the single-site accessibility-shift model'. In this model, II^{Fru} would possess a single high affinity binding site that, depending on the state of phosphorylation of the enzyme, would be exposed to alternating sides of the membrane. In the unphosphorylated enzyme, the site would only be accessible from the cytoplasmic side of the membrane whereas in the phosphorylated enzyme the site would be accessible from the periplasmic side of the membrane, and, to a much lesser extent, from the cytoplasmic side.

The kinetic characteristics of fructose phosphorylation catalyzed by ISO membrane vesicles turned out to be crucial in discriminating between the two models and seemed to favor the 'single-site accessibility-shift model'. An attractive feature of the latter model is the direct coupling of the sugar phosphorylation to the transport event (Fig. 10A). The binding site on II^{Fru} would be exposed to the cytoplasm in the unphosphorylated state. Phosphorylation of the enzyme would induce a conformational change, making the binding site accessible to the fructose in the periplasm. Fructose would bind, after which the phosphoryl group would be transferred to fructose, causing the carrier to relax to the initial conformation, with the binding site facing the cytoplasm. Dissociation of fructose-1-P into the cytoplasm would complete the catalytic cycle.

5.3. The E. coli II^{Mtl} model

An important feature of the model proposed for the *R. sphaeroides* II^{Fru} is the strong coupling between the orientation of the binding site and the state of phosphorylation of the enzyme. This is in marked contrast to the model proposed for the *E. coli* II^{Mtl} discussed in section 4 where the binding site on both the phosphorylated and unphosphorylated enzyme can change its orientation spontaneously. The latter model was based on binding studies and single turnover kinetics. Prediction of the steady-state kinetics of mannitol phosphorylation catalyzed by II^{Mtl} according to this model shows agreement, but also important differences with the model for

II^{Fru} (see Fig. 10A and B). In our laboratory, we have repeated the kinetic experiments that led to the *R. sphaeroides* II^{Fru} model and compared the data with those which led to the *E. coli* II^{Mtl} model [73,108]. Qualitatively, the results were identical except for one important difference. Steady-state kinetics of mannitol phosphorylation catalyzed by II^{Mtl} solubilized in detergent was biphasic when plotted in a Lineweaver–Burk plot as a function of the reciprocal mannitol concentration. In contrast, the identical experiment with the *R. sphaeroides* II^{Fru} showed monophasic kinetics. The difference may be easily explained by the two mechanisms in Fig. 10. Binding of mannitol to the cytoplasmic-facing binding site on the phosphorylated enzyme allows for a second pathway leading to mannitol phosphorylation that short circuits the pathway via the periplasmic-facing binding site. The latter is the only possible pathway in the case of the *R. sphaeroides* model. Comparison of the two models in Fig. 10 shows that the obligate coupling between the orientation of the binding site and the state of phosphorylation of the enzyme has to be abandoned to comply with the II^{Mtl} kinetics.

In conclusion, the steady-state kinetics of mannitol phosphorylation catalyzed by II^{Mtl} can be explained within the model shown in Fig. 8 which was based upon different types of experiments. Does this mean that the mechanisms of the *R. sphaeroides* II^{Fru} and the *E. coli* II^{Mtl} are different? Probably not. First of all, kinetically the two models are only different in that the II^{Fru} model is an extreme case of the II^{Mtl} model. The reorientation of the binding site upon phosphorylation of the enzyme is infinitely fast and complete in the former model, whereas competition between the rate of reorientation of the site and the rate of substrate binding to the site gives rise to the two pathways in the latter model. The experimental set-up may not have been adequate to detect the second pathway in case of II^{Fru}. The important differences between the two models are at the level of the molecular mechanisms. In the II^{Fru} model, the orientation of the binding site is directly linked to the state of phosphorylation of the enzyme, whereas in the II^{Mtl} model, the state of phosphorylation of the enzyme modulates the activation energy of the isomerization of the binding site between the two sides of the membrane. Steady-state kinetics by itself can never exclusively discriminate between these different models at the molecular level since a condition may be proposed where these different models show similar kinetics. The II^{Mtl} model is based upon many different types of data discussed in this chapter and the steady-state kinetics is shown to be merely consistent with the model. Therefore, the II^{Mtl} model is more likely to be representative for the mechanisms of E-IIs.

References

1. Robillard, G.T. and Lolkema, J. (1988) Biochim. Biophys. Acta 947, 493–519.
2. Saier, Jr., M.H., Yamada, M., Erni, B., Suda, K., Lengeler, J., Ebner, R., Argos, P., Rak, B.,

Schnetz, K., Lee, C.A., Stewart, G.C., Breidt, Jr., F., Waygood, E.B., Peri, K.G. and Doolittle, R.F. (1988) FASEB J. 2, 199–208.
3. Saier, Jr., M.H. and Reizer, J. (1992) J. Bacteriol. 174, 1433–1438.
4. Kundig, W. and Roseman, S. (1971) J. Biol. Chem. 246, 1407–1418.
5. Lengeler, J.W., Titgemeyer, F., Vogler, A.P. and Wohrl, B.M. (1990) Phil. Trans. R. Soc. Lond. B 326, 498–504.
6. Wu, L.F. and Saier, Jr., M.H. (1990) J. Bacteriol. 172, 7167–7178.
7. Parker, L.L. and Hall, B.G. (1990) Genetics 124, 455–471.
8. Lengeler, J.W. (1990) Biochim. Biophys. Acta 1018, 155–159.
9. Reizer, J., Reizer, A. and Saier, Jr., M.H. (1990) Res. Microbiol. 141, 1061–1067.
10. Erni, B. and Zanolari, B. (1986) J. Biol. Chem. 261, 16398–16401.
11. Yamada, M. and Saier, Jr., M.H. (1987) J. Biol. Chem. 262, 5455–5463.
12. Breidt, Jr., F., Hengstenberg, W., Finkeldei, U. and Stewart, G.C. (1987) J. Biol. Chem. 262, 16444–16449.
13. Prior, T.I. and Kornberg, H.L. (1988) J. Gen. Microbiol. 134, 2757–2765.
14. Geerse, R.H., Izzo, F. and Postma, P.W. (1989) Mol. Gen. Genet. 216, 517–525.
15. Fouet, A., Arnaud, M., Klier, A. and Rapoport, G. (1987) Proc. Natl. Acad. Sci. U.S.A. 84, 8773–8777.
16. Ebner, R. and Lengeler, J.W. (1988) Mol. Microbiol. 2, 9–17.
17. Lee, C.A. and Saier, Jr., M.H. (1983) J. Biol. Chem. 258, 10761–10767.
18. Sutrina, S.H., Reddy, P., Saier, Jr., M.H. and Reizer, J. (1990) J. Biol. Chem. 265, 18581–18589.
19. Gonzy-Treboul, G., Zagorec, M., Rain-Guion, M.C. and Steinmetz, M. (1989) Mol. Microbiol. 3, 103–112.
20. Sato, Y., Poy, F., Jacobson, G.R. and Karamitsu, H.K. (1989) J. Bacteriol. 171, 263–271.
21. Bramley, H.F. and Kornberg, H.L. (1987) Proc. Natl. Acad. Sci. U.S.A. 84, 4770–4780.
22. Schnetz, K., Toloczyki, C. and Rak, B. (1987) J. Bacteriol. 171, 263–271.
23. Peri, K.G. and Waygood, B.E. (1988) Biochemistry 27, 6054–6061.
24. Rogers, J.R., Ohgi, T., Plumbridge, J. and Soll, D. (1988) Gene 62, 197–207.
25. Erni, B., Zanolari, B. and Kocher, H.P. (1987) J. Biol. Chem. 262, 5238–5247.
26. Martin-Verstraete, I., Debarbouille, M., Klier, A. and Rapoport, G. (1990) J. Mol. Biol. 214, 657–671.
27. Reiche, B., Frank, R., Deutscher, J., Meyer, N. and Hengstenberg, W. (1988) Biochemistry 27, 6512–6516.
28. Pas, H.H. and Robillard, G.T. (1988) Biochemistry 27, 5835–5839.
29. Dorschug, M., Frank, R., Kalbitzer, H.R., Hengstenberg, W. and Deutscher, J. (1984) Eur. J. Biochem. 144, 113–119.
30. Lolkema, J.S., Swaving-Dijkstra, D., Ten Hoeve-Duurkens, R.H. and Robillard, G.T. (1990) Biochemistry 29, 10659–10663.
31. Vogler, A.P. and Lengeler, J.W. (1988) Mol. Gen. Genet. 213, 175–178.
32. Vogeler, A.P., Broekhuizen C.P., Schuitema, A., Lengeler, J.W. and Postma, P.W. (1988) Mol. Microbiol. 2, 719–726.
33. Lengeler, J., Mayer, R.J. and Schmid, K. (1982) J. Bacteriol. 151, 468–471.
34. Fischer, R., Eisermann, R., Reiche, B. and Hengstenberg, W. (1989) Gene 82, 249–257.
35. Grisafi, P.L., Scholle, A., Sugayama, J., Briggs, L., Jacobson, G.R. and Lengeler, J.W. (1989) J. Bacteriol. 171, 2719–2727.
36. Stephan, M.M., Khandekar, S.S. and Jacobson, G.R. (1989) Biochemistry 28, 7941–7946.
37. White, D.W. and Jacobson, G.R. (1990) J. Bacteriol. 172, 1509–1515.
38. Karplus, P.A. and Schmidt, G.E. (1985) Naturwissenschafter 72, 212–217.
39. Van Weeghel, R.P., Meyer, G.H., Keck, W. and Robillard, G.T. (1991) Biochemistry 30, 1774–1779.
40. Van Weeghel, R.P., Van Der Hoek, Y.Y., Pas, H.H., Elferink, M.G.L., Keck, W. and Robillard,

G.T. (1991) Biochemistry 30, 1768–1773.
41. Pas, H.H., Ten Hoeve-Duurkens, R.H. and Robillard, G.T. (1988) Biochemistry 27, 5520–5525.
42. Van Weeghel, R.P., Meyer, G., Pas, H.H., Keck, W. and Robillard, G.T. (1991) Biochemistry 30, 9478–9485.
43. Wood, B.L. and Rippon, W.B. (1991) personal communication.
44. Leonard, J.E. and Saier, Jr., M.H. (1983) J. Biol. Chem. 258, 10757–10760.
45. Robillard, G.T. and Blaauw, M. (1987) Biochemistry 26, 5796–5803.
46. Roossien, F.F. and Robillard, G.T. (1984) Biochemistry 23, 5682–5685.
47. Stephan, M.M. and Jacobson, G.R. (1986) Biochemistry 25, 4046–4051.
48. Lolkema, J.S., Ten Hoeve-Duurkens, R.H. and Robillard, G.T. (1985) Eur. J. Biochem. 149, 625–631.
49. Khandekar, S.S. and Jacobson, G.R. (1989) J. Cell Biochem. 39, 207–216.
50. Saier, Jr., M.H. (1980) J. Supramolec. Str. 14, 281–294.
51. Saier, Jr., M.H. and Leonard, J.E. (1983) In: Multifunctional Proteins (Kane, J. Ed.) pp. 11–30, CRC Press, Boca Raton.
52. Lolkema, J.S. and Robillard, G.T. (1990) Biochemistry 29, 10120–10125.
53. Pas, H.H., Ellory, C. and Robillard, G.T. (1987) Biochemistry 26, 6689–6696.
54. Erni, B. (1986) Biochemistry 25, 305–312.
55. Meins, M., Zanolari, B., Rosenbusch, J. and Erni, B. (1988) J. Biol. Chem. 263, 12986–12993.
56. Nouffer, C., Zanolari, B. and Erni, B. (1986) J. Biol. Chem. 263, 6647–6655.
57. Roossien, F.F. and Robillard, G.T. (1984) Biochemistry 23, 211–215.
58. Roossien, F.F., Van Es-Spiekman, W. and Robillard, G.T. (1986) FEBS Lett. 196, 284–290.
59. Lolkema, J.S., Ten Hoeve-Duurkens, R.H. and Robillard, G.T. (1986) Eur. J. Biochem. 154, 651–656.
60. Robillard, G.T. and Beechey, R.B. (1986) Biochemistry 25, 1346–1354.
61. Haguenauer-Tsapis, R. and Kepes, A. (1980) J. Biol. Chem. 255, 5075–5081.
62. Haguenauer-Tsapis, R. and Kepes, A. (1977) Biochim. Biophys. Acta 469, 211–215.
63. Haguenauer-Tsapis, R. and Kepes, A. (1973) Biochem. Biophys. Res. Commun. 54, 1331–1341.
64. Haguenauer-Tsapis, R. and Kepes, A. (1977) Biochim. Biophys. Acta 465, 118–130.
65. Robillard, G.T. and Konings, W.N. (1981) Biochemistry 20, 5025–5032.
66. Grenier, F.C., Waygood, E.B. and Saier, Jr., M.H. (1985) Biochemistry 24, 4872–4876.
67. Lolkema, J.S. and Walz, D. (1990) Biochemistry 29, 11180–11188.
68. Lolkema, J.S., Carrasco, N. and Kaback, H.R. (1991) Biochemistry 30, 1284–1290.
69. Lolkema, J.S., Swaving Dijkstra, D., Ten Hoeve-Duurkens, R.H. and Robillard, G.T. (1991) Biochemistry 30, 6721–6726.
70. Elferink, M.G.L., Driessen, A.J.M. and Robillard, G.T. (1990) J. Bacteriol. 172, 7119–7125.
71. Lolkema, J.S., Swaving Dijkstra, D. and Robillard, G.T. (1992) Biochemistry, in press.
72. Colowick, S.P. and Womack, F.C. (1969) J. Biol. Chem. 244, 774–777.
73. Roossien, F.F., Blaauw, M. and Robillard, G.T. (1984) Biochemistry 23, 4934–4939.
74. Roossien, F.F. and Robillard, G.T. (1984) Biochemistry 23, 4934–4939.
75. Thompson, J., Chassy, B. and Egan, W. (1985) J. Bacteriol. 162, 217–223.
76. Kornberg, H.L. and Riordan, C. (1976) J. Gen. Microbiol. 94, 75–89.
77. Postma, P.W. (1976) FEBS Lett. 61, 49–53.
78. Postma, P.W., Keizer, H.G. and Koolwijk, P. (1986) J. Bacteriol. 168, 1107–1111.
79. Postma, P.W. (1981) J. Bacteriol. 147, 382–389.
80. Ruiter, G.J.G., Postma, P.W. and van Dam, K. (1990) J. Bacteriol. 172, 4783–4789.
81. Postma, P.W. and Stock, J.B. (1980) J. Bacteriol. 141, 476–484.
82. Gachelin, G. (1970) Eur. J. Biochem, 16, 342–357.
83. Haguenauer, R. and Kepes, A. (1971) Biochemie 53, 99–107.
84. Solomon, E., Miyai, K. and Lin, E.C. (1973) J. Bacteriol. 114, 723–728.

85. Reizer, J. and Panos, C. (1980) Proc. Natl. Acad. Sci. U.S.A. 77, 5497–5501.
86. Reizer, J., Novotny, M.J., Panos, C. and Saier, Jr., M.H. (1983) J. Bacteriol. 156, 354–361.
87. Reizer, J. and Saier, Jr., M.H. (1983) J. Bacteriol. 156, 236–242.
88. Chassy, B.M. and Thompson, J. (1983) J. Bacteriol. 154, 1195–1203.
89. Hausman, S.Z., Thompson, J. and London, J. (1984) J. Bacteriol. 160, 211–215.
90. Thompson, J. and Chassy, B.M. (1985) J. Bacteriol. 156, 224–234.
91. Sutrina, S.L., Reizer, J. and Saier, Jr., M.H. (1988) J. Bacteriol. 170, 1874-1877.
92. Thompson, J. and Saier, Jr., M.H. (1981) J. Bacteriol. 146, 885-894.
93. Thompson, J. and Chassy, B.M. (1983) J. Bacteriol, 156, 70–80.
94. Lolkema, J.S., Ten Hoeve-Duurkens, R.H., Swaving Dijkstra, D. and Robillard, G.T. (1991) Biochemistry 30, 6716–6721.
95. Misset, O., Blaauw, M., Postma, P.W. and Robillard, G.T. (1983) Biochemistry 22, 6163–6167.
96. Lolkema, J.S., Ten Hoeve-Duurkens, R.H. and Robillard, G.T. (1986) Eur. J. Biochem. 154, 387–393.
97. Marquet, M., Creignou, C. and Dedonder, R. (1978) Biochimie 60, 1283–1287.
98. Huedig, H. and Hengstenberg, W. (1980) FEBS Lett. 114, 103–106.
99. Perret, J. and Gay, P. (1979) Eur. J. Biochem. 102, 237–246.
100. Simoni, R.D., Hays, J.B., Nakzawa, T. and Rosemans, S. (1973) J. Biol. Chem. 248, 957–965.
101. Erni, B., Trachsel, H., Postma, P.W. and Rosenbusch, J.P. (1982) J. Biol. Chem. 257, 13726–13730.
102. Saier, Jr., M.H., Feucht, B.U. and Roseman, S. (1971) J. Biol. Chem. 246, 7819–7821.
103. Lolkema, J.S. and Robillard, G.T. (1985) Eur. J. Biochem. 147, 69–75.
104. Perret, J. and Gay, P. (1979) Eur. J. Biochem. 102, 237–246.
105. Rephaeli, A.W. and Saier, Jr., M.H. (1980) J. Biol. Chem. 255, 8585–8591.
106. Saier, Jr., M.H., Feucht, B.U. and Mora, W.K. (1977) J. Biol. Chem. 252, 8899–8907.
107. Erni, B., Zanolari, B., Graff, P. and Kocher, H.P. (1989) J. Biol. Chem. 264, 18733–18741.
108. Lolkema, J.S., Ten Hoeve-Duurkens, R.H. and Robillard, G.T., submitted.
109. Wu, L.F., Tomich, J.M. and Saier, Jr., M.H. (1990) J. Mol. Biol. 213, 687–703.

CHAPTER 6

Mechanisms of active and passive transport in a family of homologous sugar transporters found in both prokaryotes and eukaryotes

STEPHEN A. BALDWIN

Departments of Biochemistry and Chemistry, and Protein and Molecular Biology, Royal Free Hospital School of Medicine (University of London), London NW3 2PF, U.K.

1. Introduction

Sugars play a key role in the metabolism of most organisms, ranging from bacteria to ourselves. For example, the human brain is almost entirely dependent upon glucose as an energy source. Since sugars are hydrophilic molecules their uptake into cells across the hydrophobic core of the plasma membrane requires the assistance of transport systems. A variety of such transport systems have evolved to cope with the sugar species and concentrations that may be encountered by different cells and organisms. Thus in bacteria, which may need to utilize sugars present at very low concentrations in the environment, there are many different active (energy-dependent) transport systems for the uptake of sugars against the concentration gradient. By contrast, in mammals, where blood glucose levels are kept roughly constant at 5–10 mM by complex homeostatic mechanisms, most cells take up glucose by a purely passive mechanism, driven by the concentration gradient across the membrane. Active, sodium-linked symport (co-transport) systems for glucose are found only in a few mammalian tissues, such as the small intestine and kidney, where transepithelial transport against a concentration gradient is required.

A major portion of this review will be concerned with the passive glucose transporter from the human erythrocyte. This protein (GLUT-1) belongs to a family of five homologous, passive glucose transporters so far identified by cDNA cloning in mammalian tissues. In this chapter the five isoforms will be designated GLUT-1–5, following the terminology of Fukumoto et al. [1]. Each of these isoforms has a different tissue distribution, presumably reflecting its unique physiological role (see section 6). Comparison of the amino acid sequences of these homologous proteins has identified residues which might be involved in the translocation mechanism (see section 8) and should eventually lead to a better understanding of their differing physiological roles and regulation by hormones.

Interestingly, this family of sugar transporters shows no similarity in sequence to the active, sodium-linked glucose transporter of mammalian small intestine [2]. Such a finding is not unexpected in view of the very different kinetic properties of these passive and active transporters. However, it made all the more surprising the discovery in 1987 that the mammalian passive transporters do show strong sequence similarities to a family of active sugar/H^+ symporters found in bacteria [3,4]. Subsequent studies in many laboratories have now demonstrated the existence of a large family of homologous active and passive sugar transporters with members not only in mammals and bacteria, but also in yeasts, cyanobacteria, green algae, higher plants and protozoans. The family also contains transporters whose substrates are not sugars – such as the tetracycline and citrate transporters from *Escherichia coli* and the quinate transporters of filamentous fungi. All these proteins are discussed in section 7.

The sequence similarities between the members of this large family of transporters indicate that the homologous regions of these proteins are likely to have very similar secondary and tertiary structures. It follows that the mechanisms of both active and passive transport in this family probably share many features. This chapter discusses primarily the human erythrocyte glucose transporter (GLUT-1), because it is by far the best characterized member of the family and remains the only one to have been purified in functional form (unless one considers *lac* permease from *E. coli* to be a member of the family – see section 7.4.3). However, in section 8 the information derived from the study of this protein is pooled with what we know of the characteristics of other members of the family, in an attempt to draw some general conclusions about active and passive sugar transport. Although much remains to be done, such an approach is beginning to yield clues to the molecular mechanisms of substrate translocation in this fascinating family of proteins.

2. The kinetics of sugar transport in mammalian cells

2.1. Substrate specificity

Investigation of the substrate specificity of the different mammalian sugar transporter isoforms is complicated by the frequent occurrence of more than one isoform in a single cell type. This situation is beginning to change, with the demonstration that all five known transporter isoforms can be expressed in functional form in *Xenopus* oocytes [5–10]. However, at present, most of our knowledge about the substrate specificity has come from the study of the human erythrocyte, which contains only the GLUT-1 isoform. Its specificity is rather broad, a large number of monosaccharides being substrates. Disaccharides are not transported although some, such as cellobiose and maltose, inhibit transport when present in the extracellular medium [11]. Monosaccharide substrates include both ketoses and aldoses, but the very low affinity for fructose (K_{mapp} 1.5 M) indicates that this isoform is not the normal route for cellular uptake of this metabolically important ketose *in vivo* [12]. Instead, the GLUT-2 isoform is probably responsible for fructose uptake by liver (see section 6). A wide range of aldoses, including both hexoses and pentoses, are transported by the GLUT-1 isoform (reviewed in [13]). These sugars exist in solution primarily as pyranose rings, and the demonstration by Barnett et al. [14] that 1-deoxy-D-glucose is transported confirms that sugars are transported as pyranose rings rather than in the open chain form.

The affinity of sugars for the transporter has been correlated with their stability in the $^4C_1$ chair form [13]. However, since most of the common monosaccharides exist predominantly in this form, the correlation more probably reflects the necessity for equatorial hydroxyl groups for hydrogen-bond formation with the transport protein. In addition, axial hydroxyls may sterically hinder sugar binding to the protein. Thus, of the natural hexoses and pentoses, D-glucose and D-xylose have the highest affinities for the transporter [13]. In the $^4C_1$ conformation, the β-anomers of these sugars are completely free of axial substituents. Their enantiomers L-glucose and L-xylose have extremely low affinities for the transport system [13].

The roles played by the various parts of the D-glucose molecule in its binding and translocation by the transporter have been studied using epimers, deoxy-sugars and fluoro-substituted sugars in which the fluorine atom can serve as a hydrogen-bond acceptor. The affinity of these analogues for the transporter has been investigated in a number of ways, including measurement of their transport [14,15], their inhibition of the transport of other sugars [15] and their inhibition of stereospecific D-glucose binding to erythrocyte membranes [16]. The different methodologies involved in such studies makes direct comparison of their findings difficult; for example, because the ability of a sugar to *bind* to the transporter and or inhibit the transport of other sugars does not necessarily imply that it can itself be *transported*: maltose is a good example of a non-transported sugar which nonetheless binds to the transporter. However, taken together the results of these studies have provided reasonably conclusive evidence as to the identity of the important hydrogen-bonding positions on the sugar.

The hydroxyl group at the C-2 position does not appear to be involved in hydrogen bonding, because 2-deoxy-D-glucose has an affinity for the transporter equal to or greater than that for D-glucose [14,16]. In contrast, 3-deoxy-D-glucose has a lower affinity for the transporter than glucose, but the affinity is restored by substitution of a fluorine atom for the hydroxyl group in this position, indicating that the latter accepts a hydrogen bond from the transporter protein [14,15]. The hydroxyl group at the C-1 position also appears to be an important hydrogen-bond acceptor, for the affinity of the transporter for 1-deoxy-D-glucose is much smaller than that for D-glucose [14,16]. The high affinity can be restored by replacement of one C-1 hydrogen atom with a fluorine atom in the β-position, but not in the α-position [14]. However, it is still not clear whether the transport protein discriminates between the α- and β-anomers of glucose itself: two recent studies reported no significant differences between the V_{max} and K_m for transport of the two anomers [17,18], whereas others have reported faster transport of the α-anomer [19], and the β-anomer [20], respectively.

A third important hydrogen-bonding site on the glucose molecule appears to be the ring oxygen, which must be a hydrogen-bond acceptor. Replacing this atom with sulphur in 5-thio-D-glucose results in a large loss in affinity [16]. The role of the hydroxyl groups at the C-4 and C-6 positions is less clear. D-Galactose has a lower affinity for the transporter than glucose [13–16], but this lower affinity could result either from the loss of a hydrogen bond or from steric hindrance due to the axial hydroxyl group. Unfortunately, 4-deoxy-D-glucose is not available for study. The involvement of the C-6 hydroxyl in binding is also uncertain, although there is some evidence that it functions as a hydrogen-bond acceptor [13,15,16].

In summary, the most important hydrogen-bonding sites on the glucose molecule appear to be the ring oxygen and the hydroxyls at C-1 and C-3, with a possible, although lesser, contribution from the hydroxyls at C-4 and C-6. Similar findings have been reported for studies on the rat adipocyte, which contains predominantly the GLUT-4 isoform, although the C-4 hydroxyl seems to be of less importance than for the GLUT-1 isoform [21]. However, it is not known whether all the hydrogen-bonding sites on the sugar are involved in binding to the transporter throughout the translocation process: it is possible that the two conformations of the protein (see section 5) form different sets of hydrogen bonds with the sugar during the transport. In addition, Van der Waals' contacts are likely to provide further specificity and stability to the sugar–transporter complex, as is the case for other sugar-binding proteins [22]. For example, changes in tryptophan fluorescence seen upon sugar binding to the protein suggest that one or more of these residues may be located at the binding site [23] (see section 5). In this connection it is interesting that partial stacking of aromatic residues with the pyranose ring of the substrate has been identified in both the arabinose- and galactose-binding proteins of *Escherichia coli* [22].

2.2. Specific inhibitors of transport

In addition to its inhibition by substrates and substrate analogues, D-glucose transport in the human erythrocyte and many other cell types is potently and reversibly inhibited by a number of compounds which at first sight do not appear to resemble substrates in structure. Many of these compounds bind to the transporters much more tightly than does glucose, and they have been extremely valuable tools in the investigation of the transport proteins. One of the most potent of these inhibitors is cytochalasin B (Fig. 1), a substance produced by the fungus *Helminthosporium dematioideum* [24,25]. Model building studies have suggested that this inhibitor and

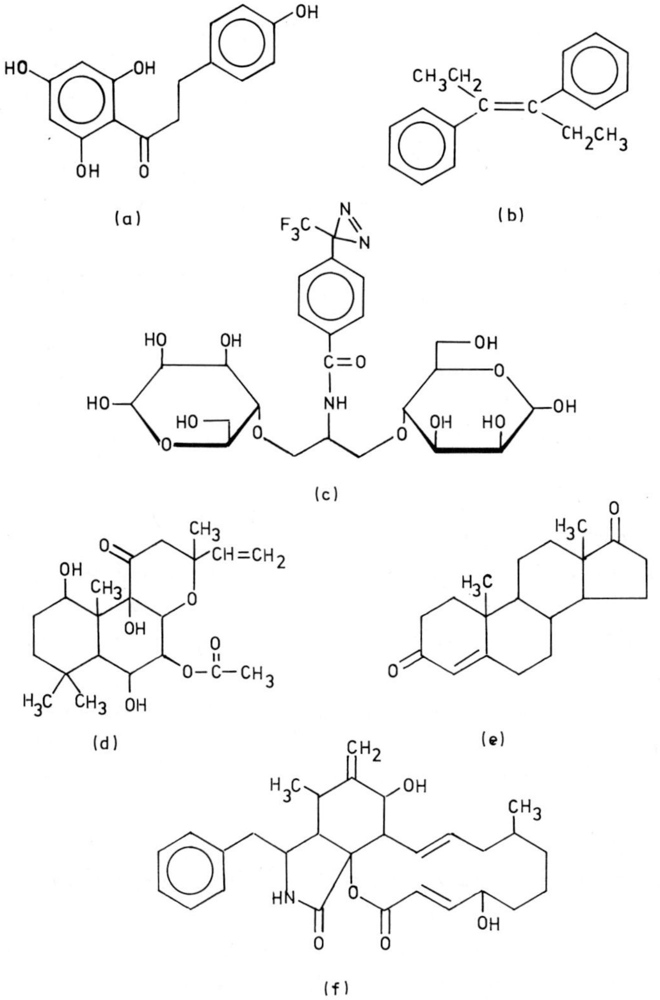

Fig. 1. The structures of sugar-transport inhibitors. (a) Phloretin, (b) diethylstilboestrol, (c) 2-N-[4-(1-azi-2,2,2-trifluoroethyl)benzoyl]-1,3-bis-(D-mannos-4-yloxy)-2-propylamine (ATB-BMPA), (d) forskolin, (e) androsten-4-ene-3,17-dione, (f) cytochalasin B.

D-glucose bind isosterically to the transport protein [26]. Cytochalasin B has been shown to inhibit 2-deoxy-D-glucose uptake by *Xenopus* oocytes expressing each of the five known mammalian transporter isoforms (GLUT-1–5) [9]. The reported K_i values for inhibition of transport in most mammalian cells, including the human erythrocyte (which contains only GLUT-1) and rat adipocyte (which contains primarily GLUT-4, see section 6), range from 50 to 500 nM [27–30]. However, cytochalasin B inhibits the GLUT-2 isoform in the hepatocyte somewhat less potently, with an apparent K_i of about 1 μM [30,31]. Direct binding measurements on human erythrocyte membranes and on rat adipocyte membranes have shown that the inhibitor binds to both the GLUT-1 and GLUT-4 isoforms with a K_d of between 100 and 200 nM, and that binding is competitively inhibited by D-glucose [32,33]. Similar studies on hepatocyte membranes have shown that it binds to the GLUT-2 isoform with a K_d of about 1 μM [30,31]. Kinetic studies on the human erythrocyte have shown that cytochalasin B binds to the transport system, in competition with glucose, exclusively at the intracellular surface of the membrane [34].

Glucose transport in the erythrocyte is also potently inhibited by a series of biphenolic compounds, including diethylstilboestrol and related compounds, and phloretin, whose structures are shown in Fig. 1. Diethylstilboestrol inhibits transport with a K_i of about 5 μM [35] and phloretin inhibits with a K_i of about 2 μM [36]. Kinetic experiments have shown that phloretin binds asymmetrically to the glucose transporter [37]. However, unlike cytochalasin B it binds exclusively to the extracellular surface of the membrane. Similar asymmetries of binding have been reported for a number of steroids that inhibit glucose transport [38]. For example, androsten-4-ene-3,17-dione (Fig. 1) which inhibits with a K_i of about 20 μM, binds almost exclusively at the inner surface of the membrane.

A more recently discovered inhibitor of the glucose transporter is the diterpene forskolin (Fig. 1). This compound inhibits glucose transport in several types of mammalian cells, including both rat adipocytes [39,40] and human erythrocytes [41]. Although it is known to be a powerful activator of adenylate cyclase [42], its effects on glucose transport in the erythrocyte have been shown to occur independently of changes in cellular cAMP levels [41], and some derivatives of forskolin which do not stimulate adenylate cyclase do inhibit transport [43]. Direct measurements of binding to the GLUT-1 isoform in human erythrocyte membranes have revealed that the inhibitor binds with a K_d of about 3 μM [44]. Even more potent inhibition of transport has recently been reported using the forskolin derivative 3-iodo-4-azidophenylamido-7-O-succinyldeacetyl-forskolin (IAPS-forskolin), which appears to have an affinity for the transporter slightly greater than that of cytochalasin B [45]. Binding of forskolin is competitively inhibited by D-glucose and Joost et al. [43] have pointed out that several of the oxygen atoms of forskolin are superimposable on hydroxyl groups of the α-D-glucose molecule, suggesting that, like cytochalasin B, forskolin binds at the substrate-binding site of the transporter. Inhibition of 3-O-methylglucose influx into erythrocytes has been found to be non-competitive, indicating that the inhibitor probably binds to the cytoplasmic surface of the membrane [41].

2.3. Kinetics of transport in the erythrocyte

2.3.1. General properties and methods of investigation

By far the most complete study of the kinetics of mammalian passive glucose transporters has been done on the GLUT-1 isoform in the human erythrocyte. The transport of glucose in this cell type is a classic example of facilitated diffusion, the permeability coefficient for D-glucose flux across the membrane being about five orders of magnitude greater than that for flux across protein-free bilayers constructed from erythrocyte membrane lipids [46]. Transport is purely passive, net flux across the membrane being driven by the concentration gradient until equilibrium is achieved, with equal glucose concentrations on both sides of the membrane. Although there is no net flux under these conditions, equal unidirectional influx and efflux of sugar continues, as can be demonstrated using isotopically labelled sugar. Glucose flux exhibits Michaelis–Menten kinetics, saturating at high concentration gradients across the membrane, and so can be characterized by a V_{max}, which is the maximal rate of flux, and a K_m value, which is the glucose concentration at which flux is half maximal. The flux rate v is then related to the glucose concentration [S] by the following equation,

$$V = \frac{V_{max}[S]}{K_m + [S]}. \tag{1}$$

V_{max} and K_m values for glucose transport in the erythrocyte have been determined in four main types of experimental situation [47].

(1) *Zero-trans experiments*. In these experiments the rate of glucose flux from the *cis* side of the membrane is measured as a function of its concentration, while the concentration on the *trans* side is kept at zero. Two types of experiment can thus be performed, zero-*trans* (zt) entry and exit, yielding V_{max} values V_{oi}^{zt} and V_{io}^{zt}, and K_m values K_{oi}^{zt} and K_{io}^{zt}, respectively, where o signifies outside and i inside the erythrocyte.

(2) *Equilibrium exchange experiments*. In this situation, the unidirectional flux of isotopically labelled glucose across the membrane is measured as a function of the glucose concentration, which is kept equal on both sides of the membrane. Because the unidirectional influx and efflux at any particular concentration are necessarily equal, the equilibrium-exchange (ee) flux is characterized by a single V_{max} value, V^{ee}, and a single K_m value, K^{ee}.

(3) *Infinite-cis experiments*. In this type of experiment, the net flux of glucose from a limitingly high concentration on the *cis* face of the membrane is measured as a function of the concentration on the *trans* face. Both entry and exit infinite-*cis* (ic) experiments can be performed, yielding two K_m values, K_{oi}^{ic} and K_{io}^{ic}, respectively. Maximal fluxes are obtained when the concentration on the *trans* side of the membrane is zero. Since this is the zero-*trans* situation, $V_{oi}^{ic} = V_{oi}^{zt}$ and $V_{io}^{ic} = V_{io}^{zt}$.

(4) *Infinite-trans experiments*. In these experiments, the unidirectional flux of isotopically labelled glucose from the *cis* side of the membrane is measured as a function of its concentration, while the concentration of unlabelled glucose on the *trans* side of the membrane is kept limitingly high. Once again, both entry and exit infinite-*trans* (it) experiments can be performed. Since maximal flux is obtained when the concentration of glucose on the *cis* face of the membrane is equal to the limitingly high concentration on the *trans* face, that is when the equilibrium exchange condition

obtains, the infinite *trans* procedure is characterized by a single V_{max} value, such that $V_{oi}^{it} = V_{io}^{it} = V^{ee}$. Two K_m values can be measured, K_{oi}^{it} and K_{io}^{it}, but these should equal K_{io}^{ic} and K_{oi}^{ic}, respectively, because in both cases the glucose concentration at the same side of the membrane is held at a limitingly high constant value, while the effect of varying the concentration at the other face is measured.

2.3.2. Transport asymmetry and the effect of cytoplasmic ATP

The results of many kinetic studies have shown that glucose transport in the erythrocyte is markedly asymmetric, the K_m and V_{max} values for zero-*trans* efflux being substantially greater than the corresponding values for influx, particularly at low

TABLE I
Comparison of experimentally determined K_m and V_{max} values for glucose transport in the erythrocyte at 0 and 20°C with values predicted by Wheeler and Whelan [65] for the simple asymmetric carrier model

Procedure	K_m (mM)		V_{max} (mM s^{-1})		Ref.
Near 0°C					
Zero-*trans* entry	0.20	(0.15)	0.0035	(0.0048)	245
	0.145	(0.15)	0.0055	(0.0048)	48
Zero-*trans* exit	1.64	(3.0)	0.071	(0.10)	48
	3.4	(3.0)	0.150	(0.10)	52
Equilibrium exchange	20	(13)	0.375	(0.43)	245
	25	(13)	0.50	(0.43)	50
	12.8	(13)	0.563	(0.43)	48
Infinite-*cis* entry	14.6	(13)	0.0051	(0.0048)	65
Infinite-*cis* exit	0.39	(0.61)	0.143	(0.10)	50
Infinite-*trans* entry	0.65	(0.61)	0.21	(0.43)	245
Infinite-*trans* exit	8.7	(13)	0.73	(0.43)	52,65
Near 20°C					
Zero-*trans* entry	1.6	(1.7)	0.6	(0.56)	245
	1.6	(1.7)	0.87	(0.56)	48
Zero-*trans* exit	5.4[a]	(6.9)	2.3	(2.3)	62
	4.6	(6.9)	2.6	(2.3)	48
Equilibrium exchange	20	(14)	4.4	(4.8)	245
	13	(14)	6.1	(4.8)	246
	17	(14)	5.9	(4.8)	48
Infinite-*cis* entry	14.3[a]	(12.6)	0.25	(0.56)	65
Infinite-*cis* exit	1.7	(3.0)	1.2	(2.3)	247
	1.9	(3.0)	3.5	(2.3)	96
Infinite-*trans* entry	1.7	(3.0)	2.7	(4.8)	245

All the results were obtained from estimates of initial rates made using fresh blood. The predicted values for K_m and V_{max} (shown in parentheses) are taken from Wheeler and Whelan [65] who fitted the data to the asymmetric carrier model by the procedures described in Wheeler [52].
[a]Data obtained at 25°C and subsequently corrected to give parameter estimates at 20°C as described by Wheeler and Whelan [65].

temperatures (the efflux values are about 12-fold greater than the influx values at 0°C [48], see Table I). Although such asymmetry has been ascribed to D-glucose binding to an intracellular component such as haemoglobin [49–51], it appears more likely that much of the asymmetry of transport arises from the properties of the transport protein itself [52]. Nevertheless, there is now considerable evidence that the asymmetric behaviour of the protein arises from the binding of cytosolic factors, which reduce the K_m and V_{max} values for zero-*trans* influx and increase the K_m value for zero-*trans* exit while leaving the V_{max} value for zero-*trans* exit unaffected. For example, Carruthers and Melchior have reported that dilution of cellular solute by lysis and resealing of cells leads to an approximately three-fold decrease in the K_m value for zero-*trans* exit of glucose [53]. They also showed that in the absence of cell solute, transport is symmetric in inside-out erythrocyte membrane vesicles [53]. Application of cellular solute (obtained by lysis of intact cells) to the external (originally cytoplasmic) surface of these vesicles decreased the K_m and V_{max} for zero-*trans* exit (from the vesicles) by up to 10-fold. Subsequent work from the same laboratory has shown that the cytosolic factor involved in transport modulation is probably adenosine triphosphate (ATP) [54]. In other studies a variety of contradictory effects of intracellular ATP on transport have been reported, ranging from inhibition through no effect to stimulation of transport (summarized in [55]). Apparent sense has recently been made of these rather disparate findings by the discovery in Carruthers' laboratory that low concentrations of Ca^{2+} profoundly inhibit sugar uptake in ATP-free ghosts, and that this inhibition is reversible only in the presence of ATP [55]. The mechanism of this inhibition is unclear but its reversal appears to require ATP hydrolysis. ATP also causes a reduction of the K_m and V_{max} for zero-*trans* sugar uptake in Ca^{2+}-free cells. The latter phenomenon is rapidly reversible and appears to be kinase-independent [54,56]. It is likely that the effects of ATP on transport seen in the absence of calcium result from direct binding to the transporter, which has been shown to bind one molecule of the nucleotide per polypeptide chain with a $K_{d(app)}$ equivalent to the concentration (40–50 μM) required for half maximal effect on transport [56,57].

2.4. Kinetic models for the transport process

In addition to the differences in zero-*trans* influx and efflux kinetic parameters that are apparent when transport is measured using intact, ATP-replete erythrocytes, there are large differences between flux rates measured under zero-*trans* as compared to equilibrium-exchange conditions. The K_m and V_{max} values for equilibrium-exchange are even larger than those for zero-*trans* efflux, again particularly at low temperature (the equilibrium-exchange values are about eight-fold higher than the zero-*trans* efflux values at 0°C [48], see Table I). In other words, there is an accelerating effect (*trans*-stimulation) when glucose is present on the *trans* side of the membrane. The first kinetic model that could account for these findings was the

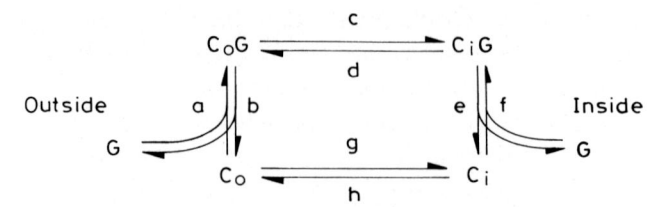

Fig. 2. The simple asymmetric carrier model for glucose transport. C denotes a sugar-binding site, which can exist in an outward-facing (C_o) or an inward-facing (C_i) conformation. Dissociation constants for sugar binding are b/a and e/f. Rate constants for carrier re-orientation are c, d, g, and h.

simple asymmetric carrier model shown in Fig. 2 [47,58]. (This model is a development of the earlier symmetric carrier model of Widdas [59].)

The characteristic feature of the model is that a *single* glucose-binding site (C) is present, that can be exposed at either the extracellular surface of the membrane (C_o) or at the cytoplasmic surface (C_i), but not simultaneously at both. Both the unloaded site (C_o, C_i) and the site with bound glucose (C_oG or C_iG) can re-orientate so as to appear on the opposite surface of the membrane, the re-orientation of the loaded site bringing about the translocation of bound glucose. Although the nature of the re-orientation process is not specified, identical kinetics would be given either by a mobile carrier or a membrane-spanning protein of fixed overall orientation in which the substrate-binding site 're-orients' by virtue of a conformational change [60,61]. The latter is the more likely situation, and so the carrier model is often alternatively named the single-site alternating conformation model. In the model the dissociation constants (b/a and e/f) for glucose binding at the two faces of the membrane can differ, and the rate constants (c,d) governing re-orientation of the loaded site are greater than those (g,h) governing re-orientation of the empty site. The latter property would account for the observation of *trans*-stimulation, i.e., that $V^{ee} > V^{zt}_{io}$ or V^{zt}_{oi}, the maximal zero-*trans* flux rates being governed by the rate of re-orientation of the unloaded sites.

Although the testing of such a model appears at first sight straightforward, determination of the necessary kinetic parameters presents a number of problems. These include the rapidity of the fluxes at physiological temperatures, coupled with the small internal volume of cells and the possibility of metabolism of intracellular glucose. Many studies have, therefore, been restricted to low temperatures, or have necessitated the use of sophisticated rapid reaction techniques [48,62]. Furthermore, for reasons which are not yet clear, uptake of sugar often exhibits 'non-ideal' behaviour, resulting in different estimates of K_m and V_{max} from initial rate measurements as compared to use of the integrated rate equation approach (see for example [48]). The major difficulty that must be faced when deciding whether to reject a kinetic model in the face of discrepancies between the values it predicts for various kinetic parameters, and those actually measured (see Table I), is, therefore, the possibility of experimental artefact. This is a particular problem in comparison of the results obtained from different laboratories, where the precise conditions used in transport assays may differ. In view of such problems, the recent demonstration by Krupka that it is possible to check a set of kinetic data for internal consistency is timely [63]. Such a test will identify at least some sets of data, which, as the result of systematic errors or of unexpected changes in the transport system in different experiments, are internally inconsistent and so cannot be used to test the adequacy of a kinetic model. Krupka has shown that for a transport system such as the glucose transporter, which obeys Michaelis–Menten kinetics, and in which the ratio of intracellular and extracellular substrate concentrations at equilibrium is 1, the following relationships between kinetic constants must exist,

$$\frac{V_{io}^{zt}}{K_{io}^{zt}} = \frac{V_{oi}^{zt}}{K_{oi}^{zt}} = \frac{V^{ee}}{K^{ee}}. \tag{2}$$

These relationships are identical to Haldane relationships, but unlike the latter, their validity does not derive from a proposed reaction scheme, but merely from the observed hyperbolic dependence of transport rates upon substrate concentration. Krupka showed that these relationships were not obeyed by the set of data previously used by Lieb [64] to reject the simple asymmetric carrier model for glucose transport. Such data therefore cannot be used either to confirm or refute the model.

The bulk of recent experimental data obtained by the use of initial rate measurements does obey Eq. (2), and appears consistent with the simple asymmetric carrier model for transport. A selection of experimentally measured values for K_m and V_{max}, determined from initial rate measurements, are compared in Table I with those predicted by Wheeler and Whelan [65] for the model. Considering the practical difficulties of measurement discussed above, the measured kinetic properties of sugar transport in the erythrocyte seem to fit the asymmetric carrier model remarkably well. However, there are a number of recent studies whose results are claimed to be irreconcilable with the simple one-site, asymmetric carrier model, even when the practical difficulties of making measurements are allowed for. Such studies have led to the formulation of alternative models, such as the two-site model proposed by Carruthers and colleagues [56,66–68]. In the latter model, two sugar-binding sites can exist simultaneously. However, given the vagaries of kinetic experiments, definitive identification of the 'correct' model will probably have to await molecular studies on the isolated transporter, rather than further kinetic studies on the erythrocyte itself. For example, direct identification of more than one sugar-binding site on the transporter would automatically exclude the simple asymmetric carrier model.

2.5. Measurements of individual rate constants for steps in the transport cycle

Assuming that the simple, four-state asymmetric carrier model does accurately describe the transport process, Lowe and Walmsley [48] have exploited the temperature-dependence of transport to determine the individual rate constants governing the re-orientation of the substrate-loaded and unloaded carrier (c and d, and g and h, respectively, in Fig. 2). Their analysis relies upon the assumption that the re-orientations of the transporter are slow in relation to the rates of sugar association with and dissociation from the carrier (i.e., the rate constants a, b, e and f are much larger than the rate constants c, d, g and h in Fig. 2). The soundness of this assumption has been confirmed by nuclear magnetic resonance (NMR) measurements [69]. Given this situation, the equations describing the V_{max} values for transport measured under differing conditions simplify to the following, where $[C]$ is the concentration of the transporter,

$$V^{ee} = \frac{[C]}{(1/c + 1/d)} \;,\quad V^{zt}_{oi} = \frac{[C]}{(1/c + 1/h)} \;,\quad V^{zt}_{io} = \frac{[C]}{(1/d + 1/g)} \;. \quad (3)$$

The V_{max} (and K_m, see below) constants determined from steady-state kinetic measurements are thus seen to be complex constants containing two or more of the individual rate constants illustrated in Fig. 2.

As had been observed by other workers, not only was transport found to be very temperature-sensitive, but the asymmetry of transport was shown to decrease with increasing temperature, so that at physiological temperatures (37°C) there was little difference in the V_{max} values measured under zero-*trans* influx, zero-*trans* efflux, or equilibrium-exchange conditions. Arrhenius plots for both zero-*trans* fluxes and equilibrium exchange were curved, a finding that was attributed to differences in the activation energies of the individual rate constants (e.g., c and d) that contribute to the composite V_{max} constants. Using non-linear least-squares procedures to fit the curves it was thus possible to estimate values for the individual rate constants and activation energies for translocation of both the loaded and the unloaded carrier (c and d, and g and h, respectively, in Fig. 2).

The rate constants and activation energies calculated from the steady-state kinetics of transport by Lowe and Walmsley are shown in Table II. Independent estimates of the values for some of these rate constants and their activation energies have more recently been obtained from pre-steady-state measurements of conformational changes in the purified transport protein using stopped-flow fluorescence procedures [19]. These measurements can be made because of differences in the intrinsic fluorescence of the outward-facing and inward-facing conformations of the transporter (C_o and C_i, respectively, in Fig. 2, see section 5 on evidence for conformational changes in the transporter). Measurement of the fluorescence transients associated with the conformational change enabled rate constant values to be determined at 10°C for steps g, h, c/K_{S_o} and d/K_{S_i} of Fig. 2, where K_{S_o} is the dissociation constant for sugar binding to the outward-facing conformation of the transporter ($\equiv b/a$) and K_{S_i} is the dissociation constant for sugar binding to the inward-facing conformation ($\equiv e/f$). Compar-

TABLE II

Rate constants governing re-orientation of the glucose transporter, and their activation energies, determined from steady-state and pre-steady-state measurements

Parameter		Steady-state		Pre-steady-state	
		Rate constant	Activation energy	Rate constant	Activation energy
g	(0°C)	12.1 s^{-1}	127.0 kJ mol^{-1}		
g	(10°C)	87 s^{-1}		9 s^{-1}	
h	(0°C)	0.726 s^{-1}	173.0 kJ mol^{-1}		118 kJ mol^{-1}
h	(10°C)	10.8 s^{-1}		5.2 s^{-1}	
c	(0°C)	1113 s^{-1}	31.7 kJ mol^{-1}		
c/K_{S_o}	(10°C)	190 mM^{-1} s^{-1}		66 mM^{-1} s^{-1}	
d	(0°C)	90.3 s^{-1}	88.0 kJ mol^{-1}		
d/K_{S_i}	(10°C)	23 mM^{-1} s^{-1}		31 mM^{-1} s^{-1}	72.8 kJ mol^{-1}

Steady-state and pre-steady-state values are taken from Lowe and Walmsley [48] and from Appleman and Lienhard [19], respectively.

ison of these values with those obtained at 10°C from steady-state analyses showed agreement within a factor of three except for g, where pre-steady-state measurements yielded a value only one-tenth that determined from steady-state measurements (Table II). Activation energies estimated by the two procedures were also very similar (Table II). Given that the pre-steady-state measurements were performed on purified transporter, in the absence of factors such as ATP that were present in the cells used for steady-state measurements, the agreement is very good.

Because the rates of sugar binding to and dissociation from the transporter are very rapid compared to the rates of transporter re-orientation, the Michaelis constants for transport by the simple asymmetric carrier model are given by the following equations,

$$K^{ee} = \frac{b(1 + g/h)}{a(1 + c/d)}, \quad K^{zt}_{oi} = \frac{b(1 + g/h)}{a(1 + c/h)}, \quad K^{zt}_{io} = \frac{e(1 + h/g)}{f(1 + d/g)}. \tag{4}$$

Using these equations, Lowe and Walmsley [48] have calculated the dissociation constants for sugar binding at the extracellular surface of the membrane ($K_{S_o} = b/a$ in Fig. 2) and at the cytoplasmic surface ($K_{S_i} = e/f = (b/a) \times [dg/ch]$) from the estimated rate constants for carrier re-orientation and the measured Michaelis constants. The dissociation constant for binding at the extracellular surface of the membrane, calculated in this way, is approximately 10 mM and is largely unaffected by temperature over the range 0–40°C. The dissociation constant for binding to the cytoplasmic side of the membrane ranges from about 13 mM at 0°C to about 23 mM at physiological temperatures (37°C). Thus, the large asymmetries in zero-*trans* Michaelis constants seen at low temperatures do not stem from a large difference in the affinity of glucose for the two conformations of the transporter. Instead, if the asymmetric carrier model is correct, they arise from the asymmetries in the rate constants for carrier re-orientation.

In contrast to the small asymmetry seen in the affinities of the two orientations of the transporter for glucose, the calculated rate constants for carrier re-orientation predict a very asymmetric distribution of both the loaded and unloaded carrier orientations at 0°C (the distributions are given by the ratios c/d and g/h, respectively). In the absence of glucose only about 6% of the transporter is predicted to be in the outward-facing conformation. This percentage rises with temperature, until at physiological temperature (37°C) about 40% of the transporters are outward-facing. A similar asymmetry is predicted for the transporter–glucose complex, the percentage of outward-facing carrier–glucose complexes rising from 7.5% at 0°C to about 60% at 37°C.

3. Characterization of the isolated human erythrocyte transporter

3.1. Purification and kinetic properties of the transporter protein

Purification of the human erythrocyte glucose transporter was first achieved in the late 1970s. This cell type was chosen because its large transport capacity suggested the presence of an abundance of transporters. In fact, there are about 5×10^5 cytochalasin B binding sites per cell, indicating that the transporter constitutes about 6% of the total membrane protein [70]. Removal of most of the major peripheral membrane proteins from erythrocytes was found to have little effect upon the transport activity [71], and so it was concluded that these components of the membrane play no obligatory part in the transport process. In contrast, solubilization of the integral membrane proteins in non-ionic detergent, followed by reconstitution in sonicated liposomes, led to the recovery of stereospecific glucose transport, albeit in low yield [72,73]. The solubilized proteins were also found to bind cytochalasin B, in a D-glucose-inhibitable fashion, after removal of detergent [33,74]. Identification of the glucose transporter amongst the other integral proteins was achieved by chromatography of Triton X-100 extracts of the membranes on diethylaminoethyl (DEAE)-cellulose [75,76]. A fraction that yielded D-glucose transport activity upon reconstitution was eluted from the chromatography column at low ionic strength. It comprised largely a single protein species that migrated as a broad band of average apparent M_r 55 000 on sodium dodecyl sulphate (SDS)/polyacrylamide gels, corresponding to erythrocyte membrane protein zone 4.5 (nomenclature of Steck [77]).

The specific activity of the purified protein for glucose transport was about 10-fold higher than that for the unfractionated Triton extract, but it was less than 1% of that expected on the basis of transport rates seen in the intact erythrocyte [75]. However, the possibility that transport activity resulted from the presence of a minor contaminant in the preparation, rather than from the major band 4.5 component, was excluded by measuring a stoichiometric function of the transporter, its ability to bind cytochalasin B. Preparations of the glucose transporter purified using Triton X-100 were found to bind 7–11 nmol cytochalasin B per mg protein [78,79]. Subsequent improvements in the purification procedure, including the use of octyl glucoside as the detergent, have increased the extent of cytochalasin B binding to 17.5 nmol per mg protein [80–82]. This figure corresponds to 0.95 molecules of cytochalasin B per polypeptide chain if the M_r of the protein is taken to be 54 117 (see section 4.1), and indicates that the transporter, or a cytochalasin B binding component thereof, constitutes the majority of the band 4.5 polypeptides in the preparation.

Although the transport activities of the isolated glucose transporter preparations were reported to be rather low, the characteristics of cytochalasin B binding and its competition by D-glucose were found to be essentially identical to those seen for the intact erythrocyte membrane. The K_d for cytochalasin B binding was 150–190 nM [78,79,82] and the K_i for D-glucose inhibition of the binding was 43 mM [78,79]. The latter finding indicated that the isolated protein retained the sugar-binding site of the native transporter, at least in part. Binding was also inhibited by other known inhibitors of glucose transport in the intact erythrocyte such as phloretin, diethylstilboestrol and maltose [79]. In fact, it now appears that the low transport activity seen in early studies of the purified transporter probably resulted from the use of non-optimal conditions for reconstitution, rather than from the lack of some essential protein component of the transport system. When the purified protein was incorporated into small phospholipid vesicles at a density of about one per vesicle, almost every transporter molecule capable of binding cytochalasin B was also found to be capable of transport [83]. However, the transport activity of each protein, measured under equilibrium-exchange conditions, was only 5% of that of the transporter in the intact erythrocyte [83]. Somewhat higher transport activities have been reported for equilibrium exchange in vesicles produced by the freeze–thaw/sonication procedure using soybean lipids, followed by separation from non-reconstituted protein [84]. In this study, a specific activity for equilibrium exchange at 23°C of 50 μmol/mg/min was obtained, about five times that calculated for the erythrocyte membrane [84]. Since the glucose transporter is known to comprise about 6% of the total membrane protein in erythrocytes [70] it follows that the specific activity of the purified transport protein is about 25% of that expected for the fully active transporter. However, in view of the known sensitivity of the reconstituted transporter activity to its lipid environment [85–88] it is not surprising that reconstitution into soybean lipids does not result in a fully active transporter.

3.2. Molecular properties of the isolated protein

3.2.1. Polypeptide composition and glycosylation state

The stoichiometry of cytochalasin B binding to the isolated glucose transporter indicates that the preparation is at least 95% pure. The most abundant contaminant is the nucleoside transporter, a glycoprotein which is very similar in its size, glycosylation and a number of other properties to the glucose transporter [80,89,90]. The purity of the preparation has been confirmed by amino acid sequencing [80]. Amino acid analysis shows that the transporter is a rather hydrophobic protein, containing only 35% polar residues [80]. Although six cysteine residues are present, disulphide bonds appear to be absent because 5.5 cysteines per chain are reactive towards 5,5'-dithiobis(2-nitrobenzoic acid) in the denatured but not-reduced protein [80].

Despite its purity, the isolated glucose transporter migrates as a very broad band of apparent M_r 55000 on SDS/polyacrylamide gels as the result of heterogeneous glycosylation [91]. The carbohydrate, which constitutes about 15% by weight of the glycoprotein [92] is N-linked and can be completely removed from the protein by treatment with endoglycosidase F. Following this treatment the protein migrates as a single sharp band of apparent M_r 46000 Da [93]. Recent investigations have shown that the heterogeneous collection of oligosaccharides attached to different transporter molecules include both high mannose-type oligosaccharides and biantennary complex-type oligosaccharides [94]. The latter have, as their outer branches, poly-N-acetyllactosamine chains containing about 16 N-acetyllactosaminyl units. The function of the oligosaccharide remains in doubt. In a recent study it was reported that removal of the carbohydrate using N-glycanase inactivated the reconstituted transporter [95]. However, previous studies have shown no effect upon transport as a result of partial deglycosylation of the transporter using endo-β-galactosidase [84] and no effect upon cytochalasin B binding as the result of complete deglycosylation of the transporter by a combination of endo-β-galactosidase and endoglycosidase F treatment [99].

3.2.2. Secondary structure

The availability of the purified transporter in large quantity has enabled investigation of its secondary structure by biophysical techniques. Comparison of the circular dichroism (CD) spectrum of the transporter in lipid vesicles with the CD spectra of water-soluble proteins of known structure indicated the presence of approximately 82% α-helix, 10% β-turns and 8% other random coil structure [97]. No β-sheet structure was detected either in this study or in a study of the protein by the same group using polarized Fourier transform infrared (FTIR) spectroscopy [98]. In our laboratory FTIR spectroscopy of the transporter has similarly revealed that the protein is predominantly α-helical [99]. However, in addition to the strong absorption bands due to α-helix, random coil and β-turn, we have also seen an absorption band at 1630 cm^{-1}, which in soluble proteins is characteristic of β-sheet. Thus, although it is clear that the transporter is predominantly α-helical, the presence of a small amount of β-sheet remains uncertain. Analysis of the linear dichroism of polarized FTIR spectra obtained using oriented multilamellar films of the reconstituted transporter indicates that the α-helices are preferentially oriented perpendicular to the plane of the lipid bilayer, and are probably tilted at less than 38° from the membrane normal [98]. The location of these helices relative to the lipid bilayer has been studied by comparing the infrared spectra of the protein before and after tryptic digestion. This study showed that both the cytoplasmic, hydrophilic regions of the transporter which are removed by digestion (see below) and the hydrophobic, membrane-embedded portions of the protein are predominantly α-helical [100].

3.2.2. Oligomeric state

It is not yet clear whether the sugar transporter functions as a monomer or as an oligomer either in its purified state or in the intact erythrocyte membrane. However, there is some evidence for the presence of a dimeric form of the protein. For example, freeze-fracture electron microscopy of the purified, reconstituted protein shows particles of diameter 62 Å, corresponding to an M_r of 110 000 [92]. Similarly, Sase et al. [101] estimated an M_r of 120 000 ± 30 000 from measurements of the frequency of intramembranous particles seen in reconstituted preparations of known protein to lipid ratio. The size of the functional glucose transporter has also been estimated from target size analysis of radiation-inactivation of cytochalasin B binding and transport activity. For the purified transporter, a target size corresponding to an M_r of 120 000 has been reported [102], whereas for erythrocyte membranes target size analyses have indicated M_r values from 124 000 to 220 000 [103–105]. The transporter may also be dimeric in its detergent-solubilized state – the results of gel filtration and sucrose gradient centrifugation experiments on Triton X-100-solubilized erythrocyte membranes yielded an estimate of 60 000–120 000 for the M_r of the cytochalasin B binding protein [33].

4. The structure of the human erythrocyte glucose transport protein

4.1. Amino acid sequence

The amino acid sequence of the human erythrocyte glucose transporter was deduced from the nucleotide sequence of a cDNA clone in 1985 [106]. Polyclonal antibodies raised against the protein were used to screen a λgt11 cDNA library prepared from the human hepatocellular carcinoma cell line HepG2. (Like many other transformed cell lines, these hepatoma cells express primarily the erythrocyte-type GLUT-1 transporter gene rather than the GLUT-2 gene which is expressed in normal hepatocytes.) A 2.8-kb cDNA clone identified in this way contained a single long open reading frame which predicted a 492-amino acid protein of M_r 54 117. A 97.6% identical sequence was subsequently deduced for a putative glucose transporter of identical size encoded by a rat brain cDNA [107]. The predicted amino acid composition of the human protein was essentially identical to that measured for the isolated human erythrocyte glucose transporter. Furthermore, fast atom bombardment mass spectroscopic mapping experiments and direct sequencing of fragments of the glucose transporter demonstrated the latter's identity to the encoded protein over at least one-quarter of the sequence. Most significantly, these protein chemical studies revealed that the native protein contained the same N- and C-terminal sequences as those predicted from the clone, indicating that the transporter does not contain a cleavable signal sequence, nor is it post-translationally cleaved in its C-terminal region. The finding that the deglycosylated protein migrates on SDS/polyacrylamide gels with a mobility corresponding to an M_r of 46 000, rather than one of 54 117, probably reflects its very hydrophobic nature. Additional proof that the cDNA clone encodes the genuine glucose transporter has been provided by the recent demonstrations that it can be expressed in functionally active form in *Xenopus* oocytes, where it exhibits kinetic properties similar to those found in the intact erythrocyte [6,7].

4.2. Arrangement in the membrane

4.2.1. Topology
The secondary structural analyses described above suggest that the transport protein spans the lipid bilayer predominantly in the form of α-helices, which are likely to be either hydrophobic or amphipathic. Analysis of the glucose transporter sequence by the algorithm of Eisenberg et al. [108] predicts the presence of 12 regions which could form such membrane-spanning segments. The predicted locations of these putative helices, together with the information discussed below, have been used to construct the model for the arrangement of the transporter polypeptide in the membrane shown in Fig. 3. In this model the hydrophilic N-terminal region (residues 1–12), C-terminal region (residues 451–492) and the large central hydrophilic region of the sequence (residues 207–271), are located on the cytoplasmic face of the mem-

---

Fig. 3. Model for the two-dimensional arrangement of the human erythrocyte glucose transporter in the membrane. Amino acid residues are identified by their single letter code. Solid bars indicate the location of introns in the transporter gene. The regions coloured black are released from the membrane upon tryptic digestion. Shaded segments indicate the probable regions photolabelled by ATB-BMPA (helix 8) and by cytochalasin B (helix 11 and the loop connecting it to helix 10). The circles with heavy outlines indicate the region labelled by IAPS-forskolin (helix 10).

187

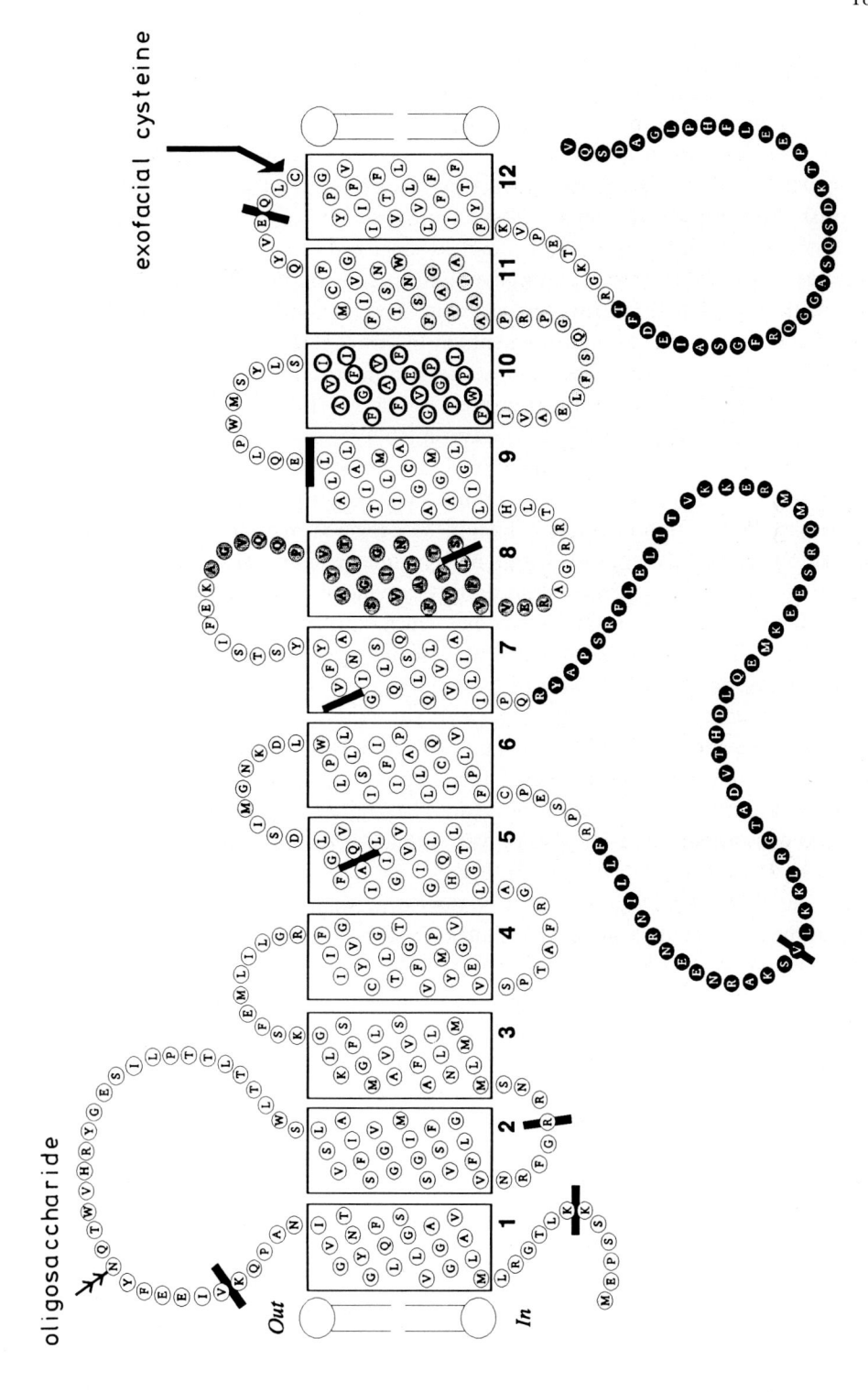

brane.

Direct evidence for the cytoplasmic location of the central and C-terminal regions has now been obtained by vectorial proteolysis experiments [100]. These have taken advantage of the observation that the extracellular domain of the membrane-bound transporter is very resistant to cleavage by a wide range of proteases [93,109]. In contrast the cytoplasmic domain is susceptible to proteases such as trypsin and chymotrypsin, which cleave the protein at sites near the mid-point and the C-terminus to yield two large, membrane-bound fragments [81,109]. Their amino acid compositions [100] and reactivity towards site-directed antibodies [110,111] indicate that the two membrane-bound fragments of the transporter contain residues 1–212 and 270–456, respectively. The intervening region (residues 213–269) and the C-terminus (residues 457–492) are recovered as water-soluble peptides (indicated in Fig. 3), confirming their cytoplasmic location [100].

Additional evidence for the cytoplasmic locations of the large, central hydrophilic region of the sequence and of the C-terminus has been provided by the use of anti-peptide polyclonal antibodies raised against these regions. We and others have shown that antibodies against segments within both regions (residues 217–272 and 450–492) bind to the cytoplasmic surface of the erythrocyte membrane [110–112]. The same is true for a number of monoclonal antibodies which have been raised against the intact transporter: all that have been investigated so far appear to recognize epitopes in these two regions [70,111,113,114].

Further evidence for the proposed topography of the transporter comes from the location of the site of glycosylation. There are only two potential sites of N-linked glycosylation, at Asn_{45} and at Asn_{411} [106]. However, when the membrane-bound transporter is cleaved by trypsin only the N-terminal fragment (residues 1–212) migrates as a broad band on SDS/polyacrylamide gels, indicative of glycosylation [100,110]. Thus Asn_{45}, located in the hydrophilic loop predicted to connect membrane-spanning helices 1 and 2, must be the extracellular site of glycosylation. The other tryptic fragment, containing residues 270–456 and the potential site of glycosylation at Asn_{411}, is not glycosylated [100,111].

Evidence for the topography of the remaining regions of the protein is less firm, and stems principally from the identification of nucleophilic amino acid side chains at the extracellular surface of the membrane. For example, many membrane-impermeant or poorly permeant sulphydryl reagents have been shown to inhibit glucose transport in the intact erythrocyte, indicating the presence of one or more exofacial cysteine residues whose modification somehow impairs the transport mechanism [115–121]. Deziel et al. identified a single exofacial sulphydryl group in the protein by treating intact erythrocytes with the membrane impermeant reagent *p*-chloromercuribenzene sulphonate to protect it from subsequent modification with *N*-ethylmaleimide [122]. This group was located on the large, C-terminal fragment (residues 270–456) produced by tryptic digestion of the transporter, and which contains three of the protein's six cysteine residues (374, 421 and 429). The precise location of the reactive exofacial cysteine remains to be definitively established, but has been tentatively identified as Cys_{429} from chemical cleavage experiments on transporter exofacially labelled with S-(bismaleimidomethyl ether)-L-[^{35}S]cysteine [123]. This residue is predicted to lie at the extracellular end of putative transmembrane helix 12 (Fig. 3) [106].

4.2.2. Three-dimensional arrangement
Little is known about the three-dimensional arrangement of the 12 putative membrane-spanning helices in the membrane. However, evidence obtained from deuterium and tritium exchange experiments suggests that they may be clustered around an aqueous channel which penetrates the membrane [99,124]. Several of the helices are predicted to be amphipathic and might play such a role [106]. From the rate of deuterium exchange as measured by FTIR it appears that more than 80% of the polypeptide backbone is readily accessible to solvent, exchanging within 1 h at 20°C in contrast to other membrane proteins such as rhodopsin and the sarcoplasmic reticulum Ca^{2+}-ATPase which exchange much more slowly [99]. Furthermore, about 20–25% of the peptide amide hydrogens appear to be free, forming a hydrogen bond to water. This value is comparable to that of water-soluble proteins, despite the fact that 50% of the protein is predicted to be membrane-spanning [124]. It follows that some of these amides are likely to lie within the membrane, lining an aqueous cavity.

4.3. Location of the substrate-binding site(s)

4.3.1. Insights from proteolytic digestion
Extensive tryptic digestion of the membrane-bound glucose transporter destroys its ability to transport sugar [125,126]. However, the digested transporter is still capable of binding cytochalasin B, albeit with reduced affinity [81,127]. Furthermore, the binding of cytochalasin is still inhibitable by D-glucose – indeed, the apparent K_i for inhibition of binding is about 10-fold lower than that seen for the undigested protein [100,127]. Because tryptic digestion removes most of the cytoplasmic, hydrophilic regions of the sequence it follows that the site(s) of glucose and inhibitor binding must be located primarily in the membrane-embedded regions of the transporter [100]. A more precise localization of the site(s) of substrate and inhibitor binding has been sought by photolabelling experiments using a range of photoactivable transport inhibitors, as detailed below.

4.3.2. Photoaffinity labelling with cytochalasin B
Although cytochalasin B normally functions as a reversible inhibitor of glucose transport, upon exposure to ultraviolet light a small proportion of the bound cytochalasin B molecules become covalently linked to the transporter protein [128–130]. Photolabelling is inhibitable by D-glucose and other transported sugars but not by

L-glucose [128–130]. The mechanism of photolabelling is not clear, but its action spectrum suggests that the reaction proceeds via photoactivation of an aromatic amino acid residue on the protein rather than by activation of the ligand itself [131]. Although the efficiency of labelling is low (6–8% at maximum [81,131]), it has been possible to exploit this phenomenon to investigate the site of binding using [^{3}H]cytochalasin B. Tryptic cleavage of the membrane-bound, photolabelled protein indicates that the site of radiolabelling lies within the C-terminal fragment of apparent M_r 18 000 (residues 270–456 – see section 4.2.1) [81,109]. Further experiments in which the labelled protein was chemically cleaved at cysteine and tryptophan residues have led to the suggestion that the site of labelling lies within the region containing residues 389–412 [81,132]. This region comprises half of helix 11 and the loop linking it to helix 10 (see Fig. 3), a location consistent with the fact that cytochalasin B binds at the cytoplasmic side of the membrane [34]. However, a firm conclusion awaits the isolation and sequencing of labelled fragments.

4.3.3. Photoaffinity labelling with bis-mannose derivatives

Cytochalasin B inhibits the glucose transporter by binding to the cytoplasmic side of the membrane. In contrast, a family of non-transported, membrane-impermeant sugar analogues synthesized by Holman's group inhibit transport by binding at the extracellular surface of the membrane [133]. These compounds are synthesized from two D-mannose moieties joined at their C-4 positions by a 2-propylamine bridge (1,3-bis(D-mannos-4-yloxy)-2-propylamine (BMPA)) (see Fig. 1) and are fairly potent inhibitors of transport (K_i for inhibition of transport in the erythrocyte of <1 mM) [133–136]. A range of photoaffinity labels for the transporter have been developed by coupling photoactivable groups to the amino group of BMPA. These include the azidosalicoyl derivative ASA-BMPA (2-N-(4-azidosalicoyl)-1,3-bis(D-mannos-4-yloxy)propylamine) [133,134] and the diazirine derivative ATB-BMPA (2-N-[4-(1-azi-2,2,2-trifluoroethyl)benzoyl]-1,3-bis-(D-mannos-4-yloxy)-2-propylamine) [136]. Both ASA-BMPA [134] and ATB-BMPA [136] were shown to label the same tryptic fragment of the transporter, of apparent M_r 18 000 that was labelled by cytochalasin B. However, although the sites of labelling by these exofacial and endofacial inhibitors are both in the C-terminal half of the protein, they are not identical. From consideration of the pattern and sizes of the labelled fragments produced by cleavage at tryptophan and cysteine residues, Holman and Rees [132] concluded that the site of labelling by ASA-BMPA lay between Cys_{347} and Trp_{388}, and tentatively assigned it to the extracellular end of helix 9. A more recent study from our laboratory, in which labelled fragments were identified by their pattern of recognition by site-directed anti-peptide antibodies, suggests that the site of labelling by ATB-BMPA lies within putative transmembrane helix 8 (Fig. 3), between residues Ala_{301} and Arg_{330} (A.F. Davies, S.A. Baldwin and G. Holman, unpublished observations).

4.3.4. Photoaffinity labelling with forskolin and its derivatives

As for cytochalasin B, U.V. irradiation of the glucose transporter/[^{3}H]forskolin complex leads to the covalent incorporation of radiolabel into the transport protein, albeit with low efficiency [137]. A much higher efficiency and specificity of labelling has been achieved using the photoactivable forskolin derivative 3-[^{125}I]iodo-4-azidophenethylamido-7-O-succinyldeacetyl-forskolin ([^{125}I]IAPS-forskolin) [45]. However, like forskolin [137] it labels the M_r 18 000 tryptic fragment derived from the C-terminal half of the transporter [45]. Peptide-mapping studies of the photolabelled transporter using cleavage at methionine and at glutamate residues suggest that the site of labelling is within transmembrane helix 10, i.e., Ile$_{369}$–Phe$_{389}$ (Fig. 3), although this conclusion remains to be confirmed by direct sequence analysis of the labelled fragments [138]. This site of labelling is thus predicted to be close to that for cytochalasin B. Since both ligands appear to bind at the cytoplasmic side of the membrane [34,41] this site probably forms a part of the cytoplasmically accessible sugar-binding site of the transporter.

4.3.5. Photoaffinity labelling with miscellaneous inhibitors

As described in section 2.3.2, ATP binds to and modulates the kinetic properties of the glucose transporter. The photoactivable ATP analogue, 8-azidoadenosine 5'-[γ-^{32}P]triphosphate (azido-ATP) has been shown to photolabel the transporter at a site within the M_r 18 000 C-terminal tryptic fragment of the transporter, but the precise site of labelling remains to be determined [57]. In addition to this nucleotide, the transporter also appears to bind nucleosides, for although these are not themselves transported (there is a completely separate nucleoside transport protein in the erythrocyte [90]), they do inhibit the transport of glucose [139]. Indeed, they seem to bind more tightly to the sugar transporter than the natural substrate, for adenosine, inosine and thymidine are more potent inhibitors of cytochalasin B binding to the erythrocyte than is D-glucose itself [139]. Furthermore, although the photoactivable nucleoside derivative, 8-azidoadenosine, is a permeant for the nucleoside transporter, it has been shown specifically to label the glucose transporter in erythrocyte membranes [140]. The site of labelling lies somewhere within the same M_r 18 000 tryptic fragment labelled by cytochalasin B, bis-mannose derivatives, forskolin and azido-ATP, i.e., in the C-terminal half of the protein [140]. Kinetic evidence suggests that nucleosides bind primarily at the cytoplasmic surface of the membrane [139], and so the site of labelling may correspond to the same cytoplasmically accessible region labelled by cytochalasin B and forskolin, but this has not yet been established by peptide mapping studies. The same is true for the transport-inhibitory steroid androsten-4-ene-3,17-dione (see section 2.2) which binds almost exclusively at the cytoplasmic side of the membrane and has been shown to be a natural photolabel for the transporter [141].

5. Conformational changes and the mechanism of transport

As described in section 2.4, much of the kinetic behaviour of the sugar transporter is compatible with a simple, asymmetric carrier model for transport. The physical counterpart of such a kinetic scheme would be a membrane-spanning transport protein with a single sugar-binding site which is alternately exposed to the extracellular and cytoplasmic sides of the membrane by virtue of a conformational change. Some of the most compelling evidence for the existence of two such asymmetric conformational states in the transporter was provided by Barnett and colleagues, who examined the ability of a variety of sugar analogues to inhibit transport in the erythrocyte [142]. They showed that non-transported sugar derivatives with a bulky substituent at the C-6 position, such as 6-O-propyl-D-galactose and 6-O-propyl-D-glucose inhibited sugar transport only when present on the outside of the cell, and not when present inside. Similar findings were made by Baker and co-workers for the hexose derivative 4,6-O-ethylidene-D-glucose [143,144]. In contrast, sugars with large substituents at the C-1 position, such as n-propyl-β-D-glucopyranoside, were found to inhibit transport only when present inside the erythrocyte [142]. In order to explain these findings Barnett and colleagues proposed that to enter the cell, the sugar first binds to an outward-facing binding site using groups in the C-1 region of the molecule. Bulky groups attached to the sugar at the C-4 or C-6 positions do not prevent this binding. However, they do prevent the subsequent conformational change to an inward-facing form of the site which occurs with transported sugars. In this inward-facing conformation the orientation of the sugar is preserved, so that the C-4 and C-6 regions of the sugar now interact with the binding site. Bulky substituents at the C-1 position do not prevent this interaction. Additional support for the presence of a single substrate-binding site came from the observation of Gorga and Lienhard [145] that ternary complexes of the transporter with cytochalasin B and either glucose or ethylidene glucose could not be detected.

Although this simple, single-site asymmetric model is attractive, some workers have proposed that multiple binding-site models are necessary to explain both the kinetics of transport and the results of ligand-binding experiments, as discussed in section 2.4. Consequently there remains some doubt about which model reflects reality. However, both single-site and multiple-site models would entail some sort of conformational change in the protein. There is now considerable evidence that the glucose transporter does indeed undergo such a change as an essential part of the catalytic mechanism. The three main lines of evidence – reactivity towards chemical reagents, the results of biophysical studies, and susceptibility to proteolytic cleavage, are outlined below.

5.1. Influence of substrates and inhibitors on reactivity towards group-specific reagents

The first evidence suggesting the involvement of a conformational change in the transport protein came from studies of the effects of the substrate glucose upon the rate of irreversible inactivation of the transporter by protein-modifying chemical reagents. Bowyer and Widdas [146] showed that glucose transport in erythrocytes is inactivated by 1-fluoro-2,4-dinitrobenzene (FDNB) and that the rate of inactivation is accelerated by the presence of D-glucose. This finding implies that sugar binding induces a conformation with increased susceptibility to reaction. Intriguingly, however, Edwards [147] found that whereas extracellular glucose accelerated inactivation, intracellular glucose retarded it. He suggested as a likely explanation that the exit of intracellular glucose during the inactivation period increased the steady-state levels of an outward-facing conformation of the transporter that is of low reactivity towards FDNB. The converse effect would occur during the entry of extracellular glucose. Additional evidence for this conclusion was provided by the finding that non-transported exofacial inhibitors of transport such as maltose, phloretin, ethylidene glucose and 6-O-alkyl-D-galactoses protected against fluorodinitrobenzene inactivation [142,148,149], presumably by stabilising the protein in an outward-facing conformation. In contrast, the non-transported endofacial inhibitor propyl β-D-glucopyranoside increased the rate of inactivation, probably as a result of causing the transporter to accumulate in the more reactive, inward-facing conformation [142].

Similar evidence for a conformational change has come from studies on the inactivation of the transporter by sulphydryl reagents. For example, Dawson and Widdas showed that the inactivation of transport by N-ethylmaleimide is potentiated by D-glucose [150]. However, the availability of membrane-impermeant thiol reagents has enabled a more thorough investigation of the nature and location of the conformationally sensitive thiol groups than has been possible for amino groups. Numerous studies have shown that cytochalasin B protects an exofacial sulphydryl group from reaction with membrane-impermeant reagents applied to the outside of intact erythrocytes [117–120]. Since cytochalasin binds to the cytoplasmic face of the membrane [34], its effect on the reactivity of the exofacial sulphydryl cannot be steric but must result from a conformational change to an inward-facing form of the transporter in which the exofacial –SH group is less accessible to modification. In contrast, maltose and phloretin, which bind to the extracellular surface of the transporter, potentiate the reaction of the –SH group with a number of reagents [117,118,151], presumably by causing the transporter to accumulate in an outward-facing conformation. Such potentiation indicates that the reactive sulphydryl is not actually located in the substrate binding site. The reaction probably inhibits transport by interfering with the translocation-associated conformational change, for modification of the exofacial sulphydryl by a number of reagents has been shown to 'lock' the transporter in an outward-facing conformation [121,151].

The conformationally sensitive sulphydryl group modified by membrane-impermeant reagents is the only exofacial sulphydryl, and as described in section 4.2.1, has been tentatively identified as Cys_{429} (Fig. 3). It is the most reactive sulphydryl in the transporter, and can be completely modified using concentrations of the membrane-permeable reagent *N*-ethylmaleimide below 1 mM [152]. However, modification of this sulphydryl does not appear to be involved in inhibition of transport by *N*-ethylmaleimide [152]. Instead, transport inhibition correlates with the modification of an endofacial sulphydryl group in the N-terminal half of the protein, which requires higher concentrations of reagent [152]. Modification of this cysteine residue is diminished by the presence of cytochalasin B [122,152]. Such diminution might be steric in origin, but since cytochalasin appears to bind in the C-terminal half of the protein, its effect is more likely to indicate that the N-terminal half of the protein also experiences a conformational change.

5.2. Biophysical studies

Infrared spectroscopy has revealed only slight changes in transporter secondary structure upon substrate or inhibitor binding [99]. The results of CD studies are contradictory, one group reporting an increase in α-helical content upon glucose binding [153], whereas another group reported a decrease [97]. Large changes in secondary structure would be unexpected, since a conformational change is more likely to involve movements in the relative positions of secondary structural elements. In this context it is interesting that polarized infrared spectroscopic data suggest that glucose binding induces a slight change in the orientation of α-helices in the protein, although it is not clear whether these are in the transmembrane or cytoplasmic domain of the protein [98].

Although ligand binding to the transporter appears to cause little change in secondary structure, large changes in accessibility of segments of the protein to solvent, indicative of a conformational change, have been detected in tritium and deuterium exchange experiments [124]. Both D-glucose and cytochalasin B were found to have large effects on the exchange characteristics of free amide hydrogens (hydrogen bonded to the solvent) and of internally structured hydrogens (those forming hydrogen bonds to other groups within the protein). In particular, the exchange rates of the latter were greatly reduced, showing that the ligands in some way stabilize the structure of the peptide backbone in the protein.

Additional evidence for conformational changes in the transporter has come from measurement of the intrinsic fluorescence of the protein tryptophan residues, of which there are six, in the presence of substrates and inhibitors of transport. The fluorescence emission spectrum of the transporter has a maximum at about 336 nm, indicating the presence of tryptophan residues in both non-polar environments (which would emit maximally at about 330 nm) and in polar environments (which would emit at 340–350 nm) [154]. The extent of quenching by the hydrophilic quencher KI indicates that more than 75% of the fluorescence is not available for quenching, and so probably stems from tryptophan residues buried within the hydrophobic interior of the protein or lipid bilayer [155]. Fluorescence is quenched by the binding of D-glucose and other transported sugars, ethylidene glucose, maltose, phloretin, and cytochalasin B but not by propylglucoside [23,56]. Quenching of fluorescence is strongest at longer wavelengths suggesting that binding of ligands reduces the quantum yield from one or more tryptophans exposed in a polar environment, possibly at the sugar binding site(s) [23]. Quenching might then result either from direct interaction with bound ligand, or from a conformational change in which a quenching group(s) of the protein is (are) more favourably oriented [23]. Probably because quenching can arise in these different ways, there is no simple correlation between the effects of ligands on protein fluorescence and their predicted effect on protein conformation (e.g., both cytochalasin B and propylglucoside should stabilize the inward-facing conformation of the transporter, but only the former causes quenching of fluorescence). Nonetheless, Appleman and Lienhard have been able to show that the outward-facing conformation of the unliganded transporter is only about 80% as fluorescent as the inward-facing conformation, and have exploited this difference to observe, and determine rate constants for, the change between these conformations as described in section 2.5 [19]. Additional evidence for structural differences between the two conformations has come from the studies of Pawagi and Deber on the effects of ligands on the efficiency of hydrophilic quenchers [155]. The exofacial ligand maltose increased the efficiency, suggesting that the relevant tryptophan residue(s) is (are) more accessible in the outward-facing conformation of the transporter. In contrast D-glucose was found to decrease the efficiency of quenching. Such a finding may stem from the presence of sugar preventing access of quenchers to tryptophan residues in the binding site, although Pawagi and Deber suggest that the effect of glucose is mediated by relocation of a tryptophan residue (possibly Trp_{388}) from an aqueous to a more hydrophobic environment within the membrane [155].

5.3. Differential susceptibility of conformers to proteolysis

The studies described above have provided evidence for the occurrence of conformational changes primarily within the hydrophobic, membrane-spanning region of the transporters. However, the hydrophilic, cytoplasmic domains of the transporter also appear to be affected by translocation-associated conformational changes. Evidence for such changes has come from examining the susceptibility of these domains to proteolytic cleavage in various circumstances [156]. It was found that the rate of tryptic cleavage of both the central and C-terminal hydrophilic regions of the purified transporter were reduced in the presence of phloretin or 4,6-O-ethylidene-D-glucose. These inhibitors bind preferentially at the extracellular side of the membrane, whereas the sites of cleavage are at the cytoplasmic side of the membrane. It follows that the effect of the inhibitors must stem from their stabilization of the protein in the outward-facing conformation in which the sites of cleavage are less accessible to trypsin, rather than from direct steric hindrance. A conformational change must also be responsible for the effect of D-glucose, which is to increase the rate of cleavage [156]. Surprisingly, ligands such as cytochalasin B and phenylglucoside, which stabilize the inward-facing conformation of the transporter, had no effect on the rate of cleavage [156]. A possible explanation might be that the purified protein exists largely in an inward-facing conformation in the absence of ligands. Alternatively, the single-site alternating conformation model might not adequately describe the mechanism of transport. Photolabelling of the transporter with bis-mannose derivatives, which bind at the exofacial surface of the membrane, has also been shown to slow the rate of tryptic cleavage and almost completely protects the transporter from cleavage with thermolysin, both in the central cytoplasmic domain and in the C-terminal region [132,136].

A recent discovery has been that tryptic digestion itself appears to alter the conformation of the transporter, stabilizing it in an inward-facing form [136]. This has been concluded because whereas the trypsinized transporter still binds and can be photolabelled by cytochalasin B [81,127], its affinity for the exofacial ligand ATB-BMPA is about 12-fold reduced [136]. Similar findings have been made for a truncated mutant of the transporter expressed in Chinese hamster ovary (CHO) cells [157]. This mutant, lacking the C-terminal 37 amino acids of the protein, was unable to transport glucose, was poorly labelled by ATB-BMPA but normally labelled by cytochalasin B. Inhibition of cytochalasin B labelling by glucose and the endofacial ligand n-propyl-β-D-glucopyranoside was normal, but there was no inhibition by the endofacial ligand, ethylidene glucose, which does inhibit labelling of the normal glucose transporter. These findings suggest that deletion of the C-terminal region, either by mutagenesis or trypsinization, somehow locks the protein into an inward-facing conformation.

6. *Homologous transporters and their distribution in mammalian tissues*

Differences in kinetics, hormonal regulation and sensitivity to inhibitors have long suggested the presence of more than one type of glucose transporter in different mammalian tissues. For example, sugar uptake into brown and white fat, skeletal and cardiac muscle is known to be rapidly and reversibly stimulated by insulin, whereas sugar uptake into the brain and liver is not. The availability since 1985 of DNA clones encoding the erythrocyte-type glucose transporter (GLUT-1) has enabled searches to be made for the putatively different transporters of other tissues, by low-stringency screening of appropriate cDNA libraries. At the time of writing, four other homologous transporter isoforms have been identified and sequenced by such procedures. All five have been shown to be genuine glucose transporters by expression in either *E. coli* (GLUT-2 [158]) or *Xenopus* oocytes (GLUT-2–5 [6,7,9,10]). The sequence similarities between the isoforms and the fact that their hydropathy plots are almost superimposable indicates that all these glucose transporters are similar in secondary and tertiary structure. It is almost certain that the five cloned isoforms only represent the tip of the iceberg, and indeed a sixth putative glucose transporter from rat liver endoplasmic reticulum has recently been purified, although its sequence has not yet been published [159]. Not only has the cloning of these proteins given us an insight into their structures and mechanisms, but it has also provided tools, in the form of isoform-specific anti-peptide antibodies and cDNA probes, with which to investigate their tissue distributions and thus their potential physiological roles. These distributions, together with the current knowledge of the structures and possible roles of these proteins, are described below.

6.1. GLUT-1

The GLUT-1 isoform of the glucose transporter has now been cloned from human [106], rat [107], mouse [160], rabbit [161] and pig [162] cDNA libraries. More than 97% of the residues in the deduced amino acid sequences are identical, and the differences involve mainly conservative replacements. This transporter isoform is, as described previously, most abundant in the human erythrocyte. The physiological significance of this abundance is unclear, because at plasma glucose concentrations of 5 mM the maximum transport capacity is 12 000 times the rate of cellular glucose utilization [163]. Furthermore, although the erythrocytes of most foetal animals have a similar high capacity for transport, in most adults, except primates, this capacity is lost soon after birth [164]. In addition to its location in the human erythrocyte, Northern blotting studies have shown that the GLUT-1 isoform is widely distributed in mammalian tissues, with particulary high levels of expression in brain (especially in the microvessels) and placenta [161,165], and in foetal tissues in general [161]. In the adult animal, immunological studies have revealed that GLUT-1 is particularly abundant in endothelial and epithelial cells that form blood–tissue barriers [166]. These barriers include the blood–brain barrier [167–170], the blood–nerve barrier [171,172], various blood–eye barriers [166,173], the choroid plexus [169] and the placenta [166,174]. Substantial immunoreactivity of the neuropil is also seen in areas of the brain known to exhibit high demands for glucose [169]. Much lower levels of expression are seen in liver, in cardiac and skeletal muscle, and in adipocytes [107,161,165]. In the latter, immunoprecipitation studies have shown that the GLUT-1 isoform probably represents only about 10% of the total glucose transporter, 90% of the protein being the GLUT-4 isoform (see below) [175]. Although GLUT-1 does undergo translocation to the plasma membrane from an intracellular source in response to insulin, the major effect of insulin on glucose transport in the adipocyte appears to be mediated by GLUT-4 (see below) [176]. Similarly, insulin has little effect on glucose transport across the blood–brain barrier, where GLUT-1 is the major, if not the sole, transporter isoform present, even though insulin receptors are present on the endothelial cells of the brain microvessels [177,178].

6.2. GLUT-2

The physiological significance of glucose transport across the hepatocyte plasma membrane differs from that in most mammalian cells because the liver can both take up and release glucose in order to maintain blood glucose levels. In contrast to fat and muscle, in liver insulin does not alter the rate of glucose transport across the plasma membrane. As discussed in section 2.2, cytochalasin B is also a much less potent inhibitor of glucose transport in the hepatocyte than in either the adipocyte or the erythrocyte, indicating the presence of a different isoform of glucose transporter. The kinetics of sugar transport in the hepatocyte are also unusual. Although estimates of the apparent K_m for glucose uptake vary (summarized in [179]), the value of 20 mM reported for uptake at physiological temperature in rat hepatocytes [180] is significantly higher than the normal blood sugar concentration and higher than the reported value of approximately 7 mM measured for zero-*trans* uptake by the GLUT-1 isoform in the human erythrocyte at 37°C [48]. This high K_m may reflect a specific adaptation of the hepatocyte glucose transporter, allowing glucose flux into or out of the hepatocyte to respond linearly to changes in glucose concentration and preventing it becoming rate-limiting for either metabolism or supply to the blood. The hepatocyte glucose transporter also differs from those of other cells in transporting fructose, with an apparent K_m of 100 mM [180].

DNA clones encoding a novel glucose transporter isoform, GLUT-2, have been isolated from human [181], rat [158] and mouse [182,183] liver cDNA libraries confirming the presence of a separate transporter isoform in the liver. The encoded proteins, predicted to contain 522, 523 and 524 residues, respectively, share more than 80% identical residues and exhibit about 55% sequence identity to GLUT-1. The principal difference between the two isoforms is the presence of an additional 32 residues in the extracellular glycosylated loop of GLUT-2, doubling its size. In addition there are extensive sequence differences at the extreme C-terminus.

Immunocytochemical studies have shown that GLUT-2 is restricted to the sinusoidal plasma membrane of the hepatocyte [184]. The isoform is also present in the insulin-producing β-cells of the pancreatic islets of Langerhans [158,185], and on the basolateral membranes of fully differentiated absorptive intestine epithelial cells and of proximal tubule cells in the kidney [184]. Its location in the kidney and small intestine suggests that it functions in tandem with the brush-border, sodium-dependent glucose transporter in the trans-epithelial transport of glucose. In the pancreatic β-cells, stimulation of insulin synthesis and secretion depends upon both the uptake and metabolism of glucose [186]. However, glucose metabolism, involving the relatively high K_m (approximately 6 mM, [187]) enzyme glucokinase, rather than uptake, is rate-limiting over the range of glucose concentrations (5–15 mM) which stimulate secretion. The function of the high K_m GLUT-2 may therefore be to enable glucokinase to act as the 'glucose sensor' by ensuring that intracellular glucose concentrations rapidly change in response to changes in extracellular sugar concentration. Support for this idea has come from studies of diabetic animals in which insulin secretion is resistant to stimulation by glucose. Thus, in both Zucker diabetic fatty rats [188] and streptozotocin-diabetic animals [189] a profound reduction in GLUT-2 expression was detected in the pancreatic β-cells, without significant alteration in its expression in the liver.

6.3. GLUT-3

Clones encoding the GLUT-3 isoform of glucose transporter were isolated from a human fetal skeletal muscle library [190]. They encoded a 496-residue protein 64.4% identical to the human GLUT-1 sequence. Characterization of the kinetics of the protein expressed in *Xenopus* oocytes revealed a K_m for equilibrium exchange of 3-O-methyl-D-glucose of about 10 mM, considerably less than the values measured for GLUT-1 and GLUT-2 which were about 20 and 40 mM, respectively [10]. Inhibition of 2-deoxyglucose transport by D-xylose was found to be more potent than for the other two isoforms, suggesting that GLUT-3 may be a xylose transporter [10]. Although isolated from a muscle library, in the adult, GLUT-3 appears to be expressed only at low levels in skeletal muscle. The highest levels of expression appear to be in brain, placenta, kidney and gall-bladder [190]. In the brain, expression is most abundant in the glial cells [191]. Elevated levels of expression have also been reported for various cancer tissues and tumour-derived cell lines [190,192].

6.4. GLUT-4

The GLUT-4 glucose-transporter isoform is found in insulin-sensitive tissues such as white and brown fat, skeletal muscle and heart. cDNA clones encoding the protein have been isolated from human [1,193], rat [5,194] and mouse [160] libraries. The human and rat proteins (509 residues) and the mouse protein (510 residues) are very similar in sequence, and about 65% identical in sequence to the GLUT-1 isoform from these species. The major differences in sequence are found at the N-terminus, where the hydrophilic, extramembranous region is about 12 residues longer than in GLUT-1, and in the central cytoplasmic loop and the C-terminus. The measured K_m value for the equilibrium exchange of 3-O-methyl-D-glucose by GLUT-4 expressed in *Xenopus* oocytes is 1.8 mM, similar to values reported for the rat adipocyte and very much lower than the value of about 21 mM measured for GLUT-1 under the same circumstances [7].

It is now clear that the GLUT-4 isoform is the major species present in insulin-sensitive cells and it is responsible for the large increase in sugar uptake seen upon exposure of the cells to insulin. Such regulation has been the subject of an enormous amount of research, which it would be inappropriate to review here. However, and in brief, a great deal of evidence now indicates that in the basal state GLUT-4 is primarily located in intracellular membranous depots including the trans-Golgi reticulum. Much, although perhaps not all, of the effect of insulin on sugar transport is brought about by translocation of the intracellular sugar transporters to the cell surface where they become inserted into the plasma membrane by an exocytotic process. Evidence for such translocation comes primarily from studies on rat adipocytes and on differentiated 3T3-L1 adipocytes, and includes subcellular fractionation experiments [5,193], immunocytochemistry [195] and cell-surface labelling experiments [176]. However, although the fact of translocation is now incontestable, the mechanism by which insulin brings it about remains unclear.

6.5. GLUT-5

The most recently cloned of the mammalian glucose transporter isoforms is human GLUT-5. This 501-residue protein exhibits only about 40% sequence identity to the other four mammalian isoforms [9]. It is expressed at highest levels in the small intestine, and at much smaller levels in kidney, skeletal muscle and adipose tissue. The role of this protein in the small intestine is not clear – GLUT-2 has been identified as a basolateral protein, presumably functioning as a means of releasing glucose into the circulation, but the subcellular location of GLUT-5 is not known. One possibility is that it corresponds to the facilitated transport system known to exist, together with the sodium-dependent transporter, in the brush-border membrane [9].

7. Homologous transporters in other organisms

The five mammalian sugar-transporter isoforms that have so far been identified by cloning are all passive transport proteins and show no sequence similarity to the sodium-dependent active sugar-transport proteins of intestine and kidney in mammals. However, since 1987 a large number of transporters, some of which are active, from lower eukaryotes and prokaryotes have been identified by genetic techniques and shown to be related in sequence to the mammalian passive transporters. All also show the presence of the 12 hydrophobic stretches of sequence that are predicted to be membrane-spanning in the mammalian glucose transporters. Clues to the mechanism of transport that can be gleaned from comparison of all these related sequences will be discussed in section 8. However, first the properties of some of these proteins will be discussed.

7.1. Fungal transporters

Yeasts contain a large number of different active and passive sugar-transport systems. The first of these to be cloned was the glucose-repressible, high-affinity passive glucose transporter of *Saccharomyces cerevisiae*, which is encoded by the *SNF3* gene

[196]. The predicted protein sequence (884 residues) is considerably longer than those of the mammalian transporters, but contains a region (residues 86–581) that is 28% identical to the human GLUT-1 sequence. A second high-affinity yeast glucose transporter is encoded by the *HXT2* gene [197]. This transporter is a 541-residue residue protein, and is 31% identical in sequence to the *SNF3* transporter, although it lacks the large hydrophilic C-terminal domain of the latter. A greater degree of sequence identity (65%) is seen in comparisons with the 574-residue passive galactose transporter of *S. cerevisiae*, encoded by the *GAL2* gene [198,199]. Very recently, the *RAG1* gene of *Kluyveromyces lactis* has also been shown to encode a glucose transporter [200]. Like the *HXT2* product, this 567-residue protein is more similar to the yeast galactose transporter than to the *SNF3*-encoded glucose transporter, exhibiting 73% and 29% sequence identity to these two proteins, respectively.

In addition to monosaccharide transporters, two disaccharide transporters from yeasts have been found to be related in sequence to the mammalian glucose transporters. These are the high-affinity maltose transporter (614 residues) of *S. cerevisiae*, encoded by the *MAL61* gene [201,202] and the lactose transporter (587 residues) of *K. lactis*, encoded by the *LAC12* gene [203]. This latter transporter is active and is probably a proton symporter [203]. Two other fungal proteins recently shown to be related in sequence to the yeast and mammalian sugar transporters are, in fact, not sugar transporters, but are responsible for the uptake of quinate in *Aspergillus nidulans* [204] and *Neurospora crassa* [205]. Although they are not sugar transporters, these proteins exhibit most of the conserved features of the sugar-transporter sequences discussed in section 8. Quinate somewhat resembles the pyranose form of a monosaccharide and presumably binds to the transporter in an analogous fashion.

7.2. Protozoan transporters

Both active and passive sugar transport systems have been detected in protozoa. The hexose transporter from the extracellular promastigote form of the parasitic protozoan *Leishmania donovanii* is a proton symporter and is inhibited by cytochalasin B [206]. Although the gene for the transporter from *L. donovanii* has not yet been identified, it may be related to a developmentally regulated gene recently cloned from *L. enriettii*, which encodes a 567-residue protein [207]. The substrate for this protein has not been established, but its sequence is about 22% identical to that of the human GLUT-1 protein, suggesting that it is a sugar transporter.

7.3. The transporters of photosynthetic organisms

Genes encoding glucose transporters have recently been cloned from three photosynthetic organisms, the prokaryotic cyanobacterium *Synechocystis* [208,209], the eukaryotic green alga *Chlorella kessleri* [210] and the higher plant *Arabidopsis thaliana* [211]. The respective transporters contain 468, 533 and 522 residues and despite the fact that they catalyze active, proton-linked transport, their amino acid sequences are about 27% identical to that of the human GLUT-1 protein.

7.4. Bacterial transporters

7.4.1. The galactose, arabinose and xylose transporters of E. coli

The bacterium *E. coli* possesses at least 7 proton-linked, active transport systems for sugars (for a recent review see [212]). Three of these transporters, which catalyze the uptake of L-arabinose, D-xylose and D-galactose by symport with protons, are related in sequence to the sugar transporters discussed above. They probably represent the best-characterized of the non-mammalian transporters, and so are discussed here in some detail.

7.4.1.1. The D-xylose/H^+ transporter. In marked contrast to the mammalian GLUT-1 transporter, the D-xylose/H^+ symporter (XylE) has a very restricted substrate specificity, the only good substrate being the pentose xylose [213,214]. Sequencing of the gene *xylE* has shown that the transporter is a 491-residue protein, about 27% identical in sequence to the human GLUT-1 protein [3,215]. However, unlike the glucose transporter, XylE is neither inhibited nor photolabelled by cytochalasin B [4,212].

7.4.1.2. The L-arabinose/H^+ transporter. L-Arabinose is the C-4 epimer of the pentose xylose. However, the L-arabinose/H^+ transporter (AraE) has a somewhat broader substrate-specificity than the xylose transporter, and will transport D-xylose, although with lower affinity than L-arabinose, indicating the importance of the configuration of the C-4 hydroxyl for binding. It will tolerate substitution of a methyl group for a hydrogen atom at the C-5 position, 6-deoxy-D-galactose (5-methyl-L-arabinose, D-fucose) being an even better substrate than arabinose. However, a hydroxymethyl group is not tolerated, D-galactose and D-glucose being very poor substrates. Furthermore, epimers with an alternative orientation of the hydroxyl group at the C-2 and C-3 are not transported [4,216,217]. A second difference from the XylE protein is that arabinose transport is inhibited by cytochalasin B (concentration required for 50% inhibition 1–3 μM) [4,212]. The inhibitor binds to the transporter with a dissociation constant of approximately 1 μM and becomes covalently attached to the protein upon U.V. irradiation [4,212]. Sequencing of the *araE* gene has revealed that the transporter is a 472-residue protein about 23% identical to the human GLUT-1 sequence and about 29% identical to XylE [218].

7.4.1.3. The D-galactose/H^+ transporter. The D-galactose/H^+ transporter (GalP) has a much broader substrate-specificity than either XylE or AraE. It also differs from AraE in that its preferred substrates are hexoses with a hydroxyl group at the C-6 position, and with the glucose configuration of the C-4 hydroxyl. In fact it rather closely resembles the mammalian GLUT-1 transporter in its substrate specificity [217,219–222]. Interestingly, although D-glucose is the best substrate, GalP is not a route of glucose entry during growth because this sugar does not induce expression of the gene *galP*.

Like the mammalian glucose transporter, GalP is inhibited by cytochalasin B (50% inhibition by 3–10 μM) [4]. The dissociation constant for binding is approximately 1 μM and cytochalasin can be used to covalently photolabel the transport protein [220]. The sequence of the gene *galP* which encodes the transporter predicts a 464-residue protein [212]. Perhaps surprisingly, in view of its similarity in specificity to the mammalian glucose transporter, it is no more similar in sequence to the latter (about 24% identical) than are AraE and Xy1E. It is more closely related to AraE than to Xy1E, sharing 63% and 32% identical residues, respectively.

7.4.2. The citrate and tetracycline transporters of E. coli

The discovery that the proton-linked arabinose and xylose transporters of *E. coli* are homologous to the mammalian glucose transporters prompted searches of the data bases for bacterial proteins of related sequence. Two additional types of transporter were identified in this way, the citrate and tetracycline transporters. The citrate transporter is a plasmid-encoded protein that confers upon the bacterium the ability to take up citrate by symport with protons [223]. Its 431-residue sequence is only about 13% identical to that of the human GLUT-1 protein [224,225]. However, alignment of the sequences using the locations of putative membrane-spanning segments as a guide reveals the occurrence of many of the conserved sequence motifs found in the sugar transporters, as discussed in section 8 (Fig. 4) [3]. In particular these include the (D/N)(R/K)XGR(R/K) motif between transmembrane helices 2 and 3, and the related (E/D)(R/K)XGR(R/K) motif between transmembrane helices 8 and 9. The major difference between the citrate transporters and the other members of the family is the shortness of the central hydrophilic region of the sequence, which at 27 residues is only half the size of the corresponding regions of the other proteins.

The bacterial tetracycline transporters are encoded by a widespread class of tetracycline resistance genes. They appear to function as antiporters, catalysing the electrically neutral efflux of an antibiotic/divalent cation complex from the cell in exchange for a proton [226,227]. Several different tetracycline genes have now been sequenced revealing a family of closely related transporters, which on average are slightly smaller than the sugar transporters (approximately 400 residues) [228–230]. Although they are not closely related to members of the sugar-transporter family they appear to contain 12 membrane-spanning segments [231,232]. As for the citrate transporter, alignment with the sugar transporters on the basis of these hydrophobic regions reveals the presence of a number of individual residues and motifs that are conserved in the sugar-transporter family (Fig. 4). In particular, there is an example of the (D/N)(R/K)XGR(R/K) motif between transmembrane helices 2 and 3.

```
                                                            Helix 1
                                                     _____
Human      GLUT1         MEPSSKKLTGRL--MLAVGGAVLGSLQFGYNTGVINAPQKVIEEFY
Human      GLUT2         MTEDKVTGTL--VFTVITAVLGSFQFGYDIGVINAPQQVIISHY
Human      GLUT3         MGTQKVTPAL--IFAITVATIGSFQFGYNTGVINAPEKIIKEFI
Human      GLUT4     MPSGFQQIGSE--DGEPPQQRVTGTL--VLAVFSAVLGSLQFGYNIGVINAPQKVIEQSY
Human      GLUT5         MEQQDQSMKEGRLTLVLA-LATLIAAFGSSFQYGYNVAAVNSPALLMQQFY
Yeast      SNF3       72.TDDISTIDDNSILFSEPP-QKQSMMMS-ICVGVFVAVGGFLFGYDTGLINSITSMNYVKS
Yeast      GAL2       44.KAGESGPEGSQSVPIEIPKKPMSEYVTVSLLCLCVAFGGFMFGWDTSTISGFVVQTDFLR
Yeast      MAL61      74...MQDAKEADESERGMPLMTALKTYPKAAAWSLLVSTTLIQEGYDTAILGAFYALPVFQK
Yeast      LAC12      47.INGVPIEDAREEVLLPGYLSKQYYKL--YGLCFITYLCATMQGYDGALMGSIY--TEDAY
E. coli    AraE          MVTINTESALTPRSLRDTRRMN-MFVSVAAAVAGLLFGLDIGVIAGALPFITDHF
E. coli    XylE              MNTQYNSSYI-FSITLVATLGGLLFGYDTAVISGTVESLNTVF
E. coli    GalP          MPDAKKQGRSNKAMT-FFVCFLAALAGLLFGLDIGVIAGALPFIADEF
Chlorella         MAGGGVVVVSGRGLSTGDYRGGLTVYVVMVAFMAACGGLLLGYDNGVTGGVVSLEAFEK
Arabidopsis       MPAGGFVVGDGQ----KAYPGKLTPFVLFTCVVAAMGGLIFGYDIGISGGVTSMPSFLK
Synechocystis           MNPSSSPSQSTANVK-FVLLISGVAALGGFLFGFDTAVINGAVAALQKHF
Leishmania         31.DDQEDAPPFMTANNARVMLVQAIGGSLNGYSIGFVGVYSTLFGYSTNCASFLQENSCTTV
Neurospora qa-y         MTLLALKEDRPTPKAVYNWRVYTCAAIASFASCMIGYDSAFIGTTLALPSFTK
E. coli    pBR322                MKSNNAL--IVILGTVTLDAVGIGLVMPVLPGLLRDIVH--
E. coli    CIT           MTQQPSRAGTFGAILRV--TSGNFLEQFDFFLFGFYATYIAKT--------
E. coli    LacY          MYYLKNTNFW-MFGLFFFYFFIMGAYFPFFPIWLHDINHISK
                                                       _____
CONSERVED                                        A  GG  FGYDTGVI

Human      GLUT1    NQTWVHRYG-------------ESILPTT-------------------------
Human      GLUT2    RHVLGVPLDDRKAINNYVINSTDELLTISYSMNPKPTPWAEEETVAA-------------
Human      GLUT3    NKT----LTDKG----------NAPPSEVL-------------------------
Human      GLUT4    NETWLGRQGP---------EGPSSIPPGT-------------------------
Human      GLUT5    NETYYGRTQ------------EFMEDFP---------------------------
Yeast      SNF3     HVAPNH---------------DSFTAQQM---------------------------
Yeast      GAL2     RFGMKHKDGTHYL---------SNVRT----------------------------
Yeast      MAL61    KYGSLNSNTGDYEI-------------------------------------------
Yeast      LAC12    LK-YYHLD-------------INS-SSGT---------------------------
E. coli    AraE     VLTS--RLQE-----------------------------------------------
E. coli    XylE     VAPQ--NLSESAANS------------------------------------------
E. coli    GalP     QITS--HTQE-----------------------------------------------
Chlorella          -FFPDVWAKKQEVHEDSPYCTYDNAKLQ------------------------------
Arabidopsis        RFFPSVYRKQQEDASTNQYCQYDSPTLT------------------------------
Synechocystis      Q--------------------TDSLLT------------------------------
Leishmania         PNADCKWFVSPTGSSYCGWPEVTCRKEYAYSSPAEMPGALARCEADSRCRWSYSDEECQN
Neurospora qa-y    --------------------EFDFASYTPGALA------------------------
E. coli    pBR322  --------SDS----------------------------------------------
E. coli    CIT     -FFP--AESEFA---------------------------------------------
E. coli    LacY    ---------------------------------------------------------
                                    _____           _____
CONSERVED                           Helix 2                  Helix 3

Human      GLUT1    ----LTTLWSLSVAIFSVGGMIGSFSVGLFVNRFGRRNSMLMMNLLAFVSAVLMGFSKLG
Human      GLUT2    -AQLITMLWSLSVSSFAVGGMTASFFGGWLGDTLGRIKAMLVANILSLVGALLMGFSKLG
Human      GLUT3    ----LTSLWSLSVAIFSVGGMIGSFSVGLFVNRFGRRNSMLIVNLLAVTGGCFMGLCKVA
Human      GLUT4    ----LTTLWSLSVAIFSVGGMISSFLIGIISQWLGRKRAMLVNNVLAVLGGSLMGLANAA
Human      GLUT5    ----LTLLWSVTVSMFPPFGIGSLLVGPLVNKFGRKGALLFNNIFSIVPAILMGCSRVA
Yeast      SNF3     ---------SILVSFLSLGTFFGALTAPFISDSYGRKPTIIFSTIFIFSIGNSLQVGAGG
Yeast      GAL2     ---------GLIVAIFNIGCAFGGIILSKGGDMYGRKKGLSIVVSVYIVGIII-QIASIN
Yeast      MAL61    ---SVSWQIGLCL-CYMAGEIVGLQVTGPSVDYMGNRYTLIMALFFLAAFIFILYFCKSL
Yeast      LAC12    ---------GLVFSIFNVGQICGAFFVPLM-DWKGRKPAILIGCLGVVIGAIISSLTTT-
E. coli    AraE     ----------WVVSSMMLGAAIGALFNGWLSFRLGRKYSLMAGAILFVLGSIGSAFATSV
E. coli    XylE     -------LLGFCVASALIGCIIGGALGGYCSNRFGRRDSLKIAAVLFFISGVGSAWPELG
E. coli    GalP     ----------WVVSSMMFGAAVGAVGSLMGFFKRGRKLSLMIGAILFVAGSLFSAAAPNV
Chlorella          ----------LFVSSLFLAGLVSCLFASWITRNWGRKVTMGIGGAFFVAGGLVNAFAQDM
Arabidopsis        ----------MFTSSLYLAALISSLVASTVTRKFGRRLSMLFGGILFCAGALINGFAKHV
Synechocystis      ---------GLSVSLALLGSALGAFGAGPIHDRHGRIKTMILAAVLFTLSSIGSGLPFTI
Leishmania         PSGYSSSESGIFAGSMIAGCLIGSVFAGPLASKIGARLSFLLVGLVGVVASVMYHASCAA
Neurospora qa-y    ------LLQSNIVSVYQAGAFFGCLFAYATSYFLGRRKSLIAFSVVFIIGAAIMLAADGQ
E. coli    pBR322  ----IASHYGVLLALYALMQFLCAPVLGALSDRFGRRPVLLASLLGATIDYAIMATTPVL
E. coli    CIT     -ALMLT----FAVFGSGFLMRPIGAVVLGAYIDRIGRRKGLMITLAIMGCGTLLIALVPGY
E. coli    LacY    ------SDTGIIFAAISLFSLLFQPLFGLLSDKLGRKYLLWIITGMLVMFAPFFIFIFG

CONSERVED              L  VS     G    IG    G       GRK        L    G
```

Fig. 4

```
                                       Helix 4
Human       GLUT1     KSF----------------EMLILGRFIIGVYCGLTTGFVPMYVGEVSPTAF--------
Human       GLUT2     PSH----------------ILIIAGRSISGLYCGLISGLVPMYIGEIAPTAL--------
Human       GLUT3     KSV----------------EMLILGRLVIGLFCGLCTGFVPMYIGEISPTAL--------
Human       GLUT4     ASY----------------EMLILGRFLIGAYSGLTSGLVPMYVGEIAPTHL--------
Human       GLUT5     TSF----------------ELIIISRLLVGICAGVSSNVVPMYLGELAPKNL--------
Yeast       SNF3      ITL----------------LIVGRVISGIGIGAISAVVPLYQAEATHKSL--------
Yeast       GAL2      KWY----------------QYFIGRIISGLGVGGIAVLCPMLISEIAPKHL--------
Yeast       MAL61     G------------------MIAVGQALCGMPWGCFQCLTVSYASEICPLAL--------
Yeast       LAC12     KS-----------------ALIGGRWFVAFFATIANAAAPTYCAEVAPAHL--------
E. coli     AraE      -------------------EMLIAARVVLGIAVGIASYTAPLYLSEMASENV--------
E. coli     XylE      FTSINPDNTVPVYLAGYVPEFVIY-RIIGGIGVGLASMLSPMYIAELAPAHI--------
E. coli     GalP      -------------------EVLILSRVLLGLAVGVASYTAPLYLSEIAPEKI--------
Chlorella             A------------------MLIVGRVLLGFGVGLGSQVVPQYLSEVAPFSH--------
Arabidopsis           W------------------MLIVGRILLGFGIGFANQAVPLYLSEMAPYKY--------
Synechocystis         WDF----------------IFW-RVLGGIGVGAASVIAPAYIAEVSPAHL--------
Leishmania            DEFW---------------VLIVGRFVIGLFLGVICVACPVYTDQNAHPKW--------
Neurospora  qa-y      GRGI---------------DPIIAGRVLAGIGVGGASNMVPIYISELAPPAV--------
E. coli     pBR322    W------------------ILYAGRIVAGIT-GATGAVAGAYIADITDGED--------
E. coli     CIT       QTIGLLAPV----------LVLVGRLLQGFSAGVELGGVSVYLSEIATPGN--------
E. coli     LacY      PLL----------------QYNILVGSIVGGIYLGFCFNAGAPA-VEAFIEKVSRRSNFEF

CONSERVED                              LI GR   G    G S  VP Y  E AP  L

                         Helix 5                                 Helix 6
Human       GLUT1     -RGALGTLHQLGIVVGILIAQVFGL-----------DSIMGNKDLWPLLLSIIFIPALLQ
Human       GLUT2     -RGALGTFHQLAIVTGILISQIIGL-----------EFILGNYDLWHILLGLSGVRAILQ
Human       GLUT3     -RGAFGTLNQLGIVVGILVAQIFGL-----------EFILGSEELWPLLLGFTILPAILQ
Human       GLUT4     -RGALGTLNQLAIVIGILIAQVLGL-----------ESLLGTASLWPLLLGLTVLPALLQ
Human       GLUT5     -RGALGVVPQLFITVGILVAQIFGL-----------RN-LLANVDGWPILLGLTGVPAALQ
Yeast       SNF3      -RGAIISTYQWAITWGLLVSSAVSQGTHARNDASS---------YRIPIGLQYVWSSFL
Yeast       GAL2      -RGTLVSCYQLMITAGIFLGYCTNYGTKSYSN----------SVQNRVPLGLCFAWSLFM
Yeast       MAL61     -RYYLTTYSNLCWTFGQLFAAGIMKNS--------QNKYANSELGYKLPFALQWIWPLPL
Yeast       LAC12     -RGKVAGLYNTLWSVGSIVAAFSTYGTNKNFPNSS--------KAFKIPLYLQMMFPGLV
E. coli     AraE      -RGKMISMYQLMVTLGIVLAFLSDTAFSYSGN-----------WRAMLGVLALPAVIL
E. coli     XylE      -RGKLVSFNQFAIIFGQLLVYCVNYFIARSGDAS-----WLNTDGWRYMFASECIPALFF
E. coli     GalP      -RGSMISMYQLMITIGILGAYLSDTAFSYTG-----------AWRWMLGVIIIPAILL
Chlorella             -RGMLNIGYQLFVTIGILIAGLVNYAVRDWEN-----------GWRLSLGLAAAPGAIL
Arabidopsis           -RGALNIGFQLSTTIGILVAEVLNYFFAKIKGGW---------GWRLSLGGAVVPALII
Synechocystis         -RGRLGSLQQLAIVSGIFIALLSNWFIALMAGGSAQNPWLFGAAAWRWMFWTELIPALLY
Leishmania            -KRTIGVMFQVFTTLGIFVAALMGLALGQSI-RFDHDGDQKVMARMQGLCVFSTLFSLLT
Neurospora  qa-y      -RGRLVGIYELGWQIGGLVGFWINYGVNTT--------MAPTRSQWLIPPFAVQLIPAGLL
E. coli     pBR322    -RARHFGLMSACFGVGMVAGPVAGGLLGA-------IS------LHAPFLAAAVLNGLNL
E. coli     CIT       -KGFYTSWQSASQQVAIVVAALIGYGLNVTLGH-DEISEW----GWRIPFFIGCMIIPLI
E. coli     LacY      GRARMFGCVGWALCASIVGIMFTIN----------------NQFVFWLGSGCALILAV

CONSERVED                RG L    QL IT GIL A                     WR  LGL   PALL

Human       GLUT1     CIVLPFCPESPRFLLINRNEENRAKSVLKKLRGTADVTH-----DLQEMKEESRQMMREK
Human       GLUT2     SLLLFFCPESPRYLYIKLDEEVKAKQSLKRLRGYDDVTK-----DINEMRKEREEASSEQ
Human       GLUT3     SAALPFCPESPRFLLINRKEENAKQIIQRLWGTQDVSQ-----DIQEMKDESARMSQEK
Human       GLUT4     LVLLPFCPESPRYLYIIQNLEGPARKSLKRLTGWADVSG-----VLAELKDEKRKLERER
Human       GLUT5     LLLLPFFPESPRYLLIQKKDEAAAKKALQTLRGWDSVDR-----EVAEIRQEDEAEKAAG
Yeast       SNF3      AIGMFFLPESPRYYVLKDKLDEAAK-SLSFLRGVP-VHDSGLLEELVEIKATYDYEASFG
Yeast       GAL2      IGALTLVPESPRYLCEVNKVEDAKRSIAKSNKVSP-EDPAV-QAELDLIMAGIEAEKLAG
Yeast       MAL61     AVGIFLAPESPWWLVKKGRIDQARRSLERILSGKPEKELLVSMELDKIKTTIEKEQKMS
Yeast       LAC12     CIFGWLIPESPRWLVGVGREKEAR-EFIIKYHLNGDRTHPLLDMEMAEIIESFHGTDLSN
E. coli     AraE      IILVVFLPNSPRWLAEKGRHIEAEEVLRMLRDTSEKARE------ELNEIRESLKL-KQG
E. coli     XylE      LMLLYTVPESPRWLMSRGKQEQAEGILRKIMGNT-LATQ------AVQEIKHSLDHGRKTG
E. coli     GalP      LIGVFFLPDSPRWFAAKRRFVDAERVLRLRDTSAEAKR-----ELDEIRESLQV--KQS
Chlorella             FLGSLVLPESPNFLVEKGKTEKGREV-LQKLRGTSEVDAEFADIVAAVEIARPITMRQSW
Arabidopsis           TIGSLVLPDTPNSMIERGQHKEEAKTK-LRRIRGVDDVSQEFDDLVAASKESQSIEHPWRN
Synechocystis         GVCAFLIPESPRYLVAQGQGEKAAAILWKV-EGG-DVPSR-----IEEIQATVSLDHKPR
Leishmania            VVLGIVTRESRAKFDGGEEGRAELNPSEYGY-----------VEMI---------
Neurospora  qa-y      FLGSFWIPESPRWLNGKREEAMKVLCWIRNLEPTDRYIVQEVSFIDADLERYTRQVGN
E. coli     pBR322    LLGCFLMQESHKG-ERRPMPLRAFNPVSSFRWARGMTIVAAL----------
E. coli     CIT       FVLRRSLQETEAFLQRKHRPDTREIFTTIAKNWR---------------
E. coli     LacY      LLFFAK-TDAPSSATVANAVGANHSAFSLKLALELFRQPKLWFLSLY------------

CONSERVED                  F PESPR L     EA    L   L G  V         EI
```

Fig. 4 (cont.)

```
                              Helix 7
Human       GLUT1     KVTILEL---FRSPA-YRQPILIAVVLQLSQQLSGINAVFYYSTSIFEKAGV----QQP-
Human       GLUT2     KVSIIQL---FTNSS-YRQPILVALMLHVAQQFSGINGIFYYSTSIFQTAGI----SKP-
Human       GLUT3     QVTVLEL---FRVSS-YRQPIIISIVLQLSQQLSGINAVFYYSTGIFKDAGV----QEP-
Human       GLUT4     PLSLLQL---LGSRT-HRQPLIIAVVLQLSQQLSGINAVFYYSTSIFETAGV----GQP-
Human       GLUT5     FISVLKL---FRMRS-LRWQLLSIIVLMGGQQLSGVNAIYYYADQIYLSAGV----PEEH
Yeast       SNF3      SSNFIDCFISSKSRPKQTLRMFTGIALQAFQQFSGINFIFYYGVNFFNKTGV----SNS-
Yeast       GAL2      NASWGELF-STKTKVF--QRLLMGVFVQMFQQLTGNNYFFYYGTVIFKSVGL----DDS-
Yeast       MAL61     DEGTYWDCVKDGINR-RRTR---IACLCWIGQCSCGASLIGYSTYFYEKAGV---STDT-
Yeast       LAC12     PLEMLDVRSLFRTRS-DRYRAMLVILMAWFGQFSGINNVCSYYLPTMLRNVGM---KSVSL
E. coli     AraE      GWALFKINRNVRRA------VFLGMLLQAMQQFTGMNIIMYYAPRIFKMAGFTT-TEQQ-
E. coli     XylE      G----------RLLMFGVGVIVIGVMLSIFQQFVGINVLYYAPEVFKTLGAST----DI-
E. coli     GalP      GWALFKENSNFRRA------VFLGMLLQQFTGMNVIMYYAPKIFELAGYTN-TTEQ-
Chlorella             A-SLF------TRR--YMPQLLTSFVIQFFQQFTGINAIIFYVPVLFSSLGSANSAA---
Arabidopsis           L----------LRRK-YRPHLTMAVMIPFFQQLTGINVIMFYAPVLFNTIGFTTDAS---
Synechocystis         FSDLL------SRRGGLLPIVWIGMGLSALQQFVGINVIFYYSSVLWRSVGF---TEEK-
Leishmania            ------------------PRLLMGCVMAGTLQLTGINAVMNYAPTIMGSLG-------LA
Neurospora  qa-y      GFWKPFLSLKQRKVQ-WRF--FLGGMLFFWQNGSGINAINYYSPTVFRSIGITG--TDTG
E. coli     pBR322    ------------------MTVFFIMQLVGQVPAALWVIFGEDRFRWSATM-------I
E. coli     CIT       ------------------IITAGTLLVAMTTTTFYFITVYTPTYGRTVLNLSARDS--
E. coli     LacY      ------------------VIGVSCTYDVFDQQFANFFTSFFATGEQGTR--------V

CONSERVED                             R        LQ  QQFSGIN I YY    IF  AG
                             Helix 8                   Helix 9
Human       GLUT1     -VYATIGSGIVNTAFTVVSLFVVERAGRRTLHLIGLAGMAGCAILMTIALALLEQLPWM-
Human       GLUT2     -VYATIGVGAVNMVFTAVSVFLVEKAGRRSLFLIGMSGMFVCAIFMSVGLVLLNKFSWM-
Human       GLUT3     -IYATIGAGVVNTIFTVVSLFLVERAGRRTLHMIGLGGMAFCSTLMTVSLLLKDNYNGM-
Human       GLUT4     -AYATIGAGVVNTVFTLVSVLLVERAGRRTLHLLGLAGMCGCAILMTVALLLLERVPAM-
Human       GLUT5     VQYVTAGTGAVNVVMTFCAVFVVELLGRRLLLLLGFSICLIACCVLTAALALQDTVSWM-
Yeast       SNF3      -YLVSFITYAVNVVFNVPGLFFVEFFGRRKVLVVGGVIMTIANFVIAIVGCSLKTVAAA-
Yeast       GAL2      -FETSIVIGVVNFASTFFSLWTVENLGRRKCLLLLGAATMMACMVIYASVGVTRLYPHGKS
Yeast       MAL61     AFTTFSIIQYCLGIAAATFVSWWASKYCGRFDLYAFGLAFQAIMFFIIGGLGCSDTHGAKM-
Yeast       LAC12     NVLMNGVYSIVTWISSICGAFFIDKIGRREGF-LGSISGAALALTGLSICTARYEKTKK-
E. coli     AraE      -MIATLVGLTFMFATFIAVFTVDKAGRKPALKIGFSVMALGTLVLGYCLMQFDNGTASS
E. coli     XylE      ALLQTIIVGVINLTFTVLAIMTVDKFGRKPLQIIGALGMAIGMFSLGTAFYTQAP-----
E. coli     GalP      -MWGTVIVGLTNVLATFIAIGLVDRWGRKPTLTLGFLVMAAGMGVLGTMMHIGIHSPSA-
Chlorella             -LLNTVVGAVNVGSTLIAVMFSDKFGRRFLLIEGGIQCCLAMLTTGVVLAIEFAKYG-T
Arabidopsis           -LMSAVVTGSVNVGATLVSIYGVDRWGRMFLFLEGGTQMLICQAVVAACIGAKFGVDG-T
Synechocystis         SLLITVITGFINILTTLVAIAFVDKFGRKPLLLMGSIGMTITLGILSVVFGGA-TVVNGQ
Leishmania            PLVGNFVVMLWNFVTTLASIPLSYVFTMRHVFLFGSIFTFCMCLFMCGIPVY-PGVSKKL
Neurospora  qa-y      FLTTGIFGVVKMVLTIIWLLWLVDLVGRRRILFIGAAGGSLCMWFIGAYIKIADPGSNKA
E. coli     pBR322    GLSLAVFGILHALAQAFVTGPATKRFGEKQAIIAGMAADALGYVLLAFATR---------
E. coli     CIT       -LVVTMLVGISNFIWLPIGGAISDRIGRRPVL-MGITLLALVTTLPVMNWLTAAP-----
E. coli     LacY      FGYVTTMGELLNASIMFFAPLIINRIGGKNALLLAGTIMSVRIIGSSFATSALE------

CONSERVED                 T   G VN    T  S   V     GRR  L     G  M
                                                 Helix 10                       Helix 11
Human       GLUT1     --------SYLSIVAIFGFVAFFEVGPGPIPWFIVAELFSQGPRPAAIAVAGFSNWTSNF
Human       GLUT2     --------SYVSMIAIFLFVSFFEIGPGPIPWFMVAEFSQGPRPAALAIAAFSNWTCNF
Human       GLUT3     --------SFVCIGAILVFVAFFEIGPGPIPWFIVAELFSQGPRPAAMAVAGCSNWTSNF
Human       GLUT4     --------SYVSIVAIFGFVAFFEIGPGPIPWFIVAELFSQGPRPAAMAVAGCSNWTSNF
Human       GLUT5     --------PYISIVCVISYVIGHALGPSPIPALLITEIFLQSSRPSAFMVGGSVHWLSNF
Yeast       SNF3      --------KVMIAFICLFIAAFSATWGGVVWVISAELYPLGVRSKCTAICAAANWLVNF
Yeast       GAL2      Q-PSSKGAGNCMIVFTCFYIFCYATTWAPVAWVITAESFPLRVKSKCMALASASNWVWGF
Yeast       MAL61     ----------GSGALLMVVAFFYNLGIAPVVFCLVSEMPSSRLRTKTIILARNAYNVIQV
Yeast       LAC12     -----KSASNGALVFIYLFGGIFSFAFTPMQSMYSTEVSTNLTRSKAQLLNFVVSGVAQF
E. coli     AraE      ------GLSWLSVGMTMMCIAGYAMSAAPVVWILCSEIQPLKCRDFGITCSTTTNWVSNM
E. coli     XylE      --------GIVALLSMLFYVAAFAMSWGPVCWVLLSEIFPNRLRGKALAIAVAAQWLANY
E. coli     GalP      --------QYFAIAMLLMFIVGFAMSAGPLIWVLCSEIQPLKGRDFGITCSTATNWIANM
Chlorella             D-PLPKAVASGILAVICIFISGFAWSWGPMGWLIPSEIFTLETRPAGTAVAVVGNFLFSF
Arabidopsis           PGELPKWYAIVVVTFICIYVAGFAWSWGPLGWLVPSEIFPLEIRSAQSITVSVNMIFTF
Synechocystis         P-TLTGAAGIIALVTANLYVFSFGFSWGPIVWVLLGEMFNNKIRAAALSVAAGVQWIANF
Leishmania            E---AKN--GVAITGILLFILGFEVCVGPCYYVLTQDMFPPSFRPRGASFTQVAQFIFNL
Neurospora  qa-y      EDAKLTSGGIAAIFFFYLWTAFYTPSWNGTPWVINSEMFDQNTRSLGQASAAANNWFWNF
E. coli     pBR322    --------GWMAFPIMILLASGGIGMPALQAMLS-RQVDDHQGQLQGSLAALTSLTSIT
E. coli     CIT       --------DFTRMTLVLLWSFSFFGMYNGAMVAALTEVMPVYVRTVGFSLAFSLATAIFG
E. coli     LacY      ------------VVILKTLHMFEVPFLLVGCF-KYITSQFEVRFSATIYLVCFCFFKQL

CONSERVED                   I    I  FV F    GP W    E F      R  AAA     NW NF
```

Fig. 4 (cont.)

```
                                                   Helix 12
                                       ─────────────────────────────────
Human         GLUT1      IVGMCFQYVEQ-----------LCGPYVFIIFTVLLVLFFIFTYVFKVPETKGRTFDEIAS
Human         GLUT2      IVALCFQYIAD-----------FCGPYVFFLFAGVLLAFTLFTFFKVPETKGKSFEEIAA
Human         GLUT3      LVGLLFPSAAH-----------YLGAYVFIIFTGFLITFLAFTFFKVPETRGRTFEDITR
Human         GLUT4      IIGMGFQYVAE-----------AMGPYVFLLFAVLLLGFFIFTFLRVPETRGRTFDQISA
Human         GLUT5      TVGLIFPFIQE-----------GLGPYSFIVFAVICLLTTIYIFLIVPETKAKTFIEINQ
Yeast         SNF3       ICALITPYIVDTGSHTS-----SLGAKIFFIWGSLNAMGVIVVYLTVYETKGLTLEEIDE
Yeast         GAL2       LIAFFTPFITS-----------AINFYYGYVFMGCLVAMFFYVFFFVPETKGLSLEEIQE
Yeast         MAL61      VVTVLIMYQLNSEKW-------NWGAKSGFFWGGFCLATLAWAVVDLPETAGRTFIEINE
Yeast         LAC12      VNQFATPKAMK-----------NIKYWFYVFYVFFDIFEFIVIYFFFVETKGRSLEELEV
E. coli       AraE       IIGATFLTLLDS----------IGAAGTFWLYTALNIAFVGITFWLIPETKNVTLEHIER
E. coli       XylE       FVSWTFPMMDKNSWLVAH----FHNGFSYWIYGCMGVLAALFMWKFVPETKGKTLEELEA
E. coli       GalP       IVGATFLTMLNT----------LGNANTFWVYAALNVLFILLTLWLVPETKHVSLEHIER
Chlorella                VIGQAFVSMLC-----------AMEYGVFLFFAGWLVIMVLCAIFLLPETKGVPIERVQA
Arabidopsis              IIAQIFLTMLC-----------HLKFGLFLVFAFFVVVMSIFVYIFLPETKGIPIKEMGQ
Synechocystis            IISTTFPPLLDT----------VGLGPAYGLYATSAAISIFFIWFFVKETKGKTLEQM   468
Leishmania               IINVCYPIATESISGGPSGNQDKGQAVAFIFFGGLGLICFVIQVFFLHPWDEERDGKKVV
Neurospora qa-y          IISRFTPQMFIK----------MEYGVYFFFASLMLLSIVFIYFFLPVTKSIPLEAMDR
E. coli       pBR322     GPLIVTAIYAASAS--------TWNGLAWIVGAALYLVCLPALRRGAWSRATST       396
E. coli       CIT        GLTPAISTALVQLTG-------DKSSPGWWLMCAALCGLAATTMLFARLSSGYQTVENKL 431
E. coli       LacY       AMIFMSVLAGNMYESIGF----QGAYLVLGLVALGFTLISVFTLSGPGPLSLLRRQVNE

CONSERVED                I      FP              F    F              F  VPETKG T EEI


Human         GLUT1      GFRQGGASQSD--KTPEELFHPLGADSQV                                 492
Human         GLUT2      EFQKKSGSAHR--PKAAVEMKFLGATETV                                 524
Human         GLUT3      AFEGQAHGADRSGKDGVMEMNSIEPAKETTTNV                             496
Human         GLUT4      AFHRTPSLLEQEVKPSTEL-EYLGPDEND                                 509
Human         GLUT5      IFKMNKVSE-VYPEKEELKELPPVTSEQ                                  501
Yeast         SNF3       LYIKSSTGVVSPKFNKDIRERALKFQYDPLQRLEDGKNTFVAKRNNFDDETPRNDFRNTI  614
Yeast         GAL2       LWEEGVLPWKSEGWIPSSRRGNNYDLEDLQHDDKPWYKAMLE                    574
Yeast         MAL61      LFRLGVPARKFKSTKVDPFAAAKAAAAEINVKDPKEDLETSVVDEGRSTPSVVNK       614
Yeast         LAC12      VFEAPNPRKASVDQAFLAQVRATLVQRNDVRVANAQNLKEQEPLKSDADHVEKLSEAESV  587
E. coli       AraE       KLMAGEKLRNIGV                                                 472
E. coli       XylE       LWEPETKKTQQTATL                                               491
E. coli       GalP       NLMKGRKLREIGAHD                                               464
Chlorella                LYARHWFWNRVMGPAAAEVIAEDEKRVAAASAIIKEEELSKAMK                  533
Arabidopsis              VWRSHWYWSRFVEDGEYGNALEMGKNSNQAGTKHV                           522
Synechocystis
Leishmania               APAIGKKELSEESIGNRAE                                           567
Neurospora qa-y          LFEIKPVQNANKNLMAELNFDRNPEREESSSLDDKDRVTQTENAV                 537
E. coli       pBR322
E. coli       CIT
E. coli       LacY       VA                                                            417
```

Fig. 4. (cont.)

Aligned sequences of 16 members of the sugar transporter family. Residues which are identical in ⩾50% of the 16 sugar-transporter sequences (excluding the quinate transporter (qa-y), the citrate transporter (CIT), the tetracycline transporter (pBR322) and *lac* permease (LacY)) are highlighted, and recorded below the sequences as 'CONSERVED'. The locations of predicted membrane-spanning helices are indicated by horizontal bars. The sequences were taken from the references cited in the text.

7.4.3. The lactose transporter of E. coli

The lactose transporter (*lac* permease) of *E. coli* is without doubt the most intensively studied and best understood of the bacterial proton-linked sugar transporters. Since its sequence was reported in 1980 [233] prodigious efforts have been made to elucidate its molecular mechanism by site-directed mutagenesis and other means. These studies have recently been reviewed elsewhere [234,235] and so will not be discussed in detail here. The important question for the present Chapter is whether the protein is related to the sugar-transporter family and so has lessons to teach us about their mechanisms. The permease is a 417-residue protein, and, like the other sugar transporters, appears to contain 12 membrane-spanning segments [236]. The only protein so far identified that is closely similar in sequence is the lactose transporter from *Klebsiella pneumoniae*, which is 60% identical [237]. However, although only about 14% of the residues in human GLUT-1 are identical with those in either of the lactose transporters, the sequences can be aligned to reveal identical residues or conservative substitutions at numerous locations where conserved residues are found in the sugar-transporter family (Fig. 4). In particular there is a DKLGLR motif between putative transmembrane segments 2 and 3, which resembles the (D/N)(R/K)XGR(R/K) motif. A similar motif (RIGGK) is also found between transmembrane segments 8 and 9. The characteristic PESP motif of the sugar transporters which follows transmembrane helix 6 is found in the two lactose transporters (*E. coli* and *K. pneumoniae*) as TDAP and PESS, respectively. Thus, although the relationship between the lactose transporters and the other proteins remains uncertain, the regions of sequence similarity suggest the possibility that they have similar three-dimensional structures.

8. Clues to the mechanism of transport from comparison of the homologous transporters

The sequence similarities of the transport proteins discussed in this review indicate that they probably have similar three-dimensional structures. It follows that at the molecular level, the mechanisms of these passive transporters, active symporters and antiporters must share many features. Comparison of their properties, and examination of their sequences for conserved motifs should, therefore, assist in the formulation of a general model for the mechanism of transport in this family of polypeptides. The aligned sequences of a selection of the family members are illustrated in Fig. 4, with those residues that are identical in ≥ 50% of the sugar transporters highlighted. Particularly striking are the regions corresponding to predicted transmembrane α-helices, which can be aligned with very few gaps. In contrast, the intervening sequences, especially those between putative helices 1 and 2, 3 and 4, 5 and 6, 6 and 7, 7 and 8, and 11 and 12, contain insertions and deletions, suggestive of a location exposed to the solvent. The locations of the introns within the human [238] and rat [239] GLUT-1 genes are also consistent with the presence of 12 membrane-spanning helices. The intron locations are identical for the two species and all occur in regions predicted to be outside or close to the membrane surface, as indicated in Fig. 3.

Studies of other members of the family have also added support to the topological model shown in the Fig. 3. In particular, chemical labelling of the native and mutated tetracycline transporter has confirmed the cytoplasmic location of the N-terminus and the loop connecting transmembrane helices 2 and 3 [231,232]. Protease digestion experiments on this protein have also provided preliminary evidence for the cytoplasmic locations of the loops connecting helices 4 and 5, and 10 and 11 [231]. Analysis of a series of 36 *lac* permease–alkaline phosphatase (*lac*Y–*pho*A) gene fusions by Calamia and Manoil [240] has also lent strong support to the 12-helix model for this protein.

Examination of the aligned sequences reveals numerous conserved residues which presumably have importance for the structure and/or function of these proteins. Proline and glycine residues are particularly well conserved. The alignments should be of great help in the design of site-directed mutagenesis experiments for testing structure/function relationships. Surprisingly, candidates for proton translocating residues present only in the active transporters are not obvious – no histidine residues are conserved, for example. However, numerous glutamine, asparagine, threonine and serine residues which could form hydrogen bonds with a sugar molecule are conserved, especially in helices 7, 8 and 11 which are predicted to be amphipathic. It is likely that such helices are associated in the membrane to form a hydrophilic cleft or channel through which the sugar moves, as predicted by tritium and deuterium exchange experiments (section 4.2.2). Polar residues of this type are known to be involved in sugar binding in water-soluble proteins [22].

Perhaps the most striking conserved features of the aligned sequences are motifs which each occur twice in each protein. For example, between helices 2 and 3 is the motif (N/D)(R/K)XGR(K/R), where X is frequently a residue with a bulky sidechain. This motif recurs between helices 8 and 9 in the form of (E/D)(R/K)XGR(R/K). Similarly, the motif PESPR which occurs after the C-terminal end of helix 6 is paralleled by the motif PETKG after helix 12. The function of these and other motifs is unknown, although the first type of motif is predicted to form a β-turn. However, they reveal that the proteins probably evolved as the result of an internal gene duplication event [3,4]. If the ancestral transporter did have only six helices, it seems reasonable to assume that the modern transporters consist of two bundles of six helices, disposed about a pseudo two-fold symmetry axis. Independent evidence for this arrangement has recently been provided by the demonstration that functional *lac* permease can be produced by the co-expression of the separate N- and C-terminal halves of the protein, suggesting that these two domains can fold independently and then associate in the membrane [241]. Furthermore, freeze-fracture electron microscopy of reconstituted *lac* permease and electron microscopy of negatively stained filamentous arrays of the protein have both revealed a cleft in the molecule separating two apparent domains [242,243]. In the latter study the dimensions of the protein parallel to the membrane plane were estimated to be approximately 5.1×2.8 nm, with the stain-filled cleft bisecting the molecule.

The information obtained from the experiments described above can be used to construct models for the three-dimensional arrangement of the membrane-spanning helices within the transport proteins. One such model, which takes the diameter of an α-helix as 1.1 nm and seems to fit the measured dimensions of *lac* permease quite nicely, is illustrated in Fig. 5, although it must be emphasized that this is only one of

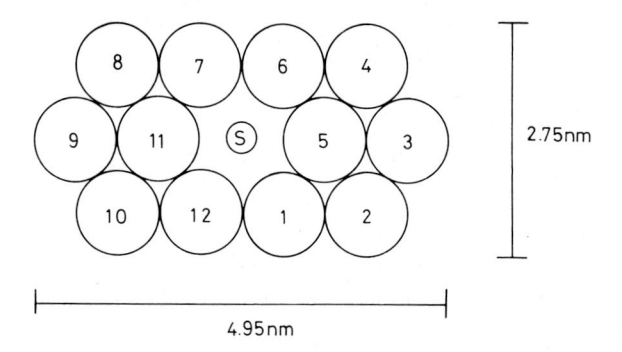

Fig. 5. A speculative model for the arrangement of the helical regions of the sugar transporters in the membrane. The helices are numbered as shown in Fig. 4. The small circle labelled 's' represents a glucose molecule.

many possibilities which must be tested by direct experimentation and molecular graphics modelling. In the model shown, the helices have been arranged so that those connected by short loops, e.g., 8 and 9 are adjacent. Helices 7 and 11 have been placed together because recent mutagenesis studies on *lac* permease have suggested that Asp_{237} and Lys_{358} of these two helices are close together, possibly forming a salt bridge [244]. Helices 7, 8 and 11, predicted to be amphipathic, might be involved in the formation of a substrate-binding cleft in this fashion in the C-terminal domain. The two 6-helix bundles have been arranged symmetrically around a central 'channel' similar to that revealed by electron microscopy, and a sphere of diameter equivalent to a glucose molecule (0.4 nm) has been included for scale. However, whether such a channel structure really exists and is part of the substrate translocation pathway is not yet clear. If the structure *is* symmetrical in the form shown, one might ask why only a single sugar-binding site exists. It is certainly intriguing that all the substrate analogues and inhibitors so far tested as photolabels of the transporter react asymmetrically, labelling only the C-terminal half of the protein (section 4). This finding suggests that the translocation pathway may be largely confined to the C-terminal domain, and raises the question of what function is played by the N-terminal domain. The mechanism by which the substrate-binding site is alternately exposed to the two sides of the membrane is also unclear, although infrared spectroscopic evidence supports the involvement of a change in helix tilt (section 5.2). Only time and experiments will give us the answers to such speculative questions, but model building may at least suggest some profitable avenues of research!

9. *Summary*

In summary, studies on the human erythrocyte glucose transporter and other members of a large family of prokaryotic and eukaryotic sugar transporters have yielded much information about the kinetics of transport, and are beginning to yield low-resolution information on the shape and arrangement of the polypeptide in the membrane. Gaining a complete understanding of the mechanism of transport at the molecular level will probably require crystallization of one of the transporters for X-ray diffraction analysis. This is likely to be a difficult and long-term process. However, in the meantime, site-directed mutagenesis, in combination with molecular modelling, may provide us with at least partial answers to the mechanism of transport in these physiologically important and fascinating proteins.

Acknowledgements

Research in the author's laboratory is supported by the SERC, the MRC and the Wellcome Trust. The author is indebted to Dr. M.T. Cairns, Dr. A. Davies, Dr. A.F. Davies and Dr. P.J.F. Henderson for many helpful discussions and access to unpublished information. Mr. R.A.J. Preston and Mrs. J. Baldwin provided invaluable help in the preparation of the figures.

References

1. Fukumoto, H., Kayano, T., Buse, J.B., Edwards, Y., Pilch, P.F., Bell, G.I. and Seino, S. (1989) J. Biol. Chem. 264, 7776–7779.
2. Hediger, M.A., Coady, M.J., Ikeda, T.S. and Wright, E.M. (1987) Nature (London) 330, 379–381.
3. Maiden, M.C.J., Davis, E.O., Baldwin, S.A., Moore, D.C.M. and Henderson, P.J.F. (1987) Nature (London) 325, 641–643.
4. Baldwin, S.A. and Henderson, P.J.F. (1989) Annu. Rev. Physiol. 51, 459–471.
5. Birnbaum, M.J. (1989) Cell 57, 305–315.
6. Gould, G.W. and Lienhard, G.E. (1989) Biochemistry 28, 9447–9452.
7. Keller, K., Strube, M. and Mueckler, M. (1989) J. Biol. Chem. 264, 18884–18889.
8. Vera, J.C. and Rosen, O.M. (1989) Mol. Cell Biol. 9, 4187–4195.
9. Kayano, T., Burant, C.F., Fukumoto, H., Gould, G.W., Fan, Y.S., Eddy, R.L., Byers, M.G., Shows, T.B., Seino, S. and Bell, G.I. (1990) J. Biol. Chem. 265, 13276–13282.
10. Gould, G.W., Thomas, H.M., Jess, T.J. and Bell, G.I. (1991) Biochemistry 30, 5139–5145.
11. Lacko, L. and Burger, M. (1962) Biochem. J. 83, 622–625.
12. LeFevre, P.G. and Marshall, J.K. (1958) Am. J. Physiol. 194, 333–337.
13. LeFevre, P.G. (1961) Pharmacol. Rev. 13, 39–70.
14. Barnett, J.E.G., Holman, G.D. and Munday, K.A. (1973) Biochem. J. 131, 211–221.
15. Riley, G.J. and Taylor, N.F. (1973) Biochem. J. 135, 773–777.
16. Kahlenberg, A. and Dolansky, D. (1972) Can. J. Biochem. 50, 638–643.
17. Carruthers, A. and Melchior, D.L. (1985) Biochemistry 24, 4244–4250.
18. Kuchel, P.W., Chapman, B.E. and Potts, J.R. (1987) FEBS Lett. 219, 5–10.
19. Appleman, J.R. and Lienhard, G.E. (1989) Biochemistry 28, 8221–8227.
20. Potts, J.R., Hounslow, A.M. and Kuchel, P.W. (1990) Biochem. J. 266, 925–928.
21. Rees, W.D. and Holman, G.D. (1981) Biochim. Biophys. Acta 646, 251–260.

22. Quiocho, F.A. (1986) Annu. Rev. Biochem. 55, 287–315.
23. Gorga, F.R. and Lienhard, G.E. (1982) Biochemistry 21, 1905–1908.
24. Carter, S.B. (1967) Nature (London) 213, 261–264.
25. Aldridge, D.C., Armstrong, J.J., Speake, R.N and Turner, W.B. (1967) J. Chem. Soc. (C) 1667–1676.
26. Griffin, J.F., Rampal, A.L. and Jung, C.Y. (1982) Proc. Natl. Acad. Sci. U.S.A. 79, 3759–3763.
27. Taverna, R.D. and Langdon, R.G. (1973) Biochim. Biophys. Acta 323, 207–219.
28. Bloch, R. (1973) Biochemistry 12, 4799–4801.
29. Jung, C.Y. and Rampal, A.L. (1977) J. Biol. Chem. 252, 5456–5463.
30. Ciaraldi, T.P., Horuk, R. and Matthaei, S. (1986) Biochem. J. 240, 115–123.
31. Axelrod, I.D. and Pilch, P.F. (1983) Biochemistry 22, 2222–2227.
32. Wardzala, L.J., Cushman, S.W. and Salans, L.B. (1978) J. Biol. Chem. 253, 8002–8005.
33. Zoccoli, M.A., Baldwin, S.A. and Lienhard, G.E. (1978) J. Biol. Chem. 253, 6923–6930.
34. Devés, R. and Krupka, R.M. (1978) Biochim. Biophys. Acta 510, 339–348.
35. Jung, C.Y., Carlson, L.M. and Whaley, D.A. (1971) Biochim. Biophys. Acta 241, 613–627.
36. LeFevre, P.G. and Marshall, J.K. (1959) J. Biol. Chem. 234, 3022–3026.
37. Krupka, R.M. (1985) J. Membr. Biol. 83, 71–80.
38. Krupka, R.M. and Devés, R. (1980) Biochim. Biophys. Acta 598, 134–144.
39. Kashiwagi, A., Huecksteadt, T.P. and Foley, J.E. (1983) J. Biol. Chem. 258, 13685–13692.
40. Joost, H.G. and Steinfelder, H.J. (1987) Mol. Pharmacol. 31, 279–283.
41. Sergeant, S. and Kim, H.D. (1985) J. Biol. Chem. 260, 14677–14682.
42. Seamon, K.B. and Daly, J.W. (1981) J. Cyclic Nucleot. Res. 7, 201–224.
43. Joost, H.G., Habberfield, A.D., Simpson, I.A., Laurenza, A. and Seamon, K.B. (1988) Mol. Pharmacol. 33, 449–453.
44. Lavis, V.R., Lee, D.P. and Shenolikar, S. (1987) J. Biol. Chem. 262, 14571–14575.
45. Wadzinski, B.E., Shanahan, M.F. and Ruoho, A.E. (1987) J. Biol. Chem. 262, 17683–17689.
46. Jung, C.Y. (1975) In: The Red Blood Cell (Surgenor, D.M., Ed.), Vol. 2, pp. 705–751, Academic Press, New York.
47. Eilam, Y. and Stein, W.D. (1974) In: Methods in Membrane Biology (Korn, E.D., Ed.), Vol. 2, pp. 283–354, Plenum Press, New York.
48. Lowe, A.G. and Walmsley, A.R. (1986) Biochim. Biophys. Acta 857, 146–154.
49. Naftalin, R.J. and Holman, G.D. (1977) In: Membrane Transport in Red Cells (Ellory, J.C. and Lew, V.L., Eds.), pp. 257–300, Academic Press, New York.
50. Baker, G.F. and Naftalin, R.J. (1979) Biochim. Biophys. Acta 550, 474–484.
51. Naftalin, R.J., Smith, P.M. and Roselaar, S.E. (1985) Biochim. Biophys. Acta 820, 235–249.
52. Wheeler, T.J. (1986) Biochim. Biophys. Acta 862, 387–398.
53. Carruthers, A. and Melchior, D.L. (1983) Biochim. Biophys. Acta 728, 254–266.
54. Hebert, D.N. and Carruthers, A. (1986) J. Biol. Chem. 261, 10093–10099.
55. Helgerson, A.L., Hebert, D.N., Naderi, S. and Carruthers, A. (1989) Biochemistry 28, 6410–6417.
56. Carruthers, A. (1986) J. Biol. Chem. 261, 11028–11037.
57. Carruthers, A. and Helgerson, A.L. (1989) Biochemistry 28, 8337–8346.
58. Geck, P. (1971) Biochim. Biophys. Acta 241, 462–472.
59. Widdas, W.F. (1952) J. Physiol. (Lond.) 118, 23–39.
60. Jardetzky, O. (1966) Nature (London) 211, 969–970.
61. Vidaver, G.A. (1966) J. Theor. Biol. 10, 301–306.
62. Brahm, J. (1983) J. Physiol. (Lond.) 339, 339–354.
63. Krupka, R.M. (1989) Biochem. J. 260, 885–891.
64. Lieb, W.R. (1982) In: Red Cell Membranes: A Methodological Approach (Ellory, J.C. and Young, J.D., Eds.), pp. 135–164, Academic Press, London.
65. Wheeler, T.J. and Whelan, J.D. (1988) Biochemistry 27, 1441–1450.
66. Carruthers, A. (1986) Biochemistry 25, 3592–3602.

67. Helgerson, A.L. and Carruthers, A. (1987) J. Biol. Chem. 262, 5464–5475.
68. Helgerson, A.L. and Carruthers, A. (1989) Biochemistry 28, 4580–4594.
69. Wang, J.-F., Falke, J.J. and Chan, S.I. (1986) Proc. Natl. Acad. Sci. U.S.A. 83, 3277–3281.
70. Allard, W.J. and Lienhard, G.E. (1985) J. Biol. Chem. 260, 8668–8675.
71. Zoccoli, M.A. and Lienhard, G.E. (1977) J. Biol. Chem. 252, 3131–3135.
72. Kasahara, M. and Hinkle, P.C. (1976) Proc. Natl. Acad. Sci. U.S.A. 73, 396–400.
73. Zala, C.A. and Kahlenberg, A. (1976) Biochem. Biophys. Res. Commun. 72, 866–874.
74. Pinkofsky, H.B., Rampal, A.L., Cowden, M.A. and Jung, C.Y. (1978) J. Biol. Chem. 253, 4930–4937.
75. Kasahara, M. and Hinkle, P.C. (1977) J. Biol. Chem. 252, 7384–7390.
76. Kahlenberg, A. and Zala, C.A. (1977) J. Supramol. Struct. 7, 287–300.
77. Steck, T.L. (1974) J. Cell Biol. 62, 1–19.
78. Baldwin, S.A., Baldwin, J.M., Gorga, F.R. and Lienhard, G.E. (1979) Biochim. Biophys. Acta 552, 183–188.
79. Sogin, D.C. and Hinkle, P.C. (1980) Biochemistry 19, 5417–5420.
80. Baldwin, S.A., Baldwin, J.M. and Lienhard, G.E. (1982) Biochemistry 21, 3836–3842.
81. Cairns, M.T., Elliot, D.A., Scudder, P.R. and Baldwin, S.A. (1984) Biochem. J. 221, 179–188.
82. Baldwin, S.A. and Lienhard, G.E. (1989) Methods Enzymol. 174, 39–50.
83. Baldwin, J.M., Gorga, J.C. and Lienhard, G.E. (1981) J. Biol. Chem. 256, 3685–3689.
84. Wheeler, T.J. and Hinkle, P.C. (1981) J. Biol. Chem. 256, 8907–8914.
85. Carruthers, A. and Melchior, D.L. (1984) Biochemistry 23, 6901–6911.
86. Connolly, T.J., Carruthers, A. and Melchior, D.L. (1985) Biochemistry 24, 2865–2873.
87. Connolly, T.J., Carruthers, A. and Melchior, D.L. (1985) J. Biol. Chem. 260, 2617–2620.
88. Tefft, R.E., Carruthers, A. and Melchior, D.L. (1986) Biochemistry 25, 3709–3718.
89. Kwong, F.Y.P., Baldwin, S.A., Scudder, P.R., Jarvis, S.M., Choy, M.Y.M. and Young, J.D. (1986) Biochem. J. 240, 349–356.
90. Kwong, F.Y.P., Davies, A., Tse, C.M., Young, J.D., Henderson, P.J.F. and Baldwin, S.A. (1988) Biochem. J. 255, 243–249.
91. Gorga, F.R., Baldwin, S.A. and Lienhard, G.E. (1979) Biochem. Biophys. Res. Commun. 91, 955–961.
92. Sogin, D.C. and Hinkle, P.C. (1978) J. Supramol. Struct. 8, 447–453.
93. Lienhard, G.E., Crabb, J.H. and Ransome, K.J. (1984) Biochim. Biophys. Acta 769, 404–410.
94. Endo, T., Kasahara, M. and Kobata, A. (1990) Biochemistry 29, 9126–9134.
95. Feugeas, J.P., Neel, D., Pavia, A.A., Laham, A., Goussault, Y. and Derappe, C. (1990) Biochim. Biophys. Acta 1030, 60–64.
96. Harris, E.J. (1964) J. Physiol. (Lond.) 173, 344–353.
97. Chin, J.J., Jung, E.K.Y., Chen, V. and Jung, C.Y. (1987) Proc. Natl. Acad. Sci. U.S.A. 84, 4113–4116.
98. Chin, J.J., Jung, E.K.Y. and Jung, C.Y. (1986) J. Biol. Chem. 261, 7101–7104.
99. Alvarez, J., Lee, D.C., Baldwin, S.A. and Chapman, D. (1987) J. Biol. Chem. 262, 3502–3509.
100. Cairns, M.T., Alvarez, J., Panico, M., Gibbs, A.F., Morris, H.R., Chapman, D. and Baldwin, S.A. (1987) Biochim. Biophys. Acta 905, 295–310.
101. Sase, S., Anraku, Y., Nagano, M., Osumi, M. and Kasahara, M. (1982) J. Biol. Chem. 257, 11100–11105.
102. Jacobs, D.B., Berenski, C.J., Spangler, R.A. and Jung, C.Y. (1987) J. Biol. Chem. 262, 8084–8087.
103. Jung, C.Y., Hsu, T.L., Hah, J.S., Cha, C. and Haas, M.N. (1980) J. Biol. Chem. 255, 361–364.
104. Cuppoletti, J., Jung, C.Y. and Green, F.A. (1981) J. Biol. Chem. 256, 1305–1306.
105. Jarvis, S.M., Ellory, J.C. and Young, J.D. (1986) Biochim. Biophys. Acta 855, 312–315.
106. Mueckler, M., Caruso, C., Baldwin, S.A., Panico, M., Blench, I., Morris, H.R., Allard, W.J., Lienhard, G.E. and Lodish, H.F. (1985) Science 229, 941-945.
107. Birnbaum, M.J., Haspel, H.C. and Rosen, O.M. (1986) Proc. Natl. Acad. Sci. U.S.A. 83, 5784–5788.

108. Eisenberg, D., Schwarz, E., Komaromy, M. and Wall, R. (1984) J. Mol. Biol. 179, 125–142.
109. Deziel, M.R. and Rothstein, A. (1984) Biochim. Biophys. Acta 776, 10–20.
110. Davies, A., Meeran, K., Cairns, M.T. and Baldwin, S.A. (1987) J. Biol. Chem. 262, 9347–9352.
111. Davies, A., Ciardelli, T.L., Lienhard, G.E., Boyle, J.M., Whetton, A.D. and Baldwin, S.A. (1990) Biochem. J. 266, 799–808.
112. Haspel, H.C., Rosenfeld, M.G. and Rosen, O.M. (1988) J. Biol. Chem. 263, 398–403.
113. Boyle, J.M., Whetton, A.D., Dexter, T.M., Meeran, K. and Baldwin, S.A. (1985) EMBO J. 4, 3093–3098.
114. Andersson, L. and Lundahl, P. (1988) J. Biol. Chem. 263, 11414–11420.
115. Van Steveninck, J., Weed, R.I. and Rothstein, A. (1965) J. Gen. Physiol. 48, 617–632.
116. Bloch, R. (1974) J. Biol. Chem. 249, 1814–1822.
117. Krupka, R.M. (1985) J. Membr. Biol. 84, 35–43.
118. May, J.M. (1989) J. Membr. Biol. 108, 227–233.
119. Batt, E.R., Abbott, R.E. and Schachter, D. (1976) J. Biol. Chem. 251, 7184–7190.
120. Roberts, S.J., Tanner, M.J.A. and Denton, R.M. (1982) Biochem. J. 205, 139–145.
121. May, J.M. (1989) Biochemistry 28, 1718–1725.
122. Deziel, M.R., Jung C.Y. and Rothstein, A. (1985) Biochim. Biophys. Acta 819, 83–92.
123. May, J.M., Buchs, A. and Carter-Su, C. (1990) Biochemistry 29, 10393–10398.
124. Jung, E.K.Y., Chin, J.J. and Jung, C.Y. (1986) J. Biol. Chem. 261, 9155–9160.
125. Masiak, S.J. and LeFevre, P.G. (1977) Biochim. Biophys. Acta 465, 371–377.
126. Baldwin, J.M., Lienhard, G.E. and Baldwin, S.A. (1980) Biochim. Biophys. Acta 599, 699–714.
127. Karim, A.R., Rees, W.D. and Holman, G.D. (1987) Biochim. Biophys. Acta 902, 402–405.
128. Shanahan, M.F. (1983) Biochemistry 22, 2750–2756.
129. Shanahan, M.F. (1982) J. Biol. Chem. 257, 7290–7293.
130. Carter-Su, C., Pessin, J.E., Mora, R., Gitomer, W. and Czech, M.P. (1982) J. Biol. Chem. 257, 5419–5425.
131. Deziel, M., Pegg, W., Mack, E., Rothstein, A. and Klip, A. (1984) Biochim. Biophys. Acta 772, 403–406.
132. Holman, G.D. and Rees, W.D. (1987) Biochim. Biophys. Acta 897, 395–405.
133. Midgley, P.J.W., Parkar, B.A. and Holman, G.D. (1985) Biochim. Biophys. Acta 812, 33–41.
134. Holman, G.D., Parkar, B.A. and Midgley, P.J.W. (1986) Biochim. Biophys. Acta 855, 115–126.
135. Holman, G.D., Karim, A.R. and Karim, B. (1988) Biochim. Biophys. Acta 946, 75–84.
136. Clark, A.E. and Holman, G.D. (1990) Biochem. J. 269, 615–622.
137. Shanahan, M.F., Morris, D.P. and Edwards, B.M. (1987) J. Biol. Chem. 262, 5978–5984.
138. Wadzinski, B.E., Shanahan, M.F., Seamon, K.B. and Ruoho, A.E. (1990) Biochem. J. 272, 151–158.
139. Jarvis, S.M. (1988) Biochem. J. 249, 383–389.
140. Jarvis, S.M., Young, J.D., Wu, J.-S.R., Belt, J.A. and Paterson, A.R.P. (1986) J. Biol. Chem. 261, 11077–11085.
141. May, J.M. and Danzo, B.J. (1988) Biochim. Biophys. Acta 943, 199–210.
142. Barnett, J.E.G., Holman, G.D., Chalkley, R.A. and Munday, K.A. (1975) Biochem. J. 145, 417–429.
143. Baker, G.F. and Widdas, W.F. (1973) J. Physiol. (Lond.) 231, 143–165.
144. Baker, G.F., Basketter, D.A. and Widdas, W.F. (1978) J. Physiol. (Lond.) 278, 377–388.
145. Gorga, F.R. and Lienhard, G.E. (1981) Biochemistry 20, 5108–5113.
146. Bowyer, F. and Widdas, W.F. (1958) J. Physiol. (Lond.) 141, 219–232.
147. Edwards, P.A.W. (1973) Biochim. Biophys. Acta 307, 415–418.
148. Krupka, R.M. (1971) Biochemistry 10, 1143–1153.
149. Baker, G.F. and Widdas, W.F. (1973) J. Physiol. (Lond.) 231, 129–142.
150. Dawson, A.C. and Widdas, W.F. (1963) J. Physiol. (Lond.) 169, 644–659.
151. May, J.M. (1988) J. Biol. Chem. 263, 13635–13640.
152. May, J.M. (1989) Biochim. Biophys. Acta 986, 207–216.

153. Pawagi, A.B. and Deber, C.M. (1987) Biochem. Biophys. Res. Commun. 145, 1087–1091.
154. Zeng, C., Suzuki, Y. and Alpert, E. (1990) Anal. Biochem. 189, 197–201.
155. Pawagi, A.B. and Deber, C.M. (1990) Biochemistry 29, 950–955.
156. Gibbs, A.F., Chapman, D. and Baldwin, S.A. (1988) Biochem. J. 256, 421–427.
157. Oka, Y., Asano, T., Shibasaki, Y., Lin, J.L., Tsukuda, K., Katagiri, H., Akanuma, Y. and Takaku, F. (1990) Nature (London) 345, 550–553.
158. Thorens, B., Sarkar, H.K., Kaback, H.R. and Lodish, H.F. (1988) Cell 55, 281–290.
159. Waddell, I.D., Scott, H., Grant, A. and Burchell, A. (1991) Biochem. J. 275, 363–367.
160. Kaestner, K.H., Christy, R.J., McLenithan, J.C., Braiterman, L.T., Cornelius, P., Pekala, P.H. and Lane, M.D. (1989) Proc. Natl. Acad. Sci. U.S.A. 86, 3150–3154.
161. Asano, T., Shibasaki, Y., Kasuga, M., Kanazawa, Y., Takaku, F., Akanuma, Y. and Oka, Y. (1988) Biochem. Biophys. Res. Commun. 154, 1204–1211.
162. Weiler Guttler, H., Zinke, H., Mockel, B., Frey, A. and Gassen, H.G. (1989) Biol. Chem. Hoppe. Seyler. 370, 467–473.
163. Jacquez, J.A. (1984) Am. J. Physiol. 246, R289–R298.
164. Widdas, W.F. (1955) J. Physiol. (Lond.) 127, 318–327.
165. Flier, J.S., Mueckler, M., McCall, A.L. and Lodish, H.F. (1987) J. Clin. Invest. 79, 657–661.
166. Takata, K., Kasahara, T., Kasahara, M., Ezaki, O. and Hirano, H. (1990) Biochem. Biophys. Res. Commun. 173, 67–73.
167. Dick, A.P.K., Harik, S.I., Klip, A. and Walker, D.M. (1984) Proc. Natl. Acad. Sci. U.S.A. 81, 7233–7237.
168. Kasanicki, M.A., Cairns, M.T., Davies, A., Gardiner, R.M. and Baldwin, S.A. (1987) Biochem. J. 247, 101–108.
169. Bagley, P.R., Tucker, S.P., Nolan, C., Lindsay, J.G., Davies, A., Baldwin, S.A., Cremer, J.E. and Cunningham, V.J. (1989) Brain Res. 499, 214–224.
170. Kalaria, R.N., Gravina, S.A., Schmidley, J.W., Perry, G. and Harik, S.I. (1988) Ann. Neurol. 24, 757–764.
171. Froehner, S.C., Davies, A., Baldwin, S.A. and Lienhard, G.E. (1988) J. Neurocytol. 17, 173–178.
172. Gerhart, D.Z. and Drewes, L.R. (1990) Brain Res. 508, 46–50.
173. Harik, S.I., Kalaria, R.N., Whitney, P.M., Andersson, L., Lundahl, P., Ledbetter, S.R. and Perry, G. (1990) Proc. Natl. Acad. Sci. U.S.A. 87, 4261–4264.
174. Barros, L.F., Baldwin, S.A., Jarvis, S.M., Cowen, T., Thrasivoulou, C., Beaumont N. and Yudilevich, D. (1991) J. Physiol. (Lond.) 446, 345P.
175. Zorzano, A., Wilkinson, W., Kotliar, N., Thoidis, G., Wadzinkski, B.E., Ruoho, A.E. and Pilch, P.F. (1989) J. Biol. Chem. 264, 12358–12363.
176. Holman, G.D., Kozka, I.J., Clark, A.E., Flower, C.J., Saltis, J., Habberfield, A.D., Simpson, I.A. and Cushman, S.W. (1990) J. Biol. Chem. 265, 18172–18179.
177. Namba, H., Lucignani, G., Nehlig, A., Patlak, C., Pettigrew, K., Kennedy, C. and Sokoloff, L. (1987) Am. J. Physiol. 252, E299–E303.
178. Pardridge, W.M., Boado, R.J. and Farrell, C.R. (1990) J. Biol. Chem. 265, 18035–18040.
179. Elliott, K.R.F. and Craik, J.D. (1982) Biochem. Soc. Trans. 10, 12–13.
180. Okuno, Y. and Gliemann, J. (1986) Biochim. Biophys. Acta 862, 329–334.
181. Fukumoto, H., Seino, S., Imura, H., Seino, Y., Eddy, R.L., Fukushima, Y., Byers, M.G., Shows, T.B. and Bell, G.I. (1988) Proc. Natl. Acad. Sci. U.S.A. 85, 5434–5438.
182. Suzue, K., Lodish, H.F. and Thorens, B. (1989) Nucleic Acids Res. 17, 10099.
183. Asano, T., Shibasaki, Y., Lin, J.L., Akanuma, Y., Takaku, F. and Oka, Y. (1989) Nucleic Acids Res. 17, 6386.
184. Thorens, B., Cheng, Z.Q., Brown, D. and Lodish, H.F. (1990) Am. J. Physiol. 259, C279–C285.
185. Orci, L., Thorens, B., Ravazzola, M. and Lodish, H.F. (1989) Science 245, 295–297.
186. Meglasson, M.D. and Matschinsky, F.M. (1986) Diabetes Metab. Rev. 2, 163–214.

187. Vischer, U., Blondel, B., Wollheim, C.B., Hoppner, E., Seitz, H.J. and Lynedjian, P.B. (1987) Biochem. J. 241, 249–255.
188. Johnson, J.H., Ogawa, A., Chen, L., Orci, L., Newgard, C.B., Alam, T. and Unger, R.H. (1990) Science 250, 546–549.
189. Thorens, B., Weir, G.C., Leahy, J.L., Lodish, H.F. and Bonner Weir, S. (1990) Proc. Natl. Acad. Sci. U.S.A. 87, 6492–6496.
190. Kayano, T., Fukumoto, H., Eddy, R.L., Fan, Y.S., Byers, M.G., Shows, T.B. and Bell, G.I. (1988) J. Biol. Chem. 263, 15245–15248.
191. Sadiq, F., Holtzclaw, L., Chundu, K., Muzzafar, A. and Devaskar, S. (1990) Endocrinology 126, 2417–2424.
192. Yamamoto, T., Seino, Y., Fukumoto, H., Koh, G., Yano, H., Inagaki, N., Yamada, Y., Inoue, K., Manabe, T. and Imura, H. (1990) Biochem. Biophys. Res. Commun. 170, 223–230.
193. James, D.E., Strube, M. and Mueckler, M. (1989) Nature (London) 338, 83–87.
194. Charron, M.J., Brosius, F.C., Alper, S.L. and Lodish, H.F. (1989) Proc. Natl. Acad. Sci. U.S.A. 86, 2535–2539.
195. Slot, J.W., Geuze, H.J., Gigengack, S., Lienhard, G.E. and James, D.E. (1991) J. Cell Biol. 113, 123–135.
196. Celenza, J.L., Marshall Carlson, L. and Carlson, M. (1988) Proc. Natl. Acad. Sci. U.S.A. 85, 2130–2134.
197. Kruckeberg, A.L. and Bisson, L.F. (1990) Mol. Cell. Biol. 10, 5903–5913.
198. Nehlin, J.O., Carlberg, M. and Ronne, H. (1989) Gene 85, 313–319.
199. Szkutnicka, K., Tschopp, J.F., Andrews, L. and Cirillo, V.P. (1989) J. Bacteriol. 171, 4486–4493.
200. Goffrini, P., Wesolowski Louvel, M., Ferrero, I. and Fukuhara, H. (1990) Nucleic Acids Res. 18, 5294.
201. Yao, B., Sollitti, P. and Marmur, J. (1989) Gene 79, 189–197.
202. Cheng, Q. and Michels, C. (1989) Genetics 123, 477–484.
203. Chang, Y.-D. and Dickson, R.C. (1988) J. Biol. Chem. 263, 16696–16703.
204. Hawkins, A.R., Lamb, H.K., Smith, M., Keyte, J.W. and Roberts, C.F. (1988) Mol. Gen. Genet. 214, 224–231.
205. Geever, R.F., Huiet, L., Baum, J.A., Tyler, B.M., Patel, V.B., Rutledge, B.J., Case, M.E. and Giles, N.H. (1989) J. Mol. Biol. 207, 15–34.
206. Zilberstein, D., Dwyer, D.M., Matthaei, S. and Horuk, R. (1986) J. Biol. Chem. 261, 15053–15057.
207. Cairns, B.R., Collard, M.W. and Landfear, S.M. (1989) Proc. Natl. Acad. Sci. U.S.A. 86, 7682–7686.
208. Zhang, C.C., Durand, M.C., Jeanjean, R. and Joset, F. (1989) Mol. Microbiol. 3, 1221–1229.
209. Schmetterer, G.R. (1990) Plant. Mol. Biol. 14, 697–706.
210. Sauer, N. and Tanner, W. (1989) FEBS Lett. 259, 43–46.
211. Sauer, N., Friedlander, K. and Graml-Wicke, U. (1990) EMBO J. 9, 3045–3050.
212. Henderson, P.J.F. (1990) J. Bioenerg. Biomembr. 22, 525–569.
213. Lam, V.M.S., Daruwalla, K.R., Henderson, P.J.F. and Jones-Mortimer, M.C. (1980) J. Bacteriol. 143, 396–402.
214. Davis, E.O. (1985) Xylose Transport in *Escherichia coli*. Ph.D. Thesis, University of Cambridge.
215. Davis, E.O. and Henderson, P.J.F. (1987) J. Biol. Chem. 262, 13928–13932.
216. Daruwalla, K.R., Paxton, A.T. and Henderson, P.J.F. (1981) Biochem. J. 200, 611–627.
217. Henderson, P.J.F. and Macpherson, A.J. (1986) Methods Enzymol. 125, 387–429.
218. Maiden, M.C.J., Jones-Mortimer, M.C. and Henderson, P.J.F. (1988) J. Biol. Chem. 263, 8003–8010.
219. Henderson, P.J.F., Giddens, R.A. and Jones-Mortimer, M.C. (1977) Biochem. J. 162, 309–320.
220. Cairns, M.T., McDonald, T.P., Horne, P., Henderson, P.J.F. and Baldwin, S.A. (1991) J. Biol. Chem. 266, 8176–8183.
221. Rotman, B., Ganesan, A.K. and Guzman, R. (1968) J. Mol. Biol. 36, 247–260.
222. Horne, P. and Henderson, P.J.F. (1983) Biochem. J. 210, 699–705.

223. Reynolds, C.H. and Silver, S. (1983) J. Bacteriol. 156, 1019–1024.
224. Sasatsu, M., Misra, T.K., Chu, L., Laddaga, R. and Silver, S. (1985) J. Bacteriol. 164, 983–993.
225. Ishiguro, N. and Sato, G. (1985) J. Bacteriol. 164, 977–982.
226. McMurry, L.M., Petrucci, R.R. and Levy, S.B. (1980) Proc. Natl. Acad. Sci. U.S.A. 77, 3974–3977.
227. Yamaguchi, A., Udagawa, T. and Sawai, T. (1990) J. Biol. Chem. 265, 4809–4813.
228. Hillen, W. and Schollmeier, K. (1983) Nucleic. Acids Res. 11, 525–539.
229. Peden, K.W. (1983) Gene 22, 277–280.
230. Waters, S.H., Rogowsky, J., Grinstead, J., Altenbuchner, J. and Schmitt, R. (1983) Nucleic Acids Res. 11, 6089–6140.
231. Eckert, B. and Beck, C.F. (1989) J. Biol. Chem. 264, 11663–11670.
232. Yamaguchi, A., Ono, N., Akasaka, T., Noumi, T. and Sawai, T. (1990) J. Biol. Chem. 265, 15525–15530.
233. Büchel, D.E., Gronenborn, B. and Müller-Hill, B. (1980) Nature (London) 283, 541–545.
234. Roepe, P.D., Consler, T.G., Menezes, M.E. and Kaback, H.R. (1990) Res. Microbiol. 141, 290–308.
235. Kaback, H.R. (1990) Biochim. Biophys. Acta 1018, 160.
236. Kaback, H.R., Bibi, E. and Roepe, P.D. (1990) Trends. Biochem. Sci. 15, 309–314.
237. McMorrow, I., Chin, D.T., Fiebig, K., Pierce, J.L., Wilson, D.M., Reeve, E.C.R. and Wilson, T.H. (1988) Biochim. Biophys. Acta 945, 315–323.
238. Fukumoto, H., Seino, S., Imura, H., Seino, Y. and Bell, G.I. (1988) Diabetes. 37, 657–661.
239. Williams, S.A. and Birnbaum, M.J. (1988) J. Biol. Chem. 263, 19513–19518.
240. Calamia, J. and Manoil, C. (1990) Proc. Natl. Acad. Sci. U.S.A. 87, 4937–4941.
241. Bibi, E. and Kaback, H.R. (1990) Proc. Natl. Acad. Sci. U.S.A. 87, 4325–4329.
242. Costello, M.J., Escaig, J., Matsushita, K., Viitanen, P.V., Menick, D.R. and Kaback, H.R. (1987) J. Biol. Chem. 262, 17072–17082.
243. Li, J. and Tooth, P. (1987) Biochemistry 26, 4816–4823.
244. King, S.C., Hansen, C.L. and Wilson, T.H. (1991) Biochim. Biophys. Acta 1062, 177–186.
245. Lacko, L., Wittke, B. and Kromphardt, H. (1972) Eur. J. Biochem. 25, 447–454.
246. Weiser, M.B., Razin, M. and Stein, W.D. (1983) Biochim. Biophys. Acta 727, 379–388.
247. Sen, A.K. and Widdas, W.F. (1962) J. Physiol. (Lond.) 160, 392–403.

CHAPTER 7

Amino acid transporters in yeast: structure, function and regulation

M. GRENSON

Université Libre de Bruxelles, Faculté des Sciences, Département de Biologie Moléculaire, Laboratoire de Physiologie Cellulaire et de Génétique des Levures, B-1050 Bruxelles, Belgium

1. Introduction

When the new term 'permease' was coined to designate bacterial membrane proteins specialized in the transport of specific metabolites [1,2], it covered a concept which was not quite new. The existence of membrane transport systems had been demonstrated in animal tissues by Cori as early as 1925 (see [3]). However, the discovery and characterization of permeases in bacteria revolutionized prospects for studying the properties of transport systems, opening the way to a new field and a very fruitful methodology.

Progress has essentially stemmed from the use of microorganisms and microbial genetics to identify transport systems, to study how they function and to analyse their regulation. It was possible to obtain a wide variety of mutations and to study their physiological effects. This, in turn, prompted a molecular approach based on recombinant DNA techniques. Such an approach is invaluable in the study of membrane transporters, where it is particularly difficult to isolate the objects to be studied. It is indeed impossible to detect transport activities in disrupted cells, so purification relies on tracing other properties of the transport systems, such as substrate affinity or immunological properties. Even this requires preliminary investigation, which is only feasible in a few selected cases. In contrast, genetics offers the possibility of labelling proteins by mutations, detecting their physiological functions and even obtaining crucial information about their structure without isolating them. This powerful tool can be applied both to permeases and to the proteins which regulate their synthesis or activity.

It was tempting to base the study of membrane transport in eukaryotic cells on similar simple principles. For this purpose, as well as for molecular biology as a whole, the yeast *Saccharomyces cerevisiae* appeared to be the best suited organism. From early times on, this yeast has occupied a privileged place for mankind. Due to its exceptional fermenting activity on sugary fruit juices or plant extracts rich in starch it was quite naturally predestined for use in the making of alcoholic beverages. In later periods, this interest favoured the study of yeast biochemistry, genetics and molecular biology.

From a genetical point of view, *Saccharomyces cerevisiae* is an ideal organism which may be considered the *Escherichia coli* of eukaryotic cells [4,5]. This is true in particular for the study of metabolic regulation and for that of membrane transport [6]. Finally, the astonishing resemblance between many yeast proteins and certain mammalian-cell proteins has seriously broadened the scope of interest. Although a few reports have appeared on amino acid transport in some other yeasts, most investigations in this field have used strains of *Saccharomyces cerevisiae*.

In this chapter, we shall focus on the molecular aspects of amino acid transport and its regulation in *Saccharomyces cerevisiae*. Kinetic, biochemical and genetic aspects of the amino acid transport systems of eukaryotic microorganisms have been reviewed earlier [7,8].

2. Physiological background: assimilation of exogenous nitrogen compounds used as a source of nitrogen or as building blocks

Yeasts are capable of utilizing the diversity of nitrogenous compounds that they find in their rich natural environment. These nitrogen-containing substances can either be used as a general source of nitrogen or they can provide ready-made metabolites. Nitrogen metabolism in *Saccharomyces cerevisiae* and degradation of organic nitrogen compounds by this and other yeasts, has been reviewed [7,9,10]. Many questions are still open. Fig. 1 summarizes what is known about the fate of nitrogenous compounds used as a source of nitrogen for biosynthesis in *Saccharomyces cerevisiae*. In this organism, many degradative pathways lead to either glutamate, or NH_4^+, or both. The ultimate nitrogen donors for biosynthesis are glutamate and glutamine. Glutamine is formed by glutamine synthetase from glutamate, NH_4^+ and ATP.

A distinctive feature of the genus *Saccharomyces* as compared to other yeasts is its inability to use nitrate or nitrite as the sole nitrogen source. Assimilation of

---

Fig. 1. Schematic representation of the main reactions involved in nitrogen utilization in *Saccharomyces cerevisiae* grown on various nitrogenous compounds. The figure is focused on the interconversion of the two main products of nitrogenous catabolism, ammonium ions and glutamate, and on the nitrogen donors for biosynthesis, glutamate and glutamine. For details see [7,9,10]. (1) Glutamate synthase; (2) anabolic glutamate dehydrogenase; (3) catabolic glutamate dehydrogenase; (4) glutamine synthetase. Due to the lack of mutants, most of the pathways involving transamination are likely but not proven. Glutamine degradation through the Ω-amidase pathway has also been described (see [101]).

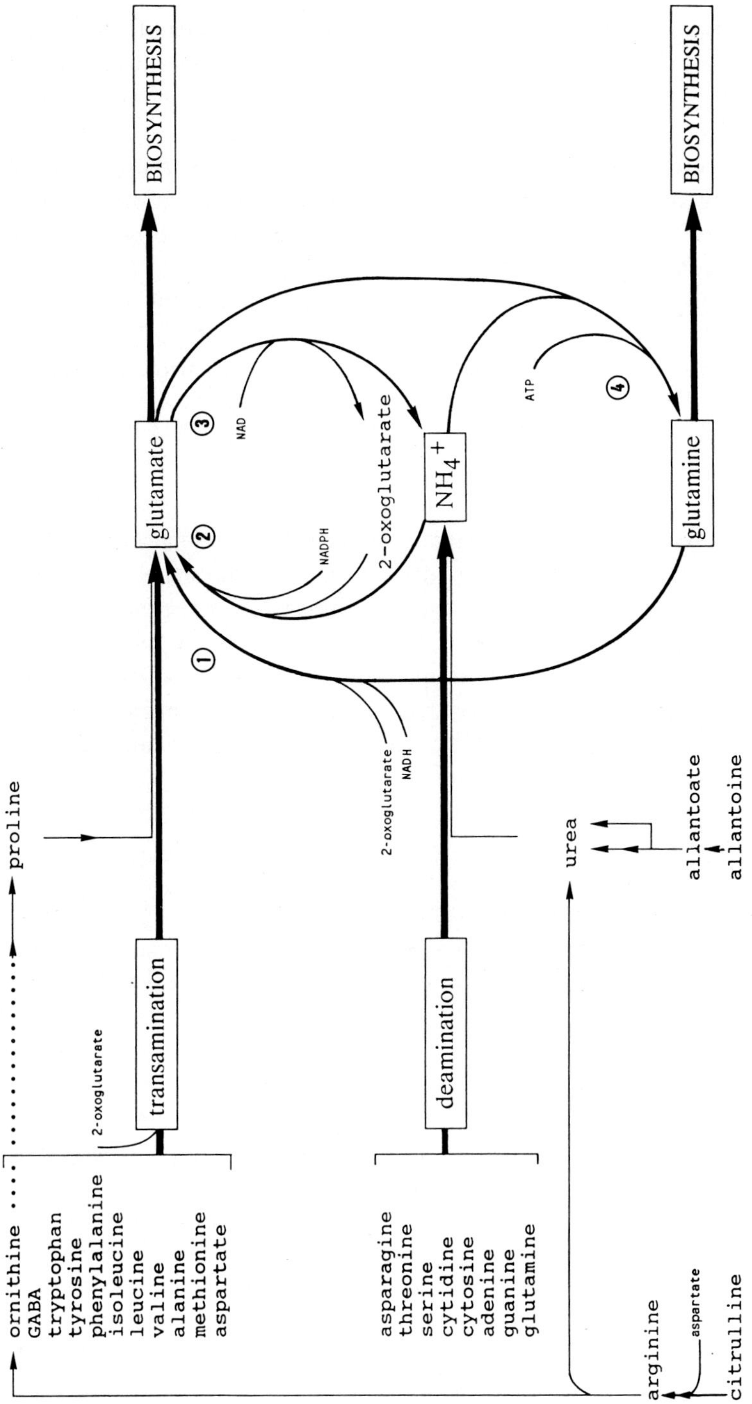
BIOSYNTHESIS
BIOSYNTHESIS
glutamate
NH4+
glutamine
2-oxoglutarate
NAD
NADPH
ATP
2-oxoglutarate
NADH
1
2
3
4
transamination
2-oxoglutarate
deamination
ornithine
GABA
tryptophan
tyrosine
phenylalanine
isoleucine
leucine
valine
alanine
methionine
aspartate
proline
asparagine
threonine
serine
cytidine
cytosine
adenine
guanine
glutamine
urea
allantoate
allantoine
arginine
aspartate
citrulline

ethylamine, cadaverine, or lysine is used as a diagnostic test for identifying species of *Saccharomyces*. For instance, *Saccharomyces cerevisiae* is unable to use any of these compounds as a source of nitrogen whereas *Saccharomyces kluyveri* and *Saccharomyces unisporus* can use them all [11].

The best known catabolic pathways of nitrogenous compounds are those of arginine, proline, allantoin and 4-aminobutyrate (GABA) degradation. Each of these is inducible under specific conditions, and all are subject to nitrogen-catabolite repression (see [7,9] and section 6.3).

3. General characteristics of amino acid transporters in Saccharomyces cerevisiae

3.1. Accumulation of amino acids

In *Saccharomyces cerevisiae*, as in most eukaryotic cells, the plasma membrane is not freely permeable to nitrogenous compounds such as amino acids. Therefore, the first step in their utilization is their catalyzed transport across the plasma membrane. Most of the transported amino acids are accumulated inside the yeast cells against a concentration gradient. When amino acids are to be used as a general source of nitrogen, this concentration is crucial because most enzymes which catalyze the first step of catabolic pathways have a low affinity for their substrates.

3.2. Multiplicity and specificity of amino acid transporters in Saccharomyces cerevisiae

In relationship to their living conditions, yeasts like *Saccharomyces cerevisiae* have developed a large number of transport systems which take up nitrogenous substances present in the external medium, and accumulate them in unmodified form (for reviews see [7,8]).

The number of transport systems in the plasma membrane of *Saccharomyces cerevisiae* is surprisingly high. In many cases a given substrate is transported by several permeases with different substrate affinities, specificities, capacities, and regulations. A number of amino acids are transported by two permeases, a specific transporter plus the general amino acid permease, but others (e.g., methionine [12], histidine [13], glutamic acid [14], lysine [15] and GABA [16]) are transported by two or three specific uptake systems with very different K_m and V_{max} values. High-affinity transporters allow the cells to scavenge even traces of amino acids from the culture medium. With such a battery of permeases, cells can take up amino acids at widely different rates over a vast range of external concentrations.

Furthermore, the multiplicity and diversity of amino acid transporters allow yeast to accumulate amino acids for both biosynthesis and catabolism under a wide variety of conditions. The regulation of these transport systems is such that only some are permanently present. These are called 'constitutive' permeases and are ready to transport amino acids for protein synthesis at any time. The additional uptake systems, which are called adaptive or inducible, develop under conditions where they may be both necessary and sufficient for cell growth or mere survival. An example is given below, in the case of GABA which is transported by three uptake systems, one of which, the UGA4 GABA-specific permease, is induced in the presence of GABA (see section 4.2).

Many nitrogen containing substances which cannot be used as metabolites (including toxic analogues of natural metabolites) are also transported, due to their likeness to natural permease substrates.

3.3. Functional specialization of amino acid transporters

Because they are not subject to competitive inhibition by other amino acids (except a few very close analogues), specific permeases are best suited for providing amino acids to be incorporated as such into proteins. In contrast, the general amino acid permease, with its broad specificity, its large capacity, and its regulation according to nitrogen availability (see section 6.3), is well adapted for taking up any available amino acid as a source of nitrogen. Such characteristics lead to functional specialization of the amino acid permeases. Similarly, functional specialization is observed in the case of several inducible catabolic pathways (e.g., GABA catabolism) where the substrate is pumped from the external medium by a specific permease which is co-regulated with the enzymes of the pathway, forming a regulon. The whole regulon is then expressed only if induced.

Despite these functional specializations, there is no exclusive use of a given permease for a specific purpose. For instance, L-arginine can be transported just as efficiently by the specific arginine permease as by the general amino acid permease, either to fulfill a specific arginine requirement in an arginine auxotroph, or for use as a general source of nitrogen. The general amino acid permease, however, due to its broad specificity and to the fact that it is regulated by nitrogen-catabolite repression (NCR, see section 6.3), can be effectively used for arginine transport only in the absence of competitors and of repressing conditions. Likewise, the proline permease can be used for GABA uptake provided there is no proline in the medium to compete with GABA (see section 4.2).

3.4. Irreversibility of amino acid accumulation

The accumulation of a number of amino acids from the external medium seems almost irreversible in *Saccharomyces cerevisiae*. The first detailed study of this phenomenon concerned histidine [13]. Histidine uptake by the specific histidine permease HIP1 is an energy dependent process which accumulates free and intact histidine against a concentration gradient. Efflux of the accumulated histidine from the cell is negligible even by exchange with added external histidine. The same has been observed for arginine, lysine, methionine, serine and threonine accumulated by their respective specific permeases, and for citrulline accumulated by the GAP1 general amino acid permease [17]. The case of proline seems rather exceptional as proline accumulated by the PUT4 proline permease can be exchanged with external proline.

3.5. Role of the vacuole in amino acid retention

In yeasts and other fungi, the vacuole is an important organelle sharing some properties with the mammalian lysosome (an acidic compartment containing a variety of hydrolytic enzymes) and with the plant cell vacuole (responsible for metabolite storage and for cytosolic ion and pH homeostasis) [18,19].

In *Saccharomyces cerevisiae*, amino acids are transported into the vacuole and stored as a nitrogen reserve in this compartment. Kinetic studies on isolated vacuolar membrane vesicles indicate the presence of seven independent H^+/amino acid antiport systems with narrow substrate specificity, all driven by a proton-motive force established by ATP hydrolysis. Their respective substrates are arginine, arginine-lysine, histidine, phenylalanine-tryptophan, tyrosine, glutamine-asparagine and isoleucine-leucine. These amino acids are taken up against a concentration gradient and accumulated in the vacuole at levels 5- to 40-fold higher than in the cytosol [20]. An arginine–histidine exchange transport system has also been detected in vacuolar membrane vesicles [21]. Acidic amino acids do not seem to accumulate in the vacuole. Since the present chapter mainly concerns plasma-membrane transporters, the reader is referred to the comprehensive reviews of Davis [22] and Klionsky et al. [23] for more details about vacuolar transport.

It is not clear to what extent vacuolar sequestration participates in the irreversibility of amino acid uptake and accumulation by the yeast cell. Once taken up by an active transport process, basic amino acids and other cations are retained in the vacuole without further expenditure of energy until they are needed. It has been thought that polyphosphates present in fungal vacuoles might play an important role in this process, since they might serve as a cation trap. In *Neurospora crassa*, however, it has been shown that polyphosphates are not required for vacuolar arginine uptake or retention (see [23]). Specific mechanisms for triggering the release of stored amino acids have been postulated and it is suggested that efflux from vacuole to cytosol requires energy (see [23]). Vacuolar sequestration might prevent exit of amino acids from the cells in the case of basic and some other amino acids, and thus at least partly account for irreversible amino acid accumulation. However, this would just displace the problem from the level of the plasma membrane to that of the vacuolar membrane. Furthermore, it does not provide a satisfactory explanation for the narrow specificity of feedback inhibition of specific plasma-membrane permeases by amino acids, which is another aspect of the irreversibility of amino acid accumulation (see section 6.1).

3.6. Efflux of amino acids

Although uptake and accumulation of most amino acids from the external medium seems to be irreversible, amino acids are excreted into the medium whenever they are overproduced above a given threshold by yeast cells [6]. This can occur under a number of specific conditions, namely in mutants with impaired regulation of amino acid biosynthesis, or in the presence of mutations preventing substrate catabolism, or when growth occurs in the presence of metabolic intermediates. It can even occur when growth is arrested under conditions where amino acid synthesis can continue.

Permeases do not seem to be directly involved in the exit of their substrates: permease inactivation by mutation does not prevent excretion of nitrogenous compounds. On the contrary, inactivation of uptake systems actually favours excretion of their substrates [6]. For instance, double mutants lacking both the general amino acid permease GAP1 and the specific arginine permease CAN1 excrete arginine into the culture medium. This suggests that efflux is mediated by mechanisms which are distinct from the characterized uptake systems. The transmembrane protein of *Saccharomyces cerevisiae* encoded by the *ATR1* gene, required for aminotriazole resistance, might be such an outward-pumping system or a component of the machinery responsible for pumping aminotriazole out of the cell [24]. When radioactive aminotriazole is added exogenously, cells with multiple copies of *ATR1* accumulate less aminotriazole than do wild-type cells, whereas cells with the *atr1* deletion retain more aminotriazole. The *ATR1* gene product appears to confer resistance to aminotriazole only. It might contain two ATP-binding sites.

Except for its narrow specificity, the *ATR1* gene product shares a number of properties with the higher eukaryotic MDR proteins responsible for multidrug resistance in tumour cells. The *MDR* gene products are also transmembrane proteins which seem to function as ATP-dependent drug-efflux pumps pumping out a variety of structurally unrelated compounds (see [25,26]).

4. Identifying transport systems

A large number of amino acid transporters have been detected by isolating mutations which selectively inactivate one permease without altering enzyme activities involving the corresponding amino acid. Competitive inhibition, kinetics and regulatory behaviour have also been used as criteria to distinguish one transport system from another (see section 4.2).

4.1. Isolating mutants affected in uptake systems

An easy method for isolating mutants with impaired amino acid transporters is based on selecting for resistance to toxic analogues which enter the cells through the same permeases as their natural equivalent. When toxic analogues are not available, mutant strains are sought, which are unable to use an amino acid either as a general source of nitrogen or as a source of a required amino acid. This is done by screening without prior selection. Since not all mutants thus isolated are affected in uptake systems, further sorting is required. Detection of uptake mutants involves measuring the uptake rate of labelled substrates and demonstrating that the low uptake rate is not a result of impaired metabolic use of the substrate (see section 6.1.1). Recently, sequence analysis of the mutated gene, showing that the gene product has the characteristic features of membrane proteins, has become one of the best ways to confirm that the mutation affects a transporter. A list of mutated amino acid permease genes is given in [7,8].

4.2. An example of transporter identification in a complex case: the three GABA transport systems of Saccharomyces cerevisiae

Recently, the GABA catabolic pathway and the GABA transporters have been receiving growing attention. Studies of this system have not only provided important insights into the regulation of transport systems, but have revealed an interesting form of transcriptional control. Analogies between the yeast system and control of the human central nervous system are not the least interesting aspect of the subject. This is why we chose GABA uptake to illustrate how complexity can be circumvented by genetics.

The yeast *Saccharomyces cerevisiae* can use GABA as its sole source of nitrogen [27]. The corresponding catabolic pathway has been established on a genetic basis, as has its requirement of a first positive regulator [28]. The pathway is inducible by GABA. A striking fact in this study was the absence of any GABA-uptake-deficient mutant, when the lack of GABA uptake should be as effective as a catabolic deficiency in preventing growth on GABA as the sole nitrogen source. The absence of uptake mutants turned out to be due to the existence of three distinct permeases capable of transporting GABA, so that only a triple mutant is devoid of GABA transport activity [16].

The three GABA transporters, as identified in the following analysis [16], include the general amino acid permease GAP1, the proline permease PUT4, and a fairly specific GABA permease UGA4. The GAP1 and PUT4 transporters have been known for a long time, but their ability to transport GABA was not recognized until recently. The identity of the three GABA transporters was established as follows. Starting with a *gap1* mutant strain (lacking general amino acid permease activity) grown in the absence of GABA, it could be shown that GABA uptake is completely inhibited by proline. This suggested that the proline permease (which is constitutive) is able to transport GABA, and that the related GABA-uptake activity is competitively inhibited by proline in the medium. This was confirmed by studying the effects of an additional mutation, *put4*, which abolishes the proline permease activity and was isolated according to Lasko and Brandriss [29]. In double *gap1*, *put4* mutant cells grown in the absence of GABA, the initial GABA uptake rate is almost undetectable, even in the absence of proline. In these double-mutant cells, a GABA-uptake activity progressively develops upon addition of GABA to the external medium, suggesting that there exists a third, inducible GABA permease. It was then possible to isolate mutants bearing a third mutation (*uga4*) causing the loss of the inducible GABA-permease activity. Triple *gap1,put4,uga4* mutant cells grow very poorly on GABA as sole nitrogen source, and their GABA-uptake rate is very low under all growth conditions tested.

The three GABA-transport-impairing mutations were then combined two by two in double mutants: *gap1,put4*; *gap1,uga4*; and *put4,uga4*. In each of these strains, the properties of the remaining active GABA-uptake system can be studied separately. This illustrates the power of genetics to dissect complex systems into their constitutive elements.

All three GABA-uptake systems are inactive in the presence of ammonium ions (see below).

5. Structure and evolution of amino acid transporters

The inactivation of transport proteins by mutations not only allows identification and functional characterization of permeases: it is also the prerequisite to isolating permease genes by functional complementation in *Saccharomyces cerevisiae*.

5.1. Molecular cloning and nucleotide sequencing of amino acid permease genes

Five structural genes for amino acid uptake systems have been cloned in *Saccharomyces cerevisiae* by functional complementation, and their putative amino acid sequences deduced from the respective nucleotide sequences (Fig. 2).

Despite the limited information available, rather clear predictions can be made about the probable structure, location, and energy coupling of the amino acid transporters of *Saccharomyces cerevisiae*, by comparing them with better known systems in both prokaryotes and eukaryotes.

In bacteria, accumulation of substrates against a concentration gradient can occur through two main classes of transport systems (see [30] for a summary). The prototype of the first class of transporters is the β-galactoside permease of *Escherichia coli* (see [31]). It is a relatively simple system involving only a single membrane-bound protein. It catalyzes a lactose–H^+ symport. Other transporters belonging to this first class are also energized by an ion gradient. They transport a wide variety of exogenous substrates across prokaryotic and eukaryotic membranes.

The second class of transporters includes more complex systems which require three or more protein components to operate. They are energized by ATP. A typical representative is the high-affinity histidine permease of *Escherichia coli* and *Salmonella typhimurium* [30]. This second class of transport system is also present in eukaryotes (see [32,33]). In eukaryotes, however, the individual protein subunits are linked into a single multidomain polypeptide. An example of this are the proteins of the MDR family, which are responsible for multidrug resistance in tumour cells. The *MDR* gene product shows strong amino acid homology with bacterial transport proteins; it seems to be a membrane glycoprotein with two consensus ATP-binding sites (see [25,26]). A family of proteins with closely related structures includes the CQR protein, responsible for chloroquine resistance in *Trypanosoma* [34], the CFTR protein, disorders of which are associated with human cystic fibrosis (see [33]), and the *STE6* gene product, which mediates export of the *a*-factor mating pheromone in *Saccharomyces cerevisiae* [35]. All of these eukaryotic proteins have two hydrophobic domains which comprise several potential membrane-spanning α-helices, and two ATP-binding domains. The amino acid sequence of the ATP-binding domains is highly homologous in all of these proteins.

The five amino acid permease genes of *Saccharomyces cerevisiae* sequenced to date seem to belong to a single class of transport systems sharing a number of common properties. The predicted gene products show hydrophobic regions which are long enough to make membrane-spanning α-helical segments. The hydrophilic and often electrically charged stretches which separate the putative transmembrane segments are very short. They do not possess any typical ATP-binding site.

The fact that no mutation outside the cloned locus is known to inactivate the same transport system also favours the idea that, in each case, the transporter is a single protein.

Hence, despite their scarcity, the available data suggest that all of these permeases belong to the first class of transporter which was mentioned above.

Another common feature of the five *Saccharomyces cerevisiae* amino acid per-

---

Fig. 2. Amino acid sequence similarities in a family of transporters of amino acids and other nitrogenous compounds. GAP1: *Saccharomyces cerevisiae* general amino acid permease [46]; HIP1: *Saccharomyces cerevisiae* histidine permease [44]; CAN1: *Saccharomyces cerevisiae* arginine permease [43]; PUT4: *Saccharomyces cerevisiae* proline permease [45]; PRNB: *Aspergillus nidulans* proline transport protein [47]; AROP: *Escherichia coli* general aromatic amino acid transport protein [48]; UGA4: *Saccharomyces cerevisiae* GABA permease [102]; CTR: *Saccharomyces cerevisiae* choline transporter [50]. Alignments were obtained by the Clustal program [103]. (−) indicates gaps introduced to optimize the alignment. Predicted membrane-associated α-helices, as determined by the algorithm of Eisenberg et al. [104], are underlined.

```
GAP1    A-GSELVGLAASESVEPRKSVPKAAKQVFWRITLFYILSLLMVGLLVPYN    344
HIP1    S-GIEMTAVSAAESKNPRETIPKAAKRTFWLITASYVTILTLIGCLVPSN    344
CAN1    Q-GTELVGITAGEAANPRKSVPRAIKKVVFRILTFYIGSLLFIGLLVPYN    346
PUT4    ILGPELVCMTSAECADQRRNIAKASRRFVWRLIFFYVLGTLAISVIVPYN    373
PRNB    IFSPELITTAAGEVEAPRRNIPKATKRFIYRVFTFYILGSLVIGVTVAYN    304
AROP    G-GLELVGITAAEADNPEQSIPKATNQVIYRILIFYIGSLAVLLSLMPW-    262

UGA4    G-SFDSCVHQSEEAKDAKKSVPIG---IISSIAVCWILGWLIIICLMACI    336
CTR     S-CLDCATHMAFEVEKPERVIPIA---IMGTVAIGFVTSFCYVIAMFFSI    316


GAP1    DKSLI-G--ASSVDAAASPFVIAI-KTHGIKGLPSVVNVVILIAVLSVGN    390
HIP1    DPRLLNG--SSSVDAASSPLVIAI-ENGGIKGLPSLMNAIILIAVVSVAN    391
CAN1    DPKLTQS--TSYV--STSPFIAI-ENSGTKVLPHIFNAVILTTIISAAN    391
PUT4    DPTLVNALAQGKPGAGSSPFVIGI-QNAGIKVLPHIINGCILTSAWSAAN    422
PRNB    DPTLEAGVESGGSGAGASPFVVAI-KTL---VLEGSTMSSMLPSGSLPGH    350
AROP    ---------TRVTADTSPFVLIF-HELGDTFVANALNIVVLTAALSVYN    301

UGA4    NPDIDSVLDSKYGFALAQIIYDSLGKKWAIAFMS-LIAFCQFLMGASITT    385
CTR     Q-DLDAVLSSTTGAPILDIYNQALGNKSGAIFLGCLILFTSFGCVIACHT    365


GAP1    SAI---YACSRTMVALAEQRFLPEIFSYVD-RKGRPLVGIAVTSAFGLIA    436
HIP1    SAV---YACSRCMVAMAHIGNLPKFLNRVD-KRGRPMNAILLTLFFGLLS    437
CAN1    SNI---YVGSRILFGLSKNKLAPKFLSRTT-KGGVPYIAVFVTAAFGALA    437
PUT4    AFM---FASTRSLLTMAQTGQAPKCLGRIN-KWGVPYVAVGVSFLCSCLA    468
PRNB    PVTHGCYAGSEKLYSLAGEGQAPKIFTRTN-RTGVPYVAVLATWTIGLLS    399
AROP    SCV---YCNSRMLFGLAQQGNAPKALASVD-KRGVPVNTILVSALVTALC    347

UGA4    AVSRQVWA-----FSRDNGLPLSKYIKRVDSKYSVPFFAILAACVGSLIL    430
CTR     WQARLCWS-----FARDNGLPLSRLWSQVNPHTGVPLNAHLMSCAWITLI    410


GAP1    FVAASKKEGEVFNWLLALSGLSSLFTWGGICICHI---------------    471
HIP1    FVAASDKQAEVFTWLSALSGLSTIFCWMAINLSHI---------------    472
CAN1    YMETSTGGDKVFEWLLNITGVAGFFAWLFISISHI---------------    472
PUT4    YLNVSSSTADVFNWFSNISTISGFLGWMCGCIAYL---------------    503
PRNB    FLNLSSSGQTVFYWFTNITTVGGFINWVLIGIAYLVCFPPSLHLNTPDQK    449
AROP    VLINYLAPESAFGLLMALVVSALVINWAMISLAHM---------------    382

UGA4    GL-LCLIDDAATDALFSLAVAGNNLAWSTPTVFRL----------TSGR    468
CTR     GL-LYLASSTAFQSLITGCIAFLLLSYIIPVICLL----------AKKR    448


GAP1    -RFRKALAAQGRGLDELSFKSPTGVWGSY-WGLFMVIIMFIAQFYVALFP    519
HIP1    -RFRQAMKVQERSLDELPFISQTGVKGSW-YGFIVLFLVLIASFW-----    515
CAN1    -RFMQALKYRGISRDELPFKAKLMPGLAY-YAATFMTIIIIQGFTAFAP    520
PUT4    -RFRKAIFYNGL-YDRLPFKTWGQPYTVW-FSLIVIGIITITNGYAIFIP    550
PRNB    QRFRKALQFHGM-LDMLPFKTPLQPYGTY-YVMFIISILTLTNGYAVFFP    497
AROP    -KFRRAKQEQGVVTRFLLLY----PLGNW-ICLLFMAAVLV--------    417

UGA4    DLFRPGPFYLG----KI--WSPIVAWTGVAFQLFIIILVMFPSQQHGITK    512
CTR     NIAH-GPFWLG----KFGFFSNIVL---LGWTVFSVVFFSFPPVLP-VTK    489


GAP1    VGDSPS-AEGFFEAYLSFPLV-MVMYIGHKIYKRNWKL-FIPAEKMDIDT    566
HIP1    ----------------TFSVP-I------RRFRSQRRI-IL---------    532
CAN1    KFNGVSFAAAYISIFL-FLAV-WILFQCIFRCRFIWKI-----GDVDIDS    563
PUT4    KYWRVA---DFIAAYITLPIF-LVLWFGHKLYTRTWRQWWLPVSEIDVTT    596
PRNB    GRFTAS---DFLVSYIVFAIF-LALYAGHKIWYRT--PWLTKVSEVDIFT    541
AROP    -------------------I-MLMTRGMGIWVYLIPVWLIVLG------    440

UGA4    S--TMNYACVIGPGIWXLAGIYYKVYK-KKYYHGPAT----NLSDD-DYT    554
CTR     D--NMNYVCVVIVGYTAYSILYWK-YKGKKEFiiALEE----SENEQAEYS    532


GAP1    GRREVDLDLLKQEIAEEKAIMATKPRWYRIWNFWW    601
HIP1    -----------------------------------    532
CAN1    DRRDIEAIVWEDH-------EPKTFWDKFWNVVA    590
PUT4    GLVEIEE--KSREIEEMR--LPPTGFKDKFLDALL    627
PRNB    GKDEIDR--LCEN--DME--RQPRNWLERVWWWIF    570
AROP    ----IGYLFKEKTAKAVKAH---------------    456

UGA4    E--------------AVGADVIDTI----MSKQEP    571
CTR     NNFDTIEDSREFSVAASDVELENEH----VPWGKK    563
```

Fig. 2

230

```
GAP1   MSNT-S---SYEKNNPDNLKHNGITI------DSEFLTQEPITIPSNGSA        40
HIP1   MPRN-PLKKEYWADVVDGFKPATSPAFENEKESTTFVTELTSKTDSAFPL        49
CAN1   MTNS---KEDADIEEKHMYNEPVTTLFHDVEASQTHHRRGSIPLKDE---        44
PUT4   MVNILPFHKN--NRHSAGVVTCADDVSGDGSGGDTKKEENVVQVTESPSS        48
PRNB   M--------------------------------------------SPPSA        6
AROP   M-------------------------------------------------        1
UGA4   MS--------------------MSSKNENKISVEQRISTDIGQAY--           25
CTR    MS--------------------IRNDNASGGYMQPDQSSNASM----           23

GAP1   VSIDETGSGSKWQDFK--DSFKRVKPIEVDPNLSEAEKVAI---------        79
HIP1   SSKDSPGINQTTNDITSSDRFRRNEDTEQED-------------------        80
CAN1   KSKELYPLRSFPTRVNGEDTFSMEDGIGDED----EGE------------        78
PUT4   GSRNNHRSDNEKDDAIRMEKISKNQSASSNGTIREDLIMDVDLEKSPSVD        98
PRNB   KSMEEGRTPS-------------------------VQYGYGDPKTLE          28
AROP   ---------------------------------------------ME         3
UGA4   ---QLQGLGSNLRSIRSKTGAGEVNYIDAAKSVNDNQLL-----------        61
CTR    ---------HKRDLRVEEEIKPLDDMDSKGAVAAD--------------        49

GAP1   --ITAQTPLKHHLKNRHLQMIAIG-GAIGTGLLVGSGTALRTGGP-SLLI       125
HIP1   ----INNTNLSGDLSVRHLLTLAVG-GAIGTGLYVNTGAALSTGGPASLVI       126
CAN1   ----VQNAEVKRELKQRHIGMIALG-GTIGTGLFIGLSTPLTNAGPVGALI       124
PUT4   GDSEPHKL-KQGLQSRHVQLIALG-GAIGTGLLVGTSSTLHTCGPAGLFI       146
PRNB   GEIEEHTATKRGLSSRQLQLLAIG-GCIGTGLFVGTSTVLTQTGPAPLLM        77
AROP   GQQHGE-QLKRGLKNRHIQLIALG-GSIGTGLFLGSASVIQSAGP-GIIL        50
UGA4   AEIGYKQELKRQFSLQVFGIAFSIMGLLPSIASVMGGGLG-GGPATLVW        110
CTR    GEV----HLRKSFSLWSILGVGFGLTNSWFGISTSMVAGISSGGPMMIVY        95

GAP1   GWGSTGTMIYAMVMALGELAVIFPI-----SGGFTTYATRFIDESFGYAN       170
HIP1   DWVIISTCLFTVINSLGELSAAFPV-----VGGFNVYSMRFIEPSFAFAV       171
CAN1   SYLFMGSLAYSVTQSLGEMATFIPV-----TSSFTVFSQRFLSPAFGAAN       169
PUT4   SYIIISAVIYPIMCALGEMVCFLPGDGSDSAGSTANLVTRYVDPSLGFAT       196
PRNB   SYIVMASIVWFVMNVLGEMTTYLPIRGV---SVPYLIGRFTEPSIGFAS       123
AROP   GYAIAGFIAFLIMRQLGEMVVEEPV-----AGSFSHFAYKYWGSFAGFAS        95
UGA4   GWFVAAFFILLVGITMAEHASSIP-----TAGGLYYWTYYYAPEGYKEII       155
CTR    GIIIVALISICIGTSLGELSSAYP-----HAGGQFWWSLKLAPPKYKRFA       140

GAP1   NFNYMLQWLVVLPLEIVAA------------------SITVNFWGTDPKY       202
HIP1   NLNYLAQWLVLLPLELVAA------------------SITIKYWNDKIN-       202
CAN1   GYMYWFSWAITFALELSVV----------------GQVIQFWTYKVPL        201
PUT4   GWNYFYCYVILVAAE-----------CTAASGV-------VEYWTTAVP-       227
PRNB   GYNYWYSFAMLLACEVSTMALLSFLSCWNPDNVGHCLGLIIEYWNPPVS-       172
AROP   GWNYWVLYVLVAMAELTAV-----------------GKYIQFWYPEIP-        126
UGA4   SFIIGCSNSLALAAGVCSIDY----GLAEEIAAAVTLTKDGNFEVTSGKL       201
CTR    AYMCGSFAYAGSVFTSASTTL-----SVATEVVGMYALT-HPEFIPKRWHI      185

GAP1   RDGFVA--LFWLAIVIINMFGVKGYGEAEFVFSFIKVITVVGFIILGIIL       250
HIP1   SDAWVA--IFYATIALANMLDVKSFGETEFVLSMIKILSIIGFTILGIVL       250
CAN1   A-AWIS--IFWVIITIMNLFPVKYYGEFEFWVASIKVLAIIGFLIYCFCM       248
PUT4   KGVWIT--IFLCVVVILNFSAVKVYGESEFWFASIKILCIVGLIILSFIL       275
PRNB   VGLWIA--IVLV-------------ESEFWFAGLKILAIIGLIILGVVL       206
AROP   --TWVSAAVFFVVINAINLTNVKVFGEMEFWFAIIKVIAVVAMIIFGGWL       174
UGA4   YGIFAGAVVVMCICTCVASGAIARLQT----LSIFANLFIIVLLFIALPI       247
CTR    FVCFELLHLFLMFFNCYGKS----LPI----ISS-SSLYISLLSFFTITI       226

GAP1   NCGGGPTGGYIGGKYWHDPGAFAG-----DTPGAKFKGVCSVFVTAAFSF       295
HIP1   SCGGGPHGGYIGGKYWHDPGAFVG-----HSSGTQFKGLCSVFVTAAFTY       295
CAN1   VCGAGVTGP-VGFRYWRNPGAWGPGIISKDKNEGRFLGWVSSLINAAFTF       297
PUT4   FWGGGPNHDRLGFRYWQHPGAFAHHLTG--GSLGNFTDIYTGIIKGAFAF       323
PRNB   FFGGGPNHERLGFRYWQDPGAFNPYLVP--GDTGKFLGFWTALIKSGFSF       254
AROP   LFSGNGGPQATVSNLW-DQGGFPLPH--------AFTGLVMMMAIIMFSF       214
UGA4   GTKH-RMGGFNDGDF-----IFGKYENLSDWNNGW-QFCLAGFMPAVWTI       290
CTR    TVLACSHGKFNDAKF-----VFATFNNETGWKNGG-IAFIVGLINPAWSF       270
```

Fig. 2 (cont.)

meases studied is their very hydrophilic amino-terminal segment, which is devoid of any typical signal sequence. Hence, they seem to belong to the membrane proteins which contain internal targeting signals [36], like the human [37], rat [38] and yeast [39] glucose transporters and other eukaryotic membrane proteins. The same is true for several other yeast transport systems, namely the uracil permease, product of the *FUR4* gene [40]; the purine-cytosine permease, product of the *FCY2* gene [41]; and the allantoate permease, product of the *UEP1/DAL5* gene [42].

In at least three cases, removing the carboxyl-terminal part of the amino acid transporter does not suppress transport activity. Thus, the 30 amino acid residues located at the carboxyl-terminal end of the arginine permease, product of the *CAN1* gene, are not essential to arginine uptake [43]. In the case of the specific GABA permease, 24 amino acids at the carboxyl-terminal end may be removed without suppressing permease activity (Hein, Jauniaux and Grenson, in preparation). This is true in the *Saccharomyces cerevisiae* glucose transporter SNF3 as well: the 87 carboxyl-terminal amino acids are not required for transporter function [39].

5.2. A family of amino acid transporters with amino acid sequence homologies

Of the deduced amino acid sequences (shown in Fig. 2) four of them reveal strong homology, namely the genes encoding the arginine permease CAN1 [43], the histidine permease HIP1 [44], the proline permease PUT4 [45], and the general amino acid permease GAP1 [46]. The four permeases present similar hydropathy profiles, and 33–40% of the amino acid residues are identical, with regions of strong identity alternating with less similar regions over nearly the entire length of the four proteins. From a total of about 600 amino acids, 81 residues are conserved in all four permeases and 185 additional positions correspond to conservative replacements. Most of the proline, tryptophan and glutamic acid residues are conserved. The major differences are located at the extremities of the proteins.

The proline transport protein prnB of *Aspergillus nidulans* [47] is very similar to the above-mentioned family of *Saccharomyces cerevisiae* amino acid transporters (about 42% identity with the *PUT4* gene product and 30% identity with the *CAN1* and the *HIP1* gene products). So is the AroP general aromatic amino acid transporter protein of *Escherichia coli* K-12, which has about 30% identity with the *HIP1* gene product [48]. Both hydrophilic ends are very different from one transporter to another (see Fig. 2).

Two additional permeases from *Saccharomyces cerevisiae* share lower homologies with this family of transporters. One of them is the UGA4 GABA-specific permease ([49], Fig. 2 and André, Hein, Grenson and Jauniaux, submitted); the second is the CTR choline transporter [50]. About 19% of the amino acid residues of these two proteins are identical. They show about 10% identity with each of the six amino acid permeases mentioned above. This identity does not include most of the amino acids which are highly conserved among the six permeases.

```
                    ┌─── GAP1
              ┌─────┤
           ┌──┤     └─── HIP1
           │  │
           │  └──────── CAN1
        ┌──┤
        │  │     ┌──── PUT4
     ┌──┤  └─────┤
     │  │        └──── PRNB
     │  │
  ───┤  └─────────── AROP
     │
     │  ┌──────────── UGA4
     └──┤
        └──────────── CTR
```

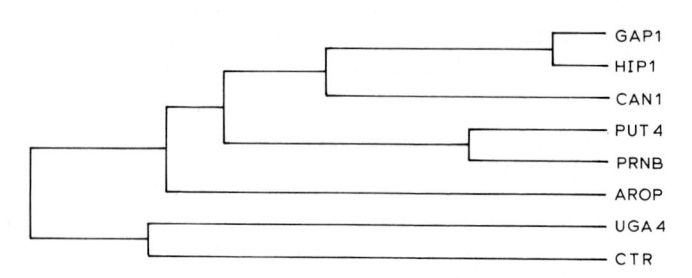

Fig. 3. Relationships within a family of transporters of nitrogenous compounds. The dendrogram is based on the alignments shown in Fig. 2. See legend of Fig. 2.

The relationships of all these transporters are illustrated by the dendrogram of the alignments shown in Fig. 3.

In contrast, the amino acid sequence of the proline permease of *Escherichia coli* [51] has no marked similarity to this family of transporters. This is also the case of the *Saccharomyces cerevisiae* purine-cytosine [41], uracil [40] and allantoate [42] permeases.

The significance of the homologies and the functions of the homologous regions have not been elucidated as yet. Also in the case of sugar uptake, a family of transporters with amino acid sequence homologies has been identified. This family includes the human [37], rat [38] and yeast SNF3 [39] glucose transporters, as well as the GAL2 galactose permease of *Saccharomyces cerevisiae* [52] and the bacterial arabinose and xylose transport proteins, AraE and XylE [53] (see comparison in [52]).

6. Regulation of amino acid transport

From a physiological point of view, uptake systems make sense to a cell only when their substrates are both available in the medium and useful to the cell. As a matter of fact, the expression of the permease structural genes and the activity of the corresponding permeases are controlled and modulated by different types of regulatory systems.

6.1. Regulation of permease activity

A rather satisfactory explanation of the irreversibility of amino acid accumulation in yeast cells is that it might result from specific regulatory mechanisms capable of immobilizing the transporters in a 'closed' position. Uptake of amino acids by a number of permeases does indeed appear to be regulated by specific, and possibly allosteric, feedback inhibition. This idea is based on the fact that a number of transport systems seem to be specifically inhibited by their internally accumulated substrates. For instance, preloading of yeast cells with histidine results in rapid inhibition of histidine uptake, whereas charging them with arginine, lysine, or other amino acids has no effect on histidine transport (although these amino acids, when pre-accumulated, do inhibit their own uptake) [13]. Likewise, threonine uptake is prevented by prior accumulation of threonine or serine, but not by preloading the cells with any other amino acid [54]. Feedback of the general amino acid permease is produced by pre-accumulation of histidine, lysine, arginine, methionine, glutamine, asparagine, aspartate, glutamate and tryptophan, i.e., probably by all of its substrates [17]. This is not competitive inhibition since the K_m value for citrulline uptake is unchanged while the V_{max} is decreased [17]. Thus, for each permease, as a rule, substrate and feedback inhibitor specificities appear to be the same. The only known exception to this rule is the proline permease which seems to be feedback-inhibited by all the amino acids [55].

Feedback inhibition of amino acid transporters by amino acids synthesized by the cells might be responsible for the well known fact that blocking protein synthesis by cycloheximide in *Saccharomyces cerevisiae* inhibits the uptake of most amino acids [56]. Indeed, under these conditions, endogenous amino acids continue to accumulate. This situation, which precludes studying amino acid transport in yeast in the presence of inhibitors of protein synthesis, is very different from that observed in bacteria, where amino acid uptake is commonly measured in the presence of chloramphenicol in order to isolate the uptake process from further metabolism of accumulated substances. In yeast, when nitrogen starvation rather than cycloheximide is used to block protein synthesis, this leads to very high uptake activity. This fact supports the 'feedback inhibition' interpretation of the observed cycloheximide effect.

Specific feedback inhibition has also been demonstrated in the case of three *Saccharomyces cerevisiae* pyrimidine-uptake systems, namely those for uracil, cytosine and uridine [57]. These compounds do not seem to be sequestered in the vacuole. The corresponding permeases are strongly and specifically inhibited by their substrates accumulating inside the cell. In this case, it could be demonstrated that endogenously produced pyrimidines are as effective as pyrimidines which accumulated from the external medium. As a result, the uptake rate for pyrimidine compounds is sharply dependent on their utilization in nucleic acid synthesis or other metabolic transformations. Hence, in *Saccharomyces cerevisiae*, simple uptake measurements cannot distinguish permease mutants from mutants affected in their utilization of pyrimidines. This must be considered a major methodological difficulty when attempting to identify a mutated protein as a putative uptake system. A complete genetic and functional analysis is needed to solve this problem. Such an analysis has been described in the case of the pyrimidine permeases [57].

The regulation of permease activity by nitrogen-catabolite inactivation will be discussed in section 6.3.

6.2. Regulation of permease synthesis

In *Saccharomyces cerevisiae*, the synthesis of amino acid permeases is controlled by several regulatory processes, some of which are general while others are more specific. The synthesis of a number of amino acid permeases is prevented by a general regulatory mechanism called nitrogen-catabolite repression (NCR), which operates in cells grown in the presence of preferred nitrogen sources such as ammonium ions, asparagine or glutamine. The prototype of this family of transporters is the GAP1 general amino acid permease. Among others, the PUT4 proline permease, the UEP1/DAL5 ureidosuccinate-allantoate permease, and the UGA4 GABA permease are subject to the same regulation. This general regulation will be described in section 6.3. A number of other permeases are practically unaffected by the presence of ammonium ions. These include specific amino acid transporters such as the arginine permease CAN1 [58], the lysine permease LYP1 [15], the methionine permease MTP1 [12], and the histidine permease HIP1 [13].

We shall first consider those regulatory processes which are distinctive for NCR and which make the expression of permease genes either 'inducible' (i.e., dependent on an inducing effector molecule) or 'constitutive' (i.e., occurring without addition of an exogenous effector).

6.2.1. Case of the NCR-insensitive amino acid permeases

Most of the permeases which are insensitive to NCR are synthesized in cells grown on minimal medium containing ammonium ions, without addition of any inducer. However, a few of them do appear to be inducible. For instance, addition to the medium of methionine, leucine valine, isoleucine and alanine, which are taken up by several distinct NCR-insensitive permeases, increases the rate of synthesis of the corresponding permeases [54]. This process involves amplification of a basal rate of permease synthesis rather than all-or-none induction. It has not been studied further at the molecular level.

6.2.2. Case of the NCR-sensitive amino acid permeases

To study the specific regulation of the synthesis of NCR-sensitive amino acid transporters, *Saccharomyces cerevisiae* cells are grown with proline or urea as the sole source of nitrogen, i.e., in the absence of NCR (see section 6.3).

6.2.2.1. Constitutive expression of permease genes. No inducer is required for synthesis of the GAP1 permease, the PUT4 permease, or the *UEP1/DAL5* ureidosuccinate-allantoate permease [59]. Some aspects of this mechanism of constitutive synthesis have been studied in the case of the UEP1/DAL5 permease. Analysis of the 5'-proximal DNA sequences of the *UEP1/DAL5* gene has led to the identification of two small regions which are required for its full expression. When multiple copies of them are inserted into the 5'-proximal region of a *lacZ* reporter gene, each of them is sufficient to confer constitutive expression to the reporter gene

[60]. The decisive sequence of each of these UASs (upstream activating sequences) is a 5'-GATAA-3' pentanucleotide, plus the directly flanking nucleotides, shown to be essential to UAS activity [61]. Several repeats of this UAS have also been detected in the 5'-proximal DNA sequences of the *GAP1* and *PUT4* genes [45,46], but their function has not been studied as yet.

Constitutive expression of the *UEP/DAL5* gene and of reporter genes located downstream from an inserted GATAA UAS requires the product of the *GLN3* gene [62]. Mutants with an affected *GLN3* gene were originally isolated as strains with low levels of glutamine synthetase; they were also shown to have lower levels of several other enzymes of nitrogen metabolism [63]. The behaviour of these mutants suggested that GLN3 might be a pleiotropic activator of the expression of several genes involved in nitrogen metabolism. Molecular analysis of the *GLN3* gene (Minehart and Magasanik cited in [64]) has shown that the deduced protein sequence contains a region rich in cysteine residues which is also found as a highly conserved structure in several regulatory proteins, including the *Aspergillus nidulans* areA gene product [65], the *Neurospora crassa* nit-2 protein [66], and a major erythroid-specific transcription factor of vertebrate cell lines (see comparison in [64]). This conserved region has DNA-binding properties *in vitro* [64,67]. Interestingly, the nit-2 protein [68] and the erythroid-specific factors of higher eukaryotes [69,70] bind to DNA sequences very similar to the GATAA UAS. There are several reported instances of proteins from lower and higher eukaryotes containing a homologous DNA-binding domain, and binding to similar DNA sequences [71]. It thus seems likely that the *GLN3* gene encodes a positive regulatory protein which binds to the GATAA UAS of several permease genes.

6.2.2.2. Inducible permeases. The GABA-specific UGA4 permease of *Saccharomyces cerevisiae* is one example of an inducible permease [16]. It is co-induced with the GABA-catabolic enzymes GABA transaminase (UGA1) and succinate semi-aldehyde dehydrogenase (UGA2). Some aspects of this induction have been elucidated and are summarized in Fig. 4. The induction process has been demonstrated at messenger level, by accumulation of *UGA4* and *UGA1* gene transcripts after addition of GABA to growing cells [49,72]. Genetic analysis has shown that induced synthesis of all three gene products requires two *trans*-acting factors, encoded by the *UGA3* [16,28,73,74] and *UGA35* [75] genes, respectively. While UGA3 behaves like a specific regulator of the GABA-inducible genes, UGA35 is pleiotropic [76,77]. It has indeed been shown that *UGA35* is identical to the *DURL* and *DAL81* genes [76,77], previously and independently identified as required for allophanate-induced expression of genes involved in urea [78], arginine [79] and allantoin [80,81] catabolism. Thus, a combination of two positive factors, one specific and one pleiotropic, is necessary for induction of *UGA4, UGA1* and *UGA2*. Molecular analysis of the *UGA3* [72] and *UGA35/DURL/DAL81* [77,82] genes has shown that the predicted amino acid sequences of both proteins are highly similar to those of DNA-binding transcriptional regulators.

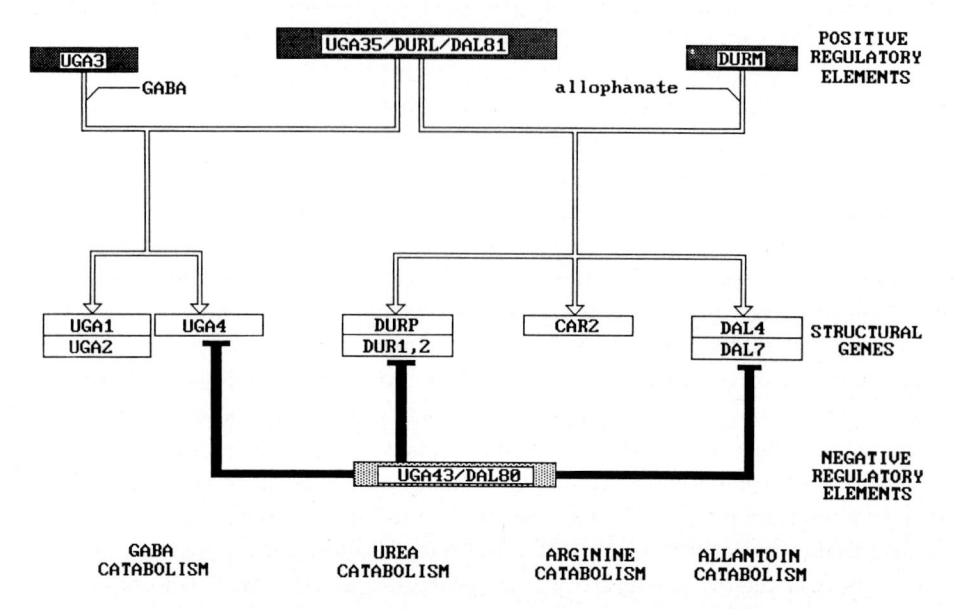

Fig. 4. Regulatory network controlling structural genes involved in the catabolism of various nitrogen sources. Construction of this figure is based on results described in [76,77,81] and on unpublished work in our laboratory. The structural genes correspond to the following enzymes or permeases: UGA1 (GABA transaminase); UGA2 (succinate semi-aldehyde dehydrogenase); UGA4 (GABA-specific permease); DURP (urea permease); DUR1,2 (urea amidolyase); CAR2 (ornithine transaminase); DAL4 (allantoin permease); DAL7 (unknown function). The identity of the *UGA43* and *DAL80* genes is likely but still hypothetical; it needs additional tests of allelism and/or sequence identity to be definitely proven.

The product of a third regulatory gene, *UGA43*, exerts a negative control on the expression of the UGA4 permease [75]. In mutants with an affected *UGA43* gene, the UGA4 permease is constitutive. The GABA-catabolic enzymes remain inducible in these mutants. The *UGA43* gene product is however pleiotropic, as it also negatively controls some of the allophanate-inducible genes [76]. The polypeptide sequence deduced from the cloned *UGA43* gene contains a cysteine-rich region which is highly similar to that found in *Neurospora crassa* nit-2, *Aspergillus nidulans* areA, and the above mentioned DNA-binding proteins of higher eukaryotes [105].

Several inducible permeases for nitrogenous compounds other than amino acids have been reported. For instance, the urea permease is induced by allophanate [83]. Induction of the urea permease, like that of UGA4, requires a combination of two *trans*-acting factors [78], one of which encoded by the *DURM* gene [78,84], is specific [76], while the other, encoded by *UGA35/DURL/DAL81*, is pleiotropic [76]. A negative control is exerted on the expression of the urea permease by the above mentioned pleiotropic *UGA43* gene product [76]. As shown independently, the expression of the urea permease as well as that of the allantoin permease (DAL4) are negatively controlled by the *DAL80* regulatory gene [81,85,86]. The *UGA43* and *DAL80* genes are probably identical. Tests of allelism or sequence identity are needed to prove it definitely.

Taken together, these results show that specific induction of several permease- as well as enzyme-genes in *Saccharomyces cerevisiae* is a complex mechanism involving distinct combinations of specific and pleiotropic, positive and negative regulatory proteins (Fig. 4).

6.3. Nitrogen-catabolite repression (NCR) and nitrogen-catabolite inactivation (NCI): two superimposed regulatory mechanisms affecting uptake systems for nitrogenous compounds

The GAP1 permease exhibits its highest activities in cells grown on poor nitrogen sources like proline or urea. It is not active in wild-type cells grown in the presence of ammonium ions. This is the result of a double control involving inhibition of permease activity (NCI) and repression of permease synthesis (NCR). These regulatory mechanisms can be lost separately as a result of mutations. Based on the behaviour of single and double mutants affected in these regulatory processes, a model has been proposed [87]. It is illustrated in Fig. 5 and commented on in section 6.3.1 and 6.3.2.

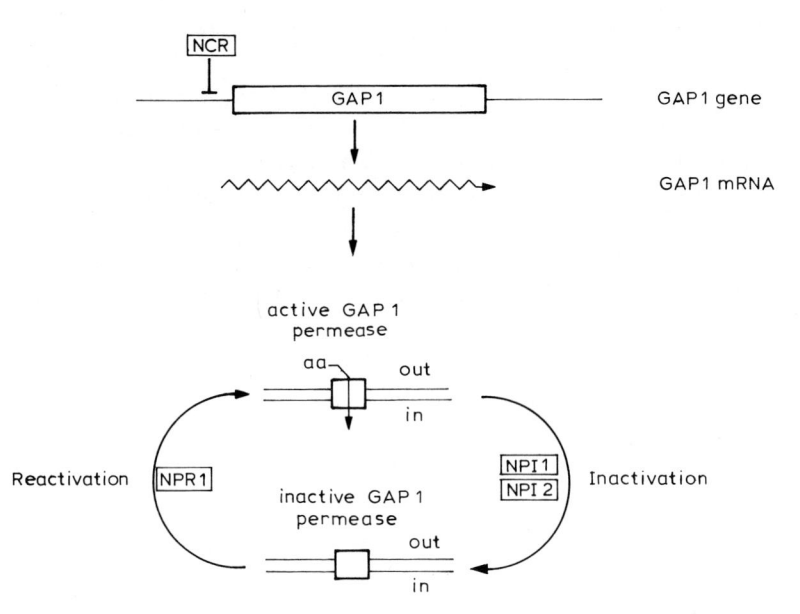

Fig. 5. Regulation of the synthesis and activity of the GAP1 permease. Repression of permease synthesis is represented in the upper part of the figure (see section 6.3.2). The lower part of the scheme features the process of inactivation and reactivation (see section 6.3.1 and [46,87,88,92]). NPI1 and NPI2 were formerly called MUT2 and MUT4, respectively.

6.3.1. Regulation of amino acid permease activity as a function of nitrogen availability
   6.3.1.1. Nitrogen catabolite inactivation (NCI): negative control of GAP1 activity. Upon addition of ammonium ions to proline grown cells, the general amino acid permease progressively undergoes rapid and complete inactivation. This inhibition of permease activity is reversible [88]. Inactivation can be suppressed by several mutations [88]. Among these, the *pgr* mutations are linked to the *GAP1* gene, whereas the *npi1* and *npi2* mutations (for nitrogen-permease inactivator) are not. NH_4^+-triggered repression of permease synthesis can easily be observed in all of these mutants, due to the absence of inactivation: whereas cell growth continues, permease activity stops increasing. As a consequence of this double regulation, the GAP1 permease is active in steady state ammonia grown cells only when both control mechanisms have been suppressed by adequate mutations.

Several other amino acid permeases are also subject to ammonium-ion-triggered inactivation. These are the PUT4 proline permease, the UEP1/DAL5 ureidosuccinate/allantoate permease [88], the GNP1 glutamine permease [89], the MEP1 and MEP2 methylamine/ammonium-ion permeases [90,89], and the UGA4 GABA permease (Vissers, André and Grenson, in preparation).

The *trans*-acting *npi1* and *npi2* mutations, which cancel the GAP1 permease inactivating process, are pleiotropic; they also release ammonia-triggered inactivation of the PUT4 and UEP1/DAL5 permeases [88], as well as that of the GNP1 glutamine permease [89]. There are other *trans*-acting mutations known which make other permeases ammonia-insensitive but they do not affect the permeases just mentioned. For instance, a *npi5* mutation formerly named *amu1* [90,89], suppresses ammonia-triggered inactivation of the MEP1 and MEP2 methylamine/ammonium ion permeases, which is not affected by the *npi1* and *npi2* mutations. Similarly, the *trans*-acting *gam1* mutation affects the UGA4 GABA permease, but not the GAP1 permease (Vissers, André and Grenson, in preparation). Although the analysis is not complete, it appears that ammonia inactivation involves a number of *trans*-acting proteins which target different spectra of ammonia-sensitive permeases.

On the other hand, mutations have been obtained which suppress ammonia inactivation of a given permease specifically. As might be expected, these mutations are linked to the corresponding structural gene of the permease. For instance, while the *pgr* mutation is located in the reading frame of the *GAP1* gene [91], the *prr* mutation is genetically linked to the *PUT4* gene, and the *gar* mutation is linked to the *UGA4* gene (Hein, Jauniaux and Grenson, in preparation). So far, only the *gar* mutated *UGA4* gene has been sequenced (Hein, Jauniaux and Grenson, in preparation). The coding sequence of the *UGA4* permease gene was modified in its carboxyl-terminal region as a result of a δ insertion. These *cis*-acting mutations are likely to affect the NCI-receptor site of the permeases.

The demonstration that a mutation affecting the permease protein makes it insensitive to ammonia inactivation reinforces the idea that ammonia inactivation involves a chemical modification of the permease.

6.3.1.2. Positive control of GAP1 activity. For the general amino acid permease to be active in derepressed cells, the integrity of a gene named *NPR1* (for nitrogen-permease reactivation) is required [87,89,90]. Mutations at the *NPR1* locus have a pleiotropic effect: they affect not only the GAP1 general amino acid permease, but also several other permeases which become partially or totally inactive. The permeases requiring the contribution of the NPR1 protein are all ammonia-sensitive. They include the PUT4 proline permease and the UEP1/DAL5 ureidosuccinate-allantoate permease [87,90], the MEP1 and MEP2 methylamine/ammonium-ion uptake systems [90], the GNP1 glutamine permease [89], the urea permease (Grenson, in preparation), and the specific GABA permease (Grenson, in preparation). It should be stressed that the target spectrum of the NPR1 molecule is wider than those of the NPI proteins. It seems, however, to be restricted to permeases, as indicated by the fact that none of the enzymes which have been tested is depressed by NPR1 deficiency [89]. The glutamate permease sensitive to NCR does not require the NPR1 product to be active; it behaves like the NCR-sensitive enzymes.

At first sight, the *NPR1* gene product might be a necessary component of these transport systems. However, it rather appears to be a regulatory protein exerting a positive control on the activity of these permeases. A remarkable effect of the *npi1* and *npi2* mutations is that they suppress the effect of *npr1* mutations. When an *NPI* gene product is inactivated by a mutation, the *NPR1* gene product is no longer required for permease activity. This strongly suggests that the positive effect of the *NPR1* gene product is required exclusively to repair or counteract the inactivating effect of the *NPI* gene products. A second reason to believe that NPR1 is a regulatory protein is provided by the information contained in the nucleotide sequence of the *NPR1* gene [92]. Indeed, the putative NPR1 protein has the characteristics of a protein kinase. There are several possibilities as to the identity of the products it might phosphorylate. The NPR1 kinase might directly phosphorylate the permeases, or more likely, it might suppress the inactivating properties of one or both *NPI* gene products.

6.3.1.3. How is GAP1 activity regulated? The available data on the regulation of GAP1 activity can be interpreted as follows (see Fig. 5). In proline grown wild-type cells, the general amino acid permease is synthesized, but it is inactivated by the NPI control mechanism acting at the PGR site of the GAP transport protein. This negative control process keeps the general amino acid permease in an inactive form as long as the NPR1 product does not intervene. In steady state wild-type cells growing on proline as their sole nitrogen source, an equilibrium is reached between the negative effect of the NPI proteins and the positive effect of the NPR1 product. Upon addition of ammonium ions, the preexisting permease is progressively inactivated by a mechanism involving the NPI proteins. Simultaneously, the NPR1 reactivating mechanism is prevented from acting, either because ammonium ions or a derivative thereof inactivate the positive-control NPR1 protein, or because they activate the NPI system which then becomes predominant.

6.3.2. Nitrogen-catabolite repression (NCR)
Nitrogen-catabolite repression of GAP1 synthesis has been studied in mutant strains containing mutations such as *npi1* or *npi2* which destroy the permease inactivation mechanism. With the help of these mutants, it could be clearly demonstrated that GAP1 synthesis is repressed in cells grown in the presence of ammonium ions.

6.3.2.1. NCR affects permease gene transcription or transcript accumulation. Three cloned structural genes for ammonia-sensitive permeases, namely the *PUT4* [93], *UEP1/DAL5* [94] and *GAP1* [46] genes, were used to test the effects of growth conditions on the production of gene transcripts. In all three cases, the amount of gene transcripts was high in proline grown wild-type cells, but very low in ammonia or asparagine grown cells [46,93,94]. Thus, NCR involves altered transcription of the three permease genes or reduced transcript stability.

6.3.2.2. Glutamine as an effector of NCR. NCR of the ammonia-sensitive permeases is observed in cells grown in the presence of ammonium ions. To trigger NCR, however, ammonium ions must be transformed into glutamine. Reduced endogenous synthesis of glutamine in a mutant with a thermosensitive glutamine synthetase results in derepression of several permeases and enzymes, even in the presence of high NH_4^+ concentrations [88]. Since repression is restored by adding glutamine to the culture medium, it would appear that glutamine is a crucial effector molecule involved in this regulation. In mutants with a thermosensitive glutamine synthetase ($gln1^{ts}$), RNA transcripts of *PUT4* [93] and *GAP1* [46] permease genes accumulate at high (non-permissive) temperatures despite the presence of ammonium ions.

6.3.2.3. The URE2/GDHCR gene product as a negative regulatory protein which participates in the repression of permease synthesis. A number of enzymes involved in the catabolism of nitrogenous compounds are repressed in yeast cells grown in the presence of ammonium ions. Mutations affecting the *URE2/GDHCR* gene render these enzymes, e.g., the catabolic NAD-linked glutamate dehydrogenase, arginase, allantoinase and urea amidolyase, insensitive to this repression (see [95]). Several ammonia-sensitive amino acid permeases behave similarly. When the accumulation of transcripts of the cloned permease genes *GAP1* [46] and *PUT4* [93] was examined under repressing conditions, it appeared in each case that the presence of a *ure2/gdhCR* mutation strongly increases transcript accumulation. Hence, the *URE2/GDHCR* gene product acts as a negative regulator which prevents transcription of permease genes in ammonia grown wild-type cells. The deduced amino acid sequence of the *URE2/GDHCR* gene product does not contain any acknowledged DNA-binding motif [96].

6.3.2.4. The GLN3 gene product as a possible target for the URE2/GDHCR gene product. Repression of the synthesis of the UEP1/DAL5 permease in ammonia-grown cells requires the product of the *URE2/GDHCR* gene [88]. On the other hand, constitutive, GLN3-dependent expression of a reporter gene under the control of the *UEP1/DAL5* GATAA/UAS is completely repressed in the presence of asparagine

[97]. These results identified the GATAA/UAS as the *cis*-acting sites through which NCR is exerted on *UEP1/DAL5* expression [97]. As *gln3* mutations are epistatic over the *ure2/gdhCR* mutations [98], NCR of transcription of this permease might involve URE2/GDHCR-dependent inhibition of the GLN3 protein as suggested for other genes [96].

6.3.2.5. Double regulation of the ammonia-sensitive permeases. It is a remarkable feature that, in wild-type cells of *Saccharomyces cerevisiae*, permeases which are regulated according to nitrogen availability are controlled, as a rule, by both regulatory mechanisms, namely the inactivation–reactivation mechanism and NCR. Although these complicated mechanisms can be distinguished by mutations which inactivate one system only, they are found together for all the permeases tested except one, namely one of the glutamate permeases, which is subject to NCR but not to inactivation [14].

7. The APF1 gene product, a common factor of unknown function which increases the activity of amino acid permeases

The *APF1-AAP1* gene has been known for a long time [99,100]. Mutations in this gene affect the activity of several distinct amino acid permeases. While the K_m of these permeases for their substrates seems unchanged, the V_{max} is lowered by a factor ranging from 3- to 50-fold or more, depending on the permease considered [100]. This behaviour suggests that the APF1 product might be a common factor used by all these permeases. Mutations affecting other genes, which are named *APF3* and *APF4*, have similar effects (Jauniaux, Vissers and Grenson, in preparation). To help elucidate the role of the 'common factor' encoded by *APF1*, the effect of the *apf1* mutations on *GAP1* transcript levels has been tested (Jauniaux, Vissers and Grenson, in preparation). Since no effect has been detected, it seems that the *APF1* gene product is not a positive regulatory protein controlling the transcription of permease genes.

8. Summary and prospects

Interesting as it is, the study of transporters for amino acids and other nitrogenous nutrients in *Saccharomyces cerevisiae* is a tricky field. Many difficulties must be circumvented to avoid trivial errors. These practical problems are linked with several features of eukaryotic uptake systems, the first being the multiplicity of permeases which transport a given substrate. In relation to this, a major point is to make certain that one is not studying more than one uptake system at a time, and this can hardly be done without genetics. Once individual uptake systems have been identified and separated with the help of genetics, a second difficulty arises, which yeast shares with many other eukaryotes. This is linked to the regulation of amino acid uptake by specific feedback inhibition of the amino acid transporters. As a consequence, the metabolic transformations of the accumulating substrate greatly influence its further uptake rate. In other words, due to feedback inhibition of the transporter, removal of the accumulated amino acid by active metabolism may increase the uptake rate, whereas blocking its metabolic transformation lowers or stops entry into the cells.

Once these difficulties are known, they can be eliminated by use of both classical and molecular genetics.

Substantial progress can be expected in the near future concerning the structure of amino acid transporters, their functional dissection, and their evolutionary filiation.

The regulation of NCR-sensitive amino acid transporters in *Saccharomyces cerevisiae* has many points in common with that of catabolic enzymes. Amino acid permeases, as well as some other transporters of nitrogenous nutrients, are integrated into the regulatory circuits, both general and specific, which control catabolic processes.

The NCR-sensitive amino acid transporters of *Saccharomyces cerevisiae* are subject to an additional regulatory mechanism which inactivates and reactivates the permeases according to nitrogen availability. This NCI regulation is effected by a group of proteins. It seems to involve covalent modifications of the permeases and/or regulatory proteins. This original NPR1-NPIs regulatory mechanism seems to be restricted to transporters of nitrogenous compounds in *Saccharomyces cerevisiae*. It has not been found to affect enzymes.

The study of the regulation of gene expression at the transcription level is progressing rapidly. A number of positive and negative regulatory proteins with different target spectra are involved. The study of their mechanism of action may help us to understand how metabolic integration is realized, including at the level of membrane transport.

Acknowledgements

I gratefully acknowledge the invaluable help of my collaborators B. André, L.A. Urrestarazu and S. Vissers throughout the preparation of this manuscript. I also thank J.-C. Jauniaux for his contribution to section 5.2 and D. Coornaert for his help in constructing the figures. I am grateful to K. Broman for reading the manuscript. The work in our laboratory was supported by Grant No. 4513.88 from the Fonds de la Recherche Fondamentale Collective and by Grant No. 86/91–88 from an Action de Recherches Concertées between the Belgian Government and the Université Libre de Bruxelles.

References

1. Rickenberg, H.V., Cohen, G.N., Buttin, G. and Monod, J. (1956) Ann. Inst. Pasteur, Paris 91, 829–857.
2. Cohen, G.N. and Monod, J. (1957) Bacteriol. Rev. 21, 169–194.
3. Wilbrandt, W. and Rosenberg, T. (1961) Pharmacol. Rev. 13, 109–183.
4. Watson, J.D., Hopkins, N.H., Roberts, J.W., Steitz, J.A. and Weiner, A.M. (1987) In: Molecular Biology of the Gene (Gillen, J.R. Ed.), pp. 550-594, The Benjamin/Cummings Publishing Company, Menlo Park, CA.
5. Sherman, F., Fink, G.R. and Hicks, J.B. (1985) Methods in Yeast Genetics, Cold Spring Harbor Laboratory, New York.
6. Grenson, M. (1973) In: Genetics of Industrial Microorganisms (Vanek, Z., Hostalek, Z. and Cudlin, J. Eds.), pp. 179–193, Academia, Prague.
7. Wiame, J.-M., Grenson, M. and Arst, H.N., Jr. (1985) Adv. Microb. Physiol. 26, 1–87.
8. Horak, J. (1986) Biochim. Biophys. Acta 864, 223–256.
9. Cooper, T.G. (1982) In: The Molecular Biology of the Yeast *Saccharomyces cerevisiae* (Strathern, J.N., Jones, E.W. and Broach, J.R. Eds.), pp. 39–99, Cold Spring Harbor Laboratory, New York.
10. Large, P.J. (1986) Yeast 2, 1–34.
11. Kreger-van Rij, N.J.W. (1984) The Yeasts, a Taxonomic Study, Elsevier, Amsterdam.
12. Gits, J.J. and Grenson, M. (1967) Biochim. Biophys. Acta 135, 507–516.
13. Crabeel, M. and Grenson, M. (1970) Eur. J. Biochem. 14, 197–204.
14. Darte, C. and Grenson, M. (1975) Biochem. Biophys. Res. Commun. 67, 1028–1033.
15. Grenson, M. (1966) Biochim. Biophys. Acta 127, 339–346.
16. Grenson, M., Muyldermans, F., Broman, K. and Vissers, S. (1987) Life Sci. Adv. 6, 35–39.
17. Crabeel, M. (1973) Etude in vivo du rétrocontrôle de perméases d'acides aminés chez la levure, Thèse, Université Libre de Bruxelles.
18. Matile, P. and Wiemken, A. (1967) Arch. Microbiol. 56, 148–155.
19. Matile, P. (1978) Annu. Rev. Plant Physiol. 29, 193–213.
20. Sato, T., Ohsumi, Y. and Anraku, Y. (1984) J. Biol. Chem. 259, 11505–11508.
21. Sato, T., Ohsumi, Y. and Anraku, Y. (1984) J. Biol. Chem. 259, 11509–11511.
22. Davis, R.H. (1986) Microbiol. Rev. 50, 280–313.
23. Klionsky, D.J., Herman, P.K. and Emr, S.D. (1990) Microbiol. Rev. 54, 266–292.
24. Kanazawa, S., Driscoll, M. and Struhl, K. (1988) Mol. Cell. Biol. 8, 664–673.
25. Gros, P., Croop, J. and Housman, D. (1986) Cell 47, 371–380.
26. Chen, C.-J., Chin, J.E., Ueda, K., Clark, D.P., Pastan, I., Gottesman, M.M. and Roninson, I.B. (1986) Cell 47, 381–389.
27. Pietruszko, R. and Fowden, L. (1961) Ann. Bot. (London) 25, 491–511.
28. Ramos, F., El Guezzar, M., Grenson, M. and Wiame, J.-M. (1985) Eur. J. Biochem. 149, 401–404.
29. Lasko, P.F. and Brandriss, M.C. (1981) J. Bacteriol. 148, 241–247.
30. Ferro-Luzi Ames, G. and Higgins, C.F. (1983) TIBS 8, 97–100.
31. Kaback, H.R., Bibi, E. and Roepe, P.D. (1990) TIBS 15, 309.
32. Higgins, C.F. (1989) Nature (London) 340, 342.
33. Higgins, C.F. (1989) Nature (London) 341, 103.
34. Foote, S.J., Thompson, J.K., Cowman, A.F. and Kemp, D.J. (1989) Cell 57, 921–930.
35. McGrath, J.P. and Varshavsky, A. (1989) Nature (London) 340, 400–404.
36. Blobel, G. (1980) Proc. Natl. Acad. Sci. U.S.A. 77, 1496–1500.
37. Mueckler, M., Caruso, C., Baldwin, S.A., Panico, M., Blench, I., Morris, H.R., Allard, W.J., Lienhard, G.E. and Lodish, H.F. (1985) Science 229, 941–945.
38. Birnbaum, M.J., Haspel, H.C. and Rosen, O.M. (1986) Proc. Natl. Acad. Sci. U.S.A. 83, 5784–5788.
39. Celenza, J.L., Marshall-Carlson, L. and Carlson, M. (1988) Proc. Natl. Acad. Sci. U.S.A. 85, 2130–

2134.
40. Jund, R., Weber, E. and Chevallier, M.R. (1988) Eur. J. Biochem. 171, 417–424.
41. Weber, E., Rodriguez, C., Chevallier, M.R. and Jund, R. (1990) Mol. Microbiol. 4, 585–596.
42. Rai, R., Genbauffe, F.S. and Cooper, T.G. (1988) J. Bacteriol. 170, 266–271.
43. Hoffmann, W. (1985) J. Biol. Chem. 260, 11831–11837.
44. Tanaka, J.-I. and Fink, G.R. (1985) Gene 38, 205–214.
45. Vandenbol, M., Jauniaux, J.-C. and Grenson, M. (1989) Gene 83, 153–159.
46. Jauniaux, J.-C. and Grenson, M. (1990) Eur. J. Biochem. 190, 39–44.
47. Sophianopoulou, V. and Schazzocchio, C. (1989) Mol. Microbiol. 3, 705–714.
48. Honoré, N. and Cole, S.T. (1990) Nucleic Acids Res. 18, 653.
49. Jauniaux, J.-C. (1990) Arch. Int. Physiol. Biochim. 98, B133.
50. Nikawa, J.-I., Hosaka, K., Tsukagoshi, Y. and Yamashita, S. (1990) J. Biol. Chem. 265, 15996–16003.
51. Nakao, T., Yamato, I. and Anraku, Y. (1987) Mol. Gen. Genet. 208, 70–75.
52. Nehlin, J.A., Carlberg, M. and Ronne, H. (1989) Gene 85, 313–319.
53. Maiden, M.C.J., Davis, E.O., Baldwin, S.A., Moore, D.C.M. and Henderson, P.J.F. (1987) Nature (London) 325, 641–643.
54. Gits, J.J. and Grenson, M. (1969) Arch. Int. Physiol. Biochim. 77, 153–154.
55. Horak, J. and Rihova, L. (1982) Biochim. Biophys. Acta 691, 144–150.
56. Grenson, M., Crabeel, M., Wiame, J.M. and Béchet, J. (1968) Biochem. Biophys. Res. Commun. 30, 414–419.
57. Grenson, M. (1969) Eur. J. Biochem. 11, 249–260.
58. Grenson, M., Mousset, M., Wiame, J.-M. and Béchet, J. (1966) Biochim. Biophys. Acta 127, 325–338.
59. Grenson, M., Hou, C. and Crabeel, M. (1970) J. Bacteriol. 103, 770–777.
60. Rai, R., Genbauffe, F.S., Sumrada, R.A. and Cooper, T.G. (1989) Mol. Cell. Biol. 9, 602–608.
61. Bysani, N., Rai, R., Daugherty, J.R. and Cooper, T.G. (1990) Yeast 6, S237.
62. Cooper, T.G., Ferguson, D., Rai, R. and Bysani, N. (1990) J. Bacteriol. 172, 1014–1018.
63. Mitchell, A.P. and Magasanik, B. (1984) Mol. Cell. Biol. 4, 2758–2766.
64. Martin, D.I.K. and Orkin, S.H. (1990) Genes Dev. 4, 1886–1898.
65. Kudla, B., Caddick, M.X., Langdon, T., Martinez-Rossi, N.M., Bennett, C.F., Sibley, S., Davies, R.W. and Arst, H.N., Jr. (1990) EMBO J. 9, 1355–1364.
66. Fu, Y.H. and Marzluf, G.A. (1990) Mol. Cell. Biol. 10, 1056–1065.
67. Fu, Y.H. and Marzluf, G.A. (1990) Mol. Microbiol. 4, 1847–1852.
68. Fu, Y.H. and Marzluf, G.A. (1990) Proc. Natl. Acad. Sci. U.S.A. 87, 5331–5335.
69. Wall, L., deBoer, E. and Grosveld, F. (1988) Genes Dev. 2, 1089–1100.
70. Evans, T., Reitman, M. and Felsenfeld, G. (1988) Proc. Natl. Acad. Sci. U.S.A. 85, 5976–5980.
71. Guarente, L. (1988) Cell 52, 303–305.
72. André, B. (1990) Mol. Gen. Genet. 220, 269–276.
73. André, B. (1990) Arch. Int. Physiol. Biochim. 98, B56.
74. André, B. and Grenson, M. (1990) Yeast 6, S323.
75. Vissers, S., André, B., Muyldermans, F. and Grenson, M. (1989) Eur. J. Biochem. 181, 357–361.
76. Vissers, S., André, B., Muyldermans, F. and Grenson, M. (1990) Eur. J. Biochem. 187, 611–616.
77. Coornaert, D., Vissers, S. and André, B. (1991) Gene 97, 163–171.
78. Jacobs, E., Dubois, E., Hennaut, C. and Wiame, J.M. (1981) Curr. Genet. 4, 13–18.
79. Hennaut, C. (1981) Curr. Genet. 4, 69–72.
80. Turoscy, V. and Cooper, T.G. (1982) J. Bacteriol. 151, 1237–1246.
81. Yoo, H.S., Genbauffe, F.S. and Cooper, T.G. (1985) Mol. Cell. Biol. 5, 2279–2288.
82. Bricmont, P.A., Daugherty, J.R. and Cooper, T.G. (1991) Mol. Cell. Biol. 11, 1161–1166.
83. Cooper, T.G. and Sumrada, R. (1975) J. Bacteriol. 121, 571–576.

84. André, B. and Jauniaux, J.-C. (1990) Nucleic Acids Res. 18, 7136.
85. Chilsholm, V.T. and Cooper, T.G. (1982) Mol. Cell. Biol. 2, 1088–1095.
86. Chisholm, G. and Cooper, T.G. (1984) Mol. Cell. Biol. 4, 947–955.
87. Grenson, M. (1983) Eur. J. Biochem. 133, 141–144.
88. Grenson, M. (1983) Eur. J. Biochem. 133, 135–139.
89. Grenson, M. and Dubois, E. (1982) Eur. J. Biochem. 121, 643–647.
90. Dubois, E. and Grenson, M. (1979) Mol. Gen. Genet. 175, 67–76.
91. Grenson, M. and Acheroy, B. (1982) Mol. Gen. Genet. 188, 261–265.
92. Vandenbol, M., Jauniaux, J.-C. and Grenson, M. (1990) Mol. Gen. Genet. 222, 393–399.
93. Jauniaux, J.-C., Vandenbol, M., Vissers, S., Broman, K. and Grenson, M. (1987) Eur. J. Biochem. 164, 601–606.
94. Rai, R., Genbauffe, F.S., Lea, H.Z. and Cooper, T.G. (1987) J. Bacteriol. 169, 3521–3524.
95. Dubois, E., Vissers, S., Grenson, M. and Wiame, J.-M. (1977) Biochem. Biophys. Res. Commun. 75, 233–239.
96. Coschigano, P.W. and Magasanik, B. (1991) Mol. Cell. Biol. 11, 822–832.
97. Cooper, T.G., Rai, R. and Yoo, H.S. (1989) Mol. Cell. Biol. 9, 5440–5444.
98. Courchesne, W.E. and Magasanik, B. (1988) J. Bacteriol. 170, 708–713.
99. Surdin, Y., Sly, W., Sire, J., Bordes, A.M. and de Robichon-Szulmajster, H. (1965) Biochim. Biophys. Acta 107, 546–566.
100. Grenson, M. and Hennaut, C. (1971) J. Bacteriol. 105, 477–482.
101. Soberon, M., Olamendi, J., Rodriguez, L. and Gonzalez, A. (1989) J. Gen. Microbiol. 135, 2693–2697.
102. Jauniaux, J.-C. and Grenson, M. (1990) Yeast 6, S564.
103. Higgins, D.G. and Sharp, P.M. (1988) Gene 73, 237–244.
104. Eisenberg, D., Schwarz, E., Komaromy, M. and Wall, R. (1984) J. Mol. Biol. 179, 125–142.
105. Coornaert, D., Vissers, S., André, B. and Grenson, M. (1992) Curr. Genet. 21, 301–307.

CHAPTER 8

Structure and function of plasma membrane Na^+/H^+ exchangers

PETER IGARASHI

Department of Medicine, Yale University School of Medicine, New Haven, CT 06510, U.S.A.

1. Introduction

1.1. The Na^+/H^+ exchanger

The plasma membranes of most eukaryotic cells contain an integral membrane transport protein, the Na^+/H^+ exchanger, which under usual physiological conditions directly couples Na^+ flux into the cell to uphill extrusion of H^+. Although not directly ATP-consuming, the process is driven by the inwardly directed electrochemical Na^+ gradient generated by the Na^+,K^+-ATPase (an example of secondary active transport). By maintaining relative intracellular alkalinization, plasma membrane Na^+/H^+ exchangers are involved in the defense of intracellular pH against acid loads. Moreover, in many cells, activation of plasma membrane Na^+/H^+ exchangers occurs following changes in cell volume and exposure to growth factors. The consequent alterations in cell Na^+ and H^+ content are responsible (or permissive) for restoration of normal cell volume or transduction of the growth factor signal. In polarized epithelia, e.g., renal proximal tubule, Na^+/H^+ exchangers play an essential role in mediating active transepithelial solute flux. Derangements of plasma membrane Na^+/H^+ exchangers may be important in some pathophysiological conditions. Perhaps most attention has been paid to a possible role in the pathogenesis of essential hypertension [1,2].

Primarily using isolated plasma membrane vesicles as an experimental preparation, the functional properties of Na^+/H^+ exchangers have been elucidated. The important kinetic properties include: (1) stoichiometry (one-for-one); (2) reversibility; (3) substrate specificity (monovalent cations Na^+, H^+, Li^+, NH_4^+, but not K^+, Rb^+, Cs^+, choline); (4) modes of operation (Na^+-for-H^+, Na^+-for-Na^+, Li^+-for-Na^+, Na^+-for-NH_4^+); (5) existence of an internal site for allosteric activation by H^+; (6) reversible inhibition by amiloride (N-amidino-5-amino-6-chloropyrazine carboxamide) and its 5-amino-substituted analogs; and (7) competitive nature of the binding of substrates and amiloride at the external transport site [3–6]. The purpose of the present chapter is to summarize progress within the last five years on the structural characterization of plasma membrane Na^+/H^+ exchangers using biochemical and molecular biological techniques, and, when possible, to correlate the structural findings with previous functional information. We will not emphasize acute regulation of Na^+/H^+ exchangers, a topic reviewed by Grinstein and Rothstein [7], nor will we consider the Na^+/H^+ exchangers of mitochondria and prokaryotes which behave differently kinetically.

1.2. Functional heterogeneity

An important development recently has been the recognition that functionally distinguishable isoforms of plasma membrane Na^+/H^+ exchangers exist in various cell types and even within the same cell. This observation has been best characterized in the porcine renal epithelial cell line (LLC-PK$_1$/Clone 4) [8] and in rabbit ileum [9]. Confluent monolayers of LLC-PK$_1$/Clone 4 cells express pharmacologically distinct Na^+/H^+ exchangers on the apical and basolateral membrane. One form is relatively sensitive to inhibition by amiloride and its 5-amino-substituted analogs (IC$_{50}$ of 44 nM for inhibition by 5-(*N*-ethyl-*N*-isopropyl)amiloride (EIPA)). This amiloride-sensitive isoform is present exclusively on the basolateral membrane of confluent LLC-PK$_1$/Clone 4 cells. The other form of Na^+/H^+ exchanger in LLC-PK$_1$/Clone 4 cells is relatively amiloride-resistant (IC$_{50}$ of 13 μM for inhibition by EIPA) and is restricted to the apical or brush border membrane. Similarly, in villus enterocytes from rabbit ileum the apical membrane form is significantly less sensitive to inhibition by amiloride (IC$_{50}$ 103 μM) than is the basolateral form (IC$_{50}$ 11 μM) [9]. Na^+/H^+ exchangers that are similarly sensitive to inhibition by amiloride as the epithelial basolateral form have been described in a wide variety of non-epithelial cells including lymphocytes and fibroblasts. The apical form of Na^+/H^+ exchanger participates in transepithelial absorption of NaCl and NaHCO$_3$. Kulanthaivel et al. [10,11] have observed that the polarized distribution of these Na^+/H^+ exchanger isoforms is reversed in human placental syncytiotrophoblast: the amiloride-sensitive form, which is also inhibited by cimetidine more than clonidine, is situated on the brush border membrane; and the amiloride-resistant form inhibited by clonidine more than cimetidine is on the basal membrane [10,11]. In addition to pharmacological differences, apical and basolateral Na^+/H^+ exchangers of epithelia differ in kinetic and regulatory properties including regulation by protein kinase C [12], sensitivity to *N*-ethylmaleimide [13], and inhibition by external K^+ [9].

Although a uniform nomenclature for Na^+/H^+ exchanger isoforms has not yet been adopted, we will refer to the amiloride-sensitive type of Na^+/H^+ exchanger that is present in the basolateral membrane of epithelia (apical membrane of placental syncytiotrophoblast) and also widely distributed in non-epithelial cells as the sensitive-type. The relatively amiloride-resistant isoform present in apical membrane of epithelia (basolateral membrane in placenta) will be referred to as the resistant-type. It is important to note that the structural basis underlying the distinction between sensitive- and resistant-types remains unknown (*vide infra*). Moreover, this classification is almost certainly an oversimplification since it is probable that multiple isoforms exist within each of these broad categories. For example, non-epithelial cells in brain express a Na^+/H^+ exchanger that is highly resistant to amiloride [14], and the apical Na^+/H^+ exchanger in ileum is affected by external potassium whereas the renal apical Na^+/H^+ exchanger is unaffected [9]. Nevertheless, in considering structure–function relationships for Na^+/H^+ exchangers it will be helpful to bear this functional classification in mind.

2. Biochemical properties of Na^+/H^+ exchangers

Biochemical studies of plasma membrane Na^+/H^+ exchangers have been directed at two major goals: (1) identification of amino acids that are involved in the transport mechanism; and (2) identification and characterization of the transport protein(s). To date, most studies have been performed on the amiloride-resistant form of Na^+/H^+ exchanger that is present in apical or brush border membrane vesicles from mammalian kidney, probably because of the relative abundance of transport activity in this starting material. However, some studies have also been performed on the amiloride-sensitive isoform present in non-epithelial cells.

2.1. 'Group-specific' modification

In contrast to amiloride and its analogs, which generally act as reversible inhibitors of the Na^+/H^+ exchanger, studies have also been performed using irreversible inhibition by covalent modifiers with the goal of identifying critical amino acid side chains that are important in the mechanism of Na^+/H^+ exchange. The general strategy has been to pretreat membranes or cells with reagents that react covalently and more or less specifically with certain amino acid side chains, then measure Na^+/H^+ exchanger activity. If pretreatment with such a 'group-specific' reagent irreversibly inactivated the Na^+/H^+ exchanger this was generally taken as evidence that the reactive group was important in the mechanism of transport. Moreover, if inactivation could be blocked in the presence of substrate or amiloride, this suggested that the critical groups were located at or near binding sites for substrate or amiloride. Such studies are subject to at least two important caveats. First, virtually all covalent modifiers have potential side reactions with groups other than the ones of interest. Second, it is important to remember that both inactivation itself and protection by substrate or amiloride could reflect 'actions at a distance' due to long-range conformational changes rather than effects on a side group directly involved in transport. Nevertheless, a similar approach has been useful for identifying critical imidazolium and sulfhydryl groups in *E. coli lac* permease [15] and for evaluating the roles of guanidinium, amino and carboxyl groups in the transport mechanism of the erythroid band 3 anion exchanger [16]. Using analogous methods, imidazolium, carboxyl, sulfhydryl and amino groups have been implicated in the transport mechanism of mammalian plasma membrane Na^+/H^+ exchangers. Effects of 'group-specific' reagents on the transport activity and regulation of plasma membrane Na^+/H^+ exchangers are summarized in Table I.

2.1.1. Imidazolium

Kinetic studies indicated that external H^+ interacted with the renal brush border Na^+/H^+ exchanger (resistant-type) at a single site with apparent pK_a 7.3–7.5. Because the imidazolium ring of histidine is the principal group that is titratable in this

TABLE I

Effects of 'group-specific' reagents on transport activity and regulation of plasma membrane Na^+/H^+ exchangers

Group	Na^+/H^+ exchanger type	Reagent[a]	Effect[b]	Protectant[c]	Ref.
Imidazolium	Lymphocyte	DEPC	I	n.d.	19
	Placental brush border	DEPC	I	Amiloride	18
	Renal brush border	DEPC, Rose bengal	I ($\downarrow V_{max}$)	Amiloride	17
Carboxyl	Lymphocyte	DCCD	None		19
	Placental brush border	DCCD	I	Amiloride	18
	Renal brush border	DCCD, EEDQ	I ($\downarrow V_{max}$ or $\uparrow K_m$)	Amilioride, Na^+, Li^+	20–24
Sulfhydryl	Lymphocyte	NEM	I	None	19
	Placental brush border	PAO	I ($\downarrow V_{max}$)	Amiloride	28
	Erythrocyte	NPM	I, R	n.d.	27
	Renal brush border	NEM	I	None	13, 22
	Renal basolateral	NEM	I	n.d.	13
Amino	Lymphocyte	PLP, DIDS	None		19
	Erythrocyte	Glutaraldehyde	R	n.d.	29
	Renal brush border	PITC	I	Amiloride	30
Carbohydrate	Renal brush border	Endo-F	I ($\downarrow V_{max}$)	n.d.	31
	Renal brush border	Endo-H	None		31

[a]Abbreviations: DEPC, diethylpyrocarbonate; DCCD, *N,N'*-dicyclohexylcarbodiimide; EEDQ, *N*-ethyoxycarbonyl-2-ethoxy-1,2-dihydroquinoline; NEM, *N*-ethylmaleimide; PAO, phenylarsine oxide; NPM, *N*-phenylmaleimide; PLP, pyridoxal phosphate; DIDS, diisothiocyanostilbene disulfonate; PITC, phenylisothiocyanate.
[b]I, inhibition of transport; R, altered regulation.
[c]n.d., not determined.

range, Grillo and Aronson examined the effects of histidine-specific reagents on transport activity [17]. Pretreatment of renal brush border membrane vesicles with diethylpyrocarbonate (DEPC) caused inactivation of Na^+/H^+ exchange that was not due to vesicle disruption or accelerated dissipation of transmembrane H^+ gradients. Reaction with DEPC followed pseudo-first-order kinetics indicating that inhibition was due to modification of a single class of imidazolium groups. Inactivation was due to a reduction in number or turnover rate of transporters (V_{max} effect) with no change in apparent affinity for external Na^+ (K_m for Na^+_o). Since DEPC reacts only with unprotonated imidazolium groups, Grillo and Aronson further evaluated the effects of pH on the reaction. Inactivation by DEPC was dependent on external but not internal pH, was blocked by amiloride (due to steric hindrance), and was enhanced by Na^+ (due to displacement of H^+). Taken together, these results indicated that an imidazolium group of histidine is the moiety that binds protons at the external transport site of the renal brush border Na^+/H^+ exchanger.

Ganapathy et al. [18] showed that the human placental brush border Na^+/H^+ exchanger (a sensitive-type) was also inactivated by DEPC, and amiloride protected against inhibition. DEPC also inactivated the sensitive-type Na^+/H^+ exchanger in thymic lymphocytes [19].

2.1.2. Carboxyl

Although a histidyl residue binds H^+ at the external transport site, such a group would not be expected to avidly bind Na^+. Anionic moieties, e.g., unprotonated carboxyl groups are more likely candidates for this role. Indeed, a number of studies have suggested a critical role of carboxyl groups in transport activity of the renal brush border Na^+/H^+ exchanger (resistant-type). Burnham et al. [20] first demonstrated inhibition of the renal brush border Na^+/H^+ exchanger by the carboxyl group-activating reagent, N-ethoxycarbonyl-2-ethoxy-1,2-dihydroquinoline (EEDQ). Inactivation by EEDQ was studied in detail by Rocco et al. [21] who observed that pretreatment with EEDQ reduced both V_{max} and apparent K_m for Na^+_o, and amiloride but not Na^+ protected against inactivation. Interestingly, 5-amino-substituted amiloride analogs which are more potent reversible inhibitors of Na^+/H^+ exchangers were less effective than amiloride itself in protecting against inactivation by EEDQ, suggesting that a free 5-amino group was necessary for protection. Alternatively, because EEDQ reacts with carboxyl groups to form a mixed carbonic anhydride that, in turn, reacts with nucleophiles, and it was not possible to distinguish whether the carboxyl group and/or an endogenous nucleophile were critical for transport activity, the free 5-amino group of amiloride might be protective by competing for reaction with a critical endogenous nucleophile.

Further evidence that carboxyl groups are important for transport activity was provided by Igarashi and Aronson [22], Friedrich et al. [23], and Kinsella et al. [24] using the carboxyl group-specific reagent, N,N'-dicyclohexylcarbodiimide (DCCD). DCCD irreversibly inactivated the brush border Na^+/H^+ exchanger in rabbit and rat kidney. Inhibition of the rabbit renal Na^+/H^+ exchanger by DCCD followed pseudo-first-order kinetics indicating that reaction was with a single class of carboxyl groups. Inhibition resulted primarily from an increase in apparent K_m for Na^+_o and was pH dependent in a manner consistent with reaction with carboxyl groups. Importantly, Na^+, Li^+ and amiloride each protected against inhibition by DCCD in a pH-dependent fashion consistent with reaction of DCCD at a carboxylate located at the external transport site where Na^+, H^+, Li^+ and amiloride compete for binding. Friedrich et al. [23] found that although other Na^+-dependent transporters were inhibited by DCCD, only the Na^+/H^+ exchanger was protected by amiloride. DCCD reacts with carboxyl groups by forming an O-acylisourea which then hydrolyzes to form the N-acylurea or is cross-linked with endogenous or exogenous nucleophiles. Thus, it was unclear whether inhibition by DCCD was due to modification of a critical carboxylate or reaction with a critical endogenous nucleophile. Igarashi and Aronson [22] found that addition of exogenous nucleophiles had no protective effect on inactivation of the rabbit renal Na^+/H^+ exchanger by DCCD (e.g., by displacing an endogenous nucleophile). Kinsella et al. [24] found that glycine methyl ester only slightly reduced the inhibition of the rat renal Na^+/H^+ exchanger (ethylenediamine had a greater protective effect but independently accelerated dissipation of the H^+ gradient). Taken together, the simplest explanation of these results is that an unprotonated carboxyl group is important for transport function and is a moiety contributing to the binding of Na^+ at the external transport site. Because binding of H^+ and Na^+ at the external transport site are known to be mutually exclusive events, Igarashi and Aronson have proposed a model in which simultaneous binding of H^+ to imidazolium and Na^+ to carboxylate would be precluded electrostatically if the groups were located in a sufficiently hydrophobic micro-environment (as may be the case since inactivation by the hydrophobic DCCD is greater than inhibition by hydrophilic carbodiimides [22–24]).

Interestingly, Grinstein et al. [19] found that DCCD did not inhibit the Na^+/H^+ exchanger in thymic lymphocytes (sensitive-type) whereas Ganapathy et al. [18] found that the placental brush border Na^+/H^+ exchanger, which is also a sensitive-type, was susceptible to DCCD, and amiloride protected from inhibition. Thus, there may be structural differences between sensitive-type exchangers from these two tissues.

2.1.3. Sulfhydryl
Several studies have described an important role for sulfhydryl groups in the mechanism of Na^+/H^+ exchange. Sulfhydryl groups are important in the transport mechanism of other proton-translocating membrane proteins [25,26]. Grinstein et al. [19] found that the Na^+/H^+ exchanger in lymphocytes (sensitive-type) was inhibited 65% by the sulfhydryl reagent, N-ethylmaleimide (NEM). Inhibition was not due to alterations in cell volume, transmembrane ionic gradients, buffering power, or cellular viability. That the the critical moiety was located on the internal

surface of the transporter was suggested by the following:

(1) reaction with NEM was affected by changes in cytoplasmic pH at constant external pH;

(2) the impermeant sulfhydryl reagent, glutathione-maleimide, was without significant effect; and

(3) extracellular Na^+ and amiloride failed to protect from inhibition.

Parker and Glosson [27] found that the Na^+/H^+ exchanger in canine erythrocytes was inhibited by the sulfhydryl reagent, N-phenylmaleimide. They also observed that lower concentrations of N-phenylmaleimide impaired regulation without affecting transport. This transporter is normally reversibly activated by hypertonic cell shrinkage. However, low concentrations ($<20 \mu mol/g$ hemoglobin) of N-phenylmaleimide caused the transporter to become irreversibly activated or inactivated, depending on the volume of the cells at the time of N-phenylmaleimide exposure. This suggests that sulfhydryl groups are involved in the mechanisms of both ion transport and switching to an activated state that accompanies hypertonic cell shrinkage.

Kulanthaivel et al. [28] found that the apical Na^+/H^+ exchanger in human placenta (sensitive-type) was sensitive to phenylarsine oxide, a reagent specific for dithiols that are situated in close proximity (vicinal dithiols). Moreover, the effect of phenylarsine oxide was to decrease V_{max} without affecting apparent affinity for Na^+_o, and was partially blocked by amiloride but not by cimetidine. Since these investigators also found that amiloride and cimetidine bound competitively with Na^+ at the external transport site of the placental brush border Na^+/H^+ exchanger, they concluded that the vicinal dithiol groups are necessary for transport function but are located at a site distinct from the external transport site.

Igarashi and Aronson [22] found that the renal brush border Na^+/H^+ exchanger (resistant-type) was inhibited 40% by 1 mM NEM, and inhibition was not blocked by 1 mM amiloride. Haggerty et al. [13] reported that both the apical and basolateral Na^+/H^+ exchangers in LLC-PK$_1$ cells were inactivated by 0.5 mM NEM, although the apical Na^+/H^+ exchanger was more sensitive to inhibition (70% inhibition compared to 20% inhibition of the basolateral transport activity).

Taken together, these results indicate that similar to other proton-translocating membrane proteins, both types of Na^+/H^+ exchangers contain critical sulfhydryl groups that are involved in the transport mechanism. These sulfhydryl groups do not appear to be present at the external transport site but may be involved in switching from an inactive to an activated state.

2.1.4. Amino

It might be unexpected for cationic groups, e.g., protonated amino groups, to be important in translocation of cations across the membrane bilayer. Indeed, Grinstein et al. [19] found that amino reagents (pyridoxal phosphate, trinitrobenzene sulfonate and diisothiocyanostilbene disulfonate) did not affect the Na^+/H^+ exchanger in lymphocytes (sensitive-type). Parker [29] also observed that the amino-reactive reagent, glutaraldehyde, did not affect the transport function of the Na^+/H^+ exchanger of dog erythrocytes, but the regulatory function was impaired. Glutaraldehyde (0.01–0.03%) had an effect similar to low concentrations of N-phenylmaleimide to permanently 'fix' the volume-sensitive exchanger in either the activate or inactive state.

A preliminary report by Huang and Warnock [30] suggests that lysyl residues may be important for transport activity of the renal brush border Na^+/H^+ exchanger (resistant-type). These investigators previously observed that external chloride blocked the inhibitory effect of amiloride suggesting that a cationic group interacted with chloride and amiloride at the external transport site. The amino-specific reagent, phenylisothiocyanate, caused 40% irreversible inactivation of the Na^+/H^+ exchanger whereas the arginine-specific reagent, phenylglyoxal, had little effect. Amiloride protected against the inactivation by phenylisothiocynate in a concentration-dependent manner. Taken together, these results suggest that amino groups may be involved in the transport mechanism of resistant-type Na^+/H^+ exchangers and the regulation (but not the transport function) of sensitive-type exchangers.

2.1.5. Carbohydrate

Yusufi et al. [31] have raised the interesting possibility that N-linked oligosaccharides may be important for the transport function of the renal brush border Na^+/H^+ exchanger. Although glycosylation is required for biological activity of certain proteins, an essential role in membrane transport had not been previously described. Administration of swainsonine to rats decreased the rate of Na^+/H^+ exchange, but did not affect the rate of other Na^+-gradient-driven transporters (Na^+/phosphate, Na^+/glucose, Na^+/proline) or amiloride-insensitive Na^+ uptake. By inhibiting α-D-mannosidase II in the medial Golgi apparatus, swainsonine inhibits formation of complex-, high-mannose-, and hybrid-type N-linked oligosaccharides. Thus, inhibition of Na^+/H^+ exchange could be explained by either impaired translocation and insertion into the membrane or decreased activity of the transporter. To distinguish between these possibilities, Yusufi et al. [31] treated brush border membrane vesicles in vitro with endo-β-N-acetylglucosaminidase F (Endo-F) or endo-β-N-acetylglucosaminidase (Endo-H). Endo-F decreased the rate of Na^+/H^+ exchange without affecting other Na^+-gradient-driven processes, whereas Endo-H was without effect. Inhibition by Endo-F was due to a decrease in V_{max} without altering apparent K_m for Na^+_o. Because of the different substrate specificities of Endo-F and Endo-H, these results suggest that complex-type N-linked oligosaccharides are essential for transport activity of the brush border Na^+/H^+ exchanger. Similar functional analyses have not been performed of the sensitive-type Na^+/H^+ exchanger in any cell.

2.2. Identification and characterization of candidate transport protein(s)

Attempts have been made to identify the protein(s) that mediate plasma membrane Na^+/H^+ exchange as a first step towards eventual purification and micro-sequencing. Two general approaches have been used: the first involves covalent labeling with group-specific reagents; the second involves interaction with amiloride analogs. These studies have identified several proteins with molecular masses between 321 and 25 kilodaltons (kDa) that were candidates for the transport protein. However, because neither the covalent modifiers nor amiloride analogs are absolutely specific for Na^+/H^+ exchange, none of these studies have unambiguously identified the transport protein. Candidates for the renal brush border Na^+/H^+ exchanger protein are listed in Table II.

2.2.1. Covalent labeling

Several laboratories have attempted to identify the amiloride-resistant Na^+/H^+ exchanger in renal brush border membranes using covalent labeling. Based on the observation that DCCD irreversibly inhibited the renal brush border Na^+/H^+ exchanger and amiloride protected against inhibition, Igarashi and Aronson [22] pretreated membrane vesicles with N,N'-dicyclohexyl[^{14}C]carbodiimide ([^{14}C]DCCD) in the presence of unlabeled amiloride. They found that a 100-kDa protein in rabbit was labeled with [^{14}C]DCCD, and labeling was reduced in the presence of amiloride. However, the concentration of amiloride required for protection against covalent labeling was higher than the concentration required for protection against kinetic inhibition. Friedrich et al. [23] performed similar studies in rat kidney brush border

TABLE II

Candidates for the renal brush border Na^+/H^+ exchanger transport protein identified by covalent labeling, affinity chromatography, or other methods

Molecular mass (kDa)	Species	Method[a]	Ref.
321	Rat	Radiation inactivation	47
178, 146	Cow	Affinity chromatography, anti-peptide antibody	40
107, 81	Pig	[^{3}H]MIA photolysis	34
100	Rabbit	[^{14}C]DCCD labeling	22
79, 72	Rabbit	[^{3}H]MIA cross-linking	36
77	Rabbit	NENMBA photolysis	37
66	Cow	^{125}I-AzPZA photolysis, anti-peptide antibody	33
65	Rat	[^{14}C]DCCD labeling, [^{14}C]Br-EIPA photolysis	23
		[^{3}H]PrBCM labeling	35

[a]Abbreviations: MIA, 5-(N-methyl-N-isobutyl)amiloride; DCCD, N,N'-dicyclohexylcarbodiimide; AzPZA, 4N-azidosalicylic-5N-piperazine-amiloride; Br-EIPA, 5-(N-ethyl-N-isopropyl)-6-bromoamiloride; PrBCM, propylbenzilylcholine mustard.

membranes. No labeling of a 100-kDa protein was observed. Rather, [^{14}C]DCCD labeled proteins of M_r 88, 65, 50, and 34 kDa, but only labeling of the 65-kDa protein was blocked by 1 mM amiloride. That the discrepancy in molecular masses between these two laboratories (100 kDa versus 65 kDa) might be due to species differences was suggested by the observation that higher DCCD concentrations were required to irreversibly inhibit the Na$^+$/H$^+$ exchanger in rabbit compared to rat, and rabbit lacked the amiloride-protectable 65-kDa protein [32].

Another approach to identify the Na$^+$/H$^+$ exchanger in renal brush border membranes has been covalent labeling with photolabile amiloride analogs. Friedrich et al. [23] observed that photoirradiation in the presence of Br-EIPA (N-amidino-3-amino-5-(N-ethyl-N-isopropyl)-6-bromopyrazine carboxamide) irreversibly inhibited the Na$^+$/H$^+$ exchanger, and Na$^+$ and Li$^+$ protected from inactivation. When these investigators photolyzed rat renal brush border membranes in the presence of ^{14}C-labeled Br-EIPA, radioactivity was primarily incorporated into a 65-kDa protein. Moreover, labeling of the 65-kDa protein was diminished in the presence of 100 mM Na$^+$ which correlated with the kinetic findings and suggested that the 65-kDa protein was a component of the transporter.

A protein of similar molecular mass (66-kDa) was identified in bovine renal brush border membranes by Ross et al. [33]. The amiloride analog 4N-azidosalicylic-5N-piperazine-amiloride (AzPZA) inhibited the Na$^+$/H$^+$ exchanger in MDCK cells with biphasic kinetics suggesting the presence of low- and high-affinity binding sites (IC$_{50}$ of 10 μM and 500 nM, respectively). Bovine brush border membranes were photolyzed with 3 nM ^{125}I-AzPZA, and a single band of 66 kDa was labeled. Labeling appeared specific since it was reduced in the presence of 500 μM 5N-piperidine-amiloride. The same workers also identified the 66-kDa protein on immunoblots of brush border membranes probed with antisera to a 20-amino acid peptide of the cytoplasmic domain of the human Na$^+$/H$^+$ exchanger (vide infra).

Wu and Lever [34] observed that the high-affinity amiloride analog 5-(N-methyl-N-[^{3}H]isobutyl)amiloride ([^{3}H]MIA) binds to porcine renal brush border membranes with a K_d of 250 nM, and binding is inhibited by amiloride, Na$^+$, and Li$^+$. When solubilized membranes were fractionated by high-pressure liquid chromatography (HPLC), [^{3}H]MIA binding activity cofractionated with Na$^+$/H$^+$ exchanger activity as measured by reconstitution into proteoliposomes (vide infra). Chromatographic fractions containing the major peak of Na$^+$/H$^+$ exchanger activity were photolyzed in the presence of [^{3}H]MIA, and radioactivity was incorporated primarily into 81- and 107-kDa proteins [34]. Moreover, labeling of the 81- and 107-kDa proteins by [^{3}H]MIA was inhibited by amiloride and Na$^+$, but not by phenamil. The latter compound was tested because one way of distinguishing Na$^+$/H$^+$ exchangers from epithelial Na$^+$ channels, Na$^+$/Ca^{2+} exchangers, and other amiloride-binding proteins is to compare the rank order of potency of inhibition by amiloride analogs. Na$^+$/H$^+$ exchange is uniquely inhibited by 5-amino-substituted amiloride analogs > amiloride > benzamil or phenamil. Epithelial Na$^+$ channels, e.g., have the opposite rank order of potency. Thus, the finding by Wu and Lever that amiloride protected better than phenamil would suggest that the 81- and 107-kDa proteins were components of the Na^+/H^+ exchanger rather than, e.g., the epithelial Na^+ channel, except that equimolar concentrations of phenamil and amiloride were not employed in their study.

An antagonist of the acetylcholine receptor (propylbenzilylcholine mustard) that irreversibly inhibits the rat renal Na^+/H^+ exchanger labeled a 65-kDa brush border protein in an amiloride-protectable manner [35]. In a preliminary report, Igarashi et al. [36], used DCCD to covalently cross-link rabbit renal brush border proteins with [^{3}H]MIA. 72- and 79-kDa proteins were identified with this technique. Initially, labeling appeared specific for the Na^+/H^+ exchanger since it was blocked by both non-radioactive MIA and Li^+, which bind competitively to the transporter. However, subsequent study revealed that benzamil protected better than amiloride itself which would not be expected given the rank order of potency of these compounds for inhibiting Na^+/H^+ exchange (*vide supra*). It was possible that this method instead labeled the Na^+/glucose transporter which has a similar molecular mass (75 kDa). A 77-kDa protein was identified in a preliminary report by Warnock et al. [37] using an amiloride analog N-ethyl-N-nitromethoxy-benzyl-amiloride (NENMBA) in which a photolabile nitromethoxy-benzene group was coupled to the 5-amino group of amiloride. NENMBA was a potent reversible inhibitor of the rabbit renal brush border Na^+/H^+ exchanger. Following photolysis, NENMBA caused irreversible inhibition of transport, and MIA protected against inactivation. Antibodies to 5-(3-aminophenyl)amiloride detected NENMBA covalently bound to a 77-kDa brush border protein, and labeling was protected by MIA better than benzamil.

Taken together, although no covalent labeling studies have unambiguously identified the renal brush border Na^+/H^+ exchanger protein, a protein of 65 kDa was most consistently identified in several species by investigators using independent methods. Because the rank order of amiloride analogs for protection from labeling of the 65-kDa protein was never evaluated, this identification is not certain. Nevertheless, the 65-kDa protein remains a possible candidate for a resistant-type Na^+/H^+ exchanger. Alternatively, the 100-kDa and/or 107-kDa proteins identified by Igarashi and Aronson, and Wu and Lever are candidates for the resistant-type apical exchanger, but could also represent contamination with membranes containing the basolateral Na^+/H^+ exchanger of similar molecular mass (*vide infra*).

2.2.2. Affinity chromatography

Huot et al. [38] used affinity chromatography to identify and partially purify an amiloride-binding protein with characteristics of the renal brush border Na^+/H^+ exchanger. The high-affinity amiloride analog A35 (5-*N*-(3-aminophenyl)amiloride) was coupled to Sepharose CL-4B through a triglycine spacer. Rabbit renal brush border membranes were solubilized with 0.6% Triton X-100, incubated with the

A35 affinity matrix, and eluted with various media. A 25-kDa protein bound to the affinity matrix and was completely eluted with 5 mM free amiloride. The abundance of the 25-kDa protein in brush border and basolateral membranes correlated closely with Na^+/H^+ exchange activity. Importantly, binding of the 25-kDa protein to the affinity matrix was blocked by MIA > amiloride > benzamil, a rank order identical to that for inhibition of Na^+/H^+ exchange activity, which suggested strongly that the 25-kDa protein was a structural component of the transporter.

Subsequently, proteolytic fragments of the rabbit renal 25-kDa amiloride-binding protein were micro-sequenced and found to have high sequence homology with rat and human NAD(P)H:quinone oxidoreductase. Indeed, enzymatic assays revealed that renal brush border membrane vesicles contain significant NADPH:quinone oxidoreductase activity. Presumably NAD(P)H:quinone oxidoreductase coincidentally binds amiloride analogs with the same rank order as the Na^+/H^+ exchanger [39].

In a preliminary report, Ross et al. [40] used affinity chromatography to identify a putative bovine renal brush border Na^+/H^+ exchanger. Brush border membranes were solubilized with Triton X-100 and chromatographed sequentially over lentil lectin Sepharose 4B and 5-(N-benzyl-N-ethyl)amiloride coupled to epoxy-activated Sepharose 6B. The eluant contained 178- and 146-kDa proteins that were susceptible to Endo-F. Moreover, the eluants reacted on dot blot immunoassays with antisera to a 20-amino acid peptide of a human Na^+/H^+ exchanger (*vide infra*). The relationship between these proteins and the 66-kDa protein previously identified by the same investigators using amiloride photolabeling is presently unclear.

2.2.3. Other

Another biochemical approach which has been considered for purification of Na^+/H^+ exchangers involves solubilization of plasma membrane proteins, chromatographic fractionation and assay of fractions containing Na^+/H^+ exchanger protein by reconstitution of amiloride-binding or transport activity. To date, although several laboratories have described solubilization and functional reconstitution of transport or amiloride-binding activities, no studies have reported substantial purification of the transport protein. LaBelle [41], Weinman et al. [42], and Shibamoto et al. [43] described reconstitution assays that involved membrane solubilization with octyl glucoside followed by addition of artificial liposomes and detergent removal by dialysis. Reconstituted Na^+/H^+ exchange activity was measured by isotopic flux [41,42] or a pH-sensitive dye [43]. LaBelle reconstituted a Na^+/H^+ exchanger from rabbit renal medulla which appears to differ from the brush border Na^+/H^+ exchanger of renal cortex (e.g., inhibited by external K^+). Moreover, he was unable to demonstrate stimulation of transport activity by an outwardly directed H^+ gradient in reconstituted proteoliposomes. In contrast, the rabbit brush border Na^+/H^+ exchanger reconstituted by Weinman et al. [44] had kinetic properties of the native transporter including stimulation by an outwardly directed H^+ gradient, saturability, apparent K_m for Na^+_o of about 4 mM and similar rank order for inhibition by amiloride analogs (although the absolute values for K_i were higher than for the native transporter). Moreover, the reconstituted Na^+/H^+ exchanger retained the capacity for regulation by cAMP-dependent protein kinase [44].

Two laboratories have reported assays measuring the binding of amiloride analogs to solubilized renal brush border membranes [34,45]. Vigne et al. [45] found that the K_d for binding of 5-(N-ethyl-N-isopropyl)amiloride to solubilized brush border membranes (45 nM) was considerably lower than the concentration required to inhibit transport in native membranes. As a possible explanation, Desir et al. [46] observed that the internal surface of brush border membrane vesicles contains a high-affinity binding site for 5-amino-substituted amiloride analogs that is distinct from the transport site of the Na^+/H^+ exchanger. Wu and Lever [34] used a binding assay for MIA and the reconstitution assay of Weinman et al. [44] to follow the Na^+/H^+ exchanger during solubilization and fractionation by HPLC. Porcine brush border vesicles were solubilized with 4% octyl glucoside and chromatographed on a Mono-Q column. Na^+/H^+ exchanger was assayed in the fractions by reconstitution with sonicated asolectin and measurement of [^{3}H]MIA binding by a centrifugation assay. Compared to starting material (unsolubilized brush border membranes), solubilization caused a two-fold increase in specific activity, and fractionation by HPLC conferred an additional seven-fold purification (15% yield). Na^+/H^+ exchange activity paralleled MIA-binding during fractionation by HPLC. As already discussed, the fractions of peak activity contained proteins that could be photolabeled with MIA in an amiloride-protectable manner.

Béliveau et al. [47] employed radiation inactivation to estimate the size of the renal brush border Na^+/H^+ exchanger *in situ* without the need for purifying the protein. Rat renal brush border membranes were frozen at $-78°C$ and γ-irradiated at 2 Mrad/h. Under these conditions there was no apparent effect on the transmembrane H^+ gradient or amiloride-insensitive Na^+ flux. However, the amiloride-sensitive Na^+ influx representing Na^+/H^+ exchange was inactivated with a single exponential permitting an estimation of the molecular size at 321 kDa. Since the molecular size of the monomer in the same membranes may be 65 kDa (*vide supra*), the authors concluded that the functional Na^+/H^+ exchanger is a multimer. This result is in agreement with Otsu et al. [48] who measured pre-steady-state kinetics of the rabbit renal Na^+/H^+ exchanger and found that at least two binding sites for Na^+ must be occupied to activate Na^+ uptake in the pre-steady state. This too was evidence that the Na^+/H^+ exchanger is oligomeric.

Whereas the above studies have attempted to identify the Na^+/H^+ exchanger in renal brush border membranes (a resistant-type), at least one study has reported possible identification of a sensitive-type transport protein [49]. The Na^+/H^+ exchanger in lymphocytes (a sensitive-type) can be activated by either 12-O-tetradecanoylphorbol 13-acetate (TPA) or osmotic shrinkage. TPA or osmotic shrinkage were each associated with phosphorylation of multiple lymphocyte membrane proteins. However, only a 60-kDa protein was significantly phosphorylated by both interventions. It was unclear whether the phosphorylated 60-kDa protein was directly involved in Na^+/H^+ exchange or was a regulatory protein. Since a human sensitive-type Na^+/H^+ exchanger protein has been identified as a 110-kDa protein (*vide infra*), the latter might seem more likely. However, it is also possible that different sensitive-type isoforms exist, one of which is a 60-kDa protein in lymphocytes.

A putative regulatory cofactor has been identified for the renal brush border Na^+/H^+ exchanger (resistant-type) [50]. Morell et al. [50] identified a 42-kDa protein that was distinct from the transporter itself and appeared to be involved in regulation by cAMP-dependent protein kinase (PKA). Evidence supporting this conclusion was:

(1) limited proteolysis of brush border membranes with trypsin eliminated regulation by PKA but not Na^+/H^+ exchange itself;

(2) an anionic fraction of solubilized brush border membranes contained reconstitutable Na^+/H^+ exchanger that was not regulated by PKA;

(3) co-reconstitution of trypsinized proteins with membrane fractions containing the 42-kDa protein restored the inhibitory effect of PKA; and

(4) the 42-kDa protein in unsolubilized brush border membranes was preferentially phosphorylated by PKA.

3. Molecular cloning of Na^+/H^+ exchangers

3.1. cDNA cloning and primary structure

Recently, recombinant DNA technology has enabled the inference of the primary structure of at least one type of plasma membrane Na^+/H^+ exchanger. Since biochemical studies failed to produce partial amino acid sequence of protein(s) that mediate plasma membrane Na^+/H^+ exchange and no antibodies were available for screening cDNA libraries, Pouysségur et al. [53], devised a novel strategy encompassing functional expression and genetic complementation for cloning a partial-length cDNA encoding a human plasma membrane Na^+/H^+ exchanger. Their approach involved the following steps:

(1) generation of mutant somatic cells deficient in plasma membrane Na^+/H^+ exchange activity;

(2) transfection of these cells with genomic DNA from another species to restore transport activity;

(3) identification of an exon-containing subgenomic fragment of the transfected DNA; and

(4) screening of a cDNA library with the subgenomic fragment.

To test whether the resultant cDNA encoded a plasma membrane Na^+/H^+ exchanger, it was re-introduced into a Na^+/H^+ exchanger-deficient cell and restoration of transport activity was evaluated. Because this approach was based on following Na^+/H^+ exchanger transport activity itself (rather than a possibly less specific measure such as amiloride binding), it is clear that the final cDNA encodes a protein that is involved in Na^+/H^+ exchange. Although it was theoretically possible that cDNAs encoded a regulatory protein rather than the transport protein itself, as will be seen below, the expected gene product does have characteristics of an integral membrane transport protein.

3.1.1. Human Na^+/H^+ exchanger cDNA

Mouse L-cells were rendered deficient in plasma membrane Na^+/H^+ exchange activity by mutagenesis and selection with a proton suicide technique. In this clever technique invented by Pouysségur et al. [51], the Na^+/H^+ exchanger is forced to operate as an acid loader in the reverse of its usual physiological mode. Under these conditions, cells expressing a Na^+/H^+ exchanger are killed, but deficient mutants survive. The Na^+/H^+ exchanger deficient mutants (called LAP1 cells) were then transfected with total human genomic DNA. Cells that acquired the human Na^+/H^+ exchanger gene (named NHE-1) were selected for by survival in acidified, HCO_3^--free medium (conditions that kill Na^+/H^+ exchanger-deficient cells). That this method selects transformants which acquire the exogenous Na^+/H^+ exchanger gene rather than restoration of an endogenous gene was demonstrated by transfection with DNA from a cell expressing a Na^+/H^+ exchanger with distinguishable kinetic properties [52].

After three sequential cycles of transformation and selection, murine L-cells which retained Na^+/H^+ exchanger activity maintained a discrete set of six *Eco*RI fragments of human genomic DNA encompassing at least 55 kilobases and presumably containing the Na^+/H^+ exchanger structural gene. A genomic library was constructed from the tertiary murine transformant, and clones containing human DNA were identified by hybridization to human repetitive DNA. In order to localize the Na^+/H^+ exchanger gene within the cloned human DNA, subgenomic restriction fragments were screened for characteristics of exon-coding sequences. An 800-bp *Pst*I fragment was selected on the basis of hybridization to single bands on Southern blots of genomic DNA from many species (indicating that it contained a single-copy evolutionarily conserved sequence perhaps encoding a Na^+/H^+ exchanger which would also be expected to be present in many species). Northern blot analysis revealed that the 800-bp *Pst*I fragment contained expressed sequences. Moreover, transcripts hybridizing to the 800-bp fragment were amplified in a mutant mouse cell overexpressing human Na^+/H^+ exchanger transport activity (ST31A cells) and no transcripts hybridized in the Na^+/H^+ exchanger-deficient LAP1 cells. Thus, the *Pst*I fragment was very likely to contain exons of the human Na^+/H^+ exchanger gene. A cDNA library constructed from a derivative of ST31A was screened with the 800-bp PstI fragment and a 3977-bp cDNA clone was isolated [53]. The 3977-bp human cDNA was expressed by transfection into mutant fibroblasts that were rendered deficient in the Na^+/H^+ exchanger. Restoration of transport activity indicated that the cDNA was very likely to encode the transport protein. When functionally expressed in recombinant baculovirus-infected cells, the human Na^+/H^+ exchanger was half-maximally inhibited by 10 nM methylpropylamiloride indicating that it represented a sensitive-type exchanger [54].

The human NHE-1 cDNA contains 407 bp of 5' untranslated region, an open reading frame of 2445 bp, and 1125 bp of 3' untranslated region terminating in a mouse B1 repeat (the latter is presumably an artifact introduced during cloning, and the actual 3' ends of human cDNAs from placenta, kidney, and breast adenocarcinoma have been recently cloned by Takaichi et al. [55,56]). The human Na^+/H^+ exchanger cDNA is predicted to encode a 91-kDa protein composed of two distinct domains, an amino-terminal hydrophobic domain (499 amino acids) consisting of 10 putative transmembrane segments and a carboxyl-terminal hydrophilic domain (316 amino acids) that is presumably cytoplasmic (see Fig. 1). This overall organization is reminiscent of other integral membrane transport proteins, viz., erythroid band 3 anion exchanger [16]. The predicted amino acid sequence is not similar to any previously described eukaryotic protein, including other Na^+-dependent transporters such as the Na^+/glucose cotransporters, nor is there similarity to the prokaryotic Na^+/H^+ exchanger which has different kinetic properties [57].

In order to identify the human Na^+/H^+ exchanger gene product, Sardet et al.

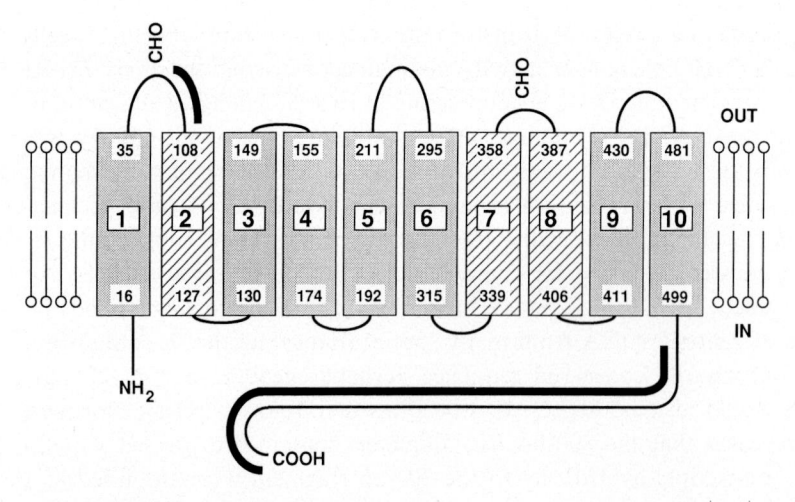

Fig. 1. Hypothetical secondary structure of a human plasma membrane Na^+/H^+ exchanger. (Adapted from Sardet et al. [53].) Shaded bars, putative transmembrane segments. Hatched bars, putative amphipathic helices (numbers at tops and bottoms of bars refer to positions of amino acids). CHO, possible site of N-linked glycosylation. Solid bars, regions of the porcine renal Na^+/H^+ exchanger used for immunolocalization in LLC-PK$_1$ cells.

[58] generated antisera to a fusion protein containing 157 amino acids from the carboxyl terminus of the predicted human Na^+/H^+ exchanger protein. These antisera recognized the human Na^+/H^+ exchanger protein expressed in Chinese hamster lung fibroblasts. Importantly, the antisera also recognized the endogenous Na^+/H^+ exchanger in various Chinese hamster lung fibroblast cell lines having differing amounts of Na^+/H^+ exchanger activity, and the abundance of immunoreactive protein paralleled transport activity, further indicating that the cloned cDNA encoded the transport protein itself. The mature human plasma membrane Na^+/H^+ exchanger had a molecular mass of 110 kDa, and was glycosylated and phosphorylated. By immunofluorescence, the gene product localized to the plasma membrane of a non-polarized cell, and the region of the hydrophilic carboxyl-terminal domain was probably cytoplasmic since permeabilization of the cells was required to obtain signal.

Plasma membrane Na^+/H^+ exchangers may be acutely activated by mitogens including thrombin, epidermal growth factor, platelet-derived growth factor, insulin, and phorbol esters. Kinetically, this stimulation reflects an increase in affinity at the internal proton modifier site [7] which may be due to protein phosphorylation. Sardet et al. [58] observed that the time course of activation of the human Na^+/H^+ exchanger by epidermal growth factor and α-thrombin paralleled the time course of phosphorylation of the 110-kDa protein, and phosphorylation occurred exclusively on serine residues [58]. Moreover, activation of the Na^+/H^+ exchanger by α-thrombin, epidermal growth factor, and okadaic acid induced phosphorylation of identical tryptic peptides indicating convergence of the tyrosine kinase and protein kinase C signal transducing pathways, perhaps via a common Na^+/H^+ exchanger kinase [59].

Recently, Pouysségur et al. [60] have reported in preliminary form mutagenesis of the cloned cDNAs to further explore the function of specific domains of the encoded protein. Mutated cDNAs were expressed in Na^+/H^+ exchanger-deficient cells and the effects on transport activity and regulation were evaluated. Proteins lacking the entire cytoplasmic domain continued to perform Na^+/H^+ exchange. However, the apparent affinity of the internal proton modifier site was altered, and activation by growth factors was abolished. In this regard, the cytoplasmic domain contains several consensus sequences for phosphorylation by protein kinase C, cAMP-dependent protein kinase, and calcium/calmodulin-dependent protein kinase II [61]. The latter may be significant since Burns et al. [62] found that stimulation of Na^+/H^+ exchange by acute intracellular acidification was ATP-dependent and inhibited by the calmodulin antagonist, W-7.

3.1.2. Other species
Because the human Na^+/H^+ exchanger cDNA was obtained from genomic DNA expressed in a murine transformant, the human tissue of origin was not known. Accordingly, several laboratories attempted to clone cDNAs corresponding to the

human Na^+/H^+ exchanger from a number of mammalian tissues. Different species were also selected for study since it was also hoped that regions which were highly conserved between species might also be functionally important. In rabbit kidney, rabbit ileum, and porcine renal epithelial cells (LLC-PK_1/Clone 4) a single transcript of about five kilobases hybridized to the human Na^+/H^+ exchanger cDNA. Corresponding cDNAs from these three sources were cloned using library screening and the polymerase chain reaction [63–65]. The composite sequence of the rabbit renal Na^+/H^+ exchanger cDNA is 3352 bp and consists of 726 bp of 5' untranslated region, an open reading frame of 2448 bp, and 178 bp of 3' untranslated region. The nucleotide sequence in each of these regions is highly similar to the human sequence (75, 92 and 83% identity). The cDNA is predicted to encode a protein composed of 816 amino acids with a molecular mass of 90 847 Da which is also highly similar to the human protein (95 and 96% amino acid identity in the membrane-associated and cytoplasmic domains). From LLC-PK_1/Clone 4 cells, Reilly et al. [65] obtained cDNAs containing 226 bp of 5' untranslated region, a 2454-bp open reading frame, and 208 bp of 3' untranslated region. This nucleotide sequence is highly similar to human (75, 92 and 85% nucleotide identity). The predicted 818-amino acid protein is 95% identical to the human protein.

Thus, partial-length cDNAs have been obtained containing the entire coding regions of rabbit and porcine renal Na^+/H^+ exchangers. The similarities of the inferred primary structures to the sensitive-type human form implies that the encoded proteins mediate Na^+/H^+ exchange and are highly evolutionarily conserved. Hydropathy plots suggest that the overall architecture is maintained and consists of an amino-terminal membrane-associated domain and a carboxyl-terminal hydrophilic domain. The high degree of identity in amino acid sequences limits possible correlations between structure and function. It is not possible, e.g., to identify histidyl or carboxyl groups that are present at the external transport site as suggested by previous kinetic studies. Rather, only general conclusions are possible. Two highly hydrophilic segments containing possible N-linked glycosylation sites (between helices 1 and 2 and between helices 7 and 8) are conserved in sequences from the three species. This permits tentative identification of these sequences as exoplasmic loops, a testable hypothesis. This finding is also consistent with the observation of Yusufi et al. [31] that at least one form of the Na^+/H^+ exchanger is sensitive to deglycosylation. Indeed, Sardet et al. [58] observed that treatment of cells with endoglycosidase shifted the electrophoretic mobility of the human Na^+/H^+ exchanger protein.

Certain transmembrane segments are predicted to form amphipathic helices and, therefore, might line an aqueous pore that conducts Na^+ or H^+ (putative transmembrane helices 2, 7, and 8). Using molecular modeling, Warnock and coworkers [66] hypothesized that three amino acids in helices 2 and 3 (His-120, Lys-116 and Glu-131) that are conserved in human, rabbit, and pig Na^+/H^+ exchangers might comprise a charge relay mechanism for proton translocation similar to that described in *lac* permease [15]. In a preliminary study, these workers found that mutating His-120 or His-349 to glycine did not affect transport activity, although mutation of His-349 did reduce affinity for amiloride [67].

Partial-length cDNAs homologous to the human Na^+/H^+ exchanger have also been cloned from rat kidney [68] and rabbit heart [69]. The 890-bp rat renal cDNA (corresponding to nucleotides 1210 to 2100 of the human sequence) encodes three putative membrane-spanning segments and part of the cytoplasmic domain. Overall, the nucleotide and amino acid sequences are greater than 90% identical to the human sequence. The 2235-bp partial-length cDNA from rabbit heart overlaps 1213 bp with the 3' end of the rabbit kidney cDNA and contains sequences corresponding principally to the cytoplasmic domain and 3' untranslated region.

3.2. Tissue and membrane localization

To confirm that the cloned rabbit renal Na^+/H^+ exchanger cDNAs encoded an amiloride-sensitive isoform that is distributed in many cells rather than an apical isoform restricted to epithelia, Hildebrandt et al. [63] studied the tissue distribution of the transcripts. Northern blot analysis was performed to determine which rabbit tissues expressed the Na^+/H^+ exchanger transcript cloned from rabbit kidney. Highest levels of expression were in stomach followed by brain, renal medulla, lung, ileum and renal cortex. Liver and skeletal muscle expressed minimal levels of the transcript. These results indicate that expression of the cloned Na^+/H^+ exchanger is tissue-specific. Moreover, the tissue distribution is consistent with the amiloride-sensitive Na^+/H^+ exchanger activity that is expressed in both epithelial and non-epithelial cells, and would not be expected for the relatively amiloride-resistant transporter activity that is limited to the apical membrane of epithelia in kidney and ileum.

Next, Reilly et al. [65] localized the Na^+/H^+ exchanger gene product in renal epithelial cells where the distributions of the kinetic isoforms was well-established. The strategy was based on the observation that the resistant- and sensitive-types are restricted to the apical and basolateral membranes, respectively, in confluent LLC-PK$_1$/Clone 4 cells [8]. Thus, if proteins encoded by the cloned cDNAs localized to the apical membrane this would indicate that they represent the resistant-type. Localization to the basolateral membrane would prove they were the sensitive-type and presence on both membranes would suggest that the two functional isoforms had identical primary structures. Na^+/H^+ exchanger proteins were localized by indirect immunofluorescence using antisera generated against a synthetic oligopeptide and a fusion protein containing most of the carboxyl-terminal hydrophilic domain (see Fig. 1). The oligopeptide was an evolutionarily conserved 17-amino acid sequence (Arg-85–Pro-101) derived from the first putative exoplasmic loop near the amino terminus of the cloned LLC-PK$_1$ Na^+/H^+ exchanger. The carboxyl-terminal domain (amino acids 514–818) was expressed as a fusion with bacterial maltose-binding protein. By indirect immunofluorescence, antisera to the oligopeptide labeled exclusively the basolateral membrane of confluent LLC-PK$_1$/Clone 4 cells. There was no labeling of the apical membrane. Labeling of the basolateral membrane was specific since it could be blocked with an excess of free peptide. Basolateral localization was confirmed by co-labeling with a monoclonal antibody against the A1-subunit of Na$^+$,K$^+$-ATPase, which is known to be restricted to the basolateral membrane in these cells.

This basolateral immunolocalization of the Na$^+$/H$^+$ exchanger was confirmed using antisera to the carboxyl-terminal domain. Affinity-purified antisera to the carboxyl-terminal domain recognized a 110-kDa protein on immunoblots of LLC-PK$_1$ cell membranes, which is in agreement with the size reported by Sardet et al. [58] for the human Na$^+$/H$^+$ exchanger. By indirect immunofluorescence, these antisera also localized solely to the basolateral membrane and not the apical membrane of confluent LLC-PK$_1$/Clone 4 cells. These experiments, using antisera to two different portions of the protein, clearly indicate that the cloned cDNAs encode a protein that is restricted to the basolateral membrane of LLC-PK$_1$ cells and must, therefore, encode a sensitive-type Na$^+$/H$^+$ exchanger. Moreover, these studies indicate that the apical resistant-type exchanger must be a different protein (*vide infra*). Tse et al. [64] examined the expression of the Na$^+$/H$^+$ exchanger in ileum and found similarly that the protein immunolocalized to the basolateral membrane of villus enterocytes where the relatively amiloride-sensitive type is located. There was no labeling of the apical membrane where the resistant-type Na$^+$/H$^+$ exchanger form is expressed.

In order to immunolocalize the Na$^+$/H$^+$ exchanger in other epithelial tissues, Biemesderfer et al. [70] raised antisera against the carboxyl-terminal 41 amino acids of the rabbit renal Na$^+$/H$^+$ exchanger (expressed as a fusion with maltose-binding protein). Indirect immunofluorescence was performed on semi-thin cryosections of kidney, stomach, ileum and colon. The affinity-purified antisera labeled the basolateral membrane of surface epithelial cells in stomach, ileum and colon. In kidney, the antisera labeled the basolateral membranes of the juxtamedullary proximal tubule, the thick ascending limb of Henle, and the collecting tubule. Importantly, no labeling of the apical membrane was observed in any tissue. Although the amiloride-sensitivity of the Na$^+$/H$^+$ exchanger in some of these epithelia has not been well studied, taken together, these results indicate that the cloned Na$^+$/H$^+$ exchanger is a sensitive-type isoform expressed *in vivo* in the basolateral membrane of polarized epithelia in ileum, stomach, colon and kidney.

In contrast to these results, Ross et al. [33] found that antisera against a 20-amino acid peptide (Ser-613–Arg-632) of the cytoplasmic domain of the human Na$^+$/H$^+$ exchanger recognized a 66-kDa protein in immunoblots of bovine renal brush border membranes. Since the purity of these membranes was not reported it is possible that this result was due to contamination with basolateral membranes (although the molecular mass would still differ from the basolateral Na$^+$/H$^+$ exchanger in LLC-

PK$_1$ cells). Alternatively there might be conservation of a micro-domain in the apical and basolateral Na$^+$/H$^+$ exchangers that was not detected with antisera to the whole cytoplasmic domain. Fliegel et al. [69] generated antisera to a 10-amino acid peptide (Arg-47–Ser-56) from near the amino terminus of the human Na$^+$/H$^+$ exchanger that recognized proteins from bovine and canine sarcolemmal vesicles of 50 and 70 kDa which also differ from the 110-kDa size reported for the human and LLC-PK$_1$ Na$^+$/H$^+$ exchangers. Whether Na$^+$/H$^+$ exchanger gene products are different in heart is unknown; as mentioned previously, there were apparent sequence differences between cDNAs from rabbit heart and kidney. It is important to note, however, that bovine and canine cDNAs corresponding to the region of the oligopeptide have not been cloned; the sequence of the oligopeptide is divergent in human and rabbit; and blocking experiments using an excess of oligopeptide were not reported.

3.3. Isoforms

Since the protein that mediates Na$^+$/H$^+$ exchange in the apical membrane of epithelia (resistant-type) has not been definitively identified, the structural features responsible for differing sensitivities to amiloride analogs remain unknown. As mentioned above, the sensitive- and resistant-types of Na$^+$/H$^+$ exchangers also differ in kinetic behavior, sensitivity to sulfhydryl modification and regulation by protein kinase C suggesting that they may be structurally different. Indeed, Haggerty et al. [71] have produced a mutant LLC-PK$_1$ cell line (PKE20) that overexpresses resistant-type Na$^+$/H$^+$ exchange with no change in sensitive activity. Another mutant (PKE6) is solely deficient in sensitive activity and has normal levels of the resistant-type (Slayman, personal communication). These results suggest that the two types of Na$^+$/H$^+$ exchangers are under separate genetic control. The immunolocalization results of Reilly et al. [65] further clarify the basis for the difference between the two types of exchangers. Antisera against the oligopeptide and fusion protein recognized the basolateral, amiloride-sensitive Na$^+$/H$^+$ exchanger. Moreover, they did not detect the apical, amiloride-resistant Na$^+$/H$^+$ exchanger. This indicates that the primary structures of the two exchangers must differ and that functional differences between the two are not due simply to variations in glycosylation, phosphorylation, or lipid environment. Rather, the resistant-type is probably encoded by a different Na$^+$/H$^+$ exchanger gene.

Although the primary structures of the sensitive- and resistant-types differ, there are likely to be some structural homologies between the two transporters which are similar in many functional respects. Accordingly, several laboratories have attempted to isolate cDNAs encoding homologs of the sensitive-type which might represent a resistant-type exchanger. Tse et al. [72] used low stringency library screening to obtain a 3.9-kilobase cDNA from rabbit ileum. The cDNA is predicted to encode an 809-amino acid protein with a molecular mass of 90 787. The hydropathy profile of the encoded protein (called NHE-2) indicates that the secondary structure is similar to NHE-1 consisting of ten putative membrane spanning segments and a large hydrophilic carboxyl-terminus. Compared with NHE-1, the overall amino acid identity is 50% (60% in the membrane-associated domain and 34% in the carboxyl-terminal domain). Putative sites of N-linked glycosylation and phosphorylation were identified. By Northern blot analysis, transcripts were expressed in rabbit ileum, kidney and adrenal. The cDNA was stably expressed in a Na^+/H^+ exchanger-deficient fibroblast (PS120) and mediated Na^+/H^+ exchange transport activity. Moreover, NHE-2 was more resistant to EIPA inhibition than NHE-1.

Recently, Tse et al. [73] and Orlowski et al. [74] have cloned a third isoform of Na^+/H^+ exchanger (named NHE-3). The inferred 832-amino acid sequence of rabbit NHE-3 is 41% identical with NHE-1, 44% identical with NHE-2, and has a similar secondary structure. In contrast to NHE-1 and NHE-2, NHE-3 is only expressed in epithelia in intestine and kidney. Moreover, administration of glucocorticoids, which stimulates transport activity of the apical Na^+/H^+ exchanger in rabbit intestine, increased levels of NHE-3 transcripts but did not affect NHE-1 or NHE-2 [75]. Taken together, these results suggest that NHE-3 may encode a resistant-type Na^+/H^+ exchanger of epithelia. A fourth Na^+/H^+ exchanger isoform (NHE-4) is preferentially expressed in stomach [74].

3.4. Genomic cloning

The rationale for cloning the Na^+/H^+ exchanger gene (*NHE-1*) was the following:

(1) Northern blot analysis revealed that expression of NHE-1 is tissue-specific with highest transcript abundance in stomach and minimal levels in liver and skeletal muscle. Moreover, there is physiological evidence that certain segments of the nephron lack a basolateral Na^+/H^+ exchanger [76]. Using RT-PCR, Krapf and Solioz [77] observed that NHE-1 transcripts are not detectable in S1 and S2 segments of superficial nephrons.

(2) Abundance of Na^+/H^+ exchanger transcripts in vascular smooth muscle and HL60 cells is increased by serum, phorbol esters, fibroblast growth factor and platelet-derived growth factor [78,79]. The increase in NHE-1 mRNA levels in PMA-treated HL60 cells is due to increased gene transcription.

(3) Stimulation of basolateral Na^+/H^+ exchanger transport activity in LLC-PK$_1$ cells during chronic metabolic acidosis is accompanied by a parallel increase in NHE-1 transcript abundance [80]. Chronic metabolic acidosis also increases the abundance of Na^+/H^+ exchanger transcripts in rat renal cortices [81] and SV40-transformed mouse proximal tubule cells [82].

It is likely that transcriptional regulation of Na^+/H^+ exchanger gene expression is responsible for some of these observations, a possibility that could be first tested by examining the structure and function of the gene promoter.

The structural gene encoding NHE-1 has been cloned in human [83] and rabbit

[84]. The human *NHE-1* gene spans 70 kilobases and the coding region is divided into 12 exons and 11 introns. Interestingly, the junctions between exons and introns do not appear to correspond to obvious functional domains of the protein, e.g., boundaries between transmembrane segments, as they do for some membrane transport proteins. Two discrete transcription initiation sites (786 and 782 bp upstream of the initiation codon) were mapped in mouse fibroblasts expressing a genomic clone of the human Na^+/H^+ exchanger (ST31A cells). The promoter/regulatory region of the human gene contains a TATA box, three GC boxes, three AP-1 sites and four half-sites for glucocorticoid receptors. The presence of a putative AP-1 site (binding site for *fos-jun* heterodimers) is particularly interesting since Rao et al. [79] showed that NHE-1 transcript abundance was increased by phorbol esters and Moe et al. [82] have hypothesized that the response to metabolic acidosis is dependent on protein kinase C activity. When expressed in NIH 3T3 fibroblasts, 1.4 kilobases of 5' flanking sequence of the human *NHE-1* gene displayed promoter activity. An influence of metabolic acidosis or growth factors on promoter activity has not yet been demonstrated. The human *NHE-1* gene has been localized to the 1p35→p36.1 region of chromosome 1 by *in situ* hybridization [85]. The gene has not yet been linked to essential hypertension [86].

In rabbit, 17 kilobases of a Na^+/H^+ exchanger gene has been isolated [84]. This clone contains six kilobases of 5' flanking sequence, the first exon (1151 bp including 351 bp of coding sequence), and a portion of the first intervening sequence. The principal transcription initiation site is located 800 bp upstream from the initiation codon. The promoter/regulatory region is (G+C)-rich, contains a TATA box, nine CACCC boxes, a single potential AP-1 site, pyrimidine-rich direct repeats, and four half-sites for steroid hormone receptors. When compared with the human sequence, the rabbit gene is similar in overall organization, location of transcription initiation sites, and DNA sequence including conservation of several potential regulatory elements.

4. Summary and future directions

Within the past five years much has been learned about the structures of plasma membrane Na^+/H^+ exchangers. Group-specific modification studies have suggested that certain amino acid side groups are important in the mechanism of transport, especially for the apical membrane isoform in kidney. Molecular cloning of cDNAs in several species has permitted inference of the primary structure of the NHE-1 amiloride-sensitive isoform, and indirectly led to immunolocalization of the gene products. It should now be possible to take the next steps of testing hypotheses about the importance of specific domains and amino acids in the mechanism of transport. The study of isoforms of NHE-1 will continue, with a particular priority of determining the membrane localizations of NHE-2, NHE-3 and NHE-4. With expression of full-length cDNAs in the appropriate system we should be capable of producing the protein in sufficient quantities to directly correlate structure and function, and elucidate the secondary and possibly the tertiary structures of this essential membrane transport protein. Regarding the Na^+/H^+ exchanger gene (*NHE-1*), attention will now focus on identifying specific regions of the gene responsible for the effects of growth factors and metabolic acidosis and delineation of the signalling pathways involved.

Acknowledgements

Work presented from the author's laboratory was supported by United States Public Health Service Grants (DK-33793, DK-42921 and AM-01423). The author was also the recipient of a New Investigator Award from the American Heart Association Connecticut Affiliate, Inc. The author is grateful to Peter S. Aronson for his mentorship and critical review of the manuscript.

References

1. Seifter, J.L. and Aronson, P.S. (1986) J. Clin. Invest. 78, 859–864.
2. Mahnensmith, R.L. and Aronson, P.S. (1985) Circ. Res. 56, 773–788.
3. Montrose, M.H. and Murer, H. (1988) In: Na^+/H^+ Exchange (Grinstein, S. Ed.), pp. 57–76, CRC Press, Boca Raton.
4. Aronson, P.S. and Igarashi, P. (1986) Curr. Top. Membr. Transp. 26, 57–75.
5. Aronson, P.S. (1985) Annu. Rev. Physiol. 47, 545–560.
6. Grinstein, S., Rotin, D. and Mason, M.J. (1989) Biochim. Biophys. Acta 988, 73–97.
7. Grinstein, S. and Rothstein, A. (1986) J. Membr. Biol. 90, 1–12.
8. Haggerty, J.G., Agarwal, N., Reilly, R.F., Adelberg, E.A. and Slayman, C.W. (1988) Proc. Natl. Acad. Sci. U.S.A. 85, 6797–6801.
9. Knickelbein, R.G., Aronson, P.S. and Dobbins, J.W. (1990) Am. J. Physiol. 259, G802–G806.
10. Kulanthaivel, P., Leibach, F.H., Mahesh, V.B., Cragoe, E.J., Jr. and Ganapathy V. (1990) J. Biol. Chem. 265, 1249–1252.
11. Kulanthaivel, P., Leibach, F.H., Mahesh, V.B., Smith, C.H., Furesz, T.C. and Ganapathy, V. (1991) FASEB J. 5, A759.
12. Casavola, V., Helmle-Kolb, C. and Murer, H. (1989) Biochem. Biophys. Res. Commun. 165, 833–837.
13. Haggerty, J.G., Reilly, R.F., Adelberg, E.A. and Slayman, C.W. (1988) J. Cell. Biol. 107, 785a.
14. Raley-Susman, K.M., Cragoe, E.J., Jr., Sapolsky, R.M. and Kopito, R.R. (1991) J. Biol. Chem. 266, 2739–2745.
15. Kaback, H.R. (1986) Annu. Rev. Biophys. Biophys. Chem. 15, 279–319.
16. Jennings, M.L. (1989) Annu. Rev. Biophys. Biophys. Chem. 18, 397–430.
17. Grillo, F.G. and Aronson, P.S. (1986) J. Biol. Chem. 261, 1120–1125.
18. Ganapathy, V., Balkovetz, D.F., Ganapathy, M.E., Mahesh, V.B., Devoe, L.D. and Leibach, F.H. (1987) Biochem. J. 245, 473–477.
19. Grinstein, S., Cohen, S. and Rothstein, A. (1985) Biochim. Biophys. Acta 812, 213–222.
20. Burnham, C., Munzesheimer, C., Rabon, E. and Sachs, G. (1982) Biochim. Biophys. Acta 685, 260–

272.
21. Rocco, V.K., Cragoe, E.J., Jr. and Warnock, D.G. (1987) Am. J. Physiol. 252, F517–F524.
22. Igarashi, P. and Aronson, P.S. (1987) J. Biol. Chem. 262, 860–868.
23. Friedrich, T., Sablotni, J. and Burckhardt, G. (1986) J. Membr. Biol. 94, 253–266.
24. Kinsella, J.L., Wehrle, J., Wilkins, N. and Sacktor, B. (1987) J. Biol. Chem. 262, 7092–7097.
25. Sokol, P.P., Holohan, P.D. and Ross, C.R. (1986) J. Biol. Chem. 261, 3282–3287.
26. Miyamoto, Y., Tiruppathi, C., Ganapathy, V. and Leibach, F.H. (1989) Biochim. Biophys. Acta 978, 25–31.
27. Parker, J.C. and Glosson, P.S. (1987) Am. J. Physiol. 253, C60–C65.
28. Kulanthaivel, P., Simon, B.J., Leibach, F.H., Mahesh, V.B. and Ganapathy, V. (1990) Biochim. Biophys. Acta 1024, 385–389.
29. Parker, J.C. (1984) J. Gen. Physiol. 84, 789–803.
30. Huang, Z.-Q. and Warnock, D.G. (1988) Kidney Int. 33, 401.
31. Yusufi, A.N.K., Szczepanska-Konkel, M. and Dousa, T.P. (1988) J. Biol. Chem. 263, 13683–13691.
32. Friedrich, T., Sablotni, J. and Burckhardt, G. (1987) Biochem. Biophys. Res. Commun. 144, 869–875.
33. Ross, W., Bertrand, W. and Morrison, A. (1990) J. Biol. Chem. 265, 5341–5344.
34. Wu, J.-S.R. and Lever, J.E. (1989) Biochemistry 28, 2980–2984.
35. Friedrich, T. and Burckhardt, G. (1988) Biochem. Biophys. Res. Comm. 157, 921–929.
36. Igarashi, P., Cragoe, E.J., Jr. and Aronson, P.S. (1987) Xth ISN Congress Satellite Symposium on Structure, Function and Regulation of Membrane Transport Proteins, Furigen, p. 86.
37. Warnock, D.G., Cragoe, E.J., Jr., Rossier, B. and Kleyman, T.R. (1989) Kidney Int. 35, 465.
38. Huot, S.J., Cassel, D., Igarashi, P., Cragoe, E.J., Jr., Slayman, C.W. and Aronson, P.S. (1989) J. Biol. Chem. 264, 683–686.
39. Huot, S.J., Slayman, C.W. and Aronson, P.S. (1990) J. Am. Soc. Nephrol. 1, 700.
40. Ross, W.R., Bertrand, W. and Morrison, A.R. (1990) Kidney Int. 37, 233.
41. LaBelle, E.F. (1984) Biochim. Biophys. Acta 770, 79–92.
42. Weinman, E.J., Shenolikar, S., Cragoe, E.J., Jr. and Dubinsky, W.P. (1988) J. Membr. Biol. 101, 1–9.
43. Shibamoto, S., Yoshida, S., Oku, N., Hayakawa, M., Hori, T. and Ito, F. (1988) J. Pharmacobiol. Dyn. 11, 669–673.
44. Weinman, E.J., Dubinsky, W.P. and Shenolikar, S. (1988) J. Membr. Biol. 101, 11–18.
45. Vigne, P., Frelin, C., Audinot, M., Borsotto, M., Cragoe, E.J., Jr. and Lazdunski, M. (1985) J. Biol. Chem. 260, 14120–14125.
46. Desir, G.V., Cragoe, E.J., Jr. and Aronson, P.S. (1991) J. Biol. Chem. 266, 2267–2271.
47. Béliveau, R., Demeule, M. and Potier, M. (1988) Biochim. Biophys. Res. Commun. 152, 484–489.
48. Otsu, K., Kinsella J., Sacktor, B. and Froehlich, J.P. (1989) Proc. Natl. Acad. Sci. U.S.A. 86, 4818–4822.
49. Grinstein, S., Goetz-Smith, J.D., Stewart, D., Beresford, B.J. and Mellors, A. (1986) J. Biol. Chem. 261, 8009–8016.
50. Morell, G., Steplock, D., Shenolikar, S. and Weinman, E.J. (1990) Am. J. Physiol. 259, F867–F871.
51. Pouysségur, J., Sardet, C., Franchi, A., L'Allemain, G. and Paris, S. (1984) Proc. Natl. Acad. Sci. U.S.A. 81, 4833–4837.
52. Franchi, A., Perucca-Lostanlen, D. and Pouysségur, J. (1986) Proc. Natl. Acad. Sci. U.S.A. 83, 9388–9392.
53. Sardet, C., Franchi, A. and Pouysségur, J. (1989) Cell 56, 271–280.
54. Fafournoux, P., Ghysdael, J., Sardet, C. and Pouysségur, J. (1991) Biochemistry 30, 9510–9515.
55. Takaichi, K., Wang, D., Chen, F. and Warnock, D.G. (1990) J. Am. Soc. Nephrol. 1, 745.
56. Takaichi, K., Wang, D., Balkovetz, D.F., Huang, Z.Q., Cragoe, E.J., Jr., Sardet, C., Pouysségur, J. and Warnock, D.G. (1990) XIth International Congress of Nephrology, Tokyo, p. 473A.
57. Karpel, R., Olami, Y., Taglicht, D., Schuldiner, S. and Padan, E. (1988) J. Biol. Chem. 263, 10408–10414.

58. Sardet, C., Counillon, L., Franchi, A. and Pouysségur, J. (1990) Science 247, 723–726.
59. Sardet, C., Fafournoux, P. and Pouysségur, J. (1991) J. Biol. Chem. 266, 19166–19171.
60. Pouysségur, J., Counillon, L., Fasournoux, P., Pages, G., Sardet, C. and Wakabayashi, S. (1991) AASLD Single Topit Conference, New York.
61. Kennelly, P.J. and Krebs, E.G. (1991) J. Biol. Chem. 266, 15555–15558.
62. Burns, K.D., Homma, T. and Harris, R.C. (1991) Am. J. Physiol. 261, F607–F616.
63. Hildebrandt, F., Pizzonia, J.H., Reilly, R.F., Rebouças, N.A., Sardet, C., Pouysségur, J., Slayman, C.W., Aronson, P.S. and Igarashi, P. (1991) Biochim. Biophys. Acta 1129, 105–108.
64. Tse, C.M., Ma, A.I., Yang, V.W., Watson, A.J.M., Levine, S., Montrose, M.H., Potter, J., Sardet, C., Pouysségur, J. and Donowitz, M. (1991) EMBO J. 10, 1957–1967.
65. Reilly, R.F., Hildebrandt, F., Biemesderfer, D., Sardet, C., Pouysségur J., Aronson, P.S., Slayman, C.W. and Igarashi, P. (1991) Am. J. Physiol. 261, F1088–F1094.
66. McDaniel, H.B., Huang, Z.-Q., Cook, W.J. and Warnock, D.G. (1990) Kidney Int. 37, 232.
67. Wang, D., Takaichi, K., Chen, F.P. and Warnock, D.G. (1991) J. Am. Soc. Nephrol. 2, 715.
68. Pearce, D., Krapf, R., Rector, F.C. and Reudelhuber, T.L. (1990) Kidney Int. 37, 233.
69. Fliegel, L., Sardet, C., Pouysségur, J. and Barr, A. (1991) FEBS Lett. 279 25–29.
70. Biemesderfer, D., Reilly, R.F., Exner, M., Hildebrandt, F., Igarashi, P. and Aronson, P.S. (1991) J. Am. Soc. Nephrol. 2, 696.
71. Haggerty, J.G., Agarwal, N., Cragoe, E.J., Jr., Slayman, C.W. and Adelberg, E.A. (1988) Am. J. Physiol. 255, C495–C501.
72. Tse, C., Watson, A.J.M., Ma, A.I., Pouysségur, J. and Donowitz, M. (1991) Gastroenteology 100, A258.
73. Tse, C.M., Brant, S.R., Walker, M.S., Pouysségur, J. and Donowitz, M. (1992) J. Biol. Chem. 267, 9340–9346.
74. Orlowski, J., Kandasamy, R.A. and Shull, G.E. (1992) J. Biol. Chem. 267, 9331–9339.
75. Yun, C., Gurubhagavatula, S., Montgomery, J., Levine, S., Walker, S., Brant, S., Pouysségur, J., Donowitz, M. and Tse, M. (1992) Gastroenterology 102, A255.
76. Geibel, J., Giebisch, G. and Boron, W.F. (1989) Am. J. Physiol. 257, F790–F797.
77. Krapf, R. and Solioz, M. (1991) J. Clin. Invest. 88, 783–788.
78. Rao, G.N., Sardet, C., Pouysségur, J. and Berk, B.C. (1990) J. Biol. Chem. 265, 19393–19396.
79. Rao, G.N., de Roux, N., Sardet, C., Pouysségur, J. and Berk, B.C. (1991) J. Biol. Chem. 266, 13485–13488.
80. Igarashi, P., Freed, M.I., Ganz, M.B. and Reilly, R.F. (1992) Am. J. Physiol., in press.
81. Krapf, R., Pearce, D., Lynch, C., Xi, C.-P., Reudelhuber, T.L., Pouysségur, J. and Rector, F.C., Jr. (1991) J. Clin. Invest. 87, 747–751.
82. Moe, O., Miller, R.T., Horie, S., Cano, A., Preisign, P.A. and Alpern, R.J. (1991) J. Clin. Invest. 88, 1703–1708.
83. Miller, R.T., Counillon, L., Pages, G., Lifton, R.P., Sardet, C. and Pouysségur, J. (1991) J. Biol. Chem. 266, 10813–10819.
84. Reboucas, N.A., Blaurock, M. and Igarashi, P. (1991) J. Am. Soc. Nephrol. 2, 710.
85. Mattei, M.G., Sardet, C., Franchi, A. and Pouysségur, J. (1988) Cytogenet. Cell. Genet. 48, 6–8.
86. Lifton, R.P., Hunt, S.C., Williams, R.R., Pouysségur, J. and Lalouel, J.-M. (1991) Hypertension 17, 8–14.

CHAPTER 9

Cl$^-$-channels

RAINER GREGER

Physiologisches Institut der Albert-Ludwigs-Universität, 7800 Freiburg i. Brsg., Germany

1. Introduction

The field of ion-channel research has met with intense interest during the past ten years. One reason for this development has been the advent of new methods such as the patch clamp technique invented by Sakmann and Neher [1] and new approaches to the cloning of the complete amino acid sequencing of ion channels as it was introduced by Numa and collaborators [2].

In comparison to cation channels (Na$^+$, K$^+$, non-selective cation, and Ca^{2+}-channels) Cl$^-$-channels had long been rather neglected. Their importance for synaptic transmission has stimulated work on the gamma-amino butyric acid-A (GABA$_A$) receptor in the laboratory of Barnard and his coworkers [3] and on the glycine-receptor channel in the laboratory of Betz and his colleagues [4]. These channels are cloned at this stage, and their function is already being studied with site-directed mutagenesis. A specific type of Cl$^-$-channel of the electrical organ of *Torpedo marmorata* has long been studied with respect to its functional properties and has now also been cloned by Jentsch et al. [5]. The relevance of the Cl$^-$-conductance for normal skeletal muscle function has been noted long ago [6] and the pathophysiology of myotonia is explained now in a fraction of the clinical cases by a defect in the Cl$^-$-conductance [6,7]. This has led to intensified efforts to characterize skeletal muscle Cl$^-$-channels and to isolate the channel proteins [8].

In previous reviews on this matter by Gögelein [9] and myself [10] it has been pointed out that the Cl$^-$-channels of the central nervous system and of skeletal muscle are distinct from those of non-excitable cells. The latter entity is in itself obviously heterogeneous with respect to its occurrence and function. In apolar as well as in polarized cells Cl$^-$-channels may be involved in volume regulation. As a simple rule gating of K$^+$- and Cl$^-$-channels is likely to occur whenever cell volume has to be down-regulated [11], as is the case in regulatory volume decrease of cell volume. A simple means to induce this phenomena is the exposure of cells to hypoosmolar solutions [12]. For example Cl$^-$-channels play an important role in histamine release by mast cells [13] or in renin release by juxtaglomerular cells of the kidney [14] and in exocytosis in general.

In many epithelia Cl^- is transported transcellularly. Cl^- is taken up by secondary or tertiary active processes such as $Na^+2Cl^-K^+$-cotransport, Na^+Cl^--cotransport, HCO_3^-–Cl^--exchange and other systems across one cell membrane and leaves the epithelial cell across the other membrane via Cl^--channels. The driving force for Cl^--exit is provided by the Cl^--uptake mechanism. The Cl^--activity, unlike that in excitable cells, is clearly above the Nernst potential [15,16], and the driving force for Cl^--exit amounts to some 20–40 mV.

The regulation of epithelial Cl^--channels has been examined in many laboratories, and these studies have been intensified with the recognition that the Cl^--channel regulation in epithelia is defective in a very common and clinically serious inherited disease, namely cystic fibrosis [17,18]. Efforts in this direction have not yet arrived at coherent concepts, and many published models [19–22] may have to be modified or revised.

The present chapter will address the following issues: (1) a very brief overview on the properties of the different types of Cl^--channels in the various mammalian cells; (2) a short summary on what is known of Cl^- channels on a molecular basis; (3) a discussion of pharmacological agents blocking the various Cl^--channels; and (4) a specific section dealing with the regulation of epithelial and maybe other Cl^--channels. This entire area has been reviewed rather extensively in the recent past. A large number of references will be provided in order to keep this text concise. The entire field of Cl^--channels in the central nervous system will only be touched upon to compare these channels to the Cl^--channels in apolar cells and epithelia.

2. Different types of Cl^--channels

A subdivision of the different types of Cl^--channels can be made on the basis of their conductance ranges. Such a subdivision is very convenient and free of bias. The single channel conductance is in fact the one parameter that is examined first by the experimenter when she or he measures the current–voltage (I/V curves for any type of unknown channel. I have recently [10] subdivided the various Cl^--channels into three categories: (1) large channels ($>100\,pS$); (2) intermediate channels (20–100 pS); and (3) small channels ($<20\,pS$). Such a subdivision is arbitrary, and in one single type of cell several different (conductance types of Cl^--channels may be present at the same time (cf. below).

In the following the Cl^--channels will be subdivided into those of the central nervous system, of muscle and *Torpedo* electroplax, of apolar non-excitable cells and of epithelia.

2.1. Cl^--channels in the nervous system

One large class of neuronal Cl^--channels are agonist controlled (receptor-operated) by gamma-amino-butyric acid ($GABA_A$-receptor channel) or by glycine. These channel families are well investigated and their amino acid sequence is known [3,4]. They share similarity in their primary structure with the acetylcholine-receptor channel [2]. Interaction of the physiological or non-physiological (e.g., muscimol) agonist with these Cl^--channels leads to their gating. Their conductance falls into the intermediate range (10–75 pS).

The consequence of the opening of these channels on membrane voltage depends on the Cl^--equilibrium voltage. In many neurones the Cl^--equilibrium voltage is hyperpolarized with respect to the resting voltage. Then GABA or some other agonist hyperpolarizes the membrane voltage [23]. In other instances, e.g., rat sympathetic neurones [24], the opposite is true. Cl^- is above equilibrium, and GABA induces a membrane depolarization. The $GABA_A$ channel which is homologous to the glycine-receptor channel possesses a phosphorylation site at the β subunit, and it has been shown recently that cAMP-dependent phosphorylation increased the glycine-induced Cl^--current in spinal trigeminal neurones [25].

Other neuronal Cl^--channels are Ca^{2+}-controlled. Increases in cytosolic Ca^{2+} enhances the probability of these channels being open [26,27]. These channels stabilize the membrane voltage by 'clamping' it towards the Cl^--equilibrium potential. Such channels have been found, e.g., in cultured mouse spinal neurones and in molluscan neurones. They subserve the repolarization phenomena and hence assist Ca^{2+}-activated K^+-channels. Their conductance is in the small to intermediate range. They are usually gated by depolarization.

The glia cells also possess Cl^--channels which are believed to play a role in the spatial buffering [28]. The channels present in astrocytes and in Schwann cells are probably different. Both are gated by depolarization, however, only the one present in astrocytes stays open over a wide range of depolarized voltages [28]. These channels probably fall into the large conductance range. They are Ca^{2+} and pH independent. Their selectivity is rather typical for Cl^--channels, with a sequence of $Cl^- \approx Br^- > SO_4^{2-}$ = isethionate > acetate > gluconate = O. In addition to these Cl^--channels, which serve to take up Cl^- when the membrane voltage is depolarized, e.g., by an increase in ambient K^+-concentration, other GABA-sensitive Cl^--channels have been found in recent studies [29]. The functional role of these channels has yet to be defined.

Cl^--channels are also present in neuroendocrine cells such as the AtT-20 pituitary cell line [30]. Ca^{2+}-dependent and voltage-activated slow Cl^--currents were found. These Cl^--currents could subserve a role comparable to that of Cl^--currents in renin-secreting cells or mast cells [13,14]. Along these lines, it has been speculated that the slow Cl^- inward currents may influence the firing rate of these neuroendocrine cells [30].

2.2. Cl^--channels of muscle and electric organ

The Cl^--conductance contributes some 50% to the total conductance of resting skeletal muscle [6]. This conductance assists the K^+-conductance in repolarizing skeletal muscle. Consequently, the absence of this conductance or its pharmacological inhibition [6,31,32] leads to myotonia. The Cl^--conductance is probably made up by several classes of Cl^--channels. Large anion conductance channels have been found in cultured skeletal muscle cells [33–35] and in adult amphibian skeletal muscle [36,37]. These channels have a conductance of some 200–700 pS. They are permeable for small anions but impermeable for large organic anions. Much smaller Cl^--channels have also been reported for adult mouse skeletal muscle [38] and for the myoballs of human skeletal muscle [39]. The Cl^--conductance of skeletal muscle can be inhibited by fairly simple organic compounds such as anthracene-9-carboxylate (A9C, cf. Fig. 2). This has been first reported by Palade and Barchi in 1977 [40], and has been reproduced by many laboratories [6]. Diphenylamine-2-carboxylate (DPC, cf. Fig. 2) a more potent inhibitor of epithelial Cl^--channels [41], is less effective in skeletal muscle [42]. The properties of Cl^--channels in heart muscle and in smooth muscle cells have also been reviewed [6,10]. The Cl^--channel of skeletal muscle has meanwhile been cloned by Jentsch and coworkers, and they have shown that the gene for the Cl^--channel of a myotonic mouse has a well defined defect [125,126]. It is worth noting that the *Torpedo* and the skeletal muscle Cl^--channels share in common large portions of their sequence.

The electric organ of *Torpedo* contains a large Cl^--conductance. The single channel equivalent of this conductance has been well studied in planar bilayers into which electric organ membrane vesicles from *Torpedo californica* have been incorporated [43]. The properties of these small to intermediate Cl^--channels (conductance 10–20 pS) have been examined kinetically. It was concluded that the channel is 'double-barrelled', that each barrel (protomeric channel) fluctuates between its open and closed state within milliseconds, and that both barrels are controlled by a common switch in their slow gating processes. This Cl^--conductance is switched on by hyperpolarization and functions, once switched on, as an outward rectifier [5,43]. The channel has been cloned from *Torpedo marmorata* (cf. below) and it has been found that it does not share similarities with the $GABA_A$- or glycine-receptor Cl^--channel. The *Torpedo* Cl^--channel is blocked by disulphonate stilbenes and DPC (Fig. 2) [5].

2.3. Cl^--channels in apolar non-excitable cells

Cl^--channels with large, intermediate, and small conductance have been found in apolar non-excitable cells. In macrophages and in fibroblasts large Cl^--channels were found [33,44]. The latter preparation, lymphocytes, monocytes and keratinocytes also contain an intermediate conductance outwardly rectifying Cl^--channel

(ICOR, cf. below) [45–48]. It has been speculated that this channel was regulated by cAMP and that its regulation was defective in cystic fibrosis [45,46]. This conclusion, as will be discussed in section 5.2, is probably not appropriate. Small Cl^--channels have been found, e.g., in mast cells and in renin-producing cells [13,14]. These channels are difficult to detect because their single channel conductance of 1–2 pS (cf. next section) makes it impossible to recognize them as single channel events. Noise analysis had to be used to determine their conductance. These latter small channels are controlled by cytosolic Ca^{2+}, but the exact mode of Ca^{2+}-control is not yet understood [13]. These channels, once gated by increasing cytosolic Ca^{2+}, determine the membrane voltage. In mast cells, for instance, the membrane voltage will be hyperpolarized by the Cl^--conductance, and this would provide extra driving force for the stimulation-induced Ca^{2+}-influx [13]. In renin-producing cells, which are juxtaglomerular specialized epitheloid smooth muscle cells, the increase in Cl^--conductance depolarizes the membrane voltage, like in other smooth muscle cells, and this leads to cell shrinkage (K^+- and Cl^--efflux). Cell shrinkage leads to an inhibition of renin release [14].

2.4. The problem of detecting small Cl^--channels

Small Cl^--channels appear to play a pivotal role in cell volume regulation and in exocytosis in general. The existence of these channels has been difficult to prove, and it is only recently that they have been identified in very many cells.

The main problem has been a methodological one. The patch clamp analysis of single channels views the 'world' of channels through a very small analytical 'window' [10]. A single channel event (opening) needs to be sufficiently long-lived and sufficiently large to be picked up within the current noise band under optimized conditions, and with the low-pass filter set to say 2 kHz. The open time needs to be close to a millisecond and the current amplitude close to 0.5 pA to permit detection.

The current amplitude cannot be increased at free will, because the clamp voltage will usually not exceed ± 100 mV. A simple calculation reveals that, given the limitations above, channels need to have a conductance of $\geqslant 10$ pS, and time constants in the millisecond range to be detected. Any channel smaller or faster than this will escape detection in this analysis.

Even beyond this, it should be clear that the large clamp voltage of say 100 mV may already lead to the inactivation of larger channels, and, still worse, the excision of the cell membrane itself may inactivate channels. It is not surprising then that the current literature gives a grossly distorted view of the world of Cl^--channels. It is full of large and intermediate channels, but much less data are available on small channels. This analytical problem can be overcome by other patch clamp techniques.

Neher and coworkers, for example, have used the whole-cell analysis of currents and have performed noise analysis to arrive at the lower limits of single channel conductance [1,13]. A modified whole cell patch clamp technique has recently been introduced by Horn and Marty [49]. They use the cation ionophore, nystatin, to perforate the cell attached patch and obtain whole-cell recordings in this fashion. This method has the advantage that larger molecules cannot escape from the cell while the whole-cell recording is being made.

I and my coworkers have recently modified this technique [50]. I diluted the nystatin in the patch pipette further until the input resistance of the cell attached membrane was just low enough to record the membrane voltage, but high enough so that the input resistance still reflected that of the cell attached patch and not of the whole cell. This method allows for the simultaneous recording of: (1) the membrane voltage; (2) the channel activity in the cell attached patch; and (3) the input resistance of the cell attached patch. If some manoeuvre induces a Cl^--current it can be detected with this method as a change in membrane voltage. If, at the same time, the conductance of the cell attached patch increases, but no single channel event appears within the membrane, these findings can be used as direct proof that the Cl^--current flows through small 'undetectable' Cl^--channels. It is to be expected that many more small Cl^--channels will be described by these new methods in the near future.

2.5. Epithelial Cl^--channels

This area of research has been reviewed by Gögelein [9] and myself [10]. Three general classes of Cl^--channels have been found.

Large conductance Cl^--channels were described for renal epithelial cells such as MDCK-cells, urinary bladder, collecting duct and A6-cells [51–54] and in pulmonary alveolar cells [55].

The intermediate conductance outwardly rectifying Cl^--channel (ICOR) has been found in many epithelia [10]. It is best described for the rectal gland of *Squalus acanthias* [56], respiratory epithelial cells [17,20,21,57–60], colonic cells [61–64], renal tubule cells [65,66] and cultured pancreatic tumor cells [67,68]. A typical recording of this channel, taken from a primary culture of cystic fibrosis respiratory cells [57], is shown in Fig. 1. The ICOR channel is characterized by a conductance of some 20–50 pS for negative clamp voltages and of 40–80 pS for positive clamp voltages. The probability of the channel to open is enhanced by depolarization and reduced by hyperpolarization. Both properties contribute to the outward rectification of the macroscopic current through these channels. In the range of physiological voltages the macroscopic current is almost voltage insensitive [57]. The kinetic properties of these ICOR channels can be described by two open and two closed time constants (states). These channels conduct halides comparably well [57]. In a detailed analysis it has been reported that I^- is the most and F^- the least permeable halide [62]. These channels are partially permeable for

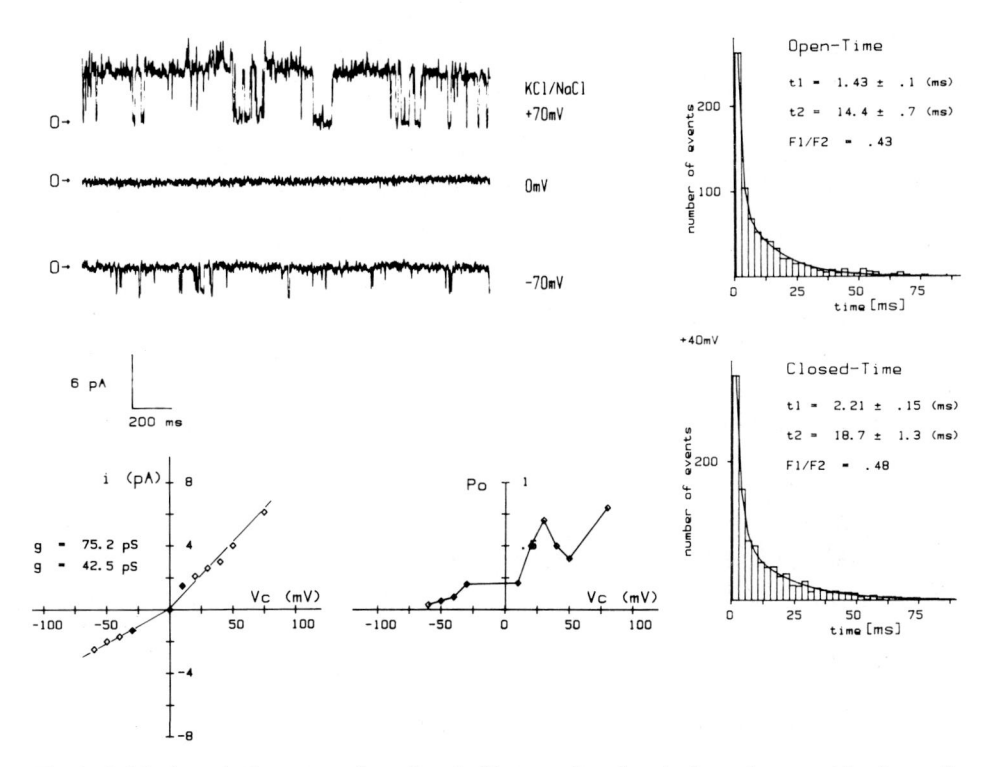

Fig. 1. Original patch clamp recording of an inside-out oriented excised membrane patch of a cystic fibrosis respiratory cell in primary culture. The methods of primary culture and patch clamp recording are described in [57]. The patch pipette is filled with a modified NaCl-Ringer's solution. The bath (on the cytosolic side) contains a modified KCl-Ringer's solution. *The upper left* panel shows typical current traces at three different clamp voltages. The zero current level for this channel is indicated by O→. Note that the channel is much more open at positive clamp voltages when compared to negative ones. In addition the open events last longer at positive voltages. *On the lower left* panel is the current–voltage (I/V) relation of this channel. Note that the conductance for positive clamp voltages is larger than that for negative ones. The open state probability (p_o) is plotted as a function of the clamp voltage (*lower half, middle*). Note that p_o is increased with depolarization. The mean current i flowing at any given time is the product of the respective current amplitude i and p_o. Since both i and p_o are higher at positive clamp voltages than at negative ones, the mean current i is strongly outwardly rectifying [57]. The time constant analysis of this channel is shown on the lower right. Two open and two closed time constants can be deduced from these histograms.

nitrate and HCO_3^- [69], but they are impermeable for large organic anions such as gluconate. The ICOR channel, unlike a previous conclusion [21], is not regulated directly by cytosolic Ca^{2+}, and it is pH-insensitive over a wide range [57,69]. The ICOR channel is blocked reversibly by a large spectrum of pharmacological agents (cf. below), amongst which 5-nitro-2-(3-phenylpropylamino)-benzoate (NPPB) [70] shows the highest affinity. The regulation of this channel will be discussed in section 5.

The physiological role of the ICOR is not clear and may be heterogeneous in the various tissues. In the thick ascending limb of the loop of Henle this channel appears to serve as the exit for Cl^- at the basal cell pole [16,65,66]. This conductive mechanism, therefore, is required for the reabsorption of Na^+ and Cl^- by this segment of the nephron [16]. In the rectal gland of *Squalus acanthias* a very similar channel is utilized for Na^+ and Cl^- secretion. In these latter cells the Cl^--channel is present in the luminal membrane and is controlled by cytosolic cAMP [15,56,71]. It has been claimed that this kind of channel is also responsible for the secretion of Cl^- in the colonic crypt cell, in colonic carcinoma cells and in respiratory epithelial cells [17,19,20,22]. Recent data have cast some doubt on this concept:

(1) The increase in the open probability of the ICOR channel induced by hormones in cell attached patches is moderate at best.

(2) The macroscopic Cl^--conductance induced by such hormones does not reflect the properties of the ICOR channel [12].

(3) The ICOR appears to be activated mostly by excision [57,72].

Cl^--channels with smaller conductance have first been noted in the rectal gland of *Squalus acanthias* by ourselves and in the colonic carcinoma cell line HT_{29} [61,73]. Later these types of 5–15 pS Cl^--channels were also found in pancreatic ducts, A6-cells and many other cells [74,75]. It is now claimed that this kind of channel is much more relevant than the ICOR for the pathophysiology of cystic fibrosis [12].

Cl^--channels with even smaller conductance have been described for the lacrimal and other exocrine glands [76,77]. These channels have a conductance of 1–2 pS. Unlike the ICOR-channel they appear to be blocked by millimolar concentrations of furosemide [77]. Most recent and only partially published data from my own laboratory obtained with the above modified nystatin technique [50,133,134] indicate that the respiratory epithelial cells and colonic carcinoma cells possess these types of small Cl^- channels, and that these channels are involved in hormonal regulation of Cl^--conductance (cf. section 5). These Cl^--channels are regulated by cytosolic Ca^{2+}. Hormonally induced increases in cytosolic Ca^{2+} lead to an abrupt increase in the probability of these small Cl^--channels being open, yet they have no effect on the ICOR-channel. Data of this kind reinforce that the physiological importance of these small Cl^--channels may have been grossly underestimated.

3. *The structure and molecular basis of Cl^--channels*

Very few of the Cl^--channels described above have been purified, and the amino acid sequence is only known for the $GABA_A$-receptor channel, the glycine-receptor channel and the *Torpedo* Cl^--channel [3–5], a Cl^--channel from skeletal muscle [125] and a Cl^--channel from MDCK cells [128]. In the following the current status will be briefly summarized.

3.1. The $GABA_A$-receptor and glycine-receptor channels

The $GABA_A$-receptor channel has been purified by virtue of its binding affinity to benzodiazepines in Barnard and his colleague's laboratory [3]. The material from bovine cerebral cortex purified in this fashion showed a high affinity to the agonist muscimol. From this material oligodeoxyribonucleotide probes have been constructed and were used to screen calf cerebral cortex cDNA libraries. The α and β subunits of the $GABA_A$-receptor, which were deduced from the DNA sequence by these techniques, have molecular masses of 44 and 55 kDa respectively. It is believed that one channel is made up of a dimeric structure ($\alpha_2\beta_2$).

A very similar cloning strategy has been used for the 48-kDa subunit of the glycine-receptor channel by Betz and collaborators [4]. Strychnine was used as an affinity ligand, and synthetic oligopeptides were constructed from the purified material. They have been used to screen cDNA libraries.

Both channels share a large homology. This was not noted at first when the sequencing work went on independently in two laboratories [3,4], whereas the fact that both channels share similarities with the nicotinic acetylcholine-receptor (ACH) channel [2] was evident immediately. Later, however, the laboratories jointly published that the channels are more closely related to each other than to the ACH channel [78]. The α and the β subunit of the $GABA_A$-receptor channel and the strychnine-binding subunit of the glycine-receptor channel are probably made up of four membrane spanning (M_1–M_4) hydrophobic domains. Both ends, the amino- and carboxy-end possess glycosylation sites, and probably face the extracellular side. Positive charges accumulate at both ends of the membrane spanning domains. This concentration of positive charges could play a role in Cl^- binding. The abundance of positive charges at the outside mouth of the channel is one important difference when compared to the nicotinic ACH channel. It is likely then that this charge accumulation at the outside mouth could act as the anion filter [3]. The entire channel is probably made up of four subunits ($\alpha_2\beta_2$ in the case of the $GABA_A$-receptor). It has been calculated that the total of 16 membrane spanning domains would form a pore with a diameter of 0.56 nm. The homology between the $GABA_A$-receptor channel and the glycine-receptor channel is largest in the M_2 segment. Both channels have probably been derived from the same ancestor and were separated late in their evolution [78].

The DNA for the $GABA_A$-receptor channel has been translated to the respective RNA by in vitro transcription of the cDNAs (for α and β) inserted into plasmids. The two RNAs, when injected together into *Xenopus* oocytes, produced GABA-sensitive Cl^--currents [3].

3.2. The Torpedo marmorata Cl^--channel

This channel has been sequenced by expression cloning by Jentsch and coworkers

[5]. As a first step the size of the mRNAs responsible for *Torpedo* Cl^--channel expression in *Xenopus* oocytes was determined. The size class of 9–10 kB was sufficient for full expression. This mRNA fraction was then used to construct a cDNA library in a plasmid vector. Single stranded DNA was used to deplete respective RNAs and hence to attenuate Cl^--channel expression. One DNA clone meeting these requirements was selected, since it hybridized with a 9–10 kB fraction in Northern blots. By additional hybrid-depletion assays and hybrid-arrest experiments using anti-sense oligonucleotides a clone with an open reading frame for 805 amino acids, encoding for a protein with a molecular mass of 89 kDa, was obtained. The mRNA derived from this cDNA elicited the expected Cl^--current in *Xenopus* oocytes. The predicted protein possesses probably 9 (up to 13) membrane spanning domains. Several glycosylation sites and possible phosphorylation sites have been identified. This protein shows no homology to the above receptor channels or the band 3 protein, which performs Cl^-–HCO^-_3-exchange [79]. In addition, this *Torpedo* Cl^--channel is not related to cation channels such as the dihydropyridine binding site of the Ca^{2+}-channel, shaker K^+-channel, or cGMP-gated cation channel [5]. The expression studies performed with the respective RNA suggest but do not prove that this cloned Cl^--channel is identical to the Cl^--channel of *Torpedo californica* studied in lipid bilayers [43].

3.3. Muscle Cl^--channels

In preliminary experiments the laboratory of Jockusch and coworkers in collaboration with Al-Awqati [32] has used indanyloxyacetic acid (IAA-94, Fig. 2, cf. below) to purify proteins between 30 and 150 kDa, which appear to be responsible for Cl^--transport and seem to be absent in one form of hereditary myotonia [8]. During the process of editing this review the Jentsch group has succeeded to clone the skeletal muscle Cl^--channel and yet another Cl^--channel [125,127].

3.4. Epithelial Cl^--channels

The current strategies to purify epithelial Cl^--channels are based on the use of affinity ligands. Landry and coworkers have assayed several potential ligands in membrane vesicles from kidney cortex and from the apical membrane of trachea [80]. They found that several compounds such as NPPB and indanyloxyacetic acid (IAA-94, Fig. 2) inhibited conductive Cl^--flux. The affinities were in the micromolar range. IAA-94 crosslinked to sepharose has then been used to isolate binding proteins [81]. The major fractions eluted from this column by excess IAA-94 had molecular masses of 27, 40, 64, and 97 kDa. Reconstitution of these proteins into liposomes led to conductive Cl^--flux, and fusion of these liposomes with planar lipid bilayers induced Cl^--currents with conductances ranging between 26 and 400 pS [81]. It has been suggested that one (or more) of these proteins resemble the epithelial Cl^--channel. It is, however, unexplained why the currents in the reconstituted system were insensitive towards the respective blockers. Regarding the source of materials in these studies it should also be noted that in the intact proximal tubule no Cl^--conductance is demonstrable in the luminal or the basolateral membrane [82,83]. However, Cl^--channels with a conductance of some 70 pS and sensitive to disulphonate stilbenes and NPPB are present in endosomal vesicles from the kidney cortex [84].

Breuer [85] has used basolateral membranes of thick ascending limb (TAL) cells to isolate the respective Cl^--channel. TAL membrane vesicles were solubilized and the proteins incorporated into liposomes. These liposomes showed conductive, disulphonate stilbene- and NPPB-sensitive Cl^--fluxes. Radioactive disulphonate stilbenes labelled 65- and 31-kDa proteins. The labelling of the 31-kDa protein could be attenuated by nonradioactive disulphonate stilbenes and NPPB. It was suggested that the 31-kDa protein was related to the Cl^--channel. One of the intrinsic problems with this approach is the poor sensitivity of the TAL Cl^--channel towards disulphonate stilbenes [86]. During the editing of this chapter, Paulmichl et al. [128] succeeded to clone a Cl^--channel from MDCK cells. They used expression cloning strategies and arrived at a Cl^--channel with about 38 kDa. A dimer of this protein might form the channel. It is worth noting that this channel is probably made up from β-leaflets rather than α-helices. This channel is outwardly rectifying and is blocked by NPPB and nucleotides. It is likely that this channel represents the ICOR channel.

4. Pharmacological modulation of Cl^--channels

The receptor-operated Cl^--channels of the central nervous system (CNS) are gated by the respective agonists GABA and glycine. Most Cl^--channels can be inhibited by disulphonate stilbenes. Muscle Cl^--channels can be inhibited by anthracene-9-carboxylate (A9C) and probably by IAA-94. The ICOR Cl^--channel is fairly sensitive to NPPB. It should be noted, however, that none of these probes, except for the GABA- and glycine-receptor Cl^--channels, is of sufficient affinity and selectivity to permit the channel identification by its use. This dilemma is one of the reasons why the purification of epithelial Cl^--channels lags behind that of the CNS Cl^--channels.

4.1. Pharmacological modulation of $GABA_A$-receptor and glycine-receptor channels

This area has been covered in a review by Eldefrawi and Eldefrawi [23]. The $GABA_A$-receptor channel is activated by GABA (Fig. 2), avermectin, muscimol, taurine (Fig. 2) and β-alanine (Fig. 2). The activation by agonists is potentiated by benzodiazepines and barbiturates. The channel is blocked by the competitive antagonist bicuculline and noncompetitive antagonists such as picrotoxin, insecticides and the disulphonate stilbene DIDS (Fig. 2). The glycine-receptor channel is also activated by taurine and β-alanine. Benzodiazepines but not barbiturates potentiate the effects of agonists. The channel is blocked non-competitively by strychnine.

4.2. Inhibition of epithelial Cl^--channels

I have reviewed this area only recently [10]. Given the heterogeneity of epithelial Cl^--channels, blockers which have been designed for some type of specific Cl^--channel will not necessarily block other Cl^--channels. The disulphonate stilbenes (SITS, DIDS, and DNDS) appear to be an exception in this respect. They block almost any type of Cl^--channel if sufficiently high concentrations are used. In fact, these substances have been designed by Cabantchik et al. [87] for the inhibition of the band 3 exchanger of the red blood cell. This carrier protein mediates the exchange of Cl^- and HCO_3^-. Its primary structure has been determined by Kopito and Lodish [79] and the binding of DIDS has been assigned to either Lys 558 or Lys 561 at the third extracellular loop. Due to the fact that DIDS inhibits Cl^--channels with entirely different structures [5,23,63,88,89] it appears likely that the DIDS-binding sites are also different in the different proteins.

Other blockers of epithelial Cl^--channels are of the aryl-amino-benzoate type or phenoxy-acetic-acid type [70]. Very few systematic surveys comparing different classes of blockers in one type of Cl^--channel are available at this stage. One such study has been performed in membrane vesicles from kidney cortex [80]. In this study IAA-94 and NPPB (cf. Fig. 2) turned out to be the most potent blocker of conductive Cl^--flux. In another systematic survey the Cl^--conductance of the sweat duct was examined, and it was found that dichloro-DPC (Fig. 2) was the most potent inhibitor of the transepithelial Cl^--conductance [90].

Several studies have been performed in the laboratory of my colleagues and I [41,70,91–93]. In one of these, starting with the structure of A9C (Fig. 2), the inhibitory effect on the Cl^--conductance in thick ascending limb (TAL) segments has been examined. It was found that DPC was much more potent than A9C [41]. In a subsequent study on the same preparation more than 100 relatives of DPC

Fig. 2. Agents controlling the opening of Cl^--channels. The structural formula, the name, the respective Cl^--channel, and an appropriate reference are provided. (A) Note the similarity of GABA, taurine, and β-alanine. Note also that DPC and its analogues such as NPPB contain a β-alanine backbone. (B) The structure of torasemide comes close to both DPC and furosemide or bumetanide. It blocks the $Na^+2Cl^-K^+$-cotransporter with very high affinity and with lesser affinity also the TAL-Cl^--channel. Note that IAA-94 is related to phenoxyacetic acids (e.g., ethacrynic acid). Amidine is related to IAA-94 but it has a positive net charge. Amidine as well as IAA-94 block the ICOR channel at around 10^{-5} mol/l [63].

Structure	Name	Target	Ref
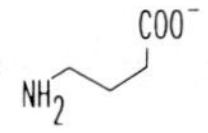	GABA	GABA-Cl-Channel	[23]
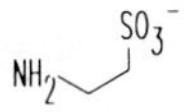	Taurine	GABA-,Glycine-Cl-Channel	[23]
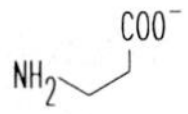	β-Alanine	GABA-,Glycine-Cl-Channel	[23]
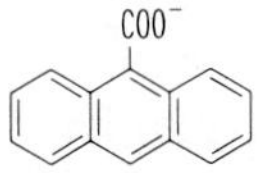	Anthracene-9-carboxylate (A9C)	Muscle-Cl-Channel	[40]
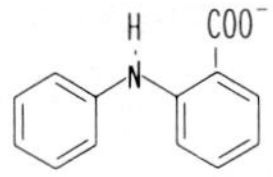	Diphenylamine-2-carboxylate (DPC)	TAL-Cl-Channel	[41]
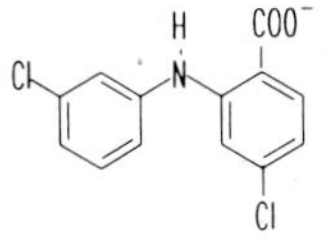	Dichlorodiphenyl-amine-2-carboxylate (DCl-DPC)	Sweat Duct-Cl-Channel	[90]
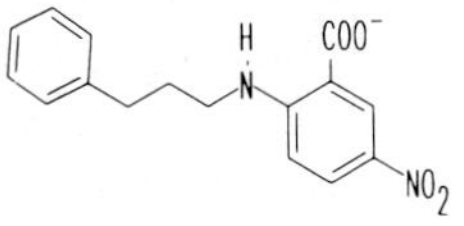	5-Nitro-2-(3-phenyl-propylamino)-benzoate (NPPB)	ICOR-Channel	[63,70]

A

Structure	Name	Target	Ref
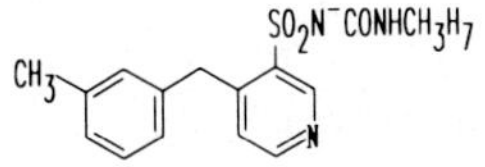	Torasemid	Na2ClK-Cotransporter	[91]
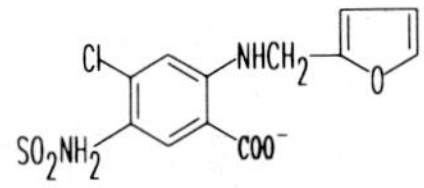	Furosemide	Na2ClK-Cotransporter / Lacrimal gl. Cl-Channel	[16] / [77]
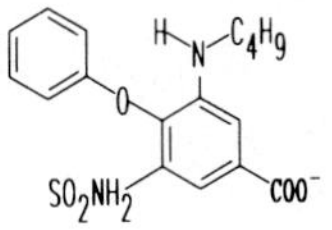	Bumetanide	Na2ClK-Cotransporter / ICOR-Channel	[16] / [unpublished]
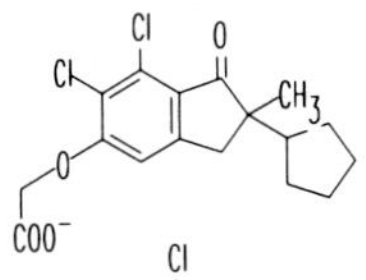	Indanyloxyacetate IAA-94	ICR-Channel	[63]
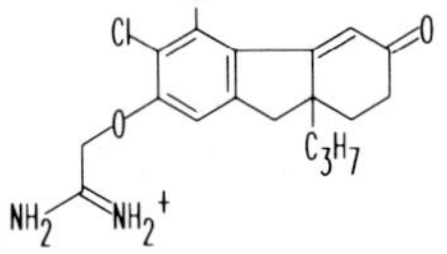	Amidine	ICOR-Channel	[63]
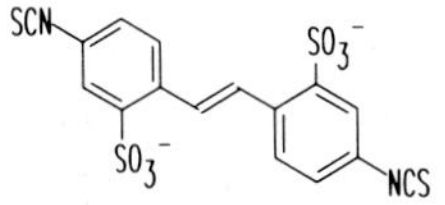	4,4'-Diisothio-cyanostilbene-2,2'-disulphonate DIDS	Band-3-Protein Cl-Channels	[5,23,63,88,89]

B

were examined, and NPPB (Fig. 2) was found to be the most potent compound in this series [70].

I have noted that NPPB is structurally related to loop diuretics of the furosemide (Fig. 2) type. These latter compounds bind to the $Na^+2Cl^-K^+$-cotransporter [16] and inhibit NaCl reabsorption in the TAL segment and NaCl secretion in epithelia such as the colonic crypt cell and rectal gland of *Squalus acanthias* [15]. We were able to show that only minor modification of the NPPB molecule on one side and of furosemide on the other led to compounds with altered selectivities [70,91–93]. One prototype of an intermediate blocker, i.e., a substance blocking both $Na^+2Cl^-K^+$-cotransport and Cl^--channels, is torasemide (Fig. 2). Hence we have performed a systematic study in order to define the constraints defining the effectiveness of this class of substances [91].

The common denominator of all these Cl^--channel blockers (Fig. 2) is their acidic function (usually carboxylate), the secondary amine with a certain spacing to the acidic group, a nitro substitution at a certain distance from the carboxylate group, and the apolar groups such as the phenyl-propyl-residue. We have suggested that the interaction with the Cl^--channel requires all of these interaction sites, and that even the apolar moiety of the blocker molecule reacts with the channel protein rather than with the membrane lipid [70].

NPPB has meanwhile been used in a lot of studies and in a large variety of different cells. Not to my surprise the results have been entirely heterogeneous. In some systems, such as the ICOR channel, NPPB turned out to be a very powerful inhibitor (reviewed by Tilmann et al. [63]). In other systems NPPB acted as a transport inhibitor but very high concentrations were necessary. In such experiments it is by no means certain that NPPB acted via its effect on Cl^--channels. Several other effects have been shown or suspected since our original study [70]. NPPB can block cyclooxygenase [94]. This is not surprising, given the similarity of NPPB to nonsteroidal antiinflammatory substances such as DPC and related molecules [95,96]. NPPB can probably act as a mitochondrial inhibitor due to its protonophoric effect [10,129]. Finally NPPB may block other channels such as non-selective cation channels and K^+-channels [97–99]. All these side effects require concentrations $> 10^{-5}$, usually $> 10^{-4}$ mol/l. If NPPB is used at lower concentrations it will still block, e.g., the ICOR channel (IC_{50} below 10^{-7} mol/l [63]) but no side effects will occur.

In a recent study on the ICOR-channel we have compared the effect of NPPB to that of the disulphonate stilbenes (Fig. 2), IAA-94 (Fig. 2) and amidine (Fig. 2). We found that all these compounds induced a flicker-type block. NPPB showed the highest affinity [63]. The results of this study were perplexing inasmuch as compounds which appear to be chemically different, and even possess opposite net charges, exert comparable effects. In further studies on the ICOR channel others and we were able to show that unsaturated fatty acids, bumetanide, the buffer HEPES and even Ca^{2+}-antagonists such as verapamil inhibited the ICOR channel reversibly [67,72,100,101]. We interpret these data as evidence for multiple binding sites at the channel protein. Several of the examined compounds (unsaturated fatty acids and verapamil) probably act via apolar interaction [72,100]. It should be noted that for most of these compounds concentrations $>10^{-4}$ mol/l are required for effective inhibition. Data of this kind should increase our alertness to the fact that so-called specific inhibitors turn into rather non-specific blockers if used at high concentrations.

For NPPB we were able to show that it blocks the ICOR channel from the extracellular side [63]. We have designed macromolecular probes (dextrans or polyethyleneglycols, molecular weight 3 kDa) to which we conjugated the NPPB molecule. These macromolecular probes were unable to penetrate the cell membrane. They inhibited the ICOR channel from the extracellular side with an IC_{50} of $<10^{-7}$ mol/l, but had no effect from the cytosolic side [63]. These probes will certainly be powerful tools for isolating and purifying the ICOR channel protein.

At this stage we appreciate the relevance of small Cl^--channels for hormonally regulated transepithelial Cl^--transport (cf. sections 2.4 and 5). Unfortunately substances such as NPPB are rather poor inhibitors of these small Cl^--channels. Therefore, new compounds specifically directed towards these channels will be required. Our search into this direction is still ongoing.

5. Regulation of epithelial Cl^--channels

The transepithelial NaCl transport is regulated by hormones. Examples are: (1) the TAL segment of the kidney where ADH increases the reabsorption of NaCl [16,102]; (2) the rectal gland of *Squalus acanthias* where the hormone rectin (a VIP analogue) increases NaCl secretion [15]; (3) the colon crypt cell where prostaglandin E_2 increases NaCl secretion [103]; or (4) the respiratory epithelial cell where catecholamines also increase the rate of NaCl secretion [103,104]. For all of these examples it has been shown that the hormone acted by increasing the respective Cl^--conductance [71,102–104]. Fig. 3 displays the general scheme from NaCl secretion. The general scheme has been elaborated for the rectal gland of *Squalus acanthias* [105] but it appears to apply also to many other epithelia [15,106]. It is interesting to note that the rate limiting step in most of these epithelia appears to be the Cl^--conductance. Once the Cl^--conductance is increased, transepithelial NaCl secretion is also enhanced. The mechanisms whereby the other membrane transport proteins are readjusted in their rates is not understood. For all of the above examples cAMP is the second messenger. It has been shown that addition of forskolin and/or dibutyryl-cAMP mimics the effect of the respective hormone.

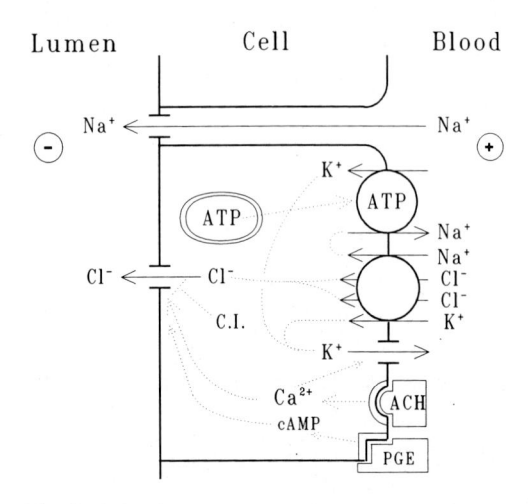

Fig. 3. Cellular model for NaCl secretion ([16] e.g., in a colonic carcinoma cell). The symbols have the following meaning: (ATP) = $(Na^+ + K^+)$-ATPase; ⟳ = $Na^+2Cl^-K^+$-cotransporter; → = ion channel; C.I. = cytosolic inhibitor [72]. Note that K^+ recycles through a basolateral conductance. Cl^- uptake is mediated by the $Na^+2Cl^-K^+$-cotransporter. The driving force for this carrier is provided by the $(Na^+ + K^+)$-ATPase. Cl^- leaves the cell through apical Cl^--channels. The polar conductances and the respective ion gradients generate the lumen negative voltage. This voltage drives the secretion of Na^+, probably through the paracellular shunt pathway. Hormones control secretion by primary regulation of the Cl^-- or K^+-conductances [16,106]. The K^+-channel regulation occurs by increases in cytosolic Ca^{2+}. The upregulation of the Cl^--channel is mediated by cAMP and Ca^{2+}. The detailed mechanisms of second messenger and Ca^{2+} interaction with the Cl^--channel are not yet clear.

5.1. The Cl^--channel defect in cystic fibrosis

This area has been reviewed by several research teams [17,18,60,107]. It goes without saying here that patients suffering from the inherited disease cystic fibrosis (CF) show multiple defects: (1) they have a reduced mucociliary clearance in the respiratory tract, predisposing for chronic lung infections; (2) they have a reduced secretion of the exocrine pancreas, leading to maldigestion and failure to thrive; (3) their ileum and colon is unable to secrete fluid, predisposing to meconium-ileus; (4) their secretory processes in the reproductive organs are defective, leading to infertility; and (5) their sweat is abnormally high in NaCl concentration due to the inability to reabsorb NaCl in the sweat ducts. Schulz and Frömter [108] were the first to recognize that the Cl^--conductance was low in the sweat ducts of CF patients many years ago. This finding has been reproduced many years later by Quinton [109], and only then was the field prepared to examine this Cl^--conductance defect further.

Several groups have shown that hormonal stimuli which lead to NaCl secretion in normal tissues fail to do so in CF tissues (reviewed in [17,18,60]). The production of the second messenger cAMP was unimpeded in these CF tissues. Hence it was concluded that either the Cl^--conductance was absent in these patients or that it was present but could not be activated by cAMP.

In respiratory epithelial (RE) cells the Cl^--conductance was attributed to the ICOR channel. In fact, it was reported by Frizzell et al. and Welsh's laboratories that catecholamines increased the incidence of ICOR channels in cell attached patches of normal RE cells but failed to do so in CF cells [110,111]. Later both laboratories presented data on excised membrane patches of RE cells in which the protein kinase A which was added to the cytosolic side produced ICOR channel activity in the normal cells but not in the CF tissues [19,20]. This finding was reproduced by Guggino and coworkers [22] for RE cells and by others for lymphocytes [46]. Protein kinase C at physiological Ca^{2+}-activities had a comparable effect in normal cells but also failed to function in CF cells [22,112].

Our own laboratory obtained different results. Not only were we unable to see a clear cut correlation between the incidence of ICOR channels in cell attached patches and the degree of hormonal stimulation [57], we were also unable to reproduce the activation studies in excised patches. In our hands, the activation of ICOR channels occurred simply by the excision, and this was equally true for the normal as for the CF cells [57]. We did note, however, that the other laboratories worked at room temperature whereas we always work at 37°C. Welsh's laboratory has shown meanwhile that excision activation of ICOR Cl^--channels is a temperature-dependent process [113]. At low temperature, excision activation is largely delayed [113] but it is immediate in our experiments at 37°C [57]. We concluded that the activation of the ICOR channels has probably little to do with phosphorylation but is rather due to the fact that the excised patch faces a new environment on the cytosolic side [57,72].

Meanwhile we have shown that the excision activation of ICOR channels is due to disinhibition [72]. The respective inhibitor, operationally named cytosolic inhibitor (CI), is present in the cytosol of placenta trophoblast cells; HT_{29}- and T_{84}-colonic carcinoma cells; and RE cells of normal and CF patients. The molecule has an apparent molecular weight of 700–1 500 Da; it is amphiphilic; heat stable; and not digested by trypsin, proteases, nucleotidases, lipases or amylase [72]. Burckhardt, Frömter and their collaborators [114] have confirmed our results and extracted a similar or identical CI from kidney cortex.

Our observations suggest an alternative hypothetical mechanism for the pathophysiology of CF. If the CI concentration was higher in CF cells, these cells would be unable to open Cl^--channels in response to increased cAMP. This concept fits well with other observations in this field. Tsui, Riordan, Collins and coworkers [115] have been able to clone the CF gene on chromosome 7. The gene product ('cystic fibrosis transmembrane conductance regulator' = CFTR) shows similarities to the multi-drug resistant (MDR) proteins but is not similar to any known ion channel. We might speculate now that CFTR might act as a regulator of CI, and a defect on CFTR, such as the 508-deletion mutation which is the most common cause of CF, might lead to an increased cytosolic concentration of CI.

Little doubt is left that a defect in CFTR is responsible for CF. This has been shown very recently in two transfection studies in which Cl^- permeability could be produced by transfecting and expressing normal CFTR [116,117]. Currently the discussion continues of whether CFTR is a regulator or the channel itself. In favour of the former is the structure itself [118]. One should keep in mind, however, that any consideration regarding the meaning of the primary sequence of CFTR is based on rather feeble arguments. Not only have we noted that Cl^--channels can have rather different structures (cf. above sections 3.1 and 3.2), we should also be aware that we still do not know which Cl^--channel is defective in CF, and apart from the MDCK-Cl^--channel [128] we do not have any primary sequence of any epithelial Cl^--channel. The latter interpretation that CFTR is the Cl^--channel is also only circumstantial and is only supported by one recent study from Riordan and coworker's laboratory [119] in which a Cl^--conductance was introduced into an insect cell after it was transfected with the mRNA for CFTR and expressed this protein. During the editing of this chapter, Bear et al. [130] have succeeded to incorporate pure CFTR into lipid bilayers and they have shown that this protein acts like a cAMP-dependent Cl^--channel.

In our patch clamp studies in excised membrane patches in which we attempted to characterize the CI we have noted that CI did not only inhibit the probability of the ICOR channel being open but we also found that the input conductance of the patch was reduced at the same time and with the same time course [72]. We have followed up on this observation and we were able to show that this reduction in input conductance is caused by an inhibition of small ($\ll 10\,pS$) Cl^--channels. Hence, we postulate that the same patches containing ICOR channels also contain small (unresolved, cf. section 2.4) Cl^--channels which are inhibited reversibly by CI. It cannot be excluded at this stage that these small Cl^--channels are responsible for the defect in CF.

5.2. Mechanisms of Cl^--channel activation in epithelia

It has been known for some time that the Cl^--conductance of epithelial cells can, in addition to its regulation via cAMP, be enhanced by increases in cytosolic Ca^{2+} (cf. Fig. 3). This has been shown with Ca^{2+}-ionophores [120,121] or with hormones increasing cytosolic Ca^{2+} such as carbachol, neurotensin, ATP, etc. [50,103,104]. Usually these agonists have dual effects. They increase the Cl^-- as well as the K^+-conductance [104]. Stutts et al. [122] have shown that CF cells still increase their Cl^--conductance in response to ATP. Another mechanism of Cl^--channel activation has been described in whole-cell patches of colonic carcinoma and RE cells [123,124]: when the cells are exposed to hypotonic media they swell and increase their Cl^--conductance. This is a rather general phenomenon which is present in a lot of cells [11]. In their effort to reduce cell volume in hypotonic media (regulatory volume decrease) the cells increase their Cl^-- and K^+-conductance and extrude KCl.

Frizzell and coworker's laboratory has examined the properties of the whole-cell currents generated by cAMP, increases in cytosolic Ca^{2+}, and increases in cell volume [12,124]. They found that the properties of all three Cl^--currents are different. The cAMP induced Cl^--current is linear and conducts Cl^- better than I^-. These findings do not fit with the properties of the ICOR channel, which is outwardly rectifying and conducts I^- better than Cl^- (cf. above). On the other hand, the Cl^--current induced by hypoosmolar solutions had characteristics similar to the ICOR channel. Comparable results have meanwhile been obtained in another laboratory (Frömter, personal communication).

We have examined the same questions utilizing the nystatin method as described above (section 2.4). We found [50] that all three pathways (cAMP, Ca^{2+} and swelling) induced Cl^--currents in HT_{29} colon carcinoma cells, but none of these manoeuvres induced the ICOR channel in cell attached patches, rather the baseline conductance increased in these patches. We interpret these data to indicate that all three pathways activate small Cl^--channels but not the ICOR channel [50 and unpublished observations]. A similar conclusion has been drawn by Tabcharani et al. [89] with respect to cAMP. Much further work will be required to characterize these small epithelial Cl^--channels. During the editing of this chapter, we [131] have shown that all three types of Cl^--conductances (cAMP-, ATP-, swelling induced) share in common all tested properties such as ion selectivity, conductance range and sensitivity towards stilbenes. In addition, the three pathways are not additive. Finally, we were able to show that the two cellular transduction pathways increase in cytosolic Ca^{2+} and cAMP converge and cooperate on the same Cl^--conductance [131]. For example, the transient nature of the Ca^{2+} mediated Cl^--conductance response can be converted into a rather stable response by cAMP. Currently we hypothesize that the Ca^{2+} signal gears the exocytosis of Cl^--channel containing vesicles, and that cAMP prevents the endocytosis of this membrane. This is supported by preliminary membrane capacitance measurements [132].

Acknowledgements

The work from the author's laboratory has been supported by DFG Gr 480/10.

References

1. Sakmann, B. and Neher, E. (1984) Annu. Rev. Physiol. 46, 455–472.
2. Noda, M., Takahashi, H., Tanabe, T., Toyosato, M., Furutani, Y., Hirose, T., Asai, M., Inayama, S., Miyata, T. and Numa, S. (1982) Nature (London) 299, 793–797.

3. Schofield, P.R., Darlison, M.G., Fujita, N., Burt, D.R., Stephenson, F.A., Rodriguez, H., Rhee, L.M., Ramachandran, J., Reale, V., Glencorse, T.A., Seeburg, P.H. and Barnard, E.A. (1987) Nature (London) 328, 221–227.
4. Grenningloh, G., Rienitz, A., Schmitt, B., Methfessel, C., Zensen, M., Beyreuther, K., Gundelfinger, E.D. and Betz, H. (1987) Nature (London) 328, 215–220.
5. Jentsch, T.J., Steinmeyer, K. and Schwarz, G. (1990) Nature (London) 348, 510–515.
6. Bretag, A.H. (1987) Physiol. Rev. 67, 618–724.
7. Rüdel, R. (1987) Pflügers Arch. 408, R3.
8. Wischmeyer, E., Weber-Schürholz, S., Jockusch, H. and Schürholz, T. (1991) Pflügers Arch. 418, R45.
9. Gögelein, H. (1988) Biochim. Biophys. Acta 947, 521–547.
10. Greger, R. (1991) In: Methods in Enzymology (Fleischer, B. and Fleischer, S., Eds.), pp. 793–810, Academic Press, Orlando.
11. Lewis, S.A. and Donaldson, P. (1990) NIPS 5, 112–119.
12. Cliff, W.H. and Frizzell, R.A. (1990) Proc. Natl. Acad. Sci. U.S.A. 87, 4956–4960.
13. Matthews, G., Neher, E. and Penner, R. (1989) J. Physiol. 418, 131–144.
14. Kurtz, A. (1990) NIPS 5, 43–46.
15. Greger, R., Schlatter, E. and Gögelein, H. (1986) NIPS 1, 134–136.
16. Greger, R. (1985) Physiol. Rev. 65, 760–797.
17. Welsh, M.J. (1990) FASEB J. 4, 2718–2725.
18. Quinton, P.M. (1990) FASEB J. 4, 2709–2717.
19. Li, M., McCann, J.D., Liedtke, C.M., Nairn, A.C., Greengard, P. and Welsh, M.J. (1988) Nature (London) 331, 358–360.
20. Schoumacher, R.A., Shoemaker, R.L., Halm, D.R., Tallant, E.A., Wallace, R.W. and Frizzell, R.A. (1987) Nature (London) 330, 752–754.
21. Frizzell, R.A., Halm, D.R., Rechkemmer, G.R. and Shoemaker, R.L. (1986) Fed. Proc. 45, 2727–2731.
22. Hwang, T.-C., Lu, L., Zeitlin, L., Gruenert, D.C., Huganir, R. and Guggino, W.B. (1989) Science 244, 1351–1353.
23. Eldefrawi, A.T. and Eldefrawi, M.E. (1987) FASEB J. 1, 262–271.
24. Ballanyi, K. and Grafe, P. (1985) J. Physiol. 365, 41–58.
25. Song, Y. and Huang, L.-Y.M. (1990) Nature (London) 348, 242–245.
26. Owen, D.G., Segal, M. and Barker, J.L. (1984) Nature (London) 311, 567–570.
27. Geletyuk, V.I. and Kazachenko, V.N. (1985) J. Membr. Biol. 86, 9–15.
28. Gray, P.T.A. and Ritchie, J.M. (1986) Proc. R. Soc. London B 228, 267–288.
29. Kettenmann, H. (1991) Pflügers Arch. 418, R3.
30. Korn, S.J. and Weight, F.F. (1987) J. Neurophysiol. 58, 1431–1451.
31. Bretag, A.H., Dawe, S.R. and Moskwa, A.G. (1980) Nature (London) 286, 625–626.
32. Weber-Schürholz, S., Zippel, M., Al-Awqati, Q., Landry, D. and Jockusch, H. (1991) Pflügers Arch. 418, R44.
33. Schwarze, W. and Kolb, H.A. (1984) Pflügers Arch. 402, 281–291.
34. Hals, G.D., Stein, P.G. and Palade, P.T. (1989) J. Gen. Physiol. 93, 385–410.
35. Blatz, A.L. and Magleby, K.L. (1983) Biophys. J. 43, 237–241.
36. Woll, K.H. and Neumcke, B. (1987) Pflügers Arch. 410, 641–647.
37. Woll, K.H., Leibowitz, M.D., Neumcke, B. and Hille, B. (1987) Pflügers Arch. 410, 632–640.
38. Brinkmeier, H. and Rüdel, R. (1991) Pflügers Arch. 418, R44.
39. Fahlke, C., Zachar, E. and Rüdel, R. (1991) Pflügers Arch. 418, R45.
40. Palade, P.T. and Barchi, R.L. (1977) J. Gen. Physiol. 69, 879–896.
41. Di Stefano, A., Wittner, M., Schlatter, E., Lang, H.J., Englert, H. and Greger, R. (1985) Pflügers Arch. 405 (Suppl. 1), S95–S100.

42. Camerino, D.C., De Luca, A.M. and Mambrini, M. (1989) J. Pharm. Pharmacol. 41, 42–45.
43. Miller, C. (1982) Philos. Trans. R. Soc. London B 299, 401–411.
44. Nobile, M. and Galietta, L.J.V. (1988) Biochem. Biophys. Res. Commun. 154, 719–726.
45. Bear, C.E. (1988) FEBS Lett. 237, 145–149.
46. Chen, J.H., Schulman, H. and Gardner, P. (1989) Science 243, 657–660.
47. Galietta, L.J.V., Barone, V., De Luca, M. and Romeo, G. (1991) Pflügers Arch. 418, 18–26.
48. Kanno, T. and Takishima, T. (1990) J. Membr. Biol. 116, 149–161.
49. Horn, R. and Marty, A. (1988) J. Gen. Physiol. 92, 145–159.
50. Greger, R., Kunzelmann, K., Lohrmann, E. and Hansen, C.P. (1991) In: Ionic Basis and Energy Metabolism of Epithelial Transport. Hot aspects from exocrine secretion (Murakami, M., Ed.), pp. 167–170, Kebun Printing, Okazaki.
51. Nelson, D.J., Tang, J.M. and Palmer, L.G. (1984) J. Membr. Biol. 80, 81–89.
52. Kolb, H.A., Brown, C.D.A. and Murer, H. (1985) Pflügers Arch. 403, 262–265.
53. Velasco, G., Prieto, M., Alvarez-Riera, J., Gascon, S. and Brros, F. (1989) Pflügers Arch. 414, 304–310.
54. Light, D.B., Schwiebert, E.M., Fejes-Toth, G., Naray-Fejes-Toth, A., Karlson, K.H., McCann, F. and Stanton, B. (1990) Am. J. Physiol. 268, F273–F280.
55. Krouse, M.E., Schneider, G.T. and Gage, P.W. (1986) Nature (London) 319, 58–60.
56. Greger, R., Schlatter, E. and Gögelein, H. (1987) Pflügers Arch. 409, 114–121.
57. Kunzelmann, K., Pavenstädt, H. and Greger, R. (1989) Pflügers Arch. 415, 172–182.
58. Jorissen, M., Vereecke, J., Carmeliet, E., Van den Berghe, H. and Cassiman, J.J. (1990) J. Membr. Biol. 117, 123–130.
59. Li, M., McCann, J.D. and Welsh, M.J. (1990) Am. J. Physiol. 259, C295–C301.
60. Halley, D.J.J., Bijman, J., De Jonge, H.R., Sinaasappel, M., Neijens, H.J. and Niermeijer, M.F. (1990) Europ. J. Pediat. 149, 670–677.
61. Hayslett, J.P., Gögelein, H., Kunzelmann, K. and Greger, R. (1987) Pflügers Arch. 410, 487–494.
62. Halm, D.R., Rechkemmer, G.R., Schoumacher, R.A. and Frizzell, R.A. (1988) Am. J. Physiol. 254, C505–C511.
63. Tilmann, M., Kunzelmann, K., Fröbe, U., Cabantchik, Z.I., Lang, H.J., Englert, H.C. and Greger, R. (1991) Pflügers Arch. 418, 556–563.
64. Diener, M., Rummel, W., Mestres, P. and Lindemann, B. (1989) J. Membr. Biol. 108, 21–30.
65. Greger, R., Bleich, M. and Schlatter, E. (1990) Renal Physiol. Biochem. 13, 37–50.
66. Paulais, M. and Teulon, J. (1990) J. Membr. Biol. 113, 253–260.
67. Hanrahan, J.W. and Tabcharani, J.A. (1990) J. Membr. Biol. 116, 65–77.
68. Schoumacher, R.A., Ram, J., Iannuzzi, M.C., Bradbury, N.A., Wallace, R.W., Tom Hon, C., Kelly, D.R., Schmid, S.M., Gelder, F.B., Rado, T.A. and Frizzell, R.A. (1990) Proc. Natl. Acad. Sci. U.S.A. 87, 4012–4016.
69. Kunzelmann, K., Gerlach, L., Fröbe, U. and Greger, R. (1991) Pflügers Arch. 417, 616–621.
70. Wangemann, P., Wittner, M., Di Stefano, A., Englert, H.C., Lang, H.J., Schlatter, E. and Greger, R. (1986) Pflügers Arch. 407 (Suppl. 2), S128–S141.
71. Greger, R., Schlatter, E. and Gögelein, H. (1985) Pflügers Arch. 403, 446–448.
72. Kunzelmann, K., Tilmann, M., Hansen, C.P. and Greger, R. (1991) Pflügers Arch. 418, 479–490.
73. Gögelein, H., Schlatter, E. and Greger, R. (1987) Pflügers Arch. 409, 122–125.
74. Gray, M.A., Pollard, C.E., Harris, A., Coleman, L., Greenwell, J.R. and Argent, B.E. (1990) Am. J. Physiol. 259, C752–C761.
75. Marunaka, Y. and Eaton, D.C. (1990) Am. J. Physiol. 258, C352–C368.
76. Findlay, I. and Petersen, O.H. (1985) Pflügers Arch. 403, 328–330.
77. Evans, M.G., Marty, A., Tan, Y.P. and Trautmann, A. (1986) Pflügers Arch. 406, 65–68.
78. Grenningloh, G., Gundelfinger, E., Schmitt, B., Betz, H., Darlison, M.G., Barnard, E.A., Schofield, P.R. and Seeburg, P.H. (1987) Nature (London) 330, 25.

79. Kopito, R.R. and Lodish, H.F. (1985) Nature (London) 316, 234–238.
80. Landry, D.W., Reitman, M., Cragoe, E.J. and Al-Awqati, Q. (1987) J. Gen. Physiol. 90, 779–798.
81. Landry, D.W., Akabas, M.H., Rehead, C., Edelman, A., Cragoe, E.J. and Al-Awqati, Q. (1989) Science 244, 1469–1472.
82. Greger, R., Bleich, M. and Schlatter, E. (1991) Kidney Int. 40:S33, S119–S124.
83. Frömter, E., Ullrich, K.J. (1980) Ann. N.Y. Acad. Sci. 77, 97–110.
84. Schmid, A., Burckhardt, G. and Gögelein, H. (1989) J. Membr. Biol. 111, 265–275.
85. Breuer, W. (1990) Biochim. Biophys. Acta 1022, 229–236.
86. Wittner, M., Weidtke, C., Schlatter, E., Di Stefano, A. and Greger, R. (1984) Pflügers Arch. 402, 52–62.
87. Cabantchik, Z.I., Knauf, P.A. and Rothstein, A. (1978) Biochim. Biophys. Acta 515, 239–302.
88. Bridges, R.J., Worrel, R.T., Frizzell, R.A. and Benos, D.J. (1989) Am. J. Physiol. 256, C902–C912.
89. Tabcharani, J.A., Low, W., Elie, D. and Hanrahan, J.W. (1990) FEBS Lett. 270, 157–164.
90. Bijman, J., Englert, H.C., Lang, H.J., Greger, R. and Frömter, E. (1987) Pflügers Arch. 408, 511–514.
91. Wittner, M., Di Stefano, A., Wangemann, P., Delarge, J., Liegeois, J.F. and Greger, R. (1987) Pflügers Arch. 408, 54–62.
92. Greger, R., Wangemann, P., Wittner, M., Di Stefano, A., Lang, H.J. and Englert, H.C. (1987) In: Diuretics: Basic, Pharmacological, and Clinical Aspects (Andreucci, V.E. and Dal Canton, A., Eds.), pp. 33–38, Martinus Nijhoff Publishing, Boston.
93. Greger, R., Lang, H.J., Englert, H.C. and Wangemann, P. (1987) In: Diuretics II (Puschett, J.B., Ed.), pp. 33–38, Elsevier, Amsterdam.
94. Breuer, W. and Skorecki, K.L. (1989) Biochem. Biophys. Res. Commun. 163, 398–405.
95. Cousin, J.L. and Motais, R. (1982) Biochim. Biophys. Acta 687, 156–164.
96. Stutts, M.J., Henke, D.C. and Boucher, R.C. (1990) Pflügers Arch. 415, 611–616.
97. Gögelein, H., Dahlem, D., Englert, H.C. and Lang, H.J. (1990) FEBS Lett. 268, 79–82.
98. Reinach, P.S. and Schoen, H.F. (1990) Biochim. Biophys. Acta 1026, 13–20.
99. Kreusel, K.M., Fischer, H., Illek, B., Clauss, W. and Hegel, U. (1991) Pflügers Arch. 418, R59.
100. Champigny, G., Verrier, B. and Lazdunski, M. (1990) Biochim. Biophys. Acta 171, 1022–1028.
101. Hwang, T.C., Guggino, S.E. and Guggino, W.B. (1990) Proc. Natl. Acad. Sci. U.S.A. 87, 5706–5709.
102. Schlatter, E. and Greger, R. (1985) Pflügers Arch. 405, 367–376.
103. Liedtke, C.M. (1989) Annu. Rev. Physiol. 51, 143–160.
104. Welsh, M.J. (1987) Physiol. Rev. 67, 1143–1184.
105. Greger, R., Schlatter, E., Wang, F. and Forrest, J.N. Jr. (1984) Pflügers Arch. 402, 376–384.
106. Petersen, O.H. (1987) NIPS 1, 92–95.
107. Berschneider, H.M., Knowles, M.R., Azizkhan, R.G., Boucher, R.C., Tobey, N.A., Orlando, R.C. and Powell, D.W. (1988) FASEB J. 2, 2625–2629.
108. Schulz, I. and Frömter, E. (1968) In: Mukoviscidose Cystische Fibrose II (Windorfer, A. and Stephan, U., Eds.), pp. 12–21, Thieme, Stuttgart.
109. Quinton, P.M. (1983) Nature (London) 301, 421–422.
110. Frizzell, R.A., Rechkemmer, G.R. and Shoemaker, R.L. (1986) Science 233, 558–560.
111. Welsh, M.J. (1986) Science 232, 1648–1649.
112. Li, M., McCann, J.D., Anderson, M.P., Clancy, J.P., Liedtke, C.M., Nairn, A.C., Greengard, P. and Welsh, M.J. (1989) Science 244, 1353–1356.
113. Welsh, M.J., Li, M. and McCann, J.D. (1989) J. Clin. Invest. 84, 2002–2007.
114. Krick, W., Disser, J., Hazama, A., Burckhardt, G. and Frömter, E. (1991) Pflügers Arch. 418, 491–499.
115. Riordan, J.R., Rommens, J.M., Kerem, B.-S., Alon, N., Rozmahel, R., Grzelczak, Z., Zielenski, J., Lok, S., Plavsic, N., Chou, J.-L., Drumm, M.L., Iannuzzi, M.C., Collins, F.S. and Tsui, L.-C. (1989) Science 245, 1066–1073.

116. Rich, D.P., Anderson, M.P., Gregory, R.J., Cheng, S.H., Paul, S., Jefferson, D.M., McCann, J.D., Klinger, K.W., Smith, A.E. and Welsh, M.J. (1990) Nature (London) 347, 358–363.
117. Drumm, M.L., Pope, H.A., Cliff, W.H., Rommens, J.M., Marvin, S.A., Tsui, L.C., Collins, F.S., Frizzell, R.A. and Wilson, J.M. (1990) Cell 62, 1227–1233.
118. Hyde, S., Emsley, P., Hartshorn, M.J., Mimmack, M.M., Gileadi, U., Pearce, S.R., Gallagher, M.P., Gill, D.R., Hubbard, R.E. and Higgins, C.F. (1990) Nature (London) 346, 362–365.
119. Kartner, N., Hanrahan, J.W., Jensen, T.J., Naismith, A.L., Sun, S., Ackerley, C.A., Reyes, E.F., Tsui, L.C., Rommens, J.M., Bear, C.E. and Riordan, J.R. (1991) Cell 64, 681–691.
120. Clancy, J.P., McCann, J.D., Li, M. and Welsh, M.J. (1990) Am. J. Physiol. 258, L25–L32.
121. Willumsen, N.J. and Boucher, R.C. (1989) Am. J. Physiol. 256, C226–C233.
122. Stutts, M.J., Boucher, R. and Knowles, M. (1990) 'Late breaking science' Arlington Symp. Cystic Fibrosis.
123. McCann, J.D., Li, M. and Welsh, M.J. (1989) J. Gen. Physiol. 94, 1015–1036.
124. Worrell, R.T., Butt, A.G., Cliff, W.H. and Frizzell, R.A. (1989) Am. J. Physiol. 256, C1111–C1119.
125. Steinmeyer, K., Ortland, C. and Jentsch, T.J. (1991) Nature (London) 354, 301–304.
126. Steinmeyer, K., Klocke, R., Ortland, C., Gronemeier, M., Jockush, H., Gründer, S. and Jentsch, T.J. (1991) Nature (London) 354, 304–308.
127. Thiemann, A., Gründer, S., Pusch, M. and Jentsch, T.J. (1992) Nature (London) 356, 57–59.
128. Paulmichl, M., Li, Y., Wickman, K., Ackerman, M., Peralta, E. and Clapham, D. (1992) Nature (London) 356, 238–241.
129. Lukacs, G.L., Nanda, A., Rotstein, O.D. and Grinstein, S. (1991) Fed. Europ. Biochem. Soc. Lett. 288, 17–20.
130. Bear, C.E., Li, C., Kartner, N., Bridges, R.J., Jensen, T.J., Ramjeesingh, M. and Riordan, J.R. (1992) Cell 68, 809–818.
131. Kubitz, R., Warth, R., Allert, N., Kunzelmann, K. and Greger, R. (1992) Pflügers Arch. in press.
132. Allert, N., Kunzelmann, K. and Greger, R. (1992) Pflügers Arch. 420, R61.
133. Kunzelmann, K., Grolik, M., Kubitz, R. and Greger, R. (1992) Pflügers Arch. in press.
134. Kunzelmann, K., Kubitz, R., Grolik, M., Warth, R. and Greger, R. (1992) Pflügers Arch. in press.

CHAPTER 10

Voltage-gated K$^+$ channels

O. PONGS

Zentrum für Molekulare Neurobiologie, 2000 Hamburg 20, Germany and Lehrstuhl für Biochemie, Ruhr-Universität, 4630 Bochum, Germany

1. Introduction

The properties of potassium channels are remarkably diverse [1]. Their distribution varies among different cellular regions and cell types displaying a spectrum of signalling capabilities [2]. Voltage-dependent potassium channels can be broadly classified according to their kinetics and pharmacological properties [1–3] as non-inactivating delayed rectifiers, inward or anomalous rectifiers rapidly inactivating transient 'A'-channels, or as channels modulated by intracellular agents such as Ca^{2+}, ATP and G-proteins. Single-channel current recordings have revealed that both delayed and transient K$^+$ currents are mediated by several subtypes of channels, which share some functional properties [4,5]. This raised the possibility that functional subtypes of potassium channels share basic structural features. The characterization of cDNAs encoding potassium channel forming proteins [6–31] has greatly advanced our knowledge of the molecular structure of potassium channels and our understanding of the molecular basis of potassium channel diversity.

The derived protein sequences indicate common features. Most notably, K$^+$ channel forming proteins possess a core region containing six potentially membrane-spanning segments flanked by highly diverse amino- and carboxyl-termini [32,33]. The six transmembrane segments may be oriented in a pseudosymmetric fashion across the membrane locating the amino- and carboxyl-termini on the intracellular side of the membrane. From structure–function analyses it may be inferred that small variations in amino acid sequences of K$^+$ channel forming proteins suffice for generating the observed diversity of voltage-gated potassium channels. This proposition will be explored in the following sections.

At this time, more then thirty K$^+$ channel DNAs have been cloned and characterized from various sources, predominantly from *Drosophila melanogaster*, mouse, rat and human cDNA/genomic libraries [6–31]. Inspection of the derived primary K$^+$ channel protein sequences indicates that voltage-gated K$^+$ channels belong to a protein family. Members of this protein family have in common several important features. The proteins are relatively small having a molecular mass between ≈ 60 and ≈ 100 kDa.

2. Structure and biophysical properties of cloned voltage-gated K^+ channels

Historically the *Shaker* (*Sh*) K^+ channel was the first K^+ channel which was cloned and characterized [6–10]. Subsequently many more K^+ channel cDNAs and genes have been isolated and studied. Yet *Sh* K^+ channels remained in the forefront of K^+ channel research. The study of *Sh* K^+ channel mutants has provided the most thorough insight into structure–function relationships of K^+ channels to date. I will first discuss in this chapter the primary sequences of voltage-gated K^+ channels. I will only use a few selected examples for discussion. As of this time, so many related K^+ channel protein sequences have been published that it is not feasible to discuss all of them. Subsequently, I will describe in detail the present knowledge about functional K^+ channel domains which are implicated in activation, inactivation and selectivity of the K^+ channel.

2.1. K^+ channels of the Shaker family

Drosophila Sh mutants were first recognized by their abnormal leg shaking behaviour under ether anaesthesia [34]. Subsequently, electrophysiological investigations revealed that *Sh* mutants have altered K^+ (A-type) currents in voltage-clamped muscle cells [35] and action potentials with delayed repolarization in cervical giant fibre interneurones [36,37]. The altered excitability of muscle and neuronal cells is caused by dysfunctional potassium channels [38]. Extensive molecular analysis of the *Sh* locus [6–12] has shown that it encodes a large transcription unit which expresses multiple A-type potassium channel subunits (Fig. 1). These subunits are translated from alternate mRNAs which are probably generated by both alternate transcription from different start sites within the K^+ channel gene and alternate splicing of primary transcript(s). The isolation of various *Sh* potassium channel cDNAs in different laboratories has created a confusing, nonsystematic nomenclature of *Sh* potassium channel subunits. Recently, a more logical nomenclature has been proposed [29] which is based on the structure of the deduced *Sh* K^+ channel proteins (Fig. 1). The proteins share a common core region which is flanked by variant amino- and carboxy-terminal sequences. The variant amino-termini are designated by a letter (A, B, D, G, H) and the variant carboxy termini by a number (1 or 2). Accordingly, *Sh*A1 is a subunit having amino-terminus A and carboxy-terminus 1 and so forth. Table I summarizes how this nomenclature is translated into the ones used by different laboratories in the *Sh* field.

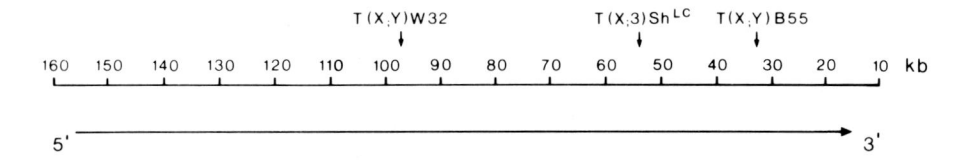

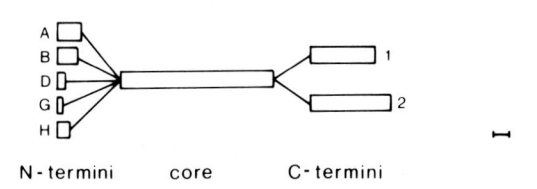

Fig. 1. Genomic map of the *Shaker* potassium channel gene. The coordinates in kb are as in [9]. Arrows indicate the location of chromosomal rearrangements in *Shaker* mutants which were used to define the gene. The gene is located on the X-chromosome at 16EF. Chromocentre is to the left. Horizontal arrow indicates the direction of transcription and the length of the longest *Shaker* primary transcript. Boxes indicate the *Shaker* K^+ channel subunit family. A common core region is flanked by variable amino-termini (A, B, D, G, H) and by one of two alternative carboxy-termini. The nomenclature is according to [29]. Code for the nomenclature of other laboratories is given in Table I. Box lengths are approximately to scale. Bar at lower right corresponds to twenty amino acids. All combinations between the core region and the amino- and carboxy-termini have been found in *Drosophila* [41].

The molecular analysis of the *Sh* K^+ channel transcription unit indicated that at least ten different K^+ channel subunits may be expressed in *Drosophila* cells. This observation raised a number of important questions; only some of them can be addressed here. Each subunit has been expressed in the *Xenopus* oocyte expression system [11,12,29,39]. These experiments indicate that each subunit expresses distinct

TABLE I
Rosetta-stone of *Shaker* K^+ channel nomenclature

From [29]	From [8,11]	From [12,39]	From [9,53]
*Sh*A1	*Sh*A	4–37	*Sh*α, *Sh*α1
*Sh*A2	*Sh*B	H4,4–4	*Sh*α2
*Sh*B1	–	29–37	*Sh*β1
*Sh*B2	*Sh*D	29–4	*Sh*β, *Sh*β2
*Sh*D1	–	–	–
*Sh*D2	–	–	*Sh*δ2
*Sh*G1	*Sh*C	–	*Sh*τ1
*Sh*G2	–	–	*Sh*τ2
*Sh*H1	–	H37,37–37	–
*Sh*H2	–	37–4	–

TABLE II
Time constants and mean single-channel open times of *Shaker* K$^+$ channel splice variants

	t_1 (ms)	t_2 (ms)	open time (ms)
*Sh*A1	103.5 ± 2.5	100 ± 10	n.d.
*Sh*A2	5.1 ± 1–3	≈ 2 000	0.7–0.9
*Sh*B1	7.3 ± 2	n.d.	n.d.
*Sh*B2	13.8 ± 3.1	90.5 ± 29	n.d.
*Sh*D2	–	≈ 2 000	3.0

Inactivation time constants are from [39,62], mean single channel open times are from [29,43]. Values were obtained at 0 mV test potentials. In the case of *Sh*A1, *Sh*A2 and *Sh*B2 ≈ 5–15% of peak current inactivates slowly, i.e., with the time constant t_2. n.d. = not determined.

voltage-gated K$^+$ channels. The properties of these K$^+$ channels vary, most notably, in their inactivation behaviour including the mean single-channel open time (Table II). This observation indicates that the expression of multiple *Sh* K$^+$ channel proteins leads to the formation of K$^+$ channels having distinct properties. This functional diversity probably reflects a mechanism for generating physiological diversity within excitable cells. Consistent with this idea is the differential distribution of splicing variants from the *Sh* locus in the *Drosophila* nervous system as well as in muscle [40,41]. An example of the biological implementation of the extensive *Sh* transcript diversity has recently been provided by a characterization of *Sh* K$^+$ channels in photoreceptor cells of *Drosophila* [42]. Single-channel properties are similar to those reported in muscle [42] with respect to channel mean open time and single-channel slope conductance (Table III). However, the photoreceptor and muscle A currents differ very significantly in the voltage-operating range of the channels. Whereas the voltage dependence of *Sh* A currents, both in muscle and in oocyte expression studies, is very similar, the activation as well as inactivation ranges of I_A in photoreceptors are shifted both by approximately 40–50 mV to more negative voltages. These differences in I_A are apparently related to the presence of different splice variants of *Sh* transcripts in photoreceptor and muscle cells [41].

TABLE III
Properties of *Shaker* K$^+$ channels in *Drosophila* muscle and photoreceptor cells

Cell type	$V^a_{1/2}$ (mV)	$V^h_{1/2}$ (mV)	Single-channel open times (ms)	Single-channel slope conductance (pS)
Muscle	+ 10	− 29	1.46 ± 0.41	11.5
Photoreceptor	− 63	− 84	1.7 ± 0.6	11

$V^a_{1/2}$ refers to the test potentials where the conductance increase has reached one half of its maximal value; $V^h_{1/2}$ to the prepulse membrane potential at which the current response to a 50 mV test potential is 50% of its maximum value. Muscle data are from [43,64], photoreceptor data from [41].

2.2. K^+ channels of the MBK/RCK/HBK family

It has been suggested that voltage-gated K^+ channel diversity evolved well before the separation of the eukaryotic world into protostomes and deuterostomes [2,14]. The general similarity among voltage-gated K^+ channels, both in structure and in electrophysiological behaviour supports this view. The characterization of *Sh* homologous K^+ channel cDNAs from *Xenopus*, mouse, rat and human cDNA libraries showed that the structure of voltage-gated K^+ channel proteins has been highly conserved during evolution [13,14,22]. Alignment of the derived primary protein sequences (Fig. 2) reveals a high degree of similarity in the structure and general architecture of *Sh*-like K^+ channel proteins. Sequence identities are 70% and higher indicating a substantial evolutionary pressure to conserve K^+ channel structure. Relatively minor variations in amino acid sequence are apparently responsible for generating distinct voltage-gated K^+ channels. This notion has been confirmed by the isolation and characterization of several MBK/RCK/HBK related cDNAs [13–15,17,18,20,23–25]. The deduced proteins assemble in the oocyte expression system into distinct voltage-gated K^+ channels. They differ in both electrophysiological

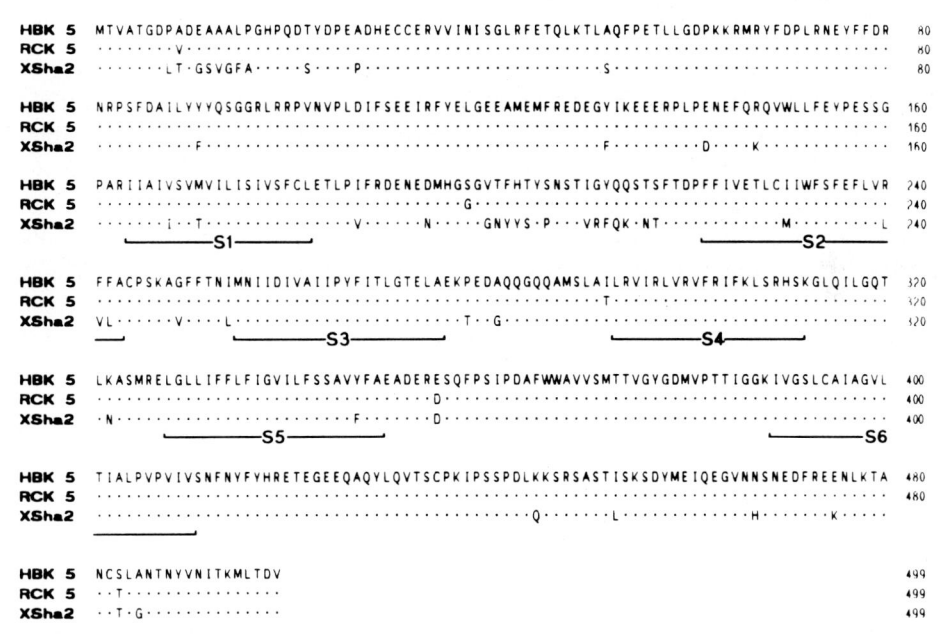

```
HBK 5   MTVATGDPADEAAALPGHPQDTYDPEADHECCERVVINISGLRFETQLKTLAQFPETLLGDPKKRMRYFDPLRNEYFFDR   80
RCK 5   ...........V...................................................................   80
XSha2   ......LT·GSVGFA····S····P·····························S·······················   80

HBK 5   NRPSFDAILYYYQSGGRLRRPVNVPLDIFSEEIRFYELGEEAMEMFREDEGYIKEEERPLPENEFQRQVWLLFEYPESSG  160
RCK 5   ................................................................................  160
XSha2   ·········F·············································F···········D····K·······  160

HBK 5   PARIIAIVSVMVILISIVSFCLETLPIFRDENEDMHGSGVTFHTYSNSTIGYQQSTSFTDPFFIVETLCIIWFSFEFLVR  240
RCK 5   ...............................G................................................  240
XSha2   ········I··T············V······N····GNYYS·P···VRFQK·NT················M·······L  240
                 ——S1——                                                      ——S2——

HBK 5   FFACPSKAGFFTNIMNIIDIVAIIPYFITLGTELAEKPEDAQQGQQAMSLAILRVIRLVRVFRIFKLSRHSKGLQILGQT  320
RCK 5   ............................................T...................................  320
XSha2   VL·····V····L··············T··G··················································  320
             ——S3——                         ——S4——

HBK 5   LKASMRELGLLIFFLFIGVILFSSAVYFAEADERESQFPSIPDAFWWAVVSMTTVGYGDMVPTTIGGKIVGSLCAIAGVL  400
RCK 5   ....................D............................................................  400
XSha2   ·N···············F····D···········································................  400
           ——S5——                                                       ——S6——

HBK 5   TIALPVPVIVSNFNYFYHRETEGEEQAQYLQVTSCPKIPSSPDLKKSRSASTISKSDYMEIQEGVNNSNEDFREENLKTA  480
RCK 5   ................................................................................  480
XSha2   ································Q······L·················H······K·····        480

HBK 5   NCSLANTNYVNITKMLTDV  499
RCK 5   ··T················  499
XSha2   ··T·G··············  499
```

Fig. 2. Sequence homologies of deduced primary sequences of subunits of voltage-gated K^+ channels of the *Shaker*/RCK family. Sequences are derived from a human (HBK5), rat (RCK5) [20] and an amphibian (XSha2) [17] K^+ channel cDNA. The related mouse sequence is identical to RCK5 and has not been listed. Identical amino acids are indicated by dots. Proposed membrane spanning segments S1–S6 are underlined by brackets. The HBK5 sequence is from Grupe (personal communication). Amino acids are numbered beginning with the methionine initiation site; numbers of the last residue in each row are given on the right-hand side.

TABLE IV
Pharmacology of K^+ channels in the oocyte expression system

K^+ channel type	4-AP (mM)	TEA (mM)	DTX (nM)	MCDP (nM)	CTX (nM)
ShA2	5.5	17.8	>200	>1000	4.0
RCK1	1.0	0.6	12	45	22
RCK2	1.5	7	20	10	1
RCK3	1.5	50	>600	>1000	1
RCK4	12.5	>100	>200	>2000	>40
RCK5	0.8	129	4	175	6
RCK7 (K_v1)	0.4	>40	>200	>600	>200

Numbers in this table refer to ID_{50} values (50% inhibition of peak current), measured at 20 mV test potential. Data for ShA2 are from [29], for RCK1, RCK3, RCK4, RCK5 from [20], for RCK2 from [22], for K_v1 from [25].

and pharmacological properties, having distinct voltage-dependent gating mechanisms, channel conductances, and toxin-binding properties (see Table IV).

2.3. K^+ channels of Shaker relatives

K^+ channel diversity may not only result from sequence variations within the Sh K^+ channel protein family, but also from the presence of an extended gene family within the genome. Three Sh relatives have been identified in Drosophila encoding the Shal, Shaw, and Shab proteins [16,28]. Like Sh gene products, the Shal, Shaw, and Shab gene products express voltage-gated K^+ channels in the oocyte expression system. These channels mediate outward currents, however, which differ markedly in their kinetic and voltage-sensitive properties from the ones of Sh currents. A detailed analysis of these K^+ channels is still missing. Therefore, a more specific comparison of Shal, Shaw, and Shab channels with the Sh K^+ channels is not yet possible.

2.4. Pharmacology of K^+ channels

Although voltage-gated K^+ channels show a striking diversity in their electrophysiological properties, they do not vary much in their pharmacological properties [1]. Thus, many more distinct K^+ channels can be measured in excitable cells than can be discerned by their pharmacological profile. To put it simply, not enough different K^+ channel blockers are available to match the diversity of K^+ channel proteins. This situation makes it very difficult at present to compare the properties of cloned K^+ channels with those of native K^+ channels. Moreover, it is not even clear whether the pharmacology of a cloned K^+ channel expressed in the oocyte system is in fact like that of the corresponding native K^+ channel in its natural environment [43]. Nevertheless, the study of binding blockers and toxins to K^+ channels has proven to be very informative for establishing structure–function relationships in the K^+ channel field. Only five different substances are in general use for characterizing voltage-gated K^+ channels. These are: (1) tetraethylammonium chloride (TEA); (2) 4-aminopyridine (4-AP); (3) dendrotoxin (DTX); (4) charybdotoxin (CTX); and (5) mast cell degranulating peptide (MCDP). TEA and 4-AP are small organic compounds, whereas DTX, CTX and MCDP are peptides varying in length between 39 and 60 amino acids [44]. TEA and CTX appear to block the open K^+ channel conformation. CTX blocks K^+ channels from the outside. TEA, on the other hand, can occlude the K^+ channel pore from both outside and inside. How 4-AP, CTX, DTX and MCDP block K^+ channels is still uncertain. The TEA binding sites on cloned K^+ channels are well-defined [45–47]; the ones for 4-AP, DTX and MCDP are not. Table IV summarizes the pharmacological profiles of some *Sh*/RCK K^+ channels. The data indicate that the observed TEA sensitivity of these channels is quite different: some channels, such as *Sh* and RCK1, are blocked by low TEA concentrations, whereas others, such as RCK4 and RCK5, are very insensitive to the presence of TEA in the external medium. Unfortunately, a comparable study for blocking K^+ channels with TEA from the inside has not been carried out. The sensitivity of K^+ channels towards DTX, CTX and MCDP is also remarkably different. Again, some channels are blocked by low toxin concentrations, whereas others are not. In contrast, the various K^+ channels vary little in their sensitivity towards 4-AP. The differences in sensitivity towards blockers could be due to local structural differences of the respective K^+ channels. Alternatively, more global structural differences might also be held responsible. The former hypothesis has been tested by site directed in vitro mutagenesis studies as well as by construction of chimeric K^+ channels. Both experimental approaches yielded comparable results which indicates that local structural changes are probably the molecular basis of different sensitivities towards K^+ channel blockers. The most consistent results have been obtained for binding TEA to the outside or inside of the K^+ channel [45–47]. Incidentally, these studies also have indicated which part of the K^+ channel protein is involved in forming the K^+ channel pore [46–48].

Chimeric K^+ channels have been constructed with DRK1 (a *Shab* relative) and NGK2 (a *Shaw* relative) [47] and independently between *Sh* and RCK1 [49]. In both cases, the loop regions between proposed membrane-spanning segments S5 and S6 were exchanged (see Fig. 3). Then, the pharmacology and single-channel behaviour of the chimeric K^+ channels were studied in the oocyte expression system. The results show that both the single-channel conductance and the blockade by external and internal TEA ions were characteristic of the donor K^+ channel. For example, the single-channel conductance of NGK2 is almost three times that of DRK1, and NGK2 is more sensitive to blockade by external TEA whereas DRK1 is more sensitive to blockade by internal TEA. Replacement of the S5–S6 loop region in NGK2 by the one of DRK1, confers to the recombinant K^+ channel the single-channel conductance and TEA sensitivity of DRK1. Vice versa, replacement of the

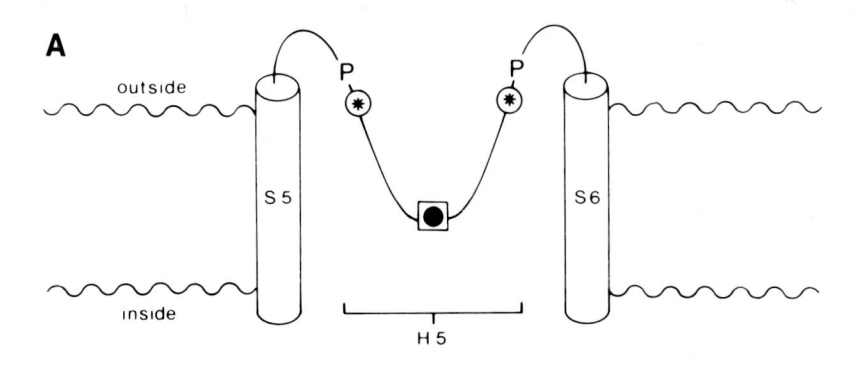

Fig. 3. (A) Model of the proposed pore forming part of K⁺ channel subunits. Segments S5 and S6 are possibly membrane-spanning helices. The helices are connected by a hydrophobic segment H5 which may be tucked into the lipid bilayer [48]. H5 is flanked by two proline residues P. Adjacent to these proline residues are amino acid side chains (*) important for external TEA binding [45,46]. Approximately halfway between these two proline residues are amino acid side chains (●) affecting internal TEA binding [46,47] and K⁺ channel selectivity [48]. (B) Mutations are indicated which affect in *Shaker* K⁺ channels external TEA (TEA$_e$) or internal TEA (TEA$_i$) binding. Concentrations of TEA for half block of the wild-type and mutant K⁺ channels are given at the right-hand side of the corresponding sequence. Data have been compiled from [45–47].

S5–S6 loop region in DRK1 by the one of NGK2 confers to the recombinant K⁺ channel the single-channel conductance and TEA sensitivity of NGK2 [47]. Thus, a region of approximately twenty amino acids in the primary sequence of K⁺ channels controls essential properties of the K⁺ channel pore [48].

A number of mutations have been introduced into this region, most notably in the primary sequence of *Sh* K⁺ channels. The effects of the mutations on inhibition by external and internal TEA, on single-channel conductance and on K⁺ channel selectivity were measured. The results of these measurements suggest that amino acids flanking the twenty amino acid region are important for external TEA binding [45]. Apparently, positively charged flanking amino acid residues decrease the binding of TEA to the external binding site. Mutation of D431 to K in *Sh* channels reduces the external TEA-affinity two- to three-fold, mutation of T449 to K at least ten-fold. V451 has not been mutated. Possibly, the presence of lysine at the equivalent position in the DRK1 sequence is responsible in DRK1 channels for low-affinity binding of external TEA. The 449 position is especially critical in determining external TEA sensitivity. Substitution of different amino acids at this position leads to TEA-affinities which vary over a 500-fold range [45]. Several pieces of evidence indicate that external TEA acts by plugging the pore. Most importantly, dissociation of TEA is enhanced by a high concentration of K^+ on the internal side of the channel, as though K^+ can enter the pore from the opposite side and expel TEA [50,51]. Therefore, the 449 position is probably very close to or even within the mouth of the channel. However, external TEA binding to K^+ channels does not appear to be voltage-dependent. This suggests that the TEA-binding site lies outside of the transmembrane electric field. Accordingly, the positive charges at positions 441, 449 and 451 in Fig. 3 may not be sensitive to the transmembrane voltage consistent with an extracellular location outside of the membrane.

Conversely, internally bound TEA is sensitive to the transmembrane voltage [50]. Also, dissociation of internally bound TEA is enhanced by a high concentration of K^+ on the external side of the channel. Finally, internal TEA gains access to the internal binding site only when the channel is open [52]. These data are consistent with an internal TEA-binding site located $\approx 15\%$ of the way into the transmembrane electric field. One particular Sh channel mutant has been found which decreases internal TEA binding ten-fold [46]. This mutation (T441S, see Fig. 3) lies within the sequence I/VTMTTV/LGY which occurs in every K^+ channel protein sequence which has been cloned to date. Most likely, T441 is part of the internal TEA binding site. DRK1 and NGK2 markedly differ in their sensitivity for internal TEA, but do not differ at the equivalent T441 position (Fig. 3) [47]. Therefore, other amino acid side chains also contribute to the internal TEA binding site. It is not known yet which amino acid residue may be responsible for the distinction between DRK1 and NGK2. An obvious candidate might be the valine–leucine replacement, which occurs two amino acid residues away from the T441 position in the DRK1/NGK2 K^+ channel proteins.

The mutation experiments indicate that amino acid residues which are only eight to eleven peptide bonds apart, affect either binding of TEA to the external or to the internal side of the channel pore. This suggests that these residues might be part of the ion conduction pore forming structures. Following this hypothesis, the sequence shown in Fig. 3 has to dip into the membrane such that D431 and T559 would lie on the outside and T441 on the inside. This proposition means that a very short stretch of eight to ten amino acids connects the two ends of the pore. Assuming that the K^+ channel pore spans the lipid bilayer, the stretch of eight to ten amino acids would have to transverse the lipid bilayer. This could only be accomplished if the peptide backbone would adopt an extended β sheet conformation.

3. Structure of K^+ channel genes

3.1. Genes in Drosophila

Four K^+ channel encoding genes have been described in *Drosophila*, namely *Shaker*, *Shab*, *Shal* and *Shaw* on the basis of cDNA analysis. Only the *Sh* gene has been characterized extensively [8–10]. A number of *Sh* mutants have been mapped such that the molecular basis of altered excitability in several *Sh* mutants is now known [38]. Mutants for *Shab*, *Shal* or *Shaw* have not been isolated yet. *Sh* represents one of the largest genes in *Drosophila* being approximately 150 kb long. So far, 24 exons have been mapped. The majority of these exons (3–22) contain part of the open reading frame (Fig. 4). Exons 2 to −2 have been mapped within the 5'-nontranslated region of the ≈10 kb long *Sh*B mRNA for which 9.8 kb of cDNA sequence information has been obtained [53]. Each alternative amino-terminus is encoded in a separate exon. The two alternative carboxy-termini are encoded in two sets of exons, 17–19 and 20–22, respectively. The chromosomal breakpoint in T(1;Y)W32 mutants interrupts the X-chromosome between exons 4 and 5. This rearrangement alters, but does not abolish I_A in W32 flies. *Sh*A and *Sh*G mRNAs, but not *Sh*D and *Sh*B mRNAs are still present in W32 flies [41]. This indicates that the W32 rearrangement disrupts the expression of some, but not all *Sh* mRNAs [8,41]. Probably, the transcription of *Sh* mRNA starts at sites which are distal as well as proximal to the W32 breakpoint in the *Sh* gene. Then, the different splice variants of *Sh* mRNA would not only be generated by alternate splicing of the primary *Sh* transcripts, but also by alternate transcription of different promoters. These have not been characterized yet. Therefore, it is still conjectural whether the tissue specific occurrence of *Sh* splice variants [40,41] implies a tissue specific regulation of transcription from different promoters of the *Sh* gene.

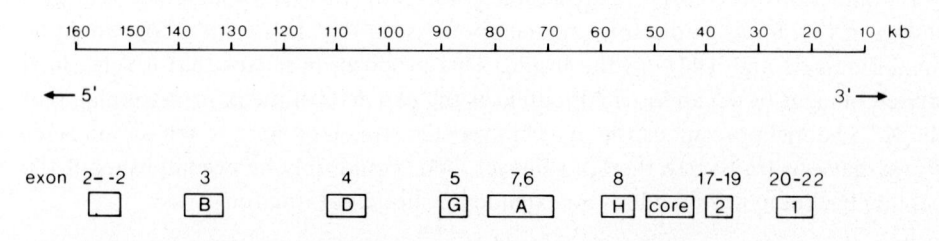

Fig. 4. Schematic organization of the *Shaker* K^+ channel gene. The coordinates of the physical map are as in [9]. The direction of transcription is indicated by arrows. Approximate location of exons is given by boxes. Open box corresponds to noncoding exons, lettered boxes to alternative amino-terminal ends of *Shaker* K^+ channel proteins and the core region, respectively, numbered boxes to the two alternative carboxy-terminal ends. Exon numbers are as in [53].

3.2. Vertebrate genes

The full characterization of a vertebrate K^+ channel gene has not been accomplished yet. Up to now, related K^+ channel mRNAs, which have been characterized in the RCK family, are not generated by alternate splicing mechanisms [20]. Instead, the mRNAs are transcribed from a set of related, but different genes. Moreover, the coding regions of RCK genes seem not to be interrupted by intervening sequences, in the coding regions [22,25,54]. However, these data do not exclude the possibility that other K^+ channel genes contain intervening sequences in their coding region and that their expression gives rise to alternate transcripts. The occurrence of many uninterrupted coding regions in vertebrate K^+ channel genes however has greatly facilitated the isolation of *Sh* related K^+ channel DNA.

4. The basis of K^+ channel diversity

4.1. Properties of homo- and heteromultimers

This chapter's brief review has already described some mechanisms from which the diversity of K^+ channel function, as evident at the level of biophysical and pharmacological studies, may arise. The multiplicity of different K^+ channel protein encoding genes probably generates some of the diversity. Alternate splicing of K^+ channel preRNAs, as observed for *Sh* RNA, possibly further increases the diversity. Whether interactions with other proteins or post-translational modifications, such as glycosylation and phosphorylation, also contribute to K^+ channel diversification, is not known. K^+ channel diversity may further arise from heteromultimeric K^+ channels [55,56]. Each of the known K^+ channel proteins resembles one of the four internally homologous repeats of Na^+ or Ca^{2+} channels [9,32,57]. By analogy, K^+ channels are multimeric membrane proteins formed by the aggregation of several independent subunits. By studying the interaction of CTX with coexpressed wild-type and toxin-insensitive mutant subunits, *Sh* K^+ channels are found to have a tetrameric structure [58]. Probably, all cloned K^+ channel subunit proteins assemble into multimeric structures of a similar stoichiometry. Thus, it seems likely that functional K^+ channels may also be formed by the aggregation of different subunits. Such heteromultimeric K^+ channels could have properties different from those of the corresponding homomultimeric channels. Two examples for the formation of heteromultimeric K^+ channels have been studied in the *Xenopus* oocyte expression system [55,56]. Coexpression of different *Sh* K^+ channel polypeptides leads to the formation of heteromultimeric K^+ channels having a kinetic behaviour distinct from those expressed by the homomultimeric K^+ channels [55]. Similarly, two vertebrate K^+ channel subunits were coexpressed either in oocytes or in HeLa cells [56]. The homomultimeric K^+ channels (RCK1 and RCK4) had distinct single-channel conductancies, kinetic behaviour and pharmacology. The formation of heteromultimeric RCK1/RCK4 K^+ channels lead to the expression of functional K^+ channels with new properties, both with respect to single-channel conductance, kinetic behaviour and pharmacology. These experiments suggest that coassembly of different K^+ channel subunits is another potential source of K^+ channel diversity.

Finally, a K^+ channel protein has been isolated from synaptic plasma membranes of bovine cerebral cortex [59]. The amino-terminal sequence of this protein is virtually identical to the amino-terminal sequence deduced from RCK5 cDNA [18,20]. RCK5 K^+ channels expressed in *Xenopus* oocytes are sensitive to DTX. The bovine protein also binds DTX with high affinity. However, the bovine protein preparation also possessed a low content of high-affinity binding sites for β-bungarotoxin. None of the cloned K^+ channels possesses a high-affinity β-bungarotoxin binding site. As the bovine RCK5-like protein copurifies with a small subunit protein it is quite possible that the binding of this small subunit protein endowes the K^+ channel with sensitivity to β-bungarotoxin. Since the small subunit protein has not been cloned yet, it is not known specifically how the small subunits affect K^+ channel function.

4.2. Functional domains in K^+ channels

When one inspects the multiple K^+ channel protein sequences that have been derived, one readily recognizes that they have related primary sequences. This suggests that they have similar three-dimensional structures. The primary sequences can be subdivided into an amino-terminal, a core and a carboxy-terminal domain (see Fig. 5). Each domain seems to contribute separately to the structure and function of a given K^+ channel [49]. Following this hypothesis, it has been possible to carry out domain swapping experiments between *Sh* and RCK proteins [49] as well as between

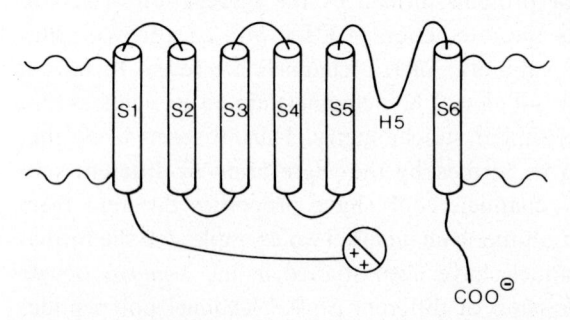

Fig. 5. Proposed topology of K^+ channel subunits inserted into the membrane. COO$^-$: carboxy-terminal. The proposed membrane-spanning segments S1–S6 in the core region of K^+ channel proteins are displayed linearly. H5 may be part of the K^+ channel pore. The amino-terminal inactivation gate is symbolized by a positively charged ball which could occlude the pore region. The extracellular side is thought to be at top and the intracellular side at bottom.

NGK2 and DRK1 proteins [47]. The results of these experiments indicate that the core domain carries the toxin-binding sites [45–47,49] as well as the K^+ channel pore [48]. The core domain comprises six hydrophobic segments which may transverse the membrane such that amino- and carboxy-terminal ends are located on the cytoplasmic side (see Fig. 5) [9]. An additional hydrophobic region, H5, tucked into the plane of the membrane between segments S5 and S6, is probably the pore-forming region of K^+ channels [48]. Rapidly inactivating K^+ channels contain an inactivation gate which is harboured within the amino-terminal sequences [29,60,61]. This inactivation gate occludes the pore of the open channel. Segment S6 represents at least in part a binding site of the inactivation gate [62]. According to the present topological model of K^+ channel subunits, this would put the inactivation gate receptor site in the vicinity of the pore forming domain H5. Finally, the carboxy-terminal domain seems to be important for the formation of active K^+ channels [38,63] but no specific function could yet be assigned to it. When major parts of the carboxy-termini of *Sh* [38] or DRK [63] K^+ channels are deleted, the resulting deletion mutants do not form active K^+ channels in the oocyte expression system.

5. General structural implications

Voltage-gated potassium channels have a complex kinetic behaviour. They occur in a closed state (C) at negative membrane potentials. Depolarization of the membrane induces a voltage-dependent transition of this closed state into another one (C^*). This transition is accompanied by a movement of charges across the electric field of the membrane. The current which is produced by this charge movement is called the gating current [2]. The sign of the current indicates that positive charges are being moved from the inside to the outside of the membrane. Alternatively, the current could also be generated by moving negative charges in the opposite direction, i.e., from the outside to the inside of the membrane. The activated closed channel (C^*) opens in a voltage-independent manner. The open state (O) is terminated by channel closure. Most potassium channels have two different modes of closure. The channels can return from the open state to the closed state (C^*) which allows them to reopen rapidly. Alternatively, they can undergo inactivation and enter into an inactivated state (I). The channels cannot reopen rapidly from the inactivated state. Two inactivated states can be discerned: a non-adsorbing (I_1) and an adsorbing inactivated state (I_2). Recovery from I_1 is faster than from I_2. Recovery from the inactivated state is accelerated by repolarization of the membrane. Taking into consideration thorough kinetic analyses of the behaviour of single-*Sh* K^+ channels [42,64], the following diagram on the different states can be formulated (Fig. 6).

Although this diagram is rather complex, it is the simplest one which would accommodate all the kinetic aspects of *Sh* K^+ channels [42]. The diagram provides

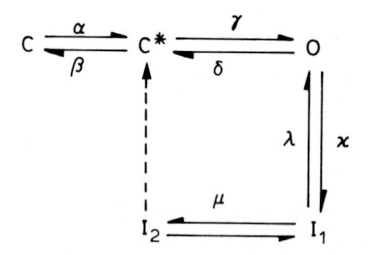

$$C \underset{\beta}{\overset{\alpha}{\rightleftarrows}} C^* \underset{\delta}{\overset{\gamma}{\rightleftarrows}} O$$

$$\text{with vertical transitions } \mu, \lambda, \kappa \text{ between } I_2, C^*, O, I_1$$

Fig. 6. Diagram on the different states of single Sh K^+ channels.

a framework, in which structural alterations of Sh K^+ channels can be correlated with functional changes. It should be noted that the diagram does not explicitly describe the pathway for recovery from inactivation. Tentatively, recovery comprises a transition from the inactivated state into the closed state C^*.

The voltage-dependent transition of C to C^* may be described by two rate constants α and β as illustrated in the diagram on the different states. Mutations in segment S4 alter the gating current of ionic channels and/or the voltage-dependence of channel activation. Although the effects of mutations in segment S4 of K^+ channels have been studied in voltage-clamp experiments [65], single-channel data are not available. From similar studies on Na^+ channel gating [66] it may be inferred that structural alterations in segment S4 affect the rate constant α. This view assigns an important role to segment S4 of K^+ channels in the voltage-dependent transition of C to C^*.

The closing rate constant β influences the stability of the activated state C^*. Accordingly, changes in β may alter the deactivation rate. The gating of Sh^5 channels was shown to be altered in this way [64]. The Sh^5 allele carries a point mutation which leads to an amino acid substitution in segment S5 of Sh channels, replacing for instance in ShA proteins F387 by I [38,67]. The voltage-dependent transition rates, into and out of the open state, were not significantly affected by the mutation. In contrast, the latencies until the channel opens following a voltage step are increased at low voltages. Thereby, the mutation increases the voltage required to activate and inactivate the Sh channel by approximately 20 mV and decreases the steepness of the voltage-dependence of steady state inactivation. Using the type of kinetic model illustrated above, an alteration in the amplitude and voltage-dependence of the deactivation rate for each subunit (β) can account for the alterations in voltage-dependent gating of Sh channels by the Sh^5 mutation [64].

The distribution of open channel times is mainly determined by the rate constants δ and κ (λ is assumed to be very small). Mutations which change the C^* to O transition (e.g., the burst size of channel opening) have not been characterized yet. However, structural alterations which affect κ and thereby the level of steady state inactivation have been described for Sh channels [29,60]. Different splice variants of Sh channels express single channels with distinct mean open times as well as with distinct inactivation kinetics [29]. These differences are due to alternate amino-termini. The kinetic analyses suggests that the amino-terminus influences the rate κ. This observation has been substantiated by a mutational analysis of the amino-terminal *Sh* protein region [60]. A region near the amino-terminus with an important role in the transition from the open state to the inactivated state has been identified. These results suggest a model where this region forms a cytoplasmic domain that interacts with the open channel to cause inactivation. The interaction requires a receptor domain to which the amino-terminus binds in order to occlude the open channel.

A mutational analysis within segment S6 of *Sh* channels has shown that a region within segment S6 may serve as a receptor for the inactivating domain [62]. Replacement of A463 to V within *Sh*A segment S6 markedly alters both the rates of inactivation and of recovery from inactivation. Incidentally, this amino acid replacement occurs in the alternative carboxyl-termini of *Sh* variants. The kinetic analysis suggests that the alteration in segment S6 leads to an accelerated transition from the non-adsorbing to the adsorbing inactivated state (I_1 to I_2 in the diagram). According to the kinetic model above, the stability of I_1 is mainly described by a single exponential with a mean duration of $1/(\lambda + \mu)$. The rate μ will influence the equilibrium between I_1 and I_2. Possibly, this rate constant is affected by sequence alterations in segment S6. As shown in Fig. 3, the putative K^+ channel pore forming region may be tucked into the membrane between segments S5 and S6. Both segments have important roles in determining the life time of the activated and inactivated channel states and therefore may be intimately linked to the K^+ channel pore.

Acknowledgement

The work of my laboratory which has been quoted in this chapter has been supported by the Deutsche Forschungsgemeinschaft, the EEC and the Fonds Chem. Industrie.

References

1. Rudy, B. (1988) Neuroscience 25, 729–749.
2. Hille, B. (1984) Ionic Channels in Excitable Membranes, Sinauer, Sunderland, MA.
3. Moczydlowski, E., Lucchesi, K. and Ravindran, A. (1988) J. Membr. Biol. 105, 95–111.
4. Marty, A. and Neher, E. (1985) J. Physiol. 367, 117–141.
5. Hoshi, T. and Aldrich, R.W. (1988) J. Gen. Physiol. 91, 73–106.
6. Papazian, D.M., Schwarz, T.L., Tempel, B.L., Jan, Y.N. and Jan, L.Y. (1987) Science 237, 749–753.
7. Baumann, A., Krah-Jentgens, I., Müller, R., Müller-Holtkamp, F., Seidel, R., Kecskemethy, N., Canal, I., Ferus, A. and Pongs, O. (1987) EMBO J. 6, 3419–3429.

8. Schwarz, T.L., Tempel, B.L., Papazian, D.M., Jan, Y.N. and Jan, L.Y. (1988) Nature (London) 331, 137–142.
9. Pongs, O., Kecskemethy, N., Müller, R., Krah-Jentgens, I., Baumann, A., Kiltz, H.H., Canal, I., Llamazares, S. and Ferrus, A. (1988) EMBO J. 7, 1087–1096.
10. Kamb, A., Tseng-Crank, J. and Tanouye, M.A. (1988) Neuron 1, 421–430.
11. Timpe, L.C., Jan, Y.N. and Jan, L.Y. (1988) Neuron 1, 659–667.
12. Iverson, L.F., Tanouye, M.A., Lester, H.A., Davidson, N. and Rudy, B. (1988) Proc. Natl. Acad. Sci. U.S.A. 85, 5723–5727.
13. Tempel, L.B., Jan, Y.N. and Jan, L.Y. (1988) Nature (London) 332, 837–839.
14. Baumann, A., Grupe, A., Ackermann, A. and Pongs, O. (1988) EMBO J. 7, 2457–2463.
15. Stühmer, W., Stocker, M., Sakmann, B., Seeburg, P., Baumann, A., Grupe, A. and Pongs, O. (1988) FEBS Lett. 242, 199–206.
16. Butler, A., Wei, A., Baker, K. and Salkoff, L. (1989) Science 243, 943–947.
17. Ribera, A. (1990) Neuron 5, 691–701.
18. McKinnon, D. (1989) J. Biol. Chem. 264, 8230–8236.
19. Frech, G.C., van Dongen, A.M.J., Schuster, G., Brown, A.M. and Joho, R.H. (1989) Nature (London) 340, 642–645.
20. Stühmer, W., Ruppersberg, J.P., Schröter, K.H., Sakmann, B., Stocker, M., Giese, K.P., Perschke, A., Baumann, A. and Pongs, O. (1989) EMBO J. 8, 3235–3244.
21. Yokoyama, S., Imoto, K., Kawamura, T., Higashida, H., Iwabe, N., Miyata, T. and Numa, S. (1989) FEBS Lett. 259, 37–42.
22. Grupe, A., Schröter, K.H., Ruppersberg, J.P., Stocker, M., Drewes, Th., Beckh, S. and Pongs, O. (1990) EMBO J. 9, 1749–1756.
23. Christie, M.J., North, R.A., Osborne, P.B., Douglass, J. and Adelman, J.P. (1990) Neuron 4, 405–411.
24. Grissmer, S., Dethlefs, B., Wasmuth, J.J., Goldin, A.L., Gutman, G.A., Cahalan, M.D. and Chandy, G. (1990) Proc. Natl. Acad. Sci. U.S.A. 87, 9411–9415.
25. Swanson, R., Marshall, J., Smith, J.S., Williams, B.J., Boyle, M.B., Folander, K., Luneau, C.J., Antanavage, J., Oliva, C., Buhrow, S.A., Bennet, C., Stein, R.B. and Kaczmarek, L. (1990) Neuron 4, 929–939.
26. Tseng-Crank, J.C., Tseng, G.N., Schwartz, A. and Tanouye, M.A. (1990) FEBS Lett. 268, 63–68.
27. Koren, G., Liman, E.R., Logothetis, D.E., Nadal-Ginard, B. and Hess, P. (1990) Neuron 2, 39–51.
28. Wei, A., Covarrubias, M., Butler, A., Baker, K., Pak, M. and Salkoff, L. (1990) Science 248, 599–603.
29. Stocker, M., Stühmer, W., Wittka, R., Wang, X., Müller, R., Ferrus, A. and Pongs, O. (1990) Proc. Natl. Acad. Sci. U.S.A. 87, 8903–8907.
30. Philipson, L.H., Hice, R.E. Schaefer, K., LaMendola, J., Bell, G., Nelson, D.J. and Steiner, D.F. (1991) Proc. Natl. Acad. Sci. U.S.A. 88, 53–57.
31. Roberds, S.L. and Tamkun, M.M. (1991) Proc. Natl. Acad. Sci. U.S.A. 88, 1798–1802.
32. Jan, L.Y. and Jan, Y.N. (1989) Cell 56, 13–25.
33. Guy, H.R. and Conti, F. (1990) Trends Neurosci. 13, 201–206.
34. Catsch, A. (1944) Z. Indukt. Abstamm. Vererbungsl. 82, 64–66.
35. Wu, C.F. and Haugland, F. (1985) J. Neurosci. 5, 2626–2640.
36. Tanouye, M.A., Ferrus, A. and Fujita, S.C. (1981) Proc. Natl. Acad. Sci. U.S.A. 78, 6548–6552.
37. Tanouye, M.A. and Ferrus, A. (1985) J. Neurogenetics 2, 253–271.
38. Lichtinghagen, R., Stocker, M., Wittka, R., Boheim, G., Stühmer, W., Ferrus, A. and Pongs, O. (1990) EMBO J. 9, 4399–4407.
39. Iverson, L.E. and Rudy, B. (1990) J. Neurosci. 10, 2903–2916.
40. Schwarz, T.L., Papazian, D.M., Carretto, R.C., Jan, Y.N. and Jan, L.Y. (1990) Neuron 2, 119–127.
41. Hardie, R.C., Voss, D., Pongs, O. and Laughlin, S.B. (1991) Neuron 6, 477–486.
42. Zagotta, W.N. and Aldrich, R.W. (1990) J. Gen. Physiol. 95, 29–60.
43. Zagotta, W.N., Germeraad, S., Garber, S.S., Hoshi, T. and Aldrich, R.W. (1989) Neuron 3, 773–782.

44. Lucchesi, K., Ravindran, A., Young, H. and Moczydlowski (1989) J. Membr. Biol. 109, 269–281.
45. MacKinnon, R. and Yellen, G. (1990) Science 250, 276–279.
46. Yellen, G., Jarman, M., Abramson, T. and MacKinnon, R. (1991) Science 251, 939–942.
47. Hartmann, H., Kirsch, G., Drewe, J., Talialatela, M., Joho, R.H. and Brown, A.M. (1991) Science 251, 942–944.
48. Yool, A.J. and Schwarz, T.L. (1991) Nature (London) 349, 700–704.
49. Stocker, M., Pongs, O., Hoth, M., Heinemann, S., Stühmer, W., Schröter, K.H. and Ruppersberg, J.P. (1991) Proc. R. Soc. London Ser. B 245, 101–107.
50. Armstrong, C.M. and Binstock, L. (1965) J. Gen. Physiol. 48, 859–873.
51. Armstrong, C.M. and Hille, B. (1972) J. Gen. Physiol. 59, 388–396.
52. Armstrong, C.M. (1966) J. Gen. Physiol. 50, 491–514.
53. Müller, R. (1990) Ph. D. Thesis, Ruhr-Universität Bochum.
54. Chandy, K.G., Williams, C.B., Spencer, R.H., Aguilar, B.A., Ghanshani, S., Tempel, B.L. and Gutman, G.A. (1990) Science 247, 973–975.
55. Isacoff, E.Y., Jan, Y.N. and Jan, L.Y. (1990) Nature (London) 345, 530–534.
56. Ruppersberg, J.P., Schröter, K.H., Sakmann, B., Stocker, M., Sewing, S. and Pongs, O. (1990) Nature (London) 345, 535–537.
57. Catterall, W. (1988) Science 242, 50–61.
58. McKinnon, R. (1991) Nature (London) 350, 232–235.
59. Scott, V.E., Parcei, D.N., Keen, J.N., Findlay, J.B. and Dolly, J.O. (1990) J. Biol. Chem. 265, 20094–20097.
60. Hoshi, T., Zagotta, W.N. and Aldrich, R.W. (1990) Science 250, 533–538.
61. Zagotta, W.N., Hoshi, T. and Aldrich, R.W. (1990) Science 250, 568–571.
62. Wittka, R., Stocker, M., Boheim, G. and Pongs, O. (1991) FEBS Lett. 286, 193–200.
63. VanDongen, A., Frech, G., Drewe, J., Joho, R. and Brown, A.M. (1990) Neuron 5, 433–443.
64. Zagotta, W.N. and Aldrich, R. (1990) J. Neurosci. 10, 1799–1810.
65. Papazian, D., Timpe, L.C., Jan, Y.N. and Jan, L.Y. (1991) Nature (London) 349, 305–310.
66. Stühmer, W., Conti, F., Suzuki, H., Wang, X., Noda, M., Yahagi, N., Kutso, H. and Numa, S. (1989) Nature (London) 339, 597–603.
67. Gautam, M. and Tanouye, M.A. (1990) Neuron 5, 67–73.

CHAPTER 11

Structure and regulation of voltage-dependent L-type calcium channels

M. MARLENE HOSEY*, REBECCA M. BRAWLEY,
CHAN FONG CHANG, LUIS M. GUTIERREZ and
CECILIA MUNDINA-WEILENMANN

Department of Pharmacology, Northwestern University Medical School, Chicago, IL 60611, U.S.A.

1. Introduction

It is well recognized that Ca^{2+} is an important regulatory element for many cellular processes, and that the major entry pathway for Ca^{2+} in many cell types is via plasma membrane Ca^{2+} channels. Ca^{2+} channels are functional pores in membranes. They exist in plasma membranes, transverse tubule membranes and in intracellular membranes such as the sarcoplasmic and endoplasmic reticulum. Ca^{2+} channels are normally closed; when opened, Ca^{2+} passively flows through the channels along the Ca^{2+} electrochemical gradient at a rate of several million Ca^{2+} ions per second.

1.1. Subtypes of Ca^{2+} channels

There are several major classes of Ca^{2+} channels: (1) receptor-operated Ca^{2+} channels in plasma membranes; (2) ligand-gated Ca^{2+} channels in intracellular membranes; and (3) voltage-dependent Ca^{2+} channels that are usually found in plasma membranes or the invaginations of the plasma membrane that are known as transverse tubule membranes. Receptor-dependent or receptor-operated Ca^{2+} channels (ROCCs) are primarily opened in response to activation of their associated receptors and, by definition, exhibit a certain amount of selectivity for Ca^{2+} over other cations. Several potentially different types of ROCCs have been characterized including ATP-sensitive channels in smooth muscle [1], mitogen and IP_3-sensitive

---

* To whom all correspondence should be addressed

channels in lymphocytes [2] and thrombin-sensitive channels in platelets [3]. One can envision that ROCCs may either be structures that have channels within the receptor itself (analogous with the nicotinic acetylcholine receptor that comprises a pore which contains a non-selective cation channel), or that they might consist of separate channel entities that are activated upon association with a receptor (see Fig. 1 [4]); this is likely to be the case for channels that are regulated by a variety of receptors that modulate the channels via GTP-binding regulatory proteins (G-proteins) or second messengers. ROCCs may exhibit some voltage-dependence but, if they are 'true' ROCCs, they should primarily be gated by their associated receptors.

The intracellular ligand-gated Ca^{2+} channels include the channels in endoplasmic and sarcoplasmic reticulum (SR) membranes that are opened upon binding of the second messenger, inositol triphosphate (IP_3). These are intracellular Ca^{2+} 'release' channels that allow Ca^{2+} to exit from intracellular stores, and consequently to increase the concentration of cytoplasmic Ca^{2+} [5]. A second type of intracellular Ca^{2+} release channel is the Ca^{2+}- and ryanodine-sensitive channel that was originally characterized and isolated from cardiac and skeletal muscle [5-7] but appears to exist in many types of cells. It has become evident that IP_3-gated channels and ryanodine-sensitive channels are structurally related but distinct proteins [8] that are present in many cell types [9]. While very interesting, time and space will not allow for further discussion of these channels.

The Ca^{2+} channels that have been the most extensively studied are the voltage-dependent Ca^{2+} channels. These channels are usually found in plasma or transverse tubule membranes. Voltage-dependent Ca^{2+} channels open in response to an appropriate membrane depolarization. Several different types of voltage-dependent Ca^{2+} channels have been described and are characterized by differences in their activation and inactivation sensitivities to voltage, their kinetic properties, and their sensitivities to activation or inhibition by a variety of pharmacological agents.

Many different voltage-dependent Ca^{2+} channels have been analyzed, and have been given several different names. The most commonly used classification system is that of Tsien and co-workers [10] who first described three different types of neuronal Ca^{2+} channels as T (conducting tiny and transient currents, and activated at low voltages); L (large, long lasting and activated at higher voltages); and N (neither, with properties intermediate between T and L) channels. This terminology has been the subject of much debate because there are channels that do not readily fall into these categories, and because there is some concern that it is too general. In other terminology T-type channels are referred to as low threshold or low voltage activated and L-type channels are referred to as high threshold or high voltage activated channels [11,12], or type I and type II channels [13].

This chapter will concentrate on the L-type or high voltage activated channels that are sensitive to dihydropyridine derivatives (see below for the description of their pharmacology). These channels are the most well characterized from many perspectives and a large amount of information has accumulated concerning the structure and regulation of L-type Ca^{2+} channels [4,14,15]. From this point on, unless otherwise noted, all references to Ca^{2+} channels will be to L-type, dihydropyridine-sensitive Ca^{2+} channels.

Within the classification of L-type Ca^{2+} channels, subtypes of channels definitely exist, and differ from one another in certain pharmacological and electrophysiological properties (e.g., compare the channels in [16–18]). In addition, there is data that indicate that other types of voltage-dependent Ca^{2+} channels may be relatives of L-type, dihydropyridine-sensitive channels (see below). The different isoforms of L channels have been termed L1, L2, etc. by Arnold Schwartz and as types 1, 2, etc. followed by a small a, b, c, etc. for alternatively spliced variants (see below) by Birnbaumer and colleagues [19]. It is more than likely that the fruits of DNA cloning strategies will lead to the identification of more members of this family and of related- and/or sub-families of different voltage-dependent Ca^{2+} channels. This has already been strongly suggested by the finding of multiple but different partial cDNA clones from mammalian brain [20] that most likely encode different but related types of Ca^{2+} channels. Furthermore, there appear to be different isoforms of channel subtypes that arise as a result of alternative splicing of the mRNA (see below for further discussion).

1.2. Functions of L-type Ca^{2+} channels

Ca^{2+} channels are important in supplying Ca^{2+} to many types of cells, particularly to excitable cells such as muscle, neurons and neurosecretory cells. In cardiac muscle cells, L-type Ca^{2+} channels play an essential role in the process of excitation–contraction coupling. Ca^{2+} must pass through cardiac L-type Ca^{2+} channels and cause Ca^{2+}-induced Ca^{2+} release in order for excitation–contraction coupling to occur in these cells (see [21] and [22]). In neurons, Ca^{2+} channels have been postulated to have several roles, including provision of Ca^{2+} for Ca^{2+}-dependent processes such as neurotransmitter release, regulation of enzyme activity, regulation of metabolism and gene expression.

A different but very interesting scenario involving L-type Ca^{2+} channels is seen in skeletal muscle, where the major component of these Ca^{2+} channels plays two roles. Skeletal muscle does not require extracellular Ca^{2+} for excitation–contraction coupling, rather it utilizes Ca^{2+} stored in the sarcoplasmic reticulum. The role of the L-type channel proteins as true Ca^{2+} channels in skeletal muscle appears to be of secondary importance, but may be to provide Ca^{2+} to the cells over longer periods of time. The main role of the L-type Ca^{2+} channel protein(s) in skeletal muscle is to act as a voltage sensor [23], a key element of excitation–contraction coupling [24,25]. In this tissue, L-type Ca^{2+} channels in the transverse tubule membranes communicate with an intracellular ryanodine-sensitive Ca^{2+} release channel in the sarcoplasmic reticulum. Upon depolarization of skeletal muscle cells the L-type Ca^{2+} channel protein(s) sense the change in voltage and communicate, almost instantaneously, with the Ca^{2+} release channel in the sarcoplasmic reticulum [24,25]. This process allows for excitation–contraction coupling; the details of the communication between the channels are still not known and this is an active area of investigation which promises to yield exciting results in the near future.

2. L-type Ca^{2+} channels

2.1. Pharmacology

The L-type Ca^{2+} channels have the richest pharmacology of all Ca^{2+} channels studied to date. Many different types of drugs have been demonstrated to bind to L-type Ca^{2+} channels and modify their ability to open [4,14,15]. Most of these drugs *block* Ca^{2+} channels and are referred to by many as Ca^{2+} channel antagonists or Ca^{2+} channel blockers. However, some of the drugs *open* Ca^{2+} channels and have sometimes been referred to as Ca^{2+} channel agonists (see [26]). In this chapter, we will refer to these drugs collectively as 'Ca^{2+} channel effectors'. Many of these agents are clinically effective drugs [27] which are currently being widely used in the treatment of cardiovascular disorders such as angina, and are currently being evaluated for use in many other disorders including treatment and prevention of strokes and migraine headaches [28].

Many of the Ca^{2+} channel effectors have been invaluable as biochemical probes to elucidate the molecular properties of Ca^{2+} channels. The most useful drugs for the biochemical characterization of Ca^{2+} channels have been the 1,4-dihydropyridine (DHP) derivatives such as ($\pm$)nitrendipine, (+)PN 200–110 and ($\pm$)Bay K 8644, which bind to receptor sites on L-type channels with dissociation constants in the pM–nM range [4]. These high affinity ligands have been particularly useful in the purification and stabilization of L-type Ca^{2+} channels. Routinely, (+)[^{3}H]PN 200–100 has been used to prelabel channels in order to allow for their detection throughout purification (e.g., see [29–31]), ($\pm$)Bay K 8644 has been used to stabilize channels that will be reconstituted after purification [32,33], and (+)[^{3}H]azidopine has been very useful as a photoaffinity ligand to identify the DHP receptor proteins [31,34]. Several interesting DHPs are the stereoisomers of Bay K 8644 and S202–791; in both cases one of the enantiomers is a channel blocker, while the other is a channel activator (e.g., see [35–38]). Other very useful drugs for the biochemical characterization of L-type Ca^{2+} channels have been the phenylalkylamines (PAA) such as verapamil or its more potent analogs D-888 and LU47781 [31,39,40], and the benzothiazepine, D-*cis*-diltiazem [41].

The use of radiolabelled Ca^{2+} channel effectors has shown that L-type Ca^{2+} channels possess several specific high affinity receptors for these agents, most of which bind the drugs with dissociation constants in the nM range [41–44]. In addition to the DHP receptor, other high affinity receptors on L channels bind the PAAs, bepridil, benzothiazepines, diphenylbutylpiperidines and indolizinsulfones [41,45–50]. There appear to be at least four (or more) distinct, but allosterically linked, drug receptors on L channels that are present in a 1:1:1:1 stoichiometry [41,45–50].

An important consideration, both for clinical use and for laboratory studies, is that some of the Ca^{2+} channel effectors exhibit a voltage-dependence of action [51–53]. This is exemplified by the DHPs, which bind with higher affinities to channels in depolarized membranes than to polarized membranes [51–53]. For example, in skeletal muscle, the drug (+)PN 200–110 blocks the calcium current with a K_d of 13 nM when the cells are held at −90 mV and with a K_d of 0.15 nM when the cells are held at the relatively depolarized potential of −65 mV [53]. Fortunately for biochemists seeking to characterize and purify the channels from isolated, and thus depolarized, membranes, the change in affinity caused by voltage is in the favorable direction. Before the voltage-dependence of binding of the DHPs was appreciated, the very high affinity of the DHPs observed in radioligand binding studies compared to their significantly lower affinity observed in intact cells, led to a concern that the biochemically identified DHP receptors might not correspond to functional Ca^{2+} channels. However, it is now generally agreed that there is a close relationship between high affinity DHP binding sites and Ca^{2+} channels.

The voltage-dependence of binding of DHPs is also important in their clinical use. In the treatment of angina, the goal is to block Ca^{2+} channels in the smooth muscle of the coronary vasculature, while minimizing effects on the channels of the already compromised cardiac muscle cells. Because smooth muscle is slightly more depolarized than cardiac muscle, the DHPs can be used at appropriate dosages to selectively target the Ca^{2+} channels in the smooth muscle of the coronary vasculature without causing much effect on the channels in the cardiac muscle cells. Thus, the voltage-dependence of action is a clinically efficacious feature of these agents.

2.2. Biochemical and molecular characterization

L-type Ca^{2+} channels are the only Ca^{2+} channels that have been extensively characterized at the biochemical level. The success in characterizing L-type channels was largely due to their rich pharmacology, and in particular to the availability of radio-labelled DHPs. However, a second major contributor to this successful characterization came from the seminal findings of Glossmann and colleagues [45] and Fosset, Lazdunski and colleagues [43] who identified skeletal muscle transverse tubule membranes as a relatively rich source of this rare membrane protein.

2.2.1. Isolation and purification of the multisubunit dihydropyridine-sensitive Ca^{2+} channels from skeletal muscle

For reasons that are not yet clear, skeletal muscle transverse (T)-tubule membranes contain 50–100-fold more high affinity DHP receptors than any other source yet identified [43,45]. Transverse tubule membranes contain 30–70 pmol/mg protein of DHP receptors that bind [^{3}H]PN 200–100 with a K_d of $\sim$0.1–0.2 nM. The strategy utilized for the purification of L-type channels was similar to that used for the purification of other high affinity ligand binding proteins, and its success was predicted from the prior use of such an approach for the purification of other ion channels [54,55]. Thus the L-type channels were purified as high affinity DHP receptors, with the anticipation that the purified component(s) would constitute functional Ca^{2+} channels.

The purified L-type channels from skeletal muscle appear to be a multisubunit protein containing the products of four different and unrelated genes. A cartoon depicting some of the features of the purified protein is shown in Fig. 1. The subunits of the channel are commonly referred to as the α_1, α_2, β, γ, and δ. These have been observed to migrate in a variety of SDS-polyacrylamide gels with apparent

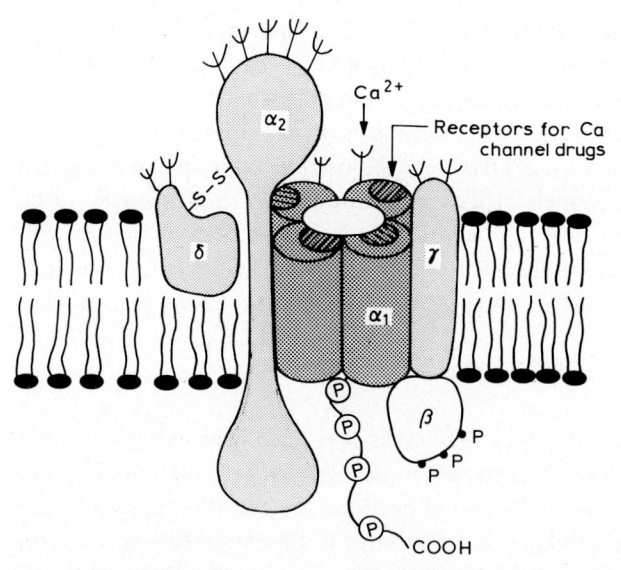

Fig. 1. Subunit structure of dihydropyridine-sensitive Ca^{2+} channels from skeletal muscle. Cartoon depicts the proposed subunit structure of Ca^{2+} channels purified from skeletal muscle T-tubule membranes. The α_1 subunit is shown as the main conducting pore with four repeated domains folded in a manner to produce a pore. Each repeated domain has six hydrophobic segments that may span the membrane [73]. The tentative arrangement of the other subunits is based on their relative hydrophobicity or hydrophilicity, their known glycosylation and/or susceptibility to phosphorylation, but is presently very speculative.

molecular weights of 165 000, 140 000, 52 000, 32 000 and ~29 000, respectively [30,39,40,56,57]. Under non-reducing conditions, the α_2/δ proteins migrate together as a 170-kDa complex. This was first noted by the use of anti-α_2 and anti-δ antibodies that cross reacted with the 140-kDa α_2 or the 29-kDa δ under reducing conditions, but which stained a single band of 170 kDa under non-reducing conditions [58]. All five polypeptides have been observed in the purified channel preparations and have been suggested to be subunits based on their co-purification in stoichiometric amounts [59], their co-immunoprecipitation with anti-α_1 and anti-β antibodies [59–61] and, most recently, by their co-migration as a large complex in non-denaturing polyacrylamide gels [62].

The α_1 polypeptide contains the receptors for the Ca^{2+} channel effectors as it can be photolabelled by DHPs and PAAs [30,39–41,56,57,63] and various other Ca^{2+} channel effectors. The other subunits do not appear to be required for proper binding of Ca^{2+} channel effectors [64], although there is some evidence from biochemical studies that the other subunits may modulate binding [65]. The α_1 and γ subunits are very hydrophobic, while the α_2/δ and β subunits are very hydrophilic [57]. The α_1 [30], α_2, δ and γ [57] peptides are glycosylated, but α_1 is much less so than α_2 [30]. The M_r of both α peptides varies slightly on different types of SDS gels [30,61] and M_r values of ~155 000–200 000 have been reported (this may have importance, see below).

2.2.2. Identification and purification of L-type channel proteins from other cells
Progress in purifying and characterizing the subunit composition of L-type Ca^{2+} channels in other cells has been hampered by the low concentration of channels. However, our laboratory has achieved a successful partial purification of L-type cardiac Ca^{2+} channels. Initially we isolated only small peptides of 60, 55 and 32 kDa [66], and others reported photoaffinity labelling with DHPs of components of 30–40 kDa [67]. Subsequently, both the 'large' α_1 and α_2 cardiac proteins were isolated [31,68]. The α_1 protein of 185–190 kDa isolated and purified from chick heart was identified by photolabelling by DHP and PAA analogs [31]; this cardiac α_1 is both structurally and immunologically different from the skeletal muscle α_1 protein. The cardiac α_1 has also been partially purified from mammalian sources, but no extensive characterization of its properties has been possible [69,70]. The 170-kDa cardiac α_2/δ protein shares many characteristics with its counterpart from skeletal muscle [68]. The peptide maps are similar and antibodies prepared against the skeletal muscle α_2 protein cross-react with the cardiac α_2 [68]. The purity of the channels isolated from cardiac muscle has not yet been sufficient to determine if cardiac channels have subunits similar to the β and γ subunits of the skeletal muscle channel. A 1:1:1 complex of the cardiac α_1 and α_2/δ peptides would yield a M_r of ~360 000, which would agree with the predictions of a hydrodynamic analysis of cardiac DHP receptors that suggested a M_r of 370 000 after correction for bound detergent [71]. The differences in structure of the cardiac and skeletal muscle α_1

polypeptides, which are the main functional units of L channels, may in part account for the known differences in properties of L channels in cardiac and skeletal muscle.

Studies with polyclonal antibodies against the α_2/δ protein have identified neuronal α_2 polypeptides [58] and other studies with anti-α_1 antibodies have identified a similar protein in brain [72]. Future studies will be aimed at further clarifying the similarities and differences in the structures of L channel proteins from various sources.

2.2.3. DNA cloning and expression of channel proteins

The DNAs for each of the subunits of the skeletal muscle channel have been cloned. From these studies there is no doubt that the α_1 subunit is the main channel forming unit of the DHP-sensitive channels (see below).

2.2.3.1. Isoforms of the α_1 subunit. The α_1 subunit of the DHP-sensitive Ca^{2+} channel from skeletal muscle was the first Ca^{2+} channel protein to be recognized [73] as a member of the superfamily of voltage-dependent ion channels that also includes the voltage-dependent Na^+ and K^+ channels [55,73]. This Ca^{2+} channel α_1 subunit has a predicted amino acid sequence that has striking homology to that of the voltage-dependent Na^+ channels [73–76]. The overall homology of 29% identical residues is found throughout the protein. The predicted amino acid sequence of the skeletal muscle α_1 peptide possesses four internally repeated units, each of which is predicted to contain six membrane-spanning regions [73]. The fourth membrane-spanning segment in each internal repeat contains clusters of basic amino acids which have been suggested to comprise the voltage sensors [73]. Two potential glycosylation sites are located in domains predicted to be extracellular. Both the hydrophilic C-terminal domain which consists of ~400 amino acids and contains several potential phosphorylation sites, and the N-terminal domain are predicted to be located intracellularly [73].

Key findings that demonstrated that the α_1 subunit is the essential component of L-type channels have come from studies of the channel activity of the expressed protein. Expression studies performed in mammalian liver fibroblasts have demonstrated that the α_1 subunit alone can form a channel [77] and contains the receptors for the DHPs, PAAs and diltiazem [64]. In very elegant studies using a mouse model of muscular dysgenesis it has been demonstrated that the α_1 subunit DNA can restore Ca^{2+} currents and the charge movement that arises from the voltage-sensing function of the channels to the mutant cells that normally lack these activities [21,78,79]. The restoration of these activities restores excitation–contraction coupling. Thus it is clear that the α_1 subunit is the major functional unit of L-type Ca^{2+} channels.

Isoforms of the α_1 subunit have now also been cloned from cardiac [80] and smooth muscle [81,82] cDNA libraries, and partial clones resembling this structure have been obtained from brain libraries [20]. In addition, a cDNA encoding a high threshold but DHP-insensitive channel has been obtained from brain [83]. In general, these isoforms are highly homologous in the membrane-spanning domains but less so in their C-terminal domains and in the intracellularly oriented hydrophilic loops connecting the repeated segments. The α_1 subunit expressed in aortic smooth muscle is essentially that expressed in cardiac muscle except that it appears to be the product of alternative splicing of the mRNA in domain 4 [81]. Alternative splicing of the α_1 transcript appears to occur in many different cells [19,81,83]. Studies utilizing the polymerase chain reaction have demonstrated that different forms of the α_1 protein arise by this mechanism [19]. Analysis of sequences in the fourth internal repeated unit demonstrated that at least three species can be identified in heart, two in skeletal muscle, two in neurons and four in neuroendocrine cells [19]. Since only a part of the channel protein was analyzed, it is possible that other forms may also exist. The functional significance of these isoforms remains to be demonstrated, although the existence of multiple forms of the channels in certain cells has been suspected for several years.

Chimeric constructs of the cardiac and skeletal muscle isoforms of the α_1 subunit have shown that the cytoplasmic linker domains are important in imparting significant differences in the properties between cardiac and skeletal muscle channels [21,79]. In addition, these studies have shown that these domains are important in the function of the skeletal muscle protein as a voltage sensor in excitation–contraction coupling. Using the dysgenic mouse model, the chimeric channels have been expressed and those with the cardiac loop that connects domains two and three but with the core of the channel from skeletal muscle exhibit cardiac type excitation–contraction coupling [79]. Thus this domain is a potential site of interaction between this protein and the other elements that contribute to excitation–contraction coupling.

2.2.3.2. Cloning of DNAs for other putative channel subunits. The cDNA for the α_2 peptide from skeletal muscle has been cloned and the results show that the α_2 peptide does not resemble any known protein [84]. The protein is very hydrophilic and contains only three short stretches that are sufficiently hydrophobic to be membrane-spanning regions. It contains 18 potential glycosylation sites, a potential signal sequence which would orient the N-terminus to the extracellular side, and a C-terminal domain that is predicted to be internal [84]. In contrast to results obtained with α_1 clones, a DNA probe from the sequence of the skeletal muscle α_2 clone cross-reacted with mRNA from many sources [84], suggesting that the α_2 subunit may be similar in different cells. This result supports the previous observations that the α_2 subunits purified from skeletal and cardiac muscle were similar in both immunological cross reactivity and in peptide mapping experiments [68]. The gene encoding the α_2 subunit also encodes the sequence of the δ subunit [85,86]. These proteins associate via disulfide bonds, but dissociate in SDS gel in the presence of reducing agents. A proposal has been made to refer to the α_2/δ complex as a single subunit [86].

The DNA encoding the β subunit of the skeletal muscle channels has been identified and encodes a unique protein that is very hydrophilic [87]. The protein contains several potential phosphorylation sites, and interestingly, contains a sequence that is homologous with an inhibitor of phosphoprotein phosphatases [87]. Whether or not the β protein might serve to inhibit dephosphorylation of the channel has not been addressed, but could be an interesting role for this protein. DNA probes from the skeletal muscle β subunit cross-reacted with mRNA from brain but not from cardiac muscle [87]. A second form of the β subunit has recently been obtained by screening a neuronal library [88]. The neuronal protein appears to be related but significantly different from the skeletal muscle β subunit. Whether other forms of the β subunit exist in other cells is currently being addressed. Finally, the γ subunit cDNA has also been identified and found to encode a very hydrophobic protein with no homology to other identified proteins [89]. A DNA probe from the skeletal muscle γ subunit did not react with mRNA from either brain or heart [89].

2.2.4. Roles of subunits of L-type Ca^{2+} channels

The question as to whether or not all the proteins discussed above are true subunits of the DHP-sensitive Ca^{2+} channel is still not completely answered, but is actively being pursued. On the one hand, a heterotetramer or heteropentamer seems very complicated, especially since the α_1 subunit itself can make a channel and the other members of this ion channel family, the Na^+ and K^+ channels [55,73], appear to be comprised predominately of a single type of subunit. On the other hand, there are compelling results that would strongly support the argument that the proteins are subunits. The co-expression of the cardiac α_1 subunit with the skeletal muscle α_2 subunit in *Xenopus* oocytes led to increased channel current [80]. It was not possible with these studies to know how this occurred (e.g., it could be due to stabilization of the mRNA or to better insertion of the newly expressed proteins into the membranes). However, our laboratory partially separated the α_1 and α_2 subunits as previously described [65] and demonstrated that the removal of the α_2/δ subunits results in a significant loss of channel activity that can be restored by adding back the α_2/δ subunits [132].

Other recent and very compelling evidence that the β subunit may also be necessary for full activity [83] comes from the study of a high threshold but DHP-insensitive brain channel that has an α_1-like channel core. The cDNA for this brain channel was recently isolated and the mRNA was expressed in *Xenopus* oocytes. The current corresponding to this channel was increased when the mRNA for either the skeletal muscle α_2 or β subunits were co-injected. However, when the brain α_1-like mRNA was co-expressed with the combination of the skeletal muscle α_2 and β mRNAs, the current was dramatically increased from 31 nA for the brain α_1 alone to 6 500 nA for the combination [83]. These striking results are the best evidence so far obtained that the β subunit has a functional role in channel activity. Although these data were obtained with a DHP-insensitive channel, they provide a strong indication that the α_2 and β peptides are important components of the L-type channels. Furthermore, they suggest that these components may play a more general role as subunits of DHP-sensitive and -insensitive channels. The future should provide exciting new information on the structure and subunit composition of various types of voltage-dependent Ca^{2+} channels. So far, a rigorous expression system for these channels has not been identified. The expression achieved in oocytes is not amenable to biochemical studies and the expression achieved in mammalian cells has been minimal, but may be enhanced by co-expression of the required subunits.

An issue related to the subunit composition of functional channels that has recently been the focus of some attention concerns the size of the functional α_1 subunit. The protein purified by most laboratories from rabbit skeletal muscle has an apparent M_r on SDS gels of approximately 165–175 K (to avoid confusion to the reader we will refer to this form as the ~ 165-kDa form). However, the cDNA for this protein predicts that a full length protein would be 212 kDa [73]. This observation has raised the question of whether or not the purified protein of ~ 165 kDa is intact or proteolyzed. To address this question, anti-peptide antibodies directed against the 18 amino acids from the C-terminal of the protein were prepared [90]. These antibodies reacted with a 210-kDa protein, but not with a smaller but 10-fold more abundant α_1 polypeptide of ~ 165-kDa (which was actually was referred to as 175 kDa in [90]). However, the ~ 165 kDa polypeptide was recognized by antibodies made against internal sequences of the α_1 subunit [90]. It was suggested that these two different size forms of the α_1 subunit might exist in skeletal muscle, and further, that the larger, 210-kDa protein might correspond to the DHP receptors that form channels, while the smaller, more abundant 165-kDa proteins might serve as voltage sensors [90]. This suggestion was supported by the previous observation that less than 5% of the purified DHP receptors could be reconstituted as active channels [32], and by data which were interpreted to indicate that only 5% of the DHP receptors in skeletal muscle may act as functional Ca^{2+} channels [91]. While these results are quite provocative, our laboratory has recently shown by reconstitution studies that the 165-kDa form of the α_1 subunit is capable of forming a channel that possesses most of the characteristics of the channels observed in intact muscle [132]. In these reconstituted preparations, there appears to be only a 165-kDa α_1 polypeptide. We have no evidence for the existence of a second, larger form in these preparations. Furthermore, as discussed below, we believe that all DHP receptors in skeletal muscle are capable of forming Ca^{2+} channels, but that they have a very low probability (0.08) of opening.

2.2.5. Reconstitution of Ca^{2+} channels

Since L-type Ca^{2+} channels were purified as high affinity DHP receptors, it was important to demonstrate that the purified preparations could act as functional DHP-sensitive Ca^{2+} channels. Purified DHP receptor preparations from skeletal muscle have been reconstituted into planar lipid bilayers by several groups and shown to exhibit Ca^{2+} channel activity [92–94]. The purified reconstituted channel possessed many characteristics in common with a DHP-sensitive Ca^{2+} channel reconstituted directly (without purification) from skeletal muscle membranes and recorded under similar conditions [93]. However, the voltage-dependence of the purified channels has not been demonstrated. As a step towards doing so, we recently demonstrated that channels reconstituted directly from skeletal muscle T-tubule membranes exhibit a voltage-dependence in bilayers that is similar to what is seen in intact cells [95]. Studies to determine the voltage-dependency of the purified channels are under way. Similar studies also should be possible to determine which subunits are necessary to obtain Ca^{2+} channel activity.

The recent studies with the skeletal muscle channels reconstituted in bilayers also established that an intrinsic property of the skeletal muscle Ca^{2+} channels is that they possess a very low probability of opening. Even under optimal conditions of voltage, etc., these channels open only 8% of the time [95]. This previously unappreciated characteristic of these channels provides an alternative explanation to an earlier study in which it was proposed that only 5% of the DHP receptors in skeletal muscle can form functional channels [91]. In that study a probability of opening of 1.0 was assumed. However, if one adjusts this to 0.08, then the very different conclusion is that all the DHP receptors can make functional channels, but that they open only a very small percentage of the time.

Biochemical studies with purified preparations incorporated into liposomes have also been performed [32,33,96–98]. Reconstituted receptors from skeletal muscle bound DHPs, PAAs and diltiazem with high affinity and in a 1:1:1 stoichiometry [97]. In general, the reconstituted proteins exhibit the characteristic pharmacological properties expected for these channels. In recent studies, our laboratory has reconstituted partially purified channels into liposomes containing the Ca^{2+}-sensitive fluorescent dye, fluo-3 [33,96]. These channels exhibit Ca^{2+} influx that is sensitive to activation by Ca^{2+} channel activators and inhibitors with affinities similar to those observed in intact cells, and the Ca^{2+} influx is dependent on the establishment of a K^+ gradient in the presence of valinomycin [132]. This assay provides a convenient and rapid approach to obtaining a 'macroscopic' picture of the activity of the channels under different conditions, while the more complex studies in lipid bilayers provide a more complete analysis of the single channel behavior.

2.3. Regulation of Ca^{2+} channels by protein phosphorylation and G-proteins

L-type Ca^{2+} channels are primarily regulated by voltage and are thus opened and closed in response to changes in membrane potential, but in addition, they are regulated by receptor-dependent processes involving protein phosphorylation and G-proteins (Fig. 2). Many lines of evidence support the hypothesis that Ca^{2+} channels are regulated by phosphorylation by several protein kinases, in particular by

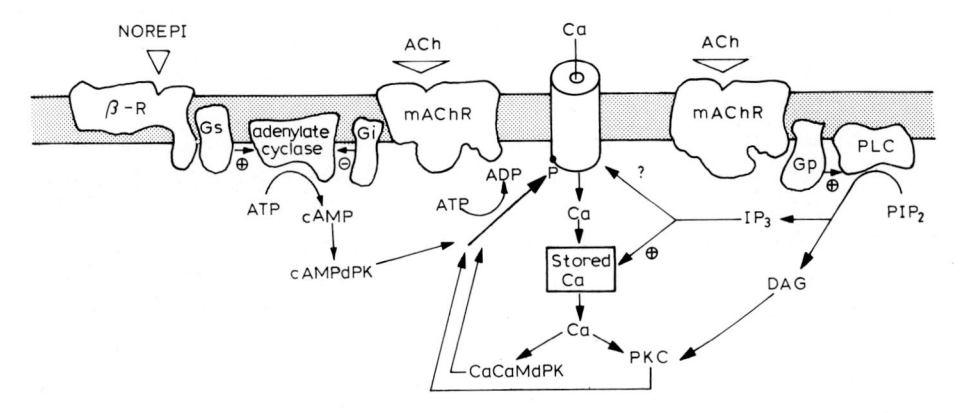

Fig. 2. Regulation of L-type Ca^{2+} channels by multiple second messenger systems. The cartoon is based on what is known or speculated about how L-type Ca^{2+} channels might be regulated by processes that involve phosphorylation by different receptor mediated pathways. The actual receptors and number of pathways may vary in different cell types. The cartoon is loosely based on events that might occur in a cardiac cell.

cAMP-dependent protein kinase (PKA) and by protein kinase C (PKC) [4,14,15]. A cartoon depicting the potential pathways of regulation by phosphorylation is given in Fig. 2. Much of this evidence has come from electrophysiological studies, but recent biochemical analyses paired with reconstitution studies have been able to elucidate the molecular events involved in these phosphorylation-dependent events.

2.3.1. Phosphorylation by cAMP-dependent protein kinase

Considerable evidence from electrophysiological studies suggests that L-type Ca^{2+} channels are regulated by a cAMP-dependent phosphorylation event (e.g., see [99]). In cardiac cells, any agent that elevates cAMP, including norepinephrine, via β-adrenergic receptors, increases the probability of opening of L channels in response to a given change in membrane potential [99]. L-type Ca^{2+} channels in other cells, including some types of neurons [100,101], skeletal muscle [102,103] and chromaffin cells [104] also are regulated by cAMP. Injection of the catalytic subunit of cAMP-dependent PKA into cardiac cells produces cAMP-like effects [105,106], and the inhibitor of this protein kinase prevents the effects [107]. In addition, the widely observed run down of Ca^{2+} channel activity over time in patch clamp experiments has been slowed, in some [100,101], but not all cases by PKA and ATP. Finally, the activity of the purified reconstituted skeletal muscle and cardiac L channels is increased by PKA [33,92,108,109]. Taken together, the results support the hypothesis that certain L channels are regulated by the cAMP-dependent phosphorylation of the channel itself or a regulatory protein.

In order to determine what phosphorylation event(s) might lead to regulation of the channels by PKA, studies of the phosphorylation of the channel subunits have been performed. Such studies showed that the skeletal muscle α_1 protein is an excellent substrate *in vitro* for PKA [108–112]. The phosphorylation occurs with incorporation of 2–2.5 mol phosphate/mol protein primarily at serine residues [111], in agreement with the predicted amino acid sequence for the α_1 peptide which indicates six serines and one threonine in seven consensus sequences for PKA phosphorylation [73]. All potential phosphorylation sites are predicted to be located intracellularly, six in a hydrophilic C-terminal cytoplasmic tail, and the seventh on a short cytoplasmic loop connecting the second and third internal repeats. The latter site (Ser 687) appears to be a preferred site for phosphorylation by PKA *in vitro* [113]. Phosphorylation of the 52-kDa β subunit has also been observed in *in vitro* studies with PKA [108,111,114,115]. However, we have found that the stoichiometry of phosphorylation of this subunit that occurs in T-tubule membranes is significantly lower than for the α_1 subunit, and only occurs to ~ 0.5 mol phosphate/mol β subunit [108].

Using reconstitution experiments, we and others have found that phosphorylation by PKA increases the activity of the skeletal muscle Ca^{2+} channels [33,108,109]. In the liposome assay with fluo-3, we found that phosphorylation of the channels with PKA leads to a two-fold increase in the rate and the extent of Ca^{2+} influx [33,108]. This result suggested that channels that are silent are activated by phosphorylation. More extensive studies using the lipid bilayer–T-tubule membrane reconstitution system have shown that phosphorylation of the skeletal muscle Ca^{2+} channels with PKA increases the probability of opening by 1.5-fold, shifts the voltage-dependence of activation ~ 7 mV to the left and slows the rate of inactivation two-fold [116]. Although the later reconstitution studies have all been performed in the presence of Bay K 8644 (to facilitate channel opening), the results agree well with most of the studies that have been performed to characterize the effect of cAMP on channels in skeletal muscle cells and fibers [102,103].

A direct demonstration that proteins associated with L-type Ca^{2+} channels are phosphorylated by PKA in intact cells has been achieved [33,117]. The results of these studies, which were achieved primarily using the technique of back phosphorylation with chick skeletal muscle [33] and cultured rat muscle cells [117], demonstrated that the α_1 subunit undergoes phosphorylation after treatment of skeletal muscle cells with the β-adrenergic agonist, isoproterenol, or the adenylyl cyclase activator, forskolin. The stoichiometry of phosphorylation in the chick muscle was ~ 1.5 mol phosphate/mol protein [33]. A 200-kDa form of the α_1 subunit also was observed to undergo phosphorylation in the rat muscle cells stimulated with isoproterenol [117]. In contrast, no phosphorylation of the β subunit was observed [33], perhaps reflecting the poorer ability of the β subunit to serve as a substrate in the *in vitro* assays. Reconstitution experiments demonstrated that the phosphorylation of the channels that occurred in the intact cells was sufficient to cause activation of the channels [33]. Taken together, the results of both *in vitro* and *in situ* studies strongly suggest that the regulation of the channels by PKA-induced phosphorylation occurs as a result of phosphorylation of the α_1 subunit. Future studies are needed to elucidate which of the putative phosphorylation sites in the α_1 subunit are phosphorylated in the intact cell and contribute to activation of the channels.

The cardiac Ca^{2+} channels have been extensively studied with electrophysiological approaches as substrates for PKA [99]. Paradoxically, the purified channels were unable to serve as substrates for phosphorylation by PKA *in vitro* under conditions in which the skeletal muscle channels were readily phosphorylated [31]. At the time that these results were obtained we proposed that the data might be explained if the phosphorylation site in the internal loop between domains 2 and 3 of the skeletal muscle channel [73,113] was not present in the heart channels and/or if the C-terminal domain was different or proteolyzed. Indeed, the cDNA cloning of the cardiac α_1 subunit showed that the serine in the loop between domains 2 and 3, which is a target of PKA in the skeletal muscle α_1 subunit is an alanine in the predicted protein from cardiac muscle [80]. In addition, the cDNA cloning predicts that the heart α_1 subunit has several potential phosphorylation sites in its C-terminal domain but these are different from those present in the skeletal muscle protein. Furthermore since the purified protein appears to have a molecular weight of $\sim 190\,kDa$ and the cDNA predicts that the intact protein should be $260\,kDa$ it is possible that the C-terminal domain of the purified protein is proteolyzed. Further studies are necessary to resolve these issues. An alternative explanation for the lack of phosphorylation observed with the purified cardiac channel protein is that phosphorylation may not occur in detergent solution with this protein. Experience with other membrane proteins [118] as well as with the skeletal muscle α_1 subunit [112] with certain protein kinases has demonstrated that certain phosphorylation reactions do not readily occur in the presence of detergent. These technical difficulties should be solved in the near future. Once achieved, it will be interesting to determine if the phosphorylation sites in different channel isoforms are similar or different, and why some L channels are regulated by cAMP while others are not.

2.3.2. Regulation of L-type channels by PKC and other protein kinases
Electrophysiological studies with activators of PKC have suggested that certain L channels are inhibited [119–121] or activated [122] by PKC, while cardiac Ca^{2+} channels have been shown to be both activated and inhibited by activators of this kinase [123]. The skeletal muscle α_1 peptide is an excellent substrate *in vitro* for PKC [112]. Phosphorylation occurs to an extent of 2 mol phosphate/mol α_1 subunit [112]. Phosphorylation of the β subunit also occurs [110] but less extensively, to 0.7 mol phosphate/mol subunit [108]. However, it is important to note that the extent of phosphorylation of the α_1 subunit is markedly dependent on the experimental conditions [112]. Full phosphorylation is only achieved when the phosphorylation is catalyzed in detergent-free conditions, and is optimum when the channel in native T-tubule membrane is used as substrate [112]. In detergent-containing reactions, very little phosphorylation of the α_1 subunit is observed, although the phosphoryla-

tion of the hydrophilic β subunit occurs as well as it does in the absence of detergent [110,112].

Using the reconstitution approaches described above, we have demonstrated that phosphorylation of the skeletal muscle Ca^{2+} channels by PKC results in activation of the channels [108]. In the fluo 3-containing liposomes, channels phosphorylated by PKC exhibited a two-fold increase in the rate and extent of Ca^{2+} influx [108]. Using the lipid bilayer–T-tubule membrane reconstitution system we are currently analyzing the effects of PKC-catalyzed phosphorylation at the single channel level [133]. The demonstration that these channels undergo phosphorylation as a result of activation of PKC in intact skeletal muscle cells has not yet been achieved.

Other studies have demonstrated that the skeletal muscle α_1 peptide can be phosphorylated in T-tubule membranes by a multifunctional Ca^{2+}/calmodulin (CaM)-dependent protein kinase [111]. Phosphorylation occurs on the α_1 subunit to an extent of 2 mol phosphate/mol subunit and on the β subunit to an extent of 0.7–1 mol phosphate/mol channel [108,111]. Phosphorylation catalyzed by the CaM-kinase on the α_1 subunit is additive to that caused by PKA and occurs on distinct sites [111]. So far, however, we have not observed any functional consequences of phosphorylation of the skeletal muscle Ca^{2+} channels by the CaM-kinase.

The skeletal muscle Ca^{2+} channels also can be phosphorylated *in vitro* by a protein kinase endogenous to the T-tubule membranes [111,115]. This kinase is neither Ca^{2+}- nor cyclic nucleotide-dependent [115], and is interesting in that it phosphorylates primarily the β subunit while the α_1 subunit is a poor substrate. However, the amount of this kinase that co-purifies with the T-tubule membranes is variable, and consequently, very few studies have been performed. So far, only low levels of phosphorylation have been obtained (no more than 0.2 mol phosphate/mol β subunit) and no functional effects of this phosphorylation have been observed in reconstitution studies.

Further studies are required, particularly at the biochemical and molecular level, to understand precisely how phosphorylation of channel subunits modifies channel function. A cartoon illustrating the probable location of the multiple sites of phosphorylation in the α_1 subunit is given in Fig. 3. Further studies are necessary to identify the amino acids that are phosphorylated and that contribute to regulation of the channel activity. Furthermore, the possibility that associated regulatory proteins are the targets of phosphorylation needs to be explored, particularly in the case of channels whose main conducting protein does not undergo phosphorylation itself. Whether or not this will be the case for the cardiac channel needs to be determined, although it is a less attractive alternative in view of the findings that the skeletal muscle α_1 subunit can be multiply phosphorylated by various protein kinases, at least *in vitro*.

2.3.3. Regulation of L-type channels by phosphoprotein phosphatases
As phosphorylation of certain Ca^{2+} channels by protein kinases has been shown to

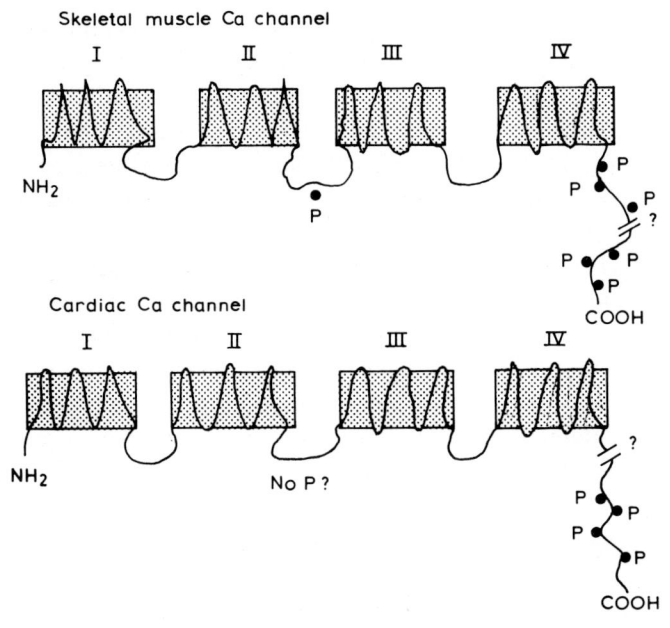

Fig. 3. Sites of potential phosphorylation of α_1 subunits of cardiac and skeletal muscle Ca^{2+} channels.

modulate channel activity, dephosphorylation by phosphoprotein phosphatases would be expected to reverse the effects. Protein phosphatases 1 and 2A and calcineurin have been shown to accelerate inactivation of Ca^{2+} channels in electrophysiological experiments, whereas potato acid phosphatase and calf intestinal alkaline phosphatase were without effect [101,124]. In biochemical studies, calcineurin dephosphorylates the skeletal muscle α_1 peptide which had been previously phosphorylated by either the PKA or the multifunctional CaM-kinase [125]. However, extensive studies characterizing this reaction have not been performed. In preliminary, but ongoing studies, protein phosphatase 1 and the catalytic subunit of protein phosphatase 2A have been found to dephosphorylate channels previously phosphorylated by PKA and PKC (Zhao, Chang and Hosey, in preparation). Taken together, these results suggest that calcineurin and phosphatases 1 and 2A may effectively reverse the effects of cAMP- and/or Ca^{2+}-dependent phosphorylation of L-type Ca^{2+} channels in intact cells. However, the effects of these dephosphorylation reactions need to be more extensively characterized and their functional effects defined.

2.3.4. Regulation of Ca^{2+} channels by G-proteins

It is appreciated that certain ion channels appear to be directly regulated by various G-proteins [126]. The cardiac and skeletal muscle L channels appear to be capable of being directly activated by G_s (the heterotrimeric G-protein that stimulates adenylate cyclase), in a manner that appears to be independent of second messenger production or activation of protein kinases [127–129]. G-protein regulation of L channels appears to occur independent of, and/or in addition to, phosphorylation-mediated regulation [127–129]. However, the study of G-protein regulation of channels has been mostly performed using electrophysiological approaches to study the channels in isolated patches of membranes where both the 'direct' G-protein pathway and the 'indirect' phosphorylation pathway may be operative. Limited studies in bilayers in which additional G_s was applied to bilayers containing T-tubule membrane vesicles resulted in increased channel activity [129]. Yet other studies have demonstrated effects of various G-proteins, notably G_o, on neuronal Ca^{2+} channels [130]. As the space allotted does not permit a detailed description of the many studies of regulation of neuronal Ca^{2+} channels, the reader is referred to the excellent review of Miller [131]. At this time, a direct demonstration that a purified Ca^{2+} channel can be activated or inhibited by a purified G-protein has not been made, but this approach is under investigation by several laboratories. It will be fascinating to sort out the relative roles of direct regulation by G-proteins versus that achieved by second messenger-catalyzed phosphorylation. As more cDNA clones of Ca^{2+} channel subunits and G-proteins become available, and as suitable expression systems are developed, new approaches to elucidating pathways of regulation should be facilitated.

Acknowledgements

We thank the members of our laboratory for their comments on this chapter, and for their efforts directed towards understanding the structure and regulation of Ca^{2+} channels. The work described from this laboratory was supported by NIH Grant HL23306 and fellowships from the American Heart Association of Metropolitan Chicago to Fong Chang, the Muscular Dystrophy Association to Cecilia Mundina-Weilenmann, the Pharmaceutical Manufacturer's Association Foundation to Rebecca M. Brawley and from the Spanish Ministry of Science to Luis M. Gutierrez.

References

1. Benham, C.D. and Tsien, R.W. (1987) Nature (London) 328, 275–278.
2. Kuno, M. and Gardner, P. (1987) Nature (London) 326, 301–304.
3. Zschauer, A., van Breeman, C., Buhler, F.R. and Nelson, M.T. (1988) Nature (London) 334, 703–705.
4. Hosey, M.M. and Lazdunski, M. (1988) J. Membr. Biol. 104, 81–105.
5. Ferris, C.D., Huganir, R.L., Supattapone, S. and Snyder, S.H. (1989) Nature (London) 342, 87–89.
6. Fleischer, S., Ogunbunni, E.M., Dixon, M.C. and Fleer, E.A.M. (1985) Proc. Natl. Sci. U.S.A. 82, 7256–7259.

7. Lai, F.A., Erickson, H.P., Rousseau, E., Liu, Q.-Y. and Meissner, G. (1988) Nature (London) 331, 315–319.
8. Migner, G.A., Sudhof, T.C., Takei, K. and DeCamilli, P. (1989) Nature (London) 342, 192–195.
9. Marks, A.R., Tempst, P., Chadwick C.C., Riviere, L., Fleischer, S. and Nadal-Ginard, B. (1990) J. Biol. Chem. 265, 20719–20722.
10. Nowycky, M.C., Fox, A.P. and Tsien, R.W. (1985) Nature (London) 316, 440–443.
11. Carbone, E. and Lux, H.D. (1984) Biophys J. 46, 413–418.
12. Fedulova, S.A., Kostyuk, P.G. and Veselovsky, N.S. (1985) J. Physiol. 359, 431–446.
13. Narahashi, R., Tsunoo, A. and Yoshii, M. (1987) J. Physiol. 383, 231–249.
14. Campbell, K.P., Leung, A.T. and Sharp, A.H. (1988) Trends Neurosci. 11, 425–530.
15. Bean, B.P. (1989) Ann. Rev. Physiol. 51, 367–384.
16. McCleskey, E.W., Fox, A.P., Feldman, D. and Tsien, R.W. (1986) J. Exp. Biol. 124, 177–190.
17. Glossmann, H., Ferry, D.R., Goll, A. and Rombusch, M. (1984) J. Cardiovasc. Pharmacol. 6, S608–S621.
18. Rosenberg, R.L., Hess, P., Reeves, J.P., Smilowitz, H. and Tsien, R.W. (1986) Science 231, 1564–1566.
19. Perez-Reyes, E., Wei, X., Castellano, and Birnbaumer, L. (1990) J. Biol. Chem. 265, 20430–20436.
20. Snutch, T.P., Leonard, J.P., Gilbert, M.M., Lester, H.A. and Davidson, N. (1990) Proc. Natl. Acad. Sci. U.S.A. 87, 3391–3395.
21. Adams, B.A., Tanabe, T., Mikami, A., Numa, S. and Beam, K.G. (1990) Nature (London) 346, 569–572.
22. Nabauer, M., Callewaert, G., Cleemann, L. and Morad, M. (1989) Science 244, 800–803.
23. Rios, E. and Brum, G. (1987) Nature (London) 325, 717–720.
24. Rios, E., Ma, J. and Gonzalez, A. (1991) J. Muscle Res. Cell Motil. 12, 127–135.
25. Catterall, W.A. (1991) Cell 64, 871–874.
26. Bechem, M., Hebisch, S. and Schramm, M. (1988) Trends Pharmacol. Sci. 9, 257–261.
27. Vanhoutte, P.M., Paoletti R. and Govoni, S. (Eds.) (1988) Calcium Antagonists: Pharmacology and Clinical Research, Ann. N.Y. Acad. Sci. vol. 522.
28. Scriabine, A., Schuurman, T. and Traber, J. (1989) FASEB J. 3, 1799–1806.
29. Borsotto, M., Barhanin, J., Fosset, M. and Lazdunski, M. (1985) J. Biol. Chem. 26 14255–14263.
30. Hosey, M.M., Barhanin, J., Schmid, A., Vandaele, S., Ptasienski, J., O'Callahan, C., Cooper, C. and Lazdunski, M. (1987) Biochem. Biophys. Res. Commun. 147, 1137–1145.
31. Chang, F.C. and Hosey, M.M. (1988) J. Biol. Chem. 263, 18929–18937.
32. Curtis, B.M. and Catterall, W.A. (1986) Biochemistry 25, 3077–3083.
33. Mundina-Weilenmann, C., Chang, C.F., Gutierrez, L.M. and Hosey, M.M. (1991) J. Biol. Chem. 266, 4067–4073.
34. Striessnig, J., Moosburger, K., Goll, A., Ferry, D.R. and Glossmann, H. (1986) Eur. J. Biochem. 161, 603–609.
35. Kongsamut, S., Kamp, T.J., Miller, R.J. and Sanguinetti, M.C. (1985) Biochem. Biophys. Res. Commun. 130, 141–148.
36. Williams, J.S., Grupp, I.L., Grupp, G., Vaghy, P.L., Dumont, A. and Schwartz, A. (1985) Biochem. Biophys. Res. Commun. 131, 13–21.
37. Kokubun, S., Prod'hom, B., Becker, C., Porzig, H. and Reuter, H. (1986) Mol. Pharmacol. 30, 571–584.
38. Hamilton, S.L., Yatani, A., Brush, K., Schwartz, A. and Brown, A.M. (1987) Mol. Pharmacol. 31, 221–231.
39. Striessnig, J., Goll, A., Moosburger, K. and Glossmann, H. (1986) FEBS Lett. 197, 204–210.
40. Vaghy, P.L., Striessnig, J., Miwa, K., Knaus, H.-G., Itagaki, K., McKenna, E., Glossmann, H. and Schwartz, A. (1987) J. Biol. Chem. 262, 14337–14342.
41. Galizzi, J.-P., Borsotto, M., Barhanin, J., Fosset, M. and Lazdunski, M. (1986) J. Biol. Chem. 261,

1393–1397.
42. Glossmann, H., Ferry, D. R., Lubbecke, F., Mewes, R. and Hofmann, F. (1982) Trends Pharmacol. Sci. 3, 431-437.
43. Fosset, M., Jaimovich, E., Delpont, E. and Lazdunski, M. (1983) J. Biol. Chem. 258, 6086–6092.
44. Bolger, G.T., Gengo, P., Klockowski, R., Luchowski, E., Siegel, H., Janis, R.A., Triggle, A.M. and Triggle, D.J. (1983) J. Pharmacol. Exp. Ther. 225, 291–309.
45. Glossmann, H., Ferry, D.R. and Boschek, C.B. (1983) Naunyn-Schmiedeberg's Arch. Pharmacol. 323, 1–11.
46. Murphy, K.M.M., Gould, R.J., Largent, B.L. and Snyder, S.H. (1983) Proc. Natl. Acad. Sci. U.S.A. 80, 860–864.
47. Ptasienski, J., McMahon, K.K. and Hosey, M.M. (1985) Biochem. Biophys. Res. Commun. 129, 1393–1397.
48. Qar, J., Galiz, J.-P., Fosset, M. and Lazdunski, M. (1987) Eur. J. Pharmacol. 141, 261–268.
49. King, V.F., Garcia, M.L., Shevell, J.L, Slaughter, R.S. and Kaczorowski, G.J. (1989) J. Biol. Chem. 264, 5633–5641.
50. Schmid, A., Romey, G., Barhanin, J. and Lazdunski, M. (1989) Mol. Pharmacol. 35, 766–773.
51. Bean, B.P. (1984) Proc. Natl. Acad. Sci. U.S.A. 81, 6388–6392.
52. Sanguinetti, M.C., Krafte, D.S. and Kass, R.S. (1986) J. Gen. Physiol. 88, 369–392.
53. Cognard, C., Romey, G., Galizzi, J.-P., Fosset, M. and Lazdunski, M. (1986) Proc. Natl. Acad. Sci. U.S.A. 83, 1518–1522.
54. Changeux, J.-P., Deviliiers-Thiery, A. and Chemouilli, P. (1984) Science 225, 1335–1345.
55. Catterall, W.A. (1988) Science 242, 50–61.
56. Sieber, M., Nastainczyk, W., Zubor, V., Wernet, W. and Hofmann, F. (1987) Eur. J. Biochem. 167, 117–122.
57. Takahashi, M., Seagar, M.J., Jones, J.F., Reber, B.F.X. and Catterall, W.A. (1987) Proc. Natl. Acad. Sci. U.S.A. 84, 5478–5482.
58. Schmid, A., Barhanin, J., Coppola, T., Borsotto, M. and Lazdunski, M. (1986) Biochemistry 25, 3492–3495.
59. Leung, A.T., Imagawa, T., Block, B., Franzini-Armstrong, C. and Campbell, K.P. (1988) J. Biol. Chem. 263, 994–1001.
60. Leung, A.T., Imagawa, T. and Campbell, K.P. (1987) J. Biol. Chem. 262, 7943–7946.
61. Morton, M.E. and Froehner, S.C. (1987) J. Biol. Chem. 262, 11904–11907.
62. Chang, C.F. and Hosey, M.M. (1990) Biochem. Biophys. Res. Commun. 172, 751–758.
63. Sharp, A.H., Imagawa, T., Leung, A.T. and Campbell, K.P. (1987) J. Biol. Chem. 262, 12309–12315.
64. Kim, H., Wei, X., Ruth, P., Perez-Reyes, E., Flockerzi, V., Hofmann, F. and Birnbaumer, L. (1990) J. Biol. Chem. 265, 11858–11863.
65. Hamilton, S.L., Hawkes, M.J., Brush, K., Cook R., Chang, R.-J. and Smilowitz, H.M. (1989) Biochemistry 28, 7820–7828.
66. Rengasamy, A., Ptasienski, J. and Hosey, M.M. (1985) Biochem. Biophys. Res. Commun. 126, 1–7.
67. Campbell, K.P., Lipshutz, G.M. and Denney, G.H. (1984) J. Biol. Chem. 259, 5384–5387.
68. Cooper, C.L., Vandaele, S., Barhanin, J., Fosset, M., Lazdunski, M. and Hosey, M.M. (1987) J. Biol. Chem. 262, 509–512.
69. Schneider, T. and Hofmann F. (1988) Eur. J. Biochem. 174, 369–375.
70. Tuana, B.S., Murphy, B.J. and Yi, Q. (1987) Mol. Cell. Biochem. 76, 173–184.
71. Horne, W.A., Weiland, G.A. and Oswald, R.E. (1986) J. Biol. Chem. 261, 3588–3594.
72. Takahashi, M. and Catterall, W.A. (1987) Science 236, 88–91.
73. Tanabe, T., Takeshima, H., Mikami, A., Flockerzi, V., Takahashi, H., Kangawa, K., Kojima, M., Matsuo, H., Hirose, T. and Numa, S. (1987) Nature (London) 328, 313–318.
74. Jan, L.Y. and Jan, Y.N. (1989) Cell 56, 13–25.
75. Noda, M., Shimizu, S., Tanabe, T., Takai, T., Kayano, T., Ikeda, T., Takahashi, H., Nakayama, H.,

Kanaoka, Y., Minamino, N., Kangawa, K., Matsuo, H., Raftery, M.A., Hirose, T., Inayama, S., Hayashida, H., Miyata, T. and Numa, S. (1984) Nature (London) 312, 121–127.
76. Noda, M., Takayuki, I., Kayano, T., Suzuki, H., Takeshima, H., Kurasaki, M., Takahashi, H. and Numa, S. (1986) Nature (London) 320, 188–192.
77. Perez, Reyes, E., Kim, H.S., Lacerda, A.E., Horne, W., Wei, X., Rampe, D., Campbell, K.P., Brown, A.M. and Birnbaumer, L. (1989) Nature (London) 340, 233–236.
78. Tanabe, T., Beam, K.G., Powell, J. and Numa, S. (1988) Nature (London) 336, 134–139.
79. Tanabe, T., Beam, K.G., Adams, B.A., Niidome, T. and Numa, S. (1990) Nature (London) 346, 567–569.
80. Mikami, A., Imoto, K, Tanabe, T., Niidome, T., Mori, Y., Takeshima, H., Narumiya, S. and Numa, S. (1989) Nature (London) 340, 230–233.
81. Koch, W.J., Ellinor, P.T. and Schwartz, A. (1990) J. Biol. Chem. 265, 17786–17791.
82. Biel, M., Ruth, P., Bosse, E., Hullin, R., Stuhmer, W., Flockerzi, V. and Hofmann, F. (1990) FEBS Lett. 269, 409–412.
83. Mori, Y., Friedrich, T., Kim, M.-S., Mikami, A., Nakai, J., Ruth, P., Bosse, E., Hofmann, F., Flockerzi, V., Furuichi, T., Mikoshiba, K., Imoto, K., Tanabe, T. and Numa, S. (1991) Nature (London) 350, 398–402.
84. Ellis, S.B., Williams, M.E., Ways, N.R., Brenner, R., Sharp, A.H., Leung, A.T., Campbell, K.P., McKenna, E., Koch, W.J., Hui, A., Schwartz, A. and Harpold, M.M. (1988) Science 241, 1661–1664.
85. DeJongh, K.S., Warner, C. and Catterall, W.A. (1990) J. Biol. Chem. 265, 14738–14741.
86. Jay, S.D., Sharp, A.H., Kahl, S.D., Vedvick, T.S., Harpold, M.M. and Campbell, K.P. (1991) J. Biol. Chem. 266, 3287–3293.
87. Ruth, P., Rohrkasten, A., Bosse, E., Regulla, S., Meyer, H.E., Flockerzi, V. and Hofmann, F. (1989) Science 245, 1115–1118.
88. Pragnell, M., Leveille, C.J., Jay, S.D. and Campbell, K.P. (1991) Biophys. J. 59, 390a.
89. Jay, S.D., Ellis, S.B., McCue, A.F., Williams, M.E., Vedvick, T.S., Harpold M.M. and Campbell, K.P. (1990) Science 248, 490–492.
90. DeJongh, K.S., Merrick, D.K. and Catterall, W.A. (1989) Proc. Natl. Acad. Sci. U.S.A. 86, 8585–8589.
91. Schwartz, L.M., McCleskey, E.W. and Almers, W. (1985) Nature (London) 314, 747–751.
92. Flockerzi, V., Oekem, H.-J., Hofmann, F., Pelzer, D., Cavalie, A. and Trautwein, W. (1986) Nature (London) 323, 66–68.
93. Smith, J.S., McKenna, E.J., Ma, J., Vilven, J., Vaghy, P.L., Schwartz, A. and Coronado, R. (1987) Biochemistry 26, 7182–7188.
94. Talvenheimo, J.A., Worley, J.F. and Nelson, M.T. (1987) Biophys. J. 52, 891–899.
95. Ma, J., Mundina-Weilenmann, C., Hosey, M.M. and Rios, E. (1991) Biophys. J. 60, 890–901.
96. Gutierrez, L.M., Chang, C.F., Mundina-Weilenmann, C. and Hosey, M.M. (1991) Biophys. J. 59, 88a.
97. Barhanin, J., Coppola, T., Schmid, A., Borsotto, M. and Lazdunski, M. (1987) Eur. J. Biochem. 164, 525–531.
98. Horne, W.A., Ghany, M.A., Racker, E., Weiland, G.A., Oswald, R.E. and Cerione, R.A. (1988) Proc. Natl. Acad. Sci. U.S.A. 85, 3718–3722.
99. Tsien, R.W., Bean, B.P., Hess, P., Lansman, J.B., Nilius, B. and Nowycky, M.C. (1986) J. Mol. Cell. Cardiol. 18, 691–710.
100. Armstrong, D. and Eckert, R. (1987) Proc. Natl. Acad. Sci. U.S.A. 84, 2518–2522.
101. Chad, J.E. and Eckert, R. (1986) J. Physiol. 378, 31–51.
102. Arreola, J., Calvo, J., Garcia, M.C. and Sanchez, J.A. (1987) J. Physiol. 393, 307–330.
103. Schmid, A., Renaud, J.-F. and Lazdunski, M. (1985) J. Biol. Chem. 260, 13041–13046.
104. Artalejo, C.R., Dahmer, M.K., Perlman, R.L. and Fox, A.P. (1991) J. Physiol. 432, 681–787.
105. Osterrieder, W., Brum, G., Hescheler, J., Trautwein, W., Flockerzi, V. and Hofmann, F. (1982)

Nature (London) 298, 576–578.
106. Brum, G., Flockerzi, V., Hofmann, F., Osterrieder, W. and Trautwein, W. (1983) Pfluegers Arch. 398, 147–154.
107. Bkaily, G. and Sperelakis, N. (1984) Am. J. Physiol. 246, H630–H634.
108. Chang, C.F., Gutierrez, L.M., Mundina-Weilenmann, C. and Hosey, M.M. (1991) Biophys. J. 59, 88a.
109. Nunoki, K., Florio, V. and Catterall, W.A. (1989) Proc. Natl. Acad. Sci. U.S.A. 86, 6816–6820.
110. Nastainczyk, W., Rohrkasten, A., Sieber, M., Rudolph, C., Schachtele, C., Marme, D. and Hofmann, F. (1987) Eur. J. Biochem. 169, 137–142.
111. O'Callahan, C.M. and Hosey, M.M. (1988) Biochemistry 27, 6071–6877.
112. O'Callahan, C.M., Ptasienski, J. and Hosey, M.M. (1988) J. Biol. Chem. 263, 17342–17349.
113. Rohrkasten, A., Meyer, H.E., Nastainczyk, W., Sieber, M. and Hofmann, F. (1988) J. Biol. Chem. 263, 15325–15329.
114. Curtis, B.M. and Catterall, W.A. (1984) Biochemistry 23, 2113–2118.
115. Imagawa, T., Leung, A.T. and Campbell, K.P. (1987) J. Biol. Chem. 262, 8333–8339.
116. Mundina-Weilenmann, C., Ma, J., Rios, E. and Hosey, M.M. (1991) Biophys. J. 60, 902–909.
117. Lai, Y., Seagar, M.J., Takahashi, M. and Catterall, W.A. (1990) J. Biol. Chem. 265, 20839–20848.
118. Kwatra, M.M., Benovic, J.L., Caron, M.G., Lefkowitz, R.J. and Hosey, M.M. (1989) Biochemistry 28, 4543–4547.
119. Rane, S.G. and Dunlap, K. (1986) Proc. Natl. Acad. Sci. U.S.A. 83, 184–188.
120. Galizzi, J.-P., Qar, J., Fosset, M., Van Renterghem, C. and Lazdunski, M. (1987) J. Biol. Chem. 262, 6947–6950.
121. Marchetti, C. and Brown, A.M. (1988) Am. J. Physiol. 254, C206–C210.
122. Dosemeci, A., Dhallan, R.S., Cohen, N.M., Lederer, W.J. and Rogers, T.B. (1988) Circ. Res. 62, 347–357.
123. Lacerda, A.E., Rampe, D. and Brown, A.M. (1989) Nature (London) 335, 249–251.
124. Heschler, J., Kameyama, M., Trautwein, W., Mieskes, G. and Soling, H.-D. (1987) Eur. J. Biochem. 165, 261–266.
125. Hosey, M.M., Borsotto, M. and Lazdunski, M. (1986) Proc. Natl. Acad. Sci. U.S.A. 83, 3733–3737.
126. Brown, A.M. and Birnbaumer, L. (1988) Am. J. Physiol. 254, H401–H410.
127. Yatani, A., Codina, J., Imoto, Y., Reeves, J.P., Birnbaumer, L. and Brown, A.M. (1987) Science 238, 1288–1292.
128. Shuba, Y.M., Hesslinger, B., Trautwein, W., McDonald, T.F. and Pelzer, D. (1990) J. Physiol. 424, 205–228.
129. Yatani, A., Imoto, Y., Codina, J., Hamilton, S.L., Brown, A.M. and Birnbaumer, L. (1988) J. Biol. Chem. 263, 9887–9895.
130. Hescheler, J., Rosenthal, W., Trautwein, W. and Schultz, G. (1987) Nature (London) 325, 445–447.
131. Miller, R.J. (1990) FASEB J. 4, 3291–3299.
132. Gutierrez, L.M., Brawley, R.M. and Hosey, M.M. (1991) J. Biol. Chem. 266, 16387–16394.
133. Ma, J., Gutierrez, L.M., Hosey, M.M. and Rios, E. (1992) Biophys. J., in press.

INDEX

5-(2-Acetamidoethyl)aminonaphtalene-1-sulfonate, *see* IAEDANS
Acetic anhydride, 94
Acetylcholine receptor, 257, 275, 281, 316
Acridine orange, 45
Adamantanyl diazirine, 7, 8
Adenosine-5′-triphosphopyridoxal, 66, 94
Adenylate cyclase, 174
Adenylate kinase, 10, 12, 94
β-Alanine, 283–285
6-*O*-alkyl-D-galactoses, 193
Allantoin permease (DAL4), 236
Allantoinase, 240
Allophanate, 236
Amidine, 285, 286
N-amidino-5-amino-6-chloropyrazine carboxamide, *see* amiloride
Amiloride, 247, 249, 251, 253–259, 265
  analogues, 248, 255–259
Amino acid efflux, 225
Amino acid permease
  general (GAP1), 223–226, 231, 234–240
Amino acid permease activity
  regulation of, 238
Amino acid transport
  regulation, 232–242
  regulation of synthesis, 234–237
  vacuolar, reviews on, 224
Amino acid transporters
  evolution, 227–232
  molecular cloning, 227
  mutants, 226
  reviews on, 220
  structure, 227–232
  yeast, 219–245
Aminochloromethoxyacridine, 119
4-Aminopyridine (4-AP), 302, 303
Aminotriazole resistance, 225
AMOG (adhesion molecule on glia), 6, 10
AMP–PCP, 97
AMP–PNP, 40
Androsten-4-ene-3, 17-dione, 173, 174, 191
8-Anilino-1-naphtalene sulfonate (ANS), 99–102
Ankyrin, 6

Anthracene-9-carboxylate (A9C), 276, 283–285
Arabinose transport proteins (AraE), 232
Arabinose transporter, 202
Arginase, 240
Arginine-histidine exchange transport system, 224
Arginine modification, 94
Arginine permease (CAN1), 223, 225, 228–231, 234
Aromatic amino acid transporter protein
  general (AroP), 228–231
ATP-imidazolidate, 98
Avermectin, 283
8-Azidoadenosine, 191
8-Azidoadenosine 5′-[γ-^{32}P]triphosphate (azido-ATP), 191
2-*N*-[4(1-azi-2,2,2-trifluoroethyl)benzoyl]-1,3-bis-(D-mannos-4-yloxy)-2-propylamine (ATB-BMPA), 173

Bacteriorhodopsin, 83, 126
Baculovirus, 262
Band 3 anion exchanger, 250, 262
Barbiturates, 283, 284
Bay K 8644, 318, 328
Benzamil, 256–258
Benzodiazepines, 283, 284
Benzothiazepines, 318, 319
Bepridil, 319
3′(2′)-*O*-biotinyl-thioinosine trisphosphate (biotinyl-S^6-ITP$_2$), 93
S-(bismaleimidomethyl ether)-L-[^{35}S]cysteine, 189
Bis-mannose derivates, 186, 190, 191, 196
Black lipid membrane, 43
Brij 56, 75
Brij 96, 75
Brij 36T, 75
Bumetamide, 285, 286
β-Bungarotoxin, 308
BY1023, 47

Ca^{2+}, 287, 290, 291
Ca^{2+}-activated K^+ channel, 275

Ca^{2+}-ATPase, 5, 6, 15, 21, 22, 29, 57–116, 126, 128, 130, 189
  ATP binding site, 81
  C-terminal ends, 59
  catalytic site, 79
  chromosome localization, 58
  classification of, 58
  conformational change mutants, 82
  crystallization, 70
  cytoplasmic domain, 89
  cytoplasmic headpiece, 65
  electron microscopy, 68, 76
  high-affinity Ca^{2+} binding site, 96
  hinge domain, 67
  isoenzymes of, 58
  location of Ca^{2+} binding sites, 79
  low-affinity Ca^{2+} binding site, 78
  molecular weights, 64
  monoclonal antibodies, 88–91
  phosphorylation domain, 66
  polyclonal antibodies, 88–91
  proteolysis, 84–88
  quantitation, 88
  review articles on, 57–116
  side chain modification, 91
  stalk region, 67, 78
  topology, 64–70
  transduction domain, 67
  transmembrane domain, 68
Ca^{2+}/calmodulin (CaM)-dependent protein kinase, 330, 331
Ca^{2+} channels, 57, 282, 307, 315–336
  ATP-sensitive, 315
  biochemical and molecular characterization, 316–319
  blockers, 318
  classification, 316
  glycosylation sites, 323
  intracellular ligand-gated, 316
  IP_3-sensitive channels, 315, 316
  α_1-isoforms, 322, 323
  mitogen-sensitive, 315
  neuronal, review on, 332
  phosphorylation, 322, 326–330
  receptor-operated (ROCCs), 315, 316
  reconstitution, 325, 326
  regulation, 326
  roles of subunits, 324, 325
  subtypes, 315–317
  thrombin-sensitive, 316
Ca^{2+}-induced Ca^{2+} release, 317

Ca^{2+}-ionophores, 290
cAMP, 280, 287–291
cAMP-dependent phosphorylation, 275
cAMP-dependent protein kinase (PKA), 30, 259, 260, 263, 287, 326–331
Cadherin, 6
Caged ATP, 18, 78
Calcineurin, 331
Calcium/calmodulin-dependent protein kinase II, 263
Calmodulin, 61, 69, 70
Calmodulin antagonist, W-7, 263
Calpain, 69
Capnophorin, 6
Carbodiimide adduct of ATP, 66, 97
Carboxyl group modification, 96
Cardiac muscles, 58
$C_{12}E_8$, 32, 34, 39, 45, 75, 90, 93
Cell adhesion molecule, 6
Cell volume regulation, 277
Cellobiose inhibition, 171
cGMP-gated cation channel, 282
Charybdotoxin (CTX), 302, 303, 307
α-[4-(N-2-chlorethyl-N-methyl-amino)]benzoyl-amide-ATP (ClR-ATP), 13
p-Chloromercuribenzene sulphonate, 188
7-Chloro-4-nitrobenzo-2-oxa-1,3-diazole (NBD-Cl), 92
Chloroquine, 228
Cholate, 45
Choline transporter (CTR), 228, 231
Chymotrypsin, 16, 18, 19, 87, 188
Cimetidine, 248, 253
Circular dichroism (CD), 19, 36, 70–77, 88, 121, 122, 184, 194
Citraconic anhydride, 29
Citrate, 203
Citrate transporter, 170
Cl^--channels, 273–294
  different types, 274–280
  epithelial, reviews on, 278
  in electric organ, 276
  in muscle, 276
  in the nervous system, 275
  intermediate conductance outwardly rectifying, see ICOR channel
  pharmacological modulation, 283–287
  reviews on, 273
  Torpedo, 280, 282
ClR-ATP, 13

Class II enzymes, 135
Clonidine, 248
Concanavalin A, 119
Conformational coupling mechanism, 98
Cr-ADP, 16
Cr-ATP, 73, 99–102
Cyanogen bromide, 32
N-cyclohexyl-N'-(4-dimethyl-amino-α-naphtyl)carbodiimide (NCD-4), 97, 99–101
Cyclooxygenase, 286
Cystic fibrosis (CF), 274, 277, 280, 288–290
Cystic fibrosis transmembrane conductance regulator (CFTR), 289, 290
Cytochalasin B, 172, 173, 182–185, 189, 191, 192, 195, 196, 201–203
Cytosine permease, 233
Cytosolic inhibitor (CI), 287, 288, 290

D-888, 318
DCCD, 17, 96, 250–252, 255–257
Decylpolyethylene glycol (decyl-PEG), 149
Dendrotoxin (DTX), 302, 303, 308
6-Deoxy-D-galactose, 202
1-Deoxy-D-glucose, 171
2-Deoxy-D-glucose, 171
3-Deoxy-D-glucose, 171
Deuterium exchange experiments, 189
Diamide, 159
Dibutyryl-cAMP, 287
Dichlorodiphenylamine-2-carboxylate (DCl-DPC), 284, 285
γ[4-(N-2-dichloroethyl-N-methylamino)]benzylamide ATP (ClrATP), 66
Dicyclohexylcarbodiimide, see DCCD
N'-dicyclohexylcarbodiimide, see DCCD
8(N-N-diethylamino)octyl trimethoxybenzoate (TMB-8), 47
Diethylpyrocarbonate (DEPC), 44, 96, 250, 251
Diethylstilboestrol, 173, 174, 183
Dihydropyridine (DHP), 282, 316–326
Diltiazem, 322
D-cis-Diltiazem, 318
Diphenylamine-2-carboxylate (DPC), 276, 285
Diphenylbutylpiperidines, 319
Disulphonate stilbene (DIDS), 250, 253, 276, 283–286
5,5'-Dithiobis(2-nitrobenzoic acid), 122, 186
Doxylstearates, 101

Dysgenic mouse, 323

E3810, 47
Electrogenic pump, 117
Electron microscopy, 3–6, 68
Emulgen, 45
Endo-β-N-acetylglucosaminidase (Endo-H), 254
Endo-β-N-acetylglucosaminidase F (Endo-F), 32, 184, 254
Endoplasmic reticulum membrane, 57–116
Enzyme II
  A domain, 140–143
  association state of, 144
  B domain, 142, 143
  C domain, 143
  coupling between transport and phosphorylation, 153–160
  coupling in vectorial phosphorylation, 158–160
  domain function, 140–143
  domain interactions, 143, 144
  domain structure, 138, 139
  equilibrium binding to, 149
  facilitated diffusion, 155–158
  kinetics of binding, 151–153
  kinetics of domain interaction, 146, 147
  orientation of binding site, 149–151
  phosphorylation, 145
  phosphorylation of carbohydrates, 154, 155
  sequence homology, 138
  steady-state-kinetics of carbohydrate phosphorylation, 160
  structure, 138, 147
Eosin, 12, 36, 40, 101
Eosin-5'-isothiocyanate, 100
Epidermal growth factor, 263
Equilibrium exchange experiments, 175–177, 180
1,N^6-ethenoadenosine-5'-diphosphate, 100
Ethoxy carbonylethoxy-dihydroquinoline (EEDQ), 41, 250, 251
1-Ethyl-3-(3 dimethylamino propyl)carbodiimide (ATP-EDC), 97
4,6-O-ethylidene-D-glucose, 192, 195
Ethylidene glucose, 193, 195
N-ethylmaleimide (MalNEt; NEM), 91, 149, 156, 188, 193, 194, 248, 250, 252, 253

N-ethyoxycarbonyl-2-ethoxy-1,2-
    dihydroquinoline (EEDQ), 41, 250,
    251
Eu^{3+}, 99, 102
Exchange activity measurements, 145
Exocytosis, 274

F_1-ATPase, 94
Fast atom bombardment mass spectroscopic
    mapping, 186
Fast-twitch skeletal muscle, 58, 64–66
Fenoctimine, 47
Fibroblast growth factor, 268
FITC, 12, 18, 29–31, 35, 40, 41, 43, 66, 72, 79,
    80, 84, 88, 93, 99–103
FITC-PE, 101, 102
Fluo-3, 326, 328, 330
Fluorescamine, 94
Fluorescein-isothiocyanate, see FITC
Fluorescence, intrinsic, 19, 101, 103, 104, 143,
    172, 180, 194
Fluorescence energy transfer, 98, 129
1-Fluoro-2,4-dinitrobenzene (FDNB), 193
Fluorosulfonylbenzoyl adenosine (FSBA), 30,
    66, 81
Fodrin, 6
Forskolin, 173, 174, 191, 287, 328
Fourier transform infrared spectroscopy,
    see FTIR spectroscopy
Fourier–Bessel reconstructions, 71
Freeze-fracture electron microscopy, 71, 73
Fructose phosphorylation, 163
FTIR spectroscopy, 68, 184, 185, 189
D-fucose, 202
Furosemide, 280, 285, 286
Furylacrylic acid, 96

G-proteins, 326, 331, 332
GABA permease (UGA4), 226–231, 234–238
$GABA_A$-receptor channel, 273, 275, 281, 283,
    285
GABA transaminase (UGA1), 235, 236
Galactose permease, 232
Galactose transporter, 202
β-Galactoside permease, 227
Gastric acid secretion, 27
Gastrin, 27
Gastritis, autoimmune, 33
Gd^{3+}, 99, 102
Glucocorticoids, 268, 269
Glucose sensor, 198

Glucose transport, 185
Glucose transporter, 169–217, 231, 232
    characterization, 182–185
    conformational changes, 192
    effect of cytoplasmic ATP, 176, 177
    expression of, 186
    glycosylation, 184
    in photosynthetic organisms, 201, 202
    kinetic models for transport, 177
    kinetics of transport, 174
    mechanism of transport, 192
    oligomeric state, 185
    secondary structure, 184
    structure, 184–191
    substrate-binding site(s), 189
    substrate specificity, 170
    three-dimensional arrangement, 189
    topology, 186
GLUT-1, 170–174, 186, 196–198, 201–203, 208
GLUT-2, 171, 173, 174, 186, 196, 198–200
GLUT-3, 196, 199
GLUT-4, 172–174, 197, 199
GLUT-5, 200
Glutamate dehydrogenase, 240
Glutamine permease (GNP1), 238, 239
Glutaraldehyde, 32, 34, 45, 66, 254
Glutathione-maleimide, 253
N-glycanase, 184
Glycine methyl ester, 252
Glycine-receptor channel, 275, 280, 281,
    283–285
Glycosylation, 33

H, K-ATPase, 5, 20
    α subunit, 28–31
    β subunit, 31–34
    ADP-sensitive intermediate, 37
    Ca^{2+} effect, 38
    cation affinity, 39
    cell distribution, 28
    conformations, 34
    cysteine residues, 33, 48
    deletion mutants, 33
    dephosphorylation, 38
    disulfide linkages, 33
    electrogenicity of ion transport, 43, 44
    exon-intron organization, 29
    H^+-ATP stoichiometry, 42
    inhibitors, 47
    interspecies homology, 29
    ion selectivity, 42

H, K-ATPase (cont'd)
  K^+-sensitive intermediate, 37
  K^+-transporting step, 43
  kinetics, 36
  lipid composition, 44
  low-affinity nucleotide site, 40
  molecular organization, 34
  monoclonal antibodies, 46
  Na^+ effects, 43
  nonhyperbolic ATP dependence, 40
  passive H^+–K^+ exchange, 46
  phospholipid requirement, 44, 45
  phosphorylation, 37
  phosphorylation capacity, 38, 39
  phosphorylation from inorganic phosphate, 41, 42
  plasma membrane insertion, 34
  reconstitution, 45
  reviews on, 27
  solubilization, 45
  tissue distribution, 28
  transport, 42–44
H^+-ATPase, 117–134
  conformational changes, 118
  cysteines, 122
  minimum functional unit, 120
  molecular mechanism, 129
  plant plasma membrane, 118
  primary structure, 121
  purification, 119
  reviews, 118
  secondary structure, 121, 122
  subunit composition, 119
  tertiary structure, 123–129
  transport, 129
  yeast, 21, 22, 118
HCO_3^-–Cl^--exchange, 274
Hg-phenyl azoferritin, 91
Histamine, 27, 274
Histidine modification, 95
Histidine permease (HIP1), 223, 228–231, 234
HOE 731, 47
Hydropathy abalysis, 29, 68
Hydroxylamine, 95
Hypertension
  essential, 247, 269

IAEDANS, 72, 92, 94, 99, 100
IAPS-forskolin, 174, 186, 191
ICOR channel, 276–280, 283, 285–290
Imidazo-[1,2a]pyridines, substituted, 48

Imidazolium groups, 250
Indanyloxyacetic acid (IAA-94), 282–286
Indolizinsulfones, 319
Infinite-*cis* experiments, 175, 176
Infinite-*trans* experiments, 175
Infrared spectroscopy, *see also* FTIR spectroscopy, 19, 122, 194, 210
Inositol trisphosphate (IP_3), 316
Insecticides, 283, 284
Insulin, 196, 199, 263
Intrinsic birefringence, 68
Iodoacetamide (IAA), 92
Iodoacetamido-fluorescein, 18, 92
Iodoacetate, 123
3-Iodo-4-azidophenylamido-7-*O*-succinyldeacetyl-forskolin, 174
Iodonaphtylazide, (INA), 7–9, 11
N-isothiocyano-phenyl-imidazole, 13

K^+ channels, 297–313, 322
  biophysical properties, 298
  functional domains, 308
  genes, 306, 307
  kinetic behaviour, 309
  nomenclature, 298
  pharmacology, 302
  shaker, 282, 298–305
  structure, 298

L-type Ca^{2+} channels, 315, 316
  pharmacology, 318, 319
Lac permease, 170, 207, 208, 250, 264, 265
Lactose-H^+ symport, 227
Lactose transporter, 201, 207, 208
Lansoprazole, 47
Lanthanide, 67, 70, 73, 79, 86
Lin-Benzo-ATP, 40
N-linked glycosylation, 30, 188, 254, 264
LU47781, 318
Lysine modification, 93
Lysine permease (LYP1), 234

Maleic anhydride, 94
Maltose inhibition, 171, 183
Maltose transporter, 201
Mannitol kinase, 159
Mannitol translocator, 159
Mast cell degranulating peptide (MCDP), 302, 303
mDAZIP, 30, 49
MDCK cells, 33

MDPQ, 36
Metabolic acidosis, 268
Methionine permease (MTP1), 234
5-Methyl-L-arabinose, 202
5-(*N*-methyl-*N*-isobutyl)amiloride (MIA), 255
Methylamine/ammonium-ion permeases, 238, 239
Methylbenzimidate, 94
3-*O*-methylglucose, 174
Michaelis–Menten kinetics, 175, 179
Multidrug resistance (MDR) proteins, 225, 228
Muscimol, 283
Muscular dysgenesis, 322
Myotonia, 276, 282

Na, K-ATPase, 1–26, 58, 66, 94, 129, 247, 266, 288
 α subunit, 7–10
 β subunit, 10, 11
 ADP sensitive phosphoenzyme, 14
 capacity for cation binding, 16
 carbohydrate residues, 31
 cation binding and occlusion, 15–18
 conformational transitions, 18
 crystals, 3, 5, 6
 cytoskeletal associations, 6
 electron microscopy of, 3–6
 enzymatic properties, 3
 isoforms, 2
 low affinity nucleotide binding site, 13
 nucleotide binding, 11–15
 phosphorylation, 11–15
 primary structure, 1, 2
 proteolytic dissection, 7, 16
 regulatory function of N-terminus, 20
 review articles on, 2
 solubilization of, 3
 transport stoichiometry, 17
Na^+–Ca^{2+} exchange, 57
Na^+ channel, 307, 322
 gating, 310
Na^+Cl^--cotransport, 274
$Na^+2Cl^-K^+$-cotransport, 274, 285–288
Na^+/H^+ exchangers, 247–272
 biochemical properties, 249
 functional heterogeneity, 248, 249
 genomic cloning, 268, 269
 glycosylation, 263, 267, 268
 isoforms, 267, 268
 kinase, 263

 molecular cloning, 260–269
 phosphorylation, 263, 267, 268
 reviews on, 248
 tissue and membrane localization, 265
NAD(P)H:quinone oxidoreductase, 258
Na^+/glucose transporter, 170, 254, 262
Na^+/phosphate transporter, 254
Na^+/proline transporter, 254
Nd^{3+}, 103
Neuron–glial adhesion, 6
Neurospora crassa, 117
Neutron diffraction, 77, 78
(±)Nitrendipine, 316
5-Nitro-2-(3-phenyl-propylamino)-benzoate (NPPB), 279, 283–287
Nitrogen-catabolite repression (NCR), 234, 237–241
Nitrogen metabolism in yeast, 220–222
p-Nitrophenylphosphatase activity, 34, 40–42
Nitrothiosulfobenzoate, 122
Noise analysis, 277, 278
Nolinium bromide, 47
Nuclear magnetic resonance, 180
Nystatin, 277, 278, 280, 291

^{18}O exchange, 35, 37, 41
Octyl-polyoxyethylene, 145
Octylglycoside, 34, 45
Okadaic acid, 263
Omeprazole, 31, 46–48
Ornithine transaminase (CAR2), 236
Ouabain, photosensitive, 7
Overhauser effect, 12

Parietal cell antigens, 33
Patch clamp technique, 273, 277, 278
Phenamil, 256, 257
o-Phenanthroline, 46
Phenylalkylamines (PAA), 318–322
Phenylarsine oxide (PAO), 250, 253
Phenylglucoside, 196
Phenylglyoxal, 94
Phenylisothiocyanate (PITC), 250
N-phenylmaleimide (NPM), 250, 253, 254
Phloretin, 173, 174, 183, 193, 195
Phorbol esters, 259, 263, 268
Phosphate acceptor, 65
Phosphoenolpyruvate-dependent carbohydrate transport system, *see* PTS
Phospholipase A_2, 44
Phospholipase C, 44

Phospholipid-sensitive region, 69
Phospholipid vesicles, 17
Phospholipids, acidic, 68
Phosphoprotein phosphatases, 323, 324, 330, 331
Phosphoryl-aspartate, 117
Phosphoryl-transfer reactions, 128
Photolabile analog of ATP, 44
Photooxidation, 95
Photosynthetic reaction center, 83, 126, 127
o-Phthalaldehyde, 94
Picoprazole, 47
Picrotoxin, 284
Ping-pong kinetics, 161
Planar lipid bilayers, 18, 276
Platelet-derived growth factor, 263, 268
PMCA (plasma membrane Ca^{2+}-ATPase) 58, 61, 62, 69
(+)PN200–110, 318–320
Polycystic kidney disease, 6
Polyethylene glycol, 45
Polyphosphates, 224
Praseodymium, 73
Proline permease (PUT4), 223, 226, 235, 238–240
Proline transport protein (prnB), 228–231
6-O-propyl-D-galactose, 192
Propyl β-D-glucopyranoside, 193, 195
N-propyl-β-D-glucopyranoside, 192, 196
6-O-propyl-D-glucose, 192
Propylbenzilylcholine mustard (PrBCM), 255, 257
Prostaglandin E_2, 287
Protease V8, 32, 87
Protein kinase C (PKC), 248, 263, 267, 269, 289, 326–331
Proteinase K, 91
Proteoliposomes, 45, 120, 123, 155
Proteolytic cleavage, 16, 18, 19, 188
Proton-translocating ATPase, 117–134
PTS, 135–167
    carbohydrate substrates, 135, 136
    components, 135, 136
    nomenclature, 136
    review on, 137
Purine-cytosine permease, 231, 232
Pyridoxal-ATP, 13
Pyridoxal phosphate (PLP), 30, 31, 94, 250, 253
Pyrimidine permeases, 233

Radiation inactivation analysis, 32, 34, 145, 255, 259
Raman spectroscopy, 19
Rectin, 287
Renin release, 274, 277
Resonance energy transfer, 103
Rhodamine-5′-isothiocyanate (RITC), 99–102
Rhodobacter spheroides, 83
Rhodopseudomonas viridis 83
Rhodopsin, 189
Ro 18–5364, 47
Ryanodine, 316, 317

S202-791, 318
Saccharomyces cerevisiae, 219–245
SCH 28080, 30, 40, 41, 48
Scopadulcic acid, 41
SERCA (Sarcoplasma reticulum Ca^{2+}-ATPase), 58–60
Single-site accessibility-shift model, 163
Site-directed antibodies, 188
Site-directed mutagenesis, 15, 67, 78, 128
Skeletal muscles
    neonatal, 58
SK&F96022, 47
Slow-twitch skeletal muscle, 58, 64, 66
Spectrin, 6
Spermine, 41
Na^+-stimulated ^{86}Rb efflux, 42
Strychnine, 281, 284
Succinate semi-aldehyde dehydrogenase (UGA2), 235, 236
Sugar transporters, 169–217
    yeast, 200
Sugar/H^+ symporters, 170
Sulphydryl groups, 187, 188
Swainsonine, 254

Taurine, 284, 285
Tb^{3+}, 99, 101
Teratoma, 62
Tetra-bromo-fluorescein, *see* Eosin
Tetrachlorosalicyl-anilide, 93
Tetracycline transporter, 170, 203, 208
Tetraethylammonium chloride (TEA), 302–305
Thermolysin, 7, 87, 91
5-Thio-D-glucose, 172
Thiomethyl-β-galactoside, 156
Thrombin, 263
Torasemide, 285, 286

Transmembrane topography, 123
Transport asymmetry, 176, 177
Trifluoperazine, 47
Trifluoromethyl-iodo-phenyldiazirine (TID), 7–9, 11, 12, 86, 123
Trinitrobenzene sulfonate, 94, 253
Trinitrophenyl-ADP (TNP-ADP), 97, 100
Trinitrophenyl-ATP (TNP-ATP), 12, 20, 35, 36, 99, 101, 102
Tritium exchange experiments, 189
Trypsin (tryptic digestion), 7, 17–19, 28, 35, 66, 67, 69, 84–86, 103, 118, 123, 188–190, 196
Tryptophan fluorescence, 19, 101, 103, 143, 172, 180, 194
Two-site affinity-shift model, 163
Tyrosine kinase, 263

Unsaturated fatty acids, 286, 287
Uracil permease, 231–233
Urea amidolyase (DUR1, 2), 236, 240
Urea permease (DURP), 236, 239
Ureidosuccinate-allantoate permease (UEP1/DAL5), 234, 238–240
Uridine permease, 233
Uvomoruylin, 6

Valinomycin, 15
Vanadate, 3, 5, 27, 40, 46, 47, 66, 67, 87
Verapamil, 47, 286, 287
Voltage-dependent, *see* Ca^{2+} channels
Voltage sensors, 322–325
Volume regulation, 273

Wheat germ agglutinin, 32

X-ray diffraction, 67, 77, 78
Xenopus oocytes, 186, 196, 199, 281, 282, 299–303, 307–309, 324
Xylose transport proteins (XylE), 232
Xylose transporters, 202

Zero-*trans* experiments, 175–177, 180, 198